Environmental Science and Engineering
Subseries: Environmental Science

Series Editors: R. Allan • U. Förstner • W. Salomons

Dan Gafta
John Akeroyd
(Eds.)

Nature Conservation

Concepts and Practice

With 113 figures, 53 in colour, and a vegetation map

Springer

EDITORS

DR. DAN GAFTA
Babeş-Bolyai University, Chair of Taxonomy and Ecology,
Republicii Street 42, 400015 Cluj-Napoca (Romania)
dgafta@grbot.ubbcluj.ro

DR. JOHN AKEROYD
Plant Talk
Lawn Cootage, West Tisbury, Salisbury
Wiltshire SP3 6SG (UK)
akeroyd@dial.pipex.com

ISSN pending

ISBN 10 3-540-47228-2 **Springer Berlin Heidelberg New York**
ISBN 13 978 3-540-47228-5 **Springer Berlin Heidelberg New York**

Library of Congress Control Number: 2006934470

Springer is a part of Springer Science+Business Media
springeronline.com

Cover design: E. Kirchner, Heidelberg
Production: A. Oelschläger
Typesetting: Camera-ready by the Editors
Printed on acid-free paper 30/2132/AO 5 4 3 2 1 0

CONTENTS

PREFACE

SHORT REVIEW AND CONCLUSIONS OF THE INTERNATIONAL SYMPOSIUM

("*Conceptions and Methods of Nature Conservation in Europe*", Cluj-Napoca, September 16–19th, 2004)

Vasile CRISTEA[1], **Dan GAFTA**[1], **John R. AKEROYD**[2]
[1] "Babeş-Bolyai" University, Department of Taxonomy and Ecology, Republicii str., 42, **400015 Cluj-Napoca (ROMANIA)**
[2] Plant Talk, Lawn Cottage, West Tisbury, **Salisbury, Wiltshire SP3 6SG, UK**

1. General review of the symposium

Organized by the Chair of Taxonomy and Ecology of the University „Babeş-Bolyai" (Cluj-Napoca, Romania), in co-operation with the Romanian Society of Phytosociology and the International Federation of Phytosociology, this international symposium was dedicated to Prof. dr. h.c. Franco PEDROTTI (University of Camerino, Italy), on the occasion of his 70th birthday.

Drawn from four continents, the 84 participants (of whom over 50% were young people under 35 years old) represented 12 countries: Belgium, Bolivia, Brazil, Germany, Italy, Japan, Poland, Romania, Russia, Slovenia, UK and USA, thus covering a large part of Earth's bio-geographical regions.

The opening ceremony took place in the University's imposing *Aula Magna* and included two welcome speeches on behalf of the Rector and the President of the Nature Monument Committee, a *Laudatio* to Prof. F. Pedrotti and a short folk music recital. The symposium programme was composed of two plenary sessions, four short communication sessions (divided into two sections: **A** – *General aspects of nature protection*; **B** – *Nature conservation in practice*) and a poster session.

The seven plenary lectures, presented by R. Pott (Hannover, Germany), E. Box (Athens, USA.), K. Fujiwara (Yokohama, Japan), N. Boşcaiu (Cluj-Napoca, Romania), R. Wittig (Frankfurt, Germany) and R. Canullo (Camerino, Italy) were much appreciated by the audience. Many discussions and debates were stimulated by the 29 oral communications, the 18 posters and the field excursion to the Suatu Natural Reserve, as well as the visit to the Village Museum and the Arcalia Centre.

Finally, we must mention the presence, during the last two days, of the Santa Lucia Choir of Magras (Val di Sole, Trentino, Italy), who delighted participants with their mountain songs.

2. Conclusions of the symposium

Without giving authors' names and the hierarchy of importance of the issues presented and debated, the conclusions of the symposium can be summarized as follows:

a. A clearer **definition** and **distinction** of the notions *protection, conservation, ecological restoration* and *ecological reconstruction* is needed both for the establishment of specific strategies and methods of biodiversity management (preservation and/or exploitation), and the use of a common and accessible language for all specialists in environment-related fields.

b. **Floristic** and **faunistic studies**, far from being obsolete, should be carried out **specifically** by habitat type and should employ – in addition – relevant **methodology** from biochemistry, molecular genetics, ethology, etc. This approach, beside its theoretical component, can be a solid scientific basis for implementing appropriate management decisions in protected areas and other important sites for biodiversity conservation.

c. The **estimation of biodiversity** should be performed using widely accepted protocols and the resulting indices should always be interpreted by taking into account environmental variables.

d. The continuation of **phytosociological investigations and syntheses** is indispensable for hierarchical classification and characterization of habitat types, as well as for the identification of new important areas worthy of protection or conservation. It has been stressed that a **diversification of approaches** to vegetation study is needed through large-scale GIS mapping and by revealing the adaptive strategies of target, key or dominant species (the latter scenario achieved by the co-operation of phytosociologists with specialists in population ecology, eco-physiology and conservation biology).

e. A multi-level approach to **land planning and assessment** through five principal components – ecological, economic, social, cultural and political – is needed. In this context, geosynphytosociology can play an important role in the delimitation of landscape units.

f. The **modern management of protected areas** cannot provide universal 'recipes', but all actions must fit within local conditions, ecosystem type, biogeographical region, as well as, in case of large reserves and national parks, with the functional zone.

g. The **monitoring** of species and ecosystems should be proceeded by complex and differentiated investigations, in order to obtain unbiased and interpretable parameters.

h. **Ecological restoration** and **reconstruction** have proved to give better results when using native species, both activities being necessary in many sites for either economic or conservation reasons.

i. An important tool for conservation actions is represented by the **elaboration**, at regional and/or local level, of the so-called '**naturalistic catalogues**', which include the most vulnerable species and their habitats along with sites that are particularly species-rich ('hot spots'). To a certain extent, the Nature 2000 network accomplishes this task.

j. For the purpose of ***ex-situ*** conservation, botanic and zoological gardens have an increasing role, but the usefulness of co-operation with different institutions to assess **effective conservation actions** has been stressed. Taking as an example threatened vernal and pre-vernal species, it should be possible to involve local people in their conservation under the guidance of specialists from botanic gardens.

k. Regarding the much-discussed future of **small reserves**, the field excursion to Suatu brought new evidence that these can accomplish the conservation role for which they were created; however, there is a need for a correct management and the inclusion of a buffer zone, in which the land should be used traditionally.

l. Universities and research institutes must change their strategies through effective co-operation with all institutions and organizations interested in or responsible for the conservation of biodiversity and/or the improvement of environment legislation. The **production of specialists** through post-graduate studies has proved to be beneficial, but it cannot be achieved without the practical experience of teachers and without effective international collaboration.

m. Starting from the celebration of contributions brought about by some pioneers of the environmental movement, it has been stressed that '**history of science**' can represent a real source of inspiration for current and future studies and actions.

n. Last but not least, the importance of scientific meetings for **consolidating inter-human relationships** was acknowledged; this is a good premise for the constitution of regional or continental teams of scientists and, implicitly, for more successful research funding.

Presenting these conclusions, we express our satisfaction at the interest and direct involvement of young people in conservation actions (as demonstrated by their considerable participation to this symposium). We are confident that the demands of all those interested in the destiny of our planet will be successfully fulfilled. The high level of similarity between the above conclusions and the aims declared in BioPlatform Romania is a proof of this convergence of conservation actions.

The publishing of these symposium proceedings is an opportunity for scientists and environmentalists to learn about a panoply of nature conservation issues (theoretical aspects, environment legislation, biodiversity evaluation, case-studies, etc.) as well as to enrich their reference database. But, maybe the most important thing is that this volume can represent for all of us a prompt **to do** something towards the preservation of the wonderful creations of nature.

The contributors are responsible for the content of their papers included in this volume. The manuscripts were received by the deadline set on December 31st, 2004.

Professor dr. ***doctor honoris causa plurime***

FRANCO PEDROTTI

LAUDATIO TO PROF. FRANCO PEDROTTI

Dan GAFTA
"Babeş-Bolyai" University, Department of Taxonomy and Ecology,
Republicii str. 42, **400015 Cluj-Napoca (ROMANIA)**

Dear colleagues,
Ladies and gentlemen,

I believe few scientists have received so many plaudits as Professor Pedrotti. I want to recall only some of them, for instance those signed by Konrad Buchwald, Fabio Cassola, Luigi Portoghesi, Piergiorgio Corbetta, Vasile Cristea and Constantin Toma. That is why my talk will be brief and focus more on facts than opinions. I am not going to emphasize Prof. Pedrotti's scientific achievements, merely his contribution to nature conservation and his human character.

During a prodigious 45-year scientific career, Prof. Pedrotti has published 336 articles, with 76 papers on land management and planning, and 80 on nature conservation. He has organized 22 symposia on vegetation science and 18 meetings on nature protection. He has acted as an associate editor of 26 journals and has been editor-in-chief of four journals (Braun-Blanquetia, Documents Phytosociologiques, L'Uomo e l'Ambiente and La Riserva di Torricchio).

His interest in natural sciences and, especially, nature protection first emerged during his time at high school, when he frequented the Natural History Museum of Trento. On these occasions, he was in contact with famous botanists such Giuseppe Dalla Fior, Benedetto Bonapace and Vittorio Marchesoni. Later, when he was a college student, he met Renzo Videsott (director of the Gran Paradiso National Park) who had a determining influence on his development as a nature conservationist.

Apart from his numerous papers, Prof. Pedrotti's commitment to the cause of nature conservation emerges from his various practical actions. He is an active member on the scientific and management committee of several national parks (Abruzzo and Sibillini) as well as regional parks in Trentino Province. Prof. Pedrotti is the founder of Torricchio Natural Reserve and Arboretum Apenninicum, but he has proposed the creation of many other protected areas. Concerning this, there is even a saying: "*the land on which Prof. Pedrotti puts his foot to perform a study becomes unavoidably a natural reserve*!" Prof. Pedrotti has also been an innovator in terms of university curriculum by creating in 1998 a post-graduate school of management of natural environment and protected areas, and more recently, a Master's degree in protected area planning and management. I prefer to make no further comment on this topic, but simply cite the words of James Sievert: "*The Titan behind the emergence of conservation history has not been a historian, but a botanist from the University of Camerino, Franco Pedrotti. More than just the dean of conservation history, Pedrotti is the high priest, housing extensive archives collected over a five-decade career as a biologist and conservationist*".

Prof. Pedrotti is one of the most unselfish scientists I have ever met. He has been always available to share his experience and to collaborate with other colleagues such as Paul Ozenda from Grenoble, Jean-Marie Géhu from Paris, Janusz Falinski from Warsaw, Maximo Liberman Cruz from La Paz, Marcello Martinelli from Sao Paolo, Nicolae Doniță and Doina Ivan from Bucharest, and Vasile Cristea from Cluj-Napoca.

It is almost impossible to quantify how much he has given to students, co-workers, colleagues, and friends in terms of time, patience, and financial, logistical and moral support. Prof. Pedrotti may be proud of the high number of pupils he has supervised. The eldest are now professors at various universities, such as Eduardo Biondi (at University of Ancona), Renato Gerdol and Filippo Piccoli (University of Ferrara), Roberto Venanzoni (University of Perugia), Salvatore Leonardi (University of Catania), and Roberto Canullo, Ettore Orsomando, Antonio dell'Uomo and Beatrice Bellomaria (University of Camerino). On the other hand, Prof. Pedrotti has hosted in Camerino many young scholars from poor countries to undertake research (from Somalia, Algeria, Albania, Bolivia, Brazil, Georgia, Bulgaria, Poland, ex-Czechoslovakia and, certainly, Romania). Prof. Pedrotti has been able to overcome any prejudice and bureaucratic impediment, probably because of his generosity and open-minded spirit. Sometimes I think Prof. Pedrotti's name – Franco – is no accident but pre-destiny, given his frankness with respect to other people.

In spite of his hidden sentimentality, Prof. Pedrotti has proved all his life to be a hard worker and most tenacious in achieving new and new challenging goals. He is such a tough person that on the occasion of centenary celebrations dedicated to the Italian Botanical Society, whose president at that time was Prof. Pedrotti, some colleagues said he would also organize the bi-centenary festivities!

It is hard to distinguish between his private life and professional career, because his existence has been dominated by **love**: love for family, friends and poor people, love for nature and his work. For his contribution to science, for his effective work on nature conservation and for everything he has given to other people and future generations, I believe Prof. Pedrotti fully deserves our respect, gratitude and warm applause.

I

PROEMINENT PIONEERS OF NATURE CONSERVATION

TRADITIONS IN ROMANIAN SOZOLOGY

Vasile CRISTEA
"Babeş-Bolyai" University, Department of Taxonomy and Ecology, Republicii str. 42,
400015 Cluj-Napoca (ROMANIA); e-mail: cristea@grbot.ubbcluj.ro

Abstract: Starting from the premise that any field requires knowledge of the ideas of its predecessors, this paper points out the contributions of three generations of representatives of the Cluj School to nature conservation. Thus, the careers of A. Borza and E. Racoviță, who laid the theoretical and practical foundations of this movement, are presented; then, the difficult mission incumbent on E. Pop under the communist regime is emphasized; and the paper ends with N. Boşcaiu, their disciple, who at the impressive age of 80 still ennobles us with his rich experience.

Introduction

A decade ago, at the United Nations Conference on Population and Development (Cairo, September 5–13, 1994), it was again demonstrated and accepted that the notion of *„planetary citizenship"* is integral to the equation: *population – environment – development*. These components cannot be considered in isolation, in the same way that the equation itself does not concern only present generations but, more especially, those of the future [30]. In fact, today we speak of a new spirit, a new collective mentality, and global and pan-European efforts in the field of sustainable development and protection [17].

Then, within the framework of the World Summit on Social Development (Copenhagen, 6–15 March 1995), it was emphasized that development, beyond economic growth, concerns the welfare of populations, and that it should first and foremost serve the cause of the whole of humanity and not just a part [31].

At the same time, it is important to be aware of the fact that nature reserves can only accomplish their role (in the current continental and global context) if they are interconnected within networks of protected areas [18], also considered at national and continental level.

This general framework also includes present concerns for the conservation of nature and biodiversity, whose value and importance no longer need to be mentioned, having been accepted already by national and international authorities.

Although *Sozology*, as a discipline in course of consolidation, has its roots in the second half of the 20th century, its foundations were laid in the first half of that century, and at present it is being enriched from a conceptual and methodological point of view. However, as in any construction, where the foundation is extremely important because depending on it one can built bigger and better, in conservation it is important to know how our predecessors, those who laid this foundation, thought and acted. And the exhortation of the strategy of sustainable development – "*Let's think globally, let's act locally!*" – also obliges us to know what has been done at international level in the field with which we are concerned, in order to act within a regionally defined plan.

These are the premises of this historical approach; this is why we should like to place the four personalities of the Cluj biological school (A. Borza, E. Racoviță, E. Pop and N. Boşcaiu) among the ‚citizens of the planet', of which Professor F. Pedrotti (to whom this Symposium is dedicated) has earned – by his work, by actions organized and by the friendship manifested towards the naturalists of the five continents – a place of great honour.

A. Borza, the true champion of the movement for nature protection in Romania

Without any doubt, none of the Romanian naturalists of yesterday and today had the vision, energy and skill of A. Borza in organizing scientific and cultural institutions or in founding a new scientific school. In order to reinforce our statement, we add here the words of the great Academician E. Pop: *"The organization achievements of professor Borza are the best known to the public, due to their concrete, permanent and spectacular character"* [22, p. 15]. We owe to A. Borza the creation of the botanical garden of Cluj-Napoca (a pearl of this part of Europe); he was the founder of the Romanian school of phytosociology; he also gathered the Romanian ethnobotanists into a movement that still breathes his own spirit; and most importantly, he organized and gave impetus to the nature protection movement in our country. Referring only to this last aspect, we shall emphasize the following contributions:

- Before World War I, he increased the awareness of the Transylvanian public, as well as that of European specialists, regarding the need to protect certain rare plant species (e.g. *Leontopodium alpinum, Saponaria bellidifolia*), centuries-old trees or „ancient nature" areas of preservation (e.g. the Craiului plain from Belioara).
- After he came to the young Romanian university of Cluj (in 1919), he became involved in the formulation of the law project of the agrarian reform, requiring that certain territories be *"completely used for the benefit of science"*. In this way he created the premises for the foundation of the first natural reserves, first in Transylvania, then throughout Romania. The political changes of that time led to the exclusion of the article concerned from the law, but a total of 27 areas remained under the custody of the Cluj Botanical Garden and became, after more than a decade, officially constituted natural reserves.

It is interesting to note that, in 1925, speaking of the study of vegetation in such reserves, A. Borza also proposed the delimitation of *"permanent squares for the exact study of succession"* [3, p. 11], an approach which today has become a rule in the ecological management of these areas.

- In addition to some popular articles, in 1924 he elaborated a real "programme manifesto" for "nature protection in Romania", considered by us as a *"synthesis rich in information, practical by its advanced proposals and visionary by the concepts presented"* [15, p. 11]. Of the multiple aspects developed by A. Borza in this "programme manifesto", we wish to emphasize the fact that, ever since that time, he saw the need to involve the international community in such protection activity, which today seems perfectly natural. Here is one of his statements, which we would strongly recommend to the politicians of the contemporary world: *"Science belongs to everyone, the right and the duty to study nature belong to the whole humanity, consequently the control of natural monuments is an international duty of everyone"* [1, p. 18].

Although published in a popular journal, another article of that year launches two important ideas for today: "architectural pollution" and "the setting up of reserve networks". In this article, the author urges us"*...not to spoil beautiful landscapes by ugly and inadequately placed buildings"* and *"...to create a network of natural reserves across the whole country ..."* [2, p. 1];

- He was aware that, without a legal base, any action for the protection of nature was destined to fail, which is why in 1925 he submitted a law project to the Ministry of Domains, and in 1927 he published an article in which he stated: *"So, we need before anything else a special law for nature protection..."* [4, p. 10]. Then, he emphasized (which we consider to be a novel concept) that the aim of setting up reserves and parks was to ensure *"...the protection of special life environments..."* [4, p. 5], in other words habitat types! His proposals also concerned some practical aspects, like the organization of a special commission at national level (with subcommissions at regional level), as well as of an "office" which *"...will be concerned with the technique and carrying out of effective protection"* [4, p. 11], etc.
- As the law proposed by him was not considered for discussion in Parliament, A. Borza organized in Cluj the first Congress of Romanian Naturalists (18–21 April 1928), convening a conference on nature protection. His exposition had the general scheme of the paper from 1924, developed with the ideas suggested by his visit to the USA (in 1926), with what he stipulated in the law project and with some new aspects, referring to: *1.* the education of the public, action in which *"The priests of all cults will assist the naturalists in this action"* [5, p. 26], and; *2.* taking into consideration of nature protection in all standard activities and actions of central and local administrations. In this sense, he stated: *"We will have to insist that all laws and actions of our State administration be impregnated with the principles of Nature protection... Thus, we will create an atmosphere of sacred respect for ancient Nature, which will be the strongest source of a well understood patriotic feeling and the feeling of universal cultural solidarity, which should animate humanity in the future"* (idem). Today, we witness such solidarity at both international and national level, but only after a delay that has significantly affected the ecological balance in many areas of Earth.
- Legislative success was accomplished in July 1930, when the first Romanian law „for the protection of natural monuments" appeared, to which new articles and two regulations were added in 1933. All this allowed A. Borza to plan

new courses of action, among which was the fulfilment of his dream to create the first national park in Romania. This was achieved in March 1935, when about 10,000 ha from the Retezat Mountains became *"...sacred land for science...a huge sanctuary of Nature..."* [6, p. 7]. Today, this national park (with an area of 38,047 ha) is included in the international Biosphere Reserves network, and some naturalists (or pseudo-naturalists!) who have not tried hard enough to inform themselves adequately have attributed (in tourist guides and maps) to other people the credit for having created this first national park. Official documents, starting in December 1923, can be consulted in the journals of the Cluj Botanical Garden (*Bul. Grăd. Bot. Muz. Bot. Univ. Cluj*, **IV**, 1924) and the Commission of Natural Monuments (*Bul. Com. Mon. Nat.*, **XI**, *1-4*, 1943).

- The resolutions of the Congress of Cracovia (December 1929), elaborated by well-known names in natural sciences from Poland, Czechoslovakia and Romania (K. Domin, W. Szafer, W. Goetel, J. Komarek, M. Siedlecki, A. Borza, etc.) propose, among others, the creation of *"transfrontier parks and reserves"*, another direction of action which is increasingly gaining ground and is regarded by many naturalists as something new.
- A. Borza did not neglect the varieties of cultivated plants, especially traditional varieties, campaigning for setting up living collections of these local varieties or land races *"...especially for future times, when, following the „standardization" of production, they will be forgotten and will disappear..."*, suggesting that, in this way *"...we keep... a living archive of our cultural history and we consider a duty of honour to do this in the interest of science"* [9, p. 6].
- More than six decades ago, A. Borza drew our attention to the danger of "aesthetic pollution" by various advertisements: *"Industrial and commercial advertisements are still rare...but they are diabolically placed where nature organically repels such inadequate settings"* [7, p. 4]. But who is now approaching this problem, except for some 'old fashioned' naturalists and architects?
- After his brutal exclusion from the Botanical Institute (in 1947), Romanian naturalists would lose the source of ideas and enthusiasm of this great naturalist, trying devious roads in the new (!) society which some were intending to build. A. Borza would only write after his partial rehabilitation (1953), but he would no longer be allowed to be directly involved in teaching and organizational activity.

Of his writings from this last period, we wish only to mention that published in 1959, in the journal *Naturaliste Canadien*. Again, he would suggest a way (considered by contemporary conservationists as a great discovery!) via the closing statements of the chapter on "nature protection": *„But now, when the danger of the rapid disappearance of these ethnobotanical notions and practices...., we become more and more preoccupied with this circle of para-botanical events, especially in combination with ethnography and economical technical research work... Then, we should combine the millenary experiences and the wisdom of our forefathers with the new methods of utilizing the endless gifts of* *the sacred earth of our country in order to grant a better and happier living to our descendants"* [8, p. 110].

E. Racoviță, a vanguard theoretician

After he travelled in the boat "Belgica" across the Antarctic seas, after he worked for almost two decades in France, falling in love with the world of caves, E. Racoviță was appointed a professor at the University of Cluj, where he organized the world's first Institute of Biospeleology. At that time a celebrated scientific personality, he presided over the Congress of Cluj (in 1928), and in 1934 offered Romanian specialists a "guide" to the definition, classification, role and management of "natural monuments". Subsequently published in French (in the Bulletin of the Society of Biogeography), this guide preceded by more than a decade that of Bourdelle (1948), considered an authoritative reference by IUCN. Whether it was known or not by specialists, Racoviță's first guide in the field of nature protection presents:

- A definition of natural monuments, considered as *"...all prehistoric stations, living beings, underground deposits and human works which, due to their scientific, artistic, picturesque and legendary interest, deserve to be conserved for the present and future use and which are considered as such by the law"* [26, p. 4-5];
- A classification of these monuments *according to their purpose and use* (natural reserves, scientific reserves, tourist reserves) and *according to their nature*: station reserves (national parks, forest reserves, ornithological shelters, etc.), reserves for geographical and geological formations, reserves for mineral, paleontological and archaeological deposits, protected species and individuals;
- A methodology regarding the declaration of an area as a reserve in which, after obligatory detailed studies, the land area will receive a temporary status of protected area, period during which the efficacy of the proposed protection measures, the social and economic impact (in relation to scientific, touristic and aesthetic benefit), and the practical possibilities of ensuring protection will be monitored;
- An extremely modern outlook, stating that *"...an ideal protection can only be achieved if a species, plant or animal is protected in an extensive reserve, in its original biotope, in other words, in its normal, consequently necessary and sufficient habitat"* [26, p. 6].

In the same year that Borza was "forbidden", E. Racoviță died, refusing special treatment compared to his fellow citizens, under conditions of cold, food deprivation and under the terror of change!

E. Pop, in the cobweb of the socialist society

A disciple of the two professors presented above, E. Pop remains in Romanian science as the founder of the school of palynology, plant cyto-physiology and, certainly, as a militant proponent of nature protection. After the political changes of 1947, for almost a decade he did not escape the attacks of those who wished to climb up the university hierarchy according to political rather than scientific criteria.

We estimate that E. Pop started to get officially involved in the field of nature protection with the publication (in 1942) of the work "The forests and our national destiny". Considering the forest not only as a "capital of richness and beauty", with decisive economic and ecological implications, but also as a "spiritual necessity" for certain peoples, he defined it in an unequalled manner: *"...a collective being, the most magnificent of all that we know, which palpitates with its own life, made up of millions of individual lives...assimilated into a harmonious appearance and a unitary spirit, which impresses in a stunning and unique way the human spirit..."* [19, p. 8]. After 39 years, he would again turn to research, protection and conservation of forests, urging us all to disturb *"... the great natural biological balance only to the extent to which its restoration is possible by the native powers of nature"* [21, p. 498].

After legislation in the field of nature protection was changed in the ‚new Romania' (1950 and 1954), and the Commission of Natural Monuments passed under the authority of the Academy, E. Pop also approached the problem of the conservation of peat bogs, both oligotrophic and eutrophic. The first category *"is not only perfectly defined from a physiognomic, economic, ecological, floristic and biocoenotic point of view, but its individuality implies a dimension that other formations do not possess: depth. The living bog forms a common body with its deposit..., which is nothing else but the complete succession of the conserved mummies of the ancestors of the present living flora..."* [20, p. 58]. Regarding the second category, E. Pop states: *"...intra-Carpathian eutrophic bogs should be considered among the most prominent relict-conserving formations that we know in the Central European floristic area* [20, p. 79].

Taking the leadership of the Commission of Natural Monuments in the early 1960s, E. Pop gave an impulse to the setting up of new natural and scientific reserves, as well as of two natural parks; he ensured rhythmicity in the publication of the specialist journal; he developed collaborative relations with foreign specialists; he initiated the organization of itinerant conferences entitled *"Nature protection on ecological grounds"*; he offered the public a fine synthesis of the Romanian reserves [24]; and he co-ordinated the realization of the project for a new law (Law for the Protection of the Environment, no. 9/1973).

He ended this activity with a swan song, entitled *"Glorification of the Carpathians"*, published posthumously in 1975, in which he draws our attention to the fact that natural monuments are distinguished from monuments of human art *"...by the monumentality of vast, immense proportions, which impresses in an inimitable way the spirit, all the more so as a natural monument is never a finished piece, but is organically integrated, without borders, by soft or wild lines, but natural by excellence, into more and more endless aspects"* [23, p. 7].

Throughout this whole activity of protection, he had to make extensive use of his diplomatic spirit and support any initiative in as a detailed a manner as possible, since the interests of our economy frequently blocked these initiatives, and the discordance between declarations and facts often made a law impracticable, despite its being conceptually adequate.

N. Boşcaiu, a disciple under three social systems

A school pupil with a passion for botany, an eminent student and a constant collaborator of Professors A. Borza and E. Pop, N. Boşcaiu would only succeed in completing his studies after six years of "re-education" in gloomy communist prisons. The reason?: his supposed participation in the anti-communist student strike in 1946! Only in 1962, at the intercession of E. Pop, would N. Boşcaiu succeed in occupying a modest position of biologist at CMN, then of botanist at the Botanical Garden, subsequently to become a librarian (!) at the Cluj branch of the Academy. Although late in the day, in 1991 he was rehabilitated completely, when he was elected a member of the Academy and when, from a librarian, he became Director of the Academy's Cluj branch library.

In spite of all these dramas in his life, N. Boşcaiu took over from E. Pop the leadership of the movement for nature protection in Romania, which he guided until the end of the 1990s, when he handed over to the ornithologist Dr. Dan Munteanu.

Among N. Boşcaiu's main contributions, we should mention the following:

- In the context of the clash between Systematics and the cutting edge of biology, expressing his *"concern that the reduction of the interest in the continuation of taxonomic work might affect the most varied fields of theoretical and applied biology* [13, p. 117], he proposed a reinvestigation of the flora on *"...the new grounds of population systematics"* [13, p. 120].

At the same time, he drew attention to the fact that: "*In the absence of a strong taxonomic basis, ecological research remain exposed to the risk of diluting into increasingly vague abstractions and generalizations detached from the real specificity of concrete biogeographical conditions*" [13, p. 119].

- By campaigning for the setting up of an ecological policy in our country 30 years ago, he pointed out that *"... contemporary realities convince us more and more that the idea of ecological policy should acquire the same importance as economic policy"* [28, p. 129].

Although in the mid-1980s, he believed that *"Today we no longer need a picture of the destruction of nature in order to increase public awareness..., but a realistic presentation of measures by means of which the environmental deadlock of our time will be overcome"* [12, p. 163], we can see that there are increasingly fewer voices to demonstrate by facts the truth he emphasized: *"Perhaps in no other area than the protection of natural resources, do the interests of the present generation interweave so closely with those of the future generations"* [12, p. 166].

- After a period of massive deforestation of the Romanian subalpine area, without the creation of the long-dreamt-of pastures for the socialist livestock sector, N. Boşcaiu demonstrated (in 1975) that the ecological role of junipers was much more important than any temporary and uncertain economic advantages, that a regeneration of these formations would only be possible at the end of a new Preboreal phase, and that *"...the legend of the junipers invading the uncared-for pastures remains a fable"* [10, p. 20];

- In the early 1980s, when documentation was initiated for setting up new national parks, N. Boşcaiu pointed out the immense importance of the Apuseni Mountains area, with a centre at Padiş, proposing the acceleration of the legalization of national park status: *"Any delay in the setting up of the national park would result in the aggravation of the deterioration processes of these national patrimony values"* [25, p. 177]. The situation of the present natural park demonstrates the truth of the words of these two conservationists!
- His 1985 article is full of innovative ideas, referring to the criteria for the setting up and ecological management of botanical reserves, stating that *"...the only real possibility of protecting species threatened to disappear remains the attempt to conserve the biological communities and ecosystems to which they belong"* [11, p. 127].

At the same time, he anticipated the practical criteria (which are used today!) in the characterization of habitats and creation of protected area networks, showing that: *"The classification of plant associations can offer in this way the ground for a detailed typology of ecosystems, intended to ensure a reference framework for the setting up of a regional reserve network"* [11, p.132].

- Finally, from 1990 he also approached the problem of the network of *"biogenetic reserves"*, not only for their holistic role in conserving genetic, specific and ecosystem diversity, but also as a symbol of the *"...increased ecological awareness of the population"* [29]. He also became involved in the elaboration of the standard acts for setting up the nine national parks proposed since the early 1980s, trying to establish a real communication with the new Ministry of the Environment, created after the revolution of December 1989.

Concluding remarks

By reviewing the main contributions of these four personalities, our intention has been to offer young Romanian people and specialists from all over the world their example of devotion, tenacity, professionalism and anticipative thinking, qualities without which research cannot advance, and nature protection cannot be achieved. If we have mentioned some constraints to which the lives of these personalities were submitted, under the vicissitudes of history, we did it to emphasize even more their achievements, to give more strength to their example.

And if we have used in the title of our paper the term "Traditions", it is because we consider that the three generations who have thought and acted for nature protection in Romania have clearly demonstrated the validity of their facts and ideas, and because this term seems to us the most appropriate to express something that has been (successfully) performed now for more than eight decades!

REFERENCES

1. Borza A., 1924a - Protecţiunea naturii în România. Bul. Grăd. Bot. Univ. Cluj, **4** (1): 3-27.
2. Borza A., 1924b - Ocrotiţi natura! Cultura poporului, Cluj, **4** (47): 1.
3. Borza A., 1925 - Dare de seamă despre starea Muzeului Botanic şi a Grădinii Botanice de la Universitatea din Cluj în anul 1925. Bul. Grăd. Bot. Univ. Cluj, **5**, app. II: 1-12.
4. Borza A., 1927 - Ne trebuie o lege pentru protecţia naturii. Tipogr. Naţională, Cluj.
5. Borza A., 1929 - Problema protecţiunii naturii în România. In: Borza A., Pop E. (eds.), I Congres al Naturaliştilor din România, Cluj: 94-127.
6. Borza A., 1933 - Retezatul, viitorul Parc Naţional al României. Carpaţii, Cluj, **1** (12): 1-7.
7. Borza A., 1939-1942 - Monumentele naturii din Ardealul central şi apusean. Apulum, Alba Iulia, **1**: 1-14.
8. Borza A., 1959 - Geobotany and allied problems in Romania. Naturaliste Canadien, Quebec, **86**: 93-111.
9. Borza A., Gürtler C., 1933 - Varietăţile de mere cultivate în Grădina Botanică din Cluj. Bul. Grăd. Bot. Univ. Cluj, **13**, app.I: 1-24.
10. Boşcaiu N., 1975 - Problema conservării vegetaţiei alpine şi subalpine. Ocrot. Nat. Med. Înconj., Bucureşti, **19** (1): 17-21.
11. Boşcaiu N., 1985a - Criterii pentru constituirea şi gestiunea ecologică a rezervaţiilor botanice. Ocrot. Nat. Med. Înconj., Bucureşti, **29** (2): 126-135.
12. Boşcaiu N., 1985b - Revoluţia în Biologie şi conştiinţa ecologică. In: Revoluţia biologică, pp. 161-167. Ed. Academiei, Bucureşti.
13. Boşcaiu N., 1986 - Importanţa sistematicii pentru protecţia florei. Ocrot. Nat. Med. Înconj., Bucureşti, **30** (2): 117-120.
14. Bourdelle E., 1948 - Essai d'unification de la nomenclature en matière de protection de la nature. Pro Natura, **1** (1): 33-36.
15. Cristea V., 1995 - La conservation de la nature en Roumanie. L'Uomo e l'Ambiente, Camerino, **18**: 1-104.
16. Cristea V., Denaeyer S., Herremans J-P., Goia I., 1996 - Ocrotirea naturii şi protecţia mediului în România. Ed. Cluj University Press, Cluj-Napoca.
17. Laliotis C., 1995 - Une lourde responsabilité. Naturopa, **77**: 10.
18. Leufeuvre J.C., 1995 - Europe: préserver la nature pour préserver le futur. Naturopa, **77**: 11-12.
19. Pop E., 1942 - Pădurile şi destinul nostru naţional. Bul. Com. Mon. Nat., Bucureşti, **9** (1-4): 1-31.
20. Pop E., 1955 - Mlaştinile noastre de turbă şi problema ocrotirii lor. Ocrot. Nat., Bucureşti, **1**: 57-105.
21. Pop E., 1971 - Perspective noi în cercetarea şi protecţia pădurilor. Rev. Păd., Bucureşti, **86** (10): 497-498.
22. Pop E., 1972 - Profesorul Alexandru Borza (1887-1971). Contr. Bot., Cluj: 7-36.
23. Pop E., 1975 - Elogiul Carpaţilor. Ocrot. Nat. Med. Înconj., Bucureşti, **19** (1): 7-9.
24. Pop E., Sălăgeanu N., 1965 - Monumente ale naturii în R.S. România. Ed. Meridiane, Bucureşti.

25. Puşcariu V., Boşcaiu N., 1981 - Viitorul Parc Naţional al Munţilor Apuseni. Ocrot. Nat. Med. Înconj., Bucureşti, **25** (2): 165-178.
26. Racoviţă E., 1934 - Monumentele naturii. Definiţie, clasificare, norme de alcătuirea legii. Cam ce trebuie făcut şi ce trebuie evitat. O expunere sumară. Bul. Com. Mon. Nat., Bucureşti, **2** (1-4): 4-7.
27. Racoviţă E., 1937 - Les monuments naturels. Soc. de Biogéographie, Paris, **5**: 15-27.
28. Toniuc N., Boşcaiu N., Filipaşcu A., 1975 - Ocrotirea naturii în prezent şi perspectivă. Ocrot. Nat. Med. Înconj., Bucureşti, **19** (2): 125-132.
29. Toniuc N., Purdelea L., Boşcaiu N., 1994 - Rezervaţiile biogenetice şi importanţa lor pentru conservarea genofondului. Ocrot. Nat. Med. Înconj., Bucureşti, **38** (2): 107-113.
30. xxx, 1994 - *Connexion*, **19** (4): 1-2.
31. xxx, 1995 - *Connexion*, **20** (1-2): 1-2.

JEAN MASSART, PIONNIER DE LA CONSERVATION DE LA NATURE EN BELGIQUE

Simone DENAEYER-DE SMET*, Jean-Paul HERREMANS, Jean VERMANDER
Institut de Gestion de l'Environnement et de l'Aménagement du Territoire (IGEAT)
Université libre de Bruxelles (ULB), avenue F.D. Roosevelt, n° 50,
1050 Bruxelles (BELGIQUE); e-mail: s.denaeyer@gmail.com

Abstract: (***Jean Massart, a pioneer of Nature Conservation in Belgium***).
Jean Massart was born in Brussels in 1865. After gaining the title Doctor of Sciences and Doctor of Medicine at ULB, the young graduate was immediately hired as a researcher in Prof. P. Héger's laboratory of human pathology. In 1892, he was offered a position as an assistant at the world-famous L. Errera Botanical Institute at ULB. From this time, he dedicated most of his time to plant biology. At the age of 32 he became Professor of Botany and started a career of field research without giving up his laboratory activities. Carrying out at the same time his functions of teacher, director of the Botanical Institute, and curator of the State Botanical Garden of Belgium (1902–1905), he started a large study of the Belgian flora, especially aimed at geobotany and plant adaptation to the environment. These investigations led to two works giving a thorough insight into the state of Nature in Belgium at the beginning of the 20th century. The first, of 166 high quality 30 x 40 cm photographic plates, shows the importance given by Massart to natural vegetation but also to rural landscapes and human activities. The second gives an outstanding synthesis of personal data collected during ten years in the field and the main geological, climatic and soil data available at this time, and concerning the 13 natural districts he indicated for the whole Belgian territory. This book points out the progressive outlook of Massart, who achieved an ecological synthesis ahead of his time. In 1912, he published an underlying book on nature conservation in Belgium, then the most industrialized country of continental Europe. His approach is multidisciplinary, taking into account the past (prehistory, history, palaeontology, etc.), present (aesthetics, welfare, natural resources protection and life conditions) and future (sustainable development). Massart can be considered a true forerunner of nature conservation because his ideas still support such European and international programmes. Massart published many other scientific papers across a range of botanical sciences, and also dedicated much time to scientific popularisation.

1. QUI ETAIT JEAN MASSART?

Des serres familiales à l'université

Jean Massart naquit le 7 mars 1865 à Etterbeek (commune de l'actuelle «Région Bruxelles-Capitale»). Ses parents exploitaient une petite entreprise horticole spécialisée dans la production de fougères et de palmiers d'appartement. Auparavant, son père, Charles Massart, avait travaillé comme ouvrier-jardinier au Jardin botanique de l'Etat à Bruxelles. C'est là qu'au contact de ses supérieurs, ce père illettré comprit toute l'importance de l'instruction. Aussi, ne ménagea-t-il aucun effort pour assurer au mieux celle de son fils, en particulier l'apprentissage des langues. Massart apprit à lire dans une école privée et accomplit l'école primaire en 4 ans; à 10 ans, il lisait Jules Vernes en anglais.

L'amour de la nature et de la vie qui imprègne toute l'œuvre de Massart est né dans les serres familiales. Dès son plus jeune âge, il y passait des heures à observer plantes et insectes; à ses petits camarades de classe qui venaient l'y rejoindre après leurs jeux, il racontait ce qu'il appelait des «histoires» pour leur faire connaître les merveilles qu'il découvrait chaque jour.

Après avoir terminé l'enseignement secondaire à 15 ans, Massart s'inscrit à la Faculté des Sciences de l'Université libre de Bruxelles. Malheureusement, la mort prématurée de son père l'oblige à assumer la gestion de l'entreprise familiale. Contraint d'abandonner l'Université, il installe un petit laboratoire de recherche dans la maison familiale et y poursuit des observations microscopiques de tissus végétaux et d'organismes unicellulaires dont il étudie expérimentalement l'irritabilité.

Conseillé par des amis, Massart reprend ses études universitaires en 1884 et brûlera les étapes. Proclamé Docteur en Sciences en 1887 (à 22 ans), il entreprend aussitôt des études de médecine car après s'être intéressé à la physiologie végétale et animale, il est attiré par la physiologie humaine. Etudiant brillant, il poursuit des recherches personnelles dans son petit laboratoire privé et ce, tout en participant encore à la gestion de l'entreprise familiale. Proclamé Docteur en Médecine quatre ans plus tard (à 26 ans), il est immédiatement engagé dans l'équipe de recherche en pathologie humaine du célèbre physiologiste, le professeur Paul Héger.

En 1892, la carrière de Massart change de cap. Le professeur Léo Errera, botaniste-physiologiste de réputation internationale, demande à son collègue Héger de lui «céder» Massart à qui il offrait une place d'assistant à l'Institut botanique qu'il venait de fonder et qu'il dirigeait. Ne pouvant assurer un avenir matériel aussi prometteur, Héger, non sans regret, se sépara de Massart dont il dira *«Massart n'était pas un collaborateur ordinaire, il animait tout le laboratoire; il communiquait son enthousiasme et sa gaîté à tous les camarades, à tous ceux qui, en même temps que lui, travaillaient avec moi»* [3].

Nommé professeur en 1897, le jeune enseignant accepte en 1902 un poste de conservateur au Jardin botanique de l'Etat pour arrondir ses fins de mois (la jeune Université libre de Bruxelles n'était pas riche et Massart avait charge de famille!). Mais, en 1905, la mort inopinée d'Errera l'obligera à renoncer à ce poste pour assumer la lourde succession de son «patron»: direction de l'Institut botanique et des recherches, enseignements de candidature et de doctorat.

Elu membre correspondant de la Classe des Sciences de l'Académie royale de Belgique en 1904, Massart en devient membre titulaire en 1911. Il était également correspondant ou membre de plusieurs Académies étrangère.

La première guerre mondiale interrompt les activités universitaires de Massart qui sera obligé de quitter la Belgique pour se réfugier à l'étranger avec toute sa famille.

Rentré au pays dès la fin des hostilités (le 11 novembre 1918), il reprend immédiatement et avec ferveur l'ensemble de ses activités à l'Université libre de Bruxelles.

Membre de la Commission royale des Monuments et des Sites, Massart y jouera un rôle décisif dans la mise sur pied d'une politique de protection de la nature. C'est aussi l'époque à laquelle il conçoit de grands projets (création d'un nouvel institut botanique dans la banlieue de Bruxelles, et d'un institut de biologie au Congo belge) que sa mort prématurée empêchera de réaliser.

Au début de l'année 1924, Massart répond à l'invitation de la *C.R.B. Educational Foundation* des USA. Il séjournera plusieurs mois aux Etats-Unis, y fera des conférences en anglais dans une dizaine d'universités et participera à plusieurs excursions dans des sites naturels célèbres. C'est au retour de ce voyage que les premiers signes de la maladie qui devait l'emporter se manifestèrent.

La mort surprit Massart pendant ses vacances, le 16 août 1925 à Houx (près de Dinant en Belgique) alors qu'il poursuivait la rédaction du compte rendu de la mission biologique belge au Brésil [32]. Il venait d'être élu Directeur de la Classe des Sciences de l'Académie royale de Belgique. Le nouvel institut botanique ne vit pas le jour. Seuls, les jardins phylogénique et éthologique qui étaient déjà tracés et partiellement aménagés purent être achevés grâce à une ASBL constituée par des amis de Massart qui, en l'honneur du disparu, le baptisèrent du nom de *«Jardin expérimental Jean Massart»*.

Massart n'était pas catholique. Ses proches se virent refuser le corbillard du village par le curé et par les instances officielles locales. Ils furent obligés de procéder à l'enterrement avant le lever du soleil [5].

Le chercheur

Massart fut un chercheur précoce. Le jour même où il obtint son premier diplôme universitaire (il a 22 ans), il exposa les résultats de recherches menées dans son petit laboratoire privé pendant ses études; ceux-ci soulevèrent l'enthousiasme de ses professeurs qui l'incitèrent à publier rapidement une note préliminaire pour prendre date. Massart n'en fit rien. Un peu plus tard, il prend connaissance des travaux du professeur W. Pfeffer et constate que le célèbre botaniste allemand a étudié les mêmes problèmes que lui et a fait les mêmes découvertes (sensibilité des Flagellates à la lumière, à certaines substances chimiques, etc.). Massart écrit alors un article qui résume les travaux de Pfeffer [7] auxquels il ajouta de nombreuses données issues de ses recherches personnelles, mais sans en revendiquer l'antériorité. Pfeffer reconnut la valeur des travaux de Massart et l'invita à travailler dans son laboratoire de Tübingen.

Etudiant de première année en Médecine, Massart se fait remarquer par un de ses professeurs (Héger), qui l'autorisera à utiliser une salle de cours occultable pendant les vacances académiques pour étudier le phototropisme de certains champignons. La même année, l'Académie royale de Belgique accepte de publier dans son *Bulletin* les résultats d'une autre recherche effectuée dans son petit laboratoire privé [6].

Etudiant au Doctorat en Médecine, Massart se voit confier une recherche qui posait problème à son Maître Héger. Cette recherche, faite en collaboration avec un autre étudiant en médecine, permit de mettre en évidence le chimiotaxisme des leucocytes aux toxines bactériennes. Cette découverte, publiée par l'Institut Pasteur de Paris [31] contribua largement au développement de la théorie naissante de l'immunité.

Nommé assistant en Botanique un an à peine après avoir été proclamé Docteur en Médecine, Massart abandonnera la recherche médicale pour se consacrer à la biologie végétale et plus particulièrement à l'étude des adaptations des végétaux à leur milieu de vie. Sans renoncer à ses recherches en laboratoire, il entame alors une carrière d'homme de terrain. En collaboration avec le professeur Charles Bommer [1], il lance une campagne d'inventaire méthodique de la flore de Belgique. Dans ce but, il fera appel à des collaborateurs volontaires pour qui il met au point une plaquette leur donnant les *«Instructions pour l'usage de la liste d'herborisation»*. Outre la localisation géographique, cette liste demande des renseignements sur le milieu de vie, c'est-à-dire sur l'autoécologie des espèces concernées. Cet inventaire très avant-gardiste (l'écologie n'est pas encore entrée en Belgique), fournira à Massart les bases d'une synthèse remarquable de la géographie botanique de la Belgique sur laquelle nous reviendrons ultérieurement.

L'explorateur et le photographe

Enthousiasmé par ses observations de terrain en Belgique, le jeune Massart désire voyager pour découvrir d'autres milieux de vie et cherchera toutes les occasions pour le faire. En 1894–1895, il obtient une bourse du gouvernement belge pour effectuer une mission scientifique en Indonésie (la Malaisie de l'époque). Pour payer le voyage, il se fait engager comme médecin à bord d'un bateau qui transporte des musulmans au pèlerinage de La Mecque. C'est la première mais aussi la dernière fois que le «Docteur Massart» exercera la médecine. Cette mission, au cours de laquelle il découvre dans leur splendeur naturelle les palmiers, fougères et autres espèces tropicales cultivées dans les serres familiales influencera profondément le rapport à la nature du jeune chercheur. Il traduira son enthousiasme débordant dans plusieurs publications [8, 9].

En 1898, il explore pendant deux mois une partie du Sahara, en compagnie de son collègue et ami, le zoologiste Auguste Lameere [10]. Il étudiera aussi la végétation alpine et visitera les fjords et les glaciers de Norvège.

Au cours de l'été 1922, il passe plusieurs mois au Brésil avec quatres jeunes universitaires belges qu'il désire initier à l'étude de la flore et de la faune tropicales et ce, en vue de la création d'un institut de biologie au Congo belge. La notoriété de Massart était telle que les membres de la très officielle «Mission biologique belge» furent reçus partout en grande pompe par les plus hautes autorités. Un monument sera érigée dans le jardin botanique de Rio de Janeiro pour commémorer cette visite.

Passionné par l'étude de la nature et des êtres vivants, Massart l'était aussi par la photographie. Il possédait un matériel photographique déjà très performant mais lourd et difficile à transporter, ce qui ne l'empêcha pas de réaliser des milliers de photos pour illustrer ses missions scientifiques à l'étranger et ses études des paysages et de la végétation en Belgique. Il révélait lui-même les photos prises sur plaques de verre gélatinées (ce qui lui posa quelques problèmes au cours de son voyage en Indonésie en raison de la chaleur qui faisait fondre la gélatine!). Ces photos utilisées par Massart pour illustrer de nombreuses publications, constituent aujourd'hui une somme impressionnante de documents d'une valeur exceptionnelle en raison de leur ancienneté (plus d'un siècle!), de la précision des légendes, du choix et de la qualité des images qui n'ont rien à envier aux photographies actuelles.

Le professeur et le vulgarisateur

Massart dispensa son enseignement universitaire à de nombreuses générations d'étudiants car à son époque, les futurs pharmaciens, médecins, chimistes, géologues et géographes suivaient le même cours de biologie générale et de botanique que les futurs biologistes. Les témoignages sont unanimes: Massart était un excellent professeur et, fait exceptionnel pour l'époque, il illustrait ses cours par des projections lumineuses dans l'amphithéâtre, par des manipulations en chaire et par ce qu'il appelait des «*démonstrations*» en salle à l'aide de matériel végétal vivant ou récolté lors de ses missions scientifiques. De plus, il organisait tous les quinze jours une herborisation dans l'une ou l'autre région du pays. C'est en évoquant ces excursions que Paul Brien, un de ses derniers élèves devenu professeur de zoologie à l'ULB écrira: *«Il* (Massart) *n'était vraiment lui-même que dans les champs, les bois, les prés, au bord des ruisseaux, sur les grèves des mers»* où il faisait découvrir par ses étudiants des choses que seuls ils n'auraient pas pu découvrir. Pour Brien, Massart était le Fabre de la Botanique [2].

Homme de terrain, Massart sera le premier en Belgique à réaliser des cultures expérimentales en plein air destinées à l'étude comparative de l'éthologie d'un certain nombre d'espèces végétales. Homme de laboratoire, il créera en 1895 le premier *laboratoire ambulant.* Destiné principalement aux étudiants mais accessible au grand public, ce laboratoire fonctionnera successivement dans différents districts botaniques du pays dans le but d'étudier *in situ* les adaptations des plantes à leur milieu de vie. Le plus célèbre fut celui installé dans les dunes littorales de Coxyde (fig. 1). Fréquenté surtout pendant les vacances académiques, Massart y accueillit de nombreux étudiants, collègues étrangers et naturalistes amateurs venus participer à ses recherches sur la végétation locale. Il avait aussi installé la résidence d'été de sa famille dans la petite cité balnéaire dont une rue porte son nom [36].

Dès la fin de la première guerre mondiale, Massart crée sur la côte belge un laboratoire de terrain (fig. 1) pour suivre une expérience grandeur nature concernant les effets de la salinité sur la végétation: la recolonisation végétale des terres dévastées par les inondations stratégiques (eau saumâtre) du front de l'Yser [28].

Massart accordait beaucoup d'importance aux collections de plantes vivantes car il y voyait le meilleur moyen pour sensibiliser l'opinion publique à l'importance des plantes dans la vie de l'homme. C'est dans ce but qu'il remania les collections du jardin botanique de l'Etat: il créa une collections phylogénique [13] pour illustrer *de visu* les phénomènes de l'évolution, et des collections éthologiques [12] pour illustrer les modes d'adaptation des plantes à leur milieu de vie. Dans le même esprit, il élabora un plan de jardin botanique à créer pour les écoles moyennes; ce plan comportait la liste et les conditions de culture de 72 espèces dont la connaissance lui paraissait indispensable [11].

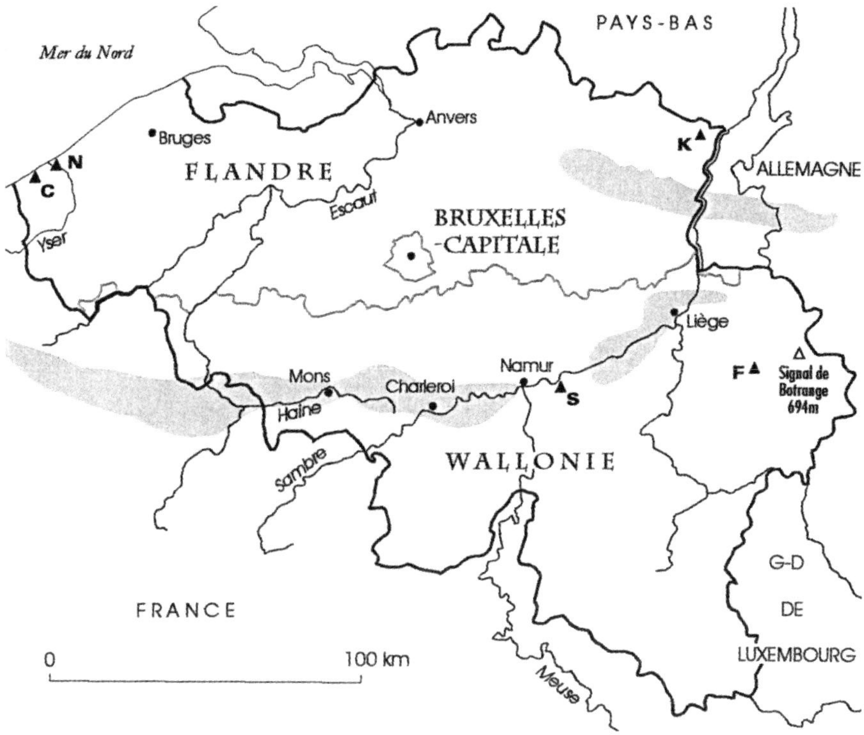

Fig. 1: Localisations du laboratoire ambulant de Jean Massart en Belgique: C – Coxyde (1902), dans les dunes littorales; **K** – Kinrooi (1895), dans une auberge du district campinien; **S** – Samson (1900), sur les bords de la Meuse, district calcaire; **F** – Francorchamps (1899), dans la maison communale, district calcaire; **N** – Laboratoire de fortune créé à Nieuport, après la première guerre mondiale pour étudier les effets des inondations stratégiques de l'Yser sur la végétation et la flore de Nieuport. En gris, bassin houiller. En trait fin: limites des trois régions actuelles de la Belgique fédérale. Signal de Botrange: point culminant du pays.

Massart aimait la science pour elle-même et désirait partager les joies de la connaissance avec le plus grand nombre. Il fut un des fondateurs et un des principaux animateurs de l'«*Extension de l'Université libre de Bruxelles*» créée en 1894 et dont le but était la diffusion de la culture scientifique dans les milieux extérieurs à l'Université (enseignants des écoles secondaire et primaire, associations diverses, grand public, etc.).

Conférencier très demandé, Massart organisait des «*conférences-promenades*» et des «*excursions scientifiques*», notamment dans les environs de Bruxelles. Les comptes-rendus détaillés, enthousiastes et pittoresques de ces excursions, rédigés par une de ses anciennes élèves, devenue professeur d'école normale, illustrent à merveille les dons d'observation et la connaissance de la nature que le maître avait l'art de transmettre aux excursionnistes [35].

Le citoyen et l'humaniste

Massart était l'antithèse du savant enfermé dans sa tour d'ivoire. C'était un citoyen à part entière, qui suivait avec attention l'évolution des conditions de vie dans un pays en plein essor industriel; il s'intéressait en particulier aux conditions de travail et au cadre de vie des plus humbles, comme en témoignent ses rapports avec une des grandes figures du socialisme belge et international, Emile Vandervelde [33, 34].

Mais ce sont surtout les années de guerre et d'après-guerre qui révélèrent ses grandes qualités morales et humaines. Quelques mois après le début des hostilités, Massart interrompt complètement ses activités à l'Université de Bruxelles car il estime que *«Ce n'est pas pendant que la géographie politique de la Terre entière est en voie de bouleversement qu'on peut s'abandonner aux spéculations de la science pure»* [24]. Il se consacrera désormais à recueillir et à diffuser un maximum de documents relatifs à l'occupation allemande de la Belgique, notamment en photographiant toutes les affiches mensongères que les Allemands faisaient placarder pour justifier leur présence dans le pays. Il assurera la diffusion de ces photos ainsi que celle d'autres informations clandestines pour soutenir le moral des combattants et des citoyens. Il tentera aussi, mais sans succès, de ramener à la raison les 93 artistes et savants allemands qui lancèrent un «*Appel au monde civilisé*» pour justifier l'invasion allemande de la Belgique. Ses actions de résistance aux Allemands menaceront rapidement sa sécurité et celle des siens et obligeront toute la famille Massart à se réfugier d'abord en Hollande, puis en Angleterre et de là, dans le sud de la France.

C'est en donnant des leçons d'anglais et en bénéficiant d'un subside de la Sorbonne pour étudier la flore méditerranéenne [21] que Massart y demeura jusqu'à la fin des hostilités. En 1916, il contribue à deux ouvrages à vocation caritative publiés à Paris [22, 23]. En 1918, il dénonce avec véhémence le manque de sens critique et la malhonnêteté intellectuelle des signataires allemands précités [24]. Rentré en Belgique dès la fin des hostilités, Massart publie un livre pour illustrer (129 photos!) les ravages de la guerre en Belgique [25].

Massart était polyglotte: outre les deux langues nationales (français et néerlandais), il connaissait l'anglais, l'allemand (ce qui lui fut très utile pendant la Grande Guerre!) et le...latin [5]; il apprit le javanais pendant son voyage en Malaisie, ce qui lui permit de traduire les paroles très poétiques de chants javanais entendus sur le bateau [8].

Il aimait aussi la musique, la littérature et la peinture, en particulier les écrivains du terroir et les peintres paysagers en raison de leur rapport à la nature et à la vie rurale. Pour initier ses étudiants aux paysages et à la végétation de la Belgique, il recommandait une visite au Musée d'Art moderne de Bruxelles, en leur conseillant les noms de 40 peintres dont les œuvres illustrent 8 des 13 districts botaniques du pays! Il recommandait aussi la lecture de certains écrivains du terroir, et ce, dans les deux langues nationales (7 francophones et 8 néerlandophones!).

Les publications et l'héritage de Jean Massart

La liste complète de ses travaux fournie par Marchal [4] traduit bien l'ampleur et la diversité des activités de Massart: une centaine de publications scientifiques, dont plusieurs comportent 50–80 pages (une d'entre elles est écrite en langue allemande), une trentaine d'écrits de vulgarisation ou d'intérêt général, un livre intitulé simplement «*Nos Arbres*» qui est un modèle d'approche globale et de vulgarisation d'un vaste sujet rendu accessible au grand public [16], un volumineux traité de biologie générale et de botanique [27, 29] et trois ouvrages dont l'intérêt scientifique et historique est exceptionnel pour la conservation de la nature en Belgique.

Malgré la modestie de son intitulé, le premier d'entre eux, l'«*Esquisse de la géographie botanique de la Belgique*» [15], est une vaste synthèse d'observations et de recherches personnelles menées pendant plus de dix ans et concernant les adaptations des plantes à leurs «conditions d'existence» comme disait Massart. Pour mieux comprendre la distribution géographique des plantes et leur origine, l'auteur ajoute à ses données personnelles de nombreuses informations concernant les conditions climatiques locales, la phénologie, l'origine géologique et la nature des sols, sans oublier l'influence des établissements humains successifs. Il aboutit ainsi à la définition d' unités géobotaniques plus pragmatiques que celles proposées par ses prédécesseurs car elles tiennent compte à la fois de la qualité du milieu naturel et du degré d'intervention de l'homme. Le territoire de la Belgique est ainsi divisé en treize districts naturels caractérisés par une physionomie propre illustrant la grande diversité des sols, des conditions climatiques locales, des paysages et de la végétation et ce, malgré la faible étendue du pays. Par l'approche scientifique pluridisciplinaire et humaniste qui imprègne cet ouvrage, Massart peut, à juste titre, être considéré comme le précurseur de l'approche écologique de la géographie botanique.

Le deuxième ouvrage consacré aux «*Aspects de la végétation en Belgique*» se présente d'une manière inhabituelle en raison de ses dimensions et surtout en raison de son contenu: deux volumes grand in-folio, quelques pages de texte seulement mais un ensemble de cent soixante six planches photographiques de grandes dimensions (30 x 40 cm) réalisées par Massart et représentant les paysages et les aspects de la végétation les plus typiques de quatre districts naturels de la Belgique [14–17]. Trois autres volumes et quatre cents autres planches de mêmes dimensions étaient prévus pour couvrir l'ensemble du territoire. Malheureusement, le coût élevé de tels documents et la situation précaire de l'après-guerre ne permirent pas leur réalisation. Les deux premiers volumes constituent néanmoins une source d'informations particulièrement intéressantes concernant la vie rurale de l'époque car elles nous fournissent un «état des lieux» précis d'une partie de la Belgique qui, quelques années plus tard, allait subir des modifications brutales (inondations et pilonnages lors de la Grande Guerre, pollution ou disparition d'espaces verts liées au développement industriel et à l'urbanisation concomitante). De plus, le choix judicieux des sujets photographiés et les informations fournies par les légendes dépassent le cadre habituel de la géographie botanique «scientifique» de l'époque. En effet, les paysages de Massart nous renseignent à la fois sur les formations végétales naturelles, sur les cultures, sur les pratiques agricoles et sur les autres activités humaines qui en découlent. C'est pourquoi, de tels documents présentent aussi un grand intérêt pour les historiens, les ethnologues, les spécialistes des paysages, etc.

Le troisième ouvrage que Massart qualifie de «brochure» ou encore de «notice» mais qui comporte néanmoins 308 pages, fera de lui le père-fondateur de la protection de la nature en Belgique [18]. Nous lui consacrerons les pages qui suivent.

2. JEAN MASSART ET LA PROTECTION DE LA NATURE EN BELGIQUE

La situation particulière de la Belgique au début du 20ᵉ siècle

En temps qu'Etat indépendant, la Belgique n'existe que depuis 1830. C'est un des plus petits pays d'Europe (32.500 km^2), bordé au nord par 66 km de côte sablonneuse et limité vers l'intérieur par une frontière curieusement «festonnée» car totalement artificielle. Le pays est traversé par deux grands fleuves, l'Escaut et la Meuse, dont il ne possède ni les sources, ni les embouchures (fig.1). Le relief est modéré et s'élève vers le sud-est, de la plaine de Flandre au plateau de l'Ardenne (altitude maximum: 694 m). Le climat est océanique, doux et humide mais présente néanmoins des différences relativement importantes lorsqu'on s'éloigne de la côte. Malgré l'exiguïté de son territoire, la Belgique est traversée par un grand nombre de couches géologiques différentes à l'origine de la diversité des sols, des paysages et du couvert végétal.

Comme le montrent fort bien les nombreuses photos qui illustrent les travaux de Massart que nous venons d'évoquer, la Belgique de l'époque possède encore des paysages très diversifiés et des espaces verts peu ou non perturbés par les activités humaines mais cependant très menacés. En effet, depuis la fin du 19ᵉ siècle, le pays connaît un essor industriel sans précédent. En Wallonie, l'exploitation des mines de charbon du sillon Haine-Sambre-Meuse (fig.1) ne cesse d'augmenter et favorise le développement des industries métallurgiques et sidérurgiques. Des usines chimiques sont implantées dans ce même sillon industriel. Des usines de production de zinc fonctionnent à plein rendement . De nouveaux gisements de houille ont été découverts dans les profondeurs du sol de la Campine; leur exploitation est imminente. L'utilisation d'explosifs facilite et multiplie l'exploitation des carrières dans tout le pays (calcaire, marbre, porphyre).

Des sols improductifs sont améliorés par l'adjonction d'engrais chimiques produits par l'industrie. L'agriculture et l'élevage s'intensifient. Les voies de communications (chemins de fer, canaux, etc.) se développent. Les salaires et les niveaux de vie s'améliorent. Le tourisme côtier se développe. Bref, l'économie se porte bien mais la nature va mal.

Les littérateurs et les artistes sont les premiers à dénoncer les atteintes portées aux paysages et à la nature. Ils sont bientôt relayés par des sociétés artistiques ou savantes intéressées, elles aussi, au maintien des paysages naturels dans tout le pays. Des hommes politiques influents interpellent le gouvernement belge pour que des mesures de protection soient prises sans tarder. La Société royale de Botanique crée un *«Comité pour la Protection de la Nature en Belgique»*. C'est sous les auspices de ce comité que Massart publie en 1912 *«Pour la Protection de la Nature en Belgique»*. Dans l'avant-propos de l'ouvrage [18], l'auteur rend hommage aux artistes et aux sociétés *«qui se sont donné la mission de lutter avec vigueur contre de nouveaux actes de vandalisme»* et il ajoute: *«nombreux sont les paysages que leur intervention judicieuse a sauvé de la dévastation»*.

Massart n'était donc ni le premier ni le seul à défendre la nature dans son pays mais il fut capable de proposer très rapidement un programme de protection concret et réaliste permettant de poursuivre à grande échelle les actions déjà entreprises par des associations locales *«pour défendre la nature contre les empiètements de l'industrialisme»*. Le ton est donné; il n'est pas toujours académique.

Selon Massart, la situation de la Belgique est particulièrement préoccupante car malgré l'exiguïté de son territoire, elle est devenue le pays le plus industrialisé de l'Europe continentale, et la densité de sa population est la plus élevée au monde (227 habitants/km^2 en 1900) et *«Il est donc fort compréhensible que la destruction de la nature y soit fort avancée»*.

Pour illustrer son propos, Massart décrit les graves dégâts provoqués par les établissements qui grillent les minerais de zinc: *«Chacun d'eux s'entoure d'une ceinture de déserts: les bruyères, les herbes, les arbres, tout meurt dans la large zone où s'abattent les fumées»*. Quant aux charbonnages de Wallonie, il constate qu' *«ils sont le siège d'une industrie tellement dense et active que loin d'y rencontrer des coins de nature, c'est à peine si l'on peut y voir un champ ou une prairie, resserrée entre un terril et un chemin de fer; encore l'herbe y est-elle noire de poussière»*. Massart dénonce aussi la disparition de territoires sauvages et de leur flore et faune naturelles provoquée par la «mise en valeur» de terrains «improductifs»: marécages transformés en prairies de fauche par drainage artificiel, landes à bruyères humides transformées en cultures, landes à bruyères sèches labourées, fertilisées et transformées en prairies ou plantées de pins sylvestres (*Pinus sylvestris*), fonds tourbeux drainés et plantés d'épicéas (*Picea abies*) fournisseurs rapides de bois de mine. Il conclut, non sans humour: *«Une terre doit être extraordinairement maigre, rocheuse ou marécageuse pour que le Belge ne réussisse pas à lui faire produire quelque chose»*! Ce que Massart déplore surtout, c'est que *«presque rien n'a été fait en Belgique pour protéger la nature»* alors que le pays est devenu une des principales puissances économiques européennes et que dans les pays industrialisés voisins, plusieurs mesures de protection sont déjà effectives.

L'organisation à Paris, en 1909, du «1er Congrès international pour la Protection des Paysages» témoigne de l'importance internationale de cet aspect de la protection de la nature dès le début de 20e siècle.

Le programme de protection de la nature proposé par Massart pour la Belgique

C'est un programme inspiré par l'urgence pour protéger la nature dans un très petit pays où *«...les cultures, les chemins de fer, les carrières, les villas...auront bientôt tout envahi ...»*. Massart propose deux types d'actions: quelques mesures simples et d'application immédiate (du moins si la volonté existe) d'une part, et la création de divers types de réserves naturelles d'autres part.

Comme mesures simples, Massart propose notamment: l'abandon de certaines pratiques archaïques telles que l'étrépage de la bruyère humide pour enlever les mottes qui serviront de combustibles, l'arrêt des boisements des landes à bruyères qui présentent un grand intérêt pour les zoologistes, l'arrêt de l'assèchement systématique des fonds humides pour planter des épicéas, ou, plus simple encore, le rafraîchissement d'une coupe géologique à l'origine d'un nouvel étage, et souvent envahie par la végétation, l'évitement du comblement de carrières comportant des témoins préhistoriques, comme par exemple l'exploitation d'un gisement de silex, que Massart décrit par un schéma!), l'arrêt de la destruction systématique d'animaux très menacés tels le grand corbeau. Il préconise aussi une mesure purement administrative: le maintien des noms de rue ou de lieux-dits après leurs modifications (tracé de nouvelles rues par exemple) afin de perpétuer le souvenir des sites, à défaut des sites eux-mêmes [20].

Pour la création de réserves, Massart propose:

1. La délimitation de ce qu'il appelle des parcs naturels *«dans les contrées où la physionomie originelle du paysage est restée à peu près intacte»* et dans lesquelles existent encore des localités *«qui ont été fort peu défigurées par l'homme, et d'une façon toute superficielle»*. Ce qu'il faut protéger dans ce type de réserves, ce ne sont pas les espèces rares mais au contraire les espèces banales et abondantes qui composent les formations végétales et déterminent ainsi la physionomie des paysages. Massart insiste sur l'étendue de ce type de réserve qui doit être considérable. D'abord, pour que celui qui s'y promène puisse avoir jusqu'à un certain point *«l'illusion de l'espace et ne pas se heurter tout de suite à la civilisation»*. Ensuite, pour que les populations animales et végétales puissent lutter librement pour la vie, dans des conditions d'existence à l'abri des modifications induites par la proximité de cultures, d'industries polluantes, de drainages, etc.

Massart ne parle pas de protection des écosystèmes parce que le vocable n'a pas encore été inventé (il sera proposé par Tansley en 1935), mais il en perçoit déjà nettement le concept. Citons un exemple: la création d'un grand parc naturel en Campine, vaste région sablonneuse riche en landes et étangs mais futur «pays noir» à cause de l'exploitation imminente du charbon avec son cortège de terrils et ses corons (cités ouvrières).

2. La mise en réserve d'espaces verts de dimensions plus modestes, ayant gardé plus ou moins intacte leur allure primitive et qui existent encore dans les régions agricoles et industrielles et au voisinage des grandes villes. Le but de ces «coins» comme les appelle Massart est de renseigner sur l'aspect primitif des sites et non sur leur flore et faune originelles. Ces «coins» constitueront des lieux d'excursions pédagogiques de proximité et des sources de matériel didactique vivant (récolte de plantes, chasse aux insectes) pour les écoles et pour ... les universités. Citons un exemple: la protection en région bruxelloise d'un vallon dont la végétation ne comporte pas d'espèces rares mais permet de se faire une idée de la flore colonisant jadis tous les endroits humides comparables.

3. La protection d'habitats (Massart utilisait le terme d'habitations) d'espèces rares dont l'intérêt est principalement scientifique: espèces reliques, témoins des conditions géologiques et climatiques anciennes ou de mouvements de flore récents, espèces indicatrices de sols particuliers, espèces potentiellement utiles pour la recherche fondamentale (il rappelle le cas de l'*Amphioxus*, petit organisme marin, qui a permis d'éclaircir l'embryogenèse humaine), etc.

4. Le maintien de certaines coupes géologiques fournies par les tranchées de chemins de fer et de carrières abandonnées présentant un intérêt scientifique, la protection des grottes pour éviter que *«leurs exploitants transplantent des stalactites et des stalagmites afin de ménager des effets plus pittoresques»*, la protection des fossiles et des témoins archéologiques exhumés lors du creusement de canaux et de travaux de dragage pour éviter qu'ils ne soient vendus, donnés ou tout simplement jetés par ignorance.

Pour concrétiser ses propositions, Massart établit une liste de 75 sites prioritaires qu'il connaît personnellement ou sur lesquels il possède suffisamment de renseignements; il précise toutefois que cette liste est «loin d'être complète». Ces sites sont répartis dans les treize districts naturels de la Belgique de manière à conserver au mieux l'étonnante diversité des paysages, de la végétation et des sols si bien mise en évidence par Massart lui-même dans son *Esquisse de la géographie botanique de la Belgique* [15]. Chacun des sites fait l'objet d'une description précise, de la justification de son choix et est illustré par une ou plusieurs photos souvent explicitées par un schéma ou une carte géologique ou géographique.

Pour assurer la protection des espaces verts convertis en réserves, Massart exige l'interdiction totale de la chasse et de la pêche. Pour les autres activités humaines susceptibles d'y être exercées, il préconise d'appliquer les mesures de protection avec souplesse, en les adaptant au cas par cas, comme par exemple, prévoir des indemnisations pour les communes lorsque des terrains ayant toujours servi de vaine pâture seront rendus inaccessibles aux troupeaux.

Massart veut aussi lutter contre l'ignorance parfois responsable de la disparition de certains sites naturels. Il cite deux exemples tristement célèbres en Belgique: (*i*) l'assèchement, au sein d'un bois de deux mille ha, d'un marécage de trois hectares dont le propriétaire ignorait la composition botanique exceptionnelle et (*ii*) la destruction d'un vaste pré salé établi sur un type d'alluvion sablo-argileuse rare et peut-être unique (selon Massart) pour y installer un terrain de golf. Seul remède selon Massart: une éducation à la nature pour tous les acteurs de la société.

Comme pour mieux convaincre de l'urgence des mesures à prendre en Belgique, Massart consacre dans l'ouvrage précité [18] une quarantaine de pages à «Ce qui a été fait ailleurs» en Europe et dans le monde. Une fois de plus, il nous éblouit par l'abondance et la qualité des informations et de l'illustration photographique (clichés prêtés par des collègues étrangers) qui montrent à quel point au début du 20e siècle, la protection de la nature préoccupait déjà le monde politique et la communauté scientifique.

Faute de place, nous ne pouvons rappeler toutes les mesures de protection rapportées par Massart. Les premières sont d'ordre purement esthétique comme par exemple en Prusse, où une loi contre la défiguration des paysages par les affiches est en vigueur depuis 1902 et où une autre loi édictée en 1907 permet aux autorités locales d'empêcher l'érection de bâtiments qui risqueraient d'enlaidir les paysages ou de leur enlever leur caractère naturel. Dans d'autres pays européens, les mesures de protection sont à la fois d'ordre esthétique et scientifique; elles concernent souvent des paysages mais aussi certaines espèces végétales ou animales menacées par le tourisme (flore alpine), par la mode (les aigrettes pour leurs plumes) ou par les chasseurs et pêcheurs (oiseaux de proie, martin-pêcheurs, etc.), des arbres remarquables, des blocs erratiques, etc. Il existe aussi des projets ou déjà quelques créations de parcs naturels. Massart évoque aussi les mesures prises pour éviter l'anéantissement des grands animaux des savanes africaines (Convention de Londres 1900), des mammifères à fourrure et des oiseaux à beau plumage; concernant ces derniers, Massart rapporte qu'en 1907, 19.742 dépouilles d'oiseaux de paradis ont été vendues sur le marché de Londres pour garnir les chapeaux des dames [18].

La relation de l'homme à la nature

Pour Massart, la protection de la nature est indispensable au développement harmonieux de l'homme au plan esthétique, socioculturel et scientifique. Suivant les traces des premiers protecteurs de la nature, Massart considère que les beautés de la nature sont des monuments naturels qui, au même titre que les monuments de pierre, méritent respect et protection. Pour lui, *«un paysage est un sujet de tableau qui se suffit à lui-même»*. Il nous en fournit de nombreuses preuves, comme par exemple, un superbe étang à nénuphars photographié en 1904, que n'aurait probablement pas dédaigné Monet (dont les célèbres nénuphars ont été peints la même année!). Massart attache une importance particulière aux paysages, non seulement en tant que cadre de vie quotidien mais aussi, pour certains d'entre eux, comme témoins du passé historique, économique, agricole, etc.

Mais Massart est avant tout un homme de science et comme il le déclare dans l'avant-propos de son célèbre ouvrage: *«La Science, à la poursuite de la Vérité, a droit aux mêmes égards que l'Art, à la poursuite de la Beauté»*. Il prône une véritable culture scientifique car son approche de la protection de la nature est pluridisciplinaire: les sites à protéger concernent autant la botanique que la zoologie, la géologie, la géographie, la préhistoire et l'archéologie. Il recommande aussi le recours à la toponymie et à la linguistique (exemples à l'appui) pour découvrir des animaux disparus (castors, par exemple), des végétations et pratiques agricoles anciennes, etc.

Mais ce qui préoccupe surtout Massart en ce début de 20ᵉ siècle, c'est la primauté du pouvoir économique sur la nature, sur la science, et sur les êtres vivants. Face à l'urbanisation galopante du territoire et à la disparition concomitante des espaces naturels, Massart interroge: *«L'utilisation du territoire doit-elle aller jusqu'aux plus extrêmes limites; faut-il que l'industrie et la culture prennent possession des moindres parcelles du sol?»*. Et il répond: *«... nous ne devons pas - nous ne pouvons pas - permettre que les derniers coins de nature qui nous restent encore s'effacent devant l'artificiel (car) nous porterions vis-à-vis des générations futures une responsabilité par trop lourde, si nous ne leur laissions pas la faculté de constater* de visu, *ne fut-ce qu'en un petit nombre de points, quel était l'état physique de notre pays avant son entière dénaturation»*. Cette référence aux générations futures se retrouvera soixante ans plus tard dans le rapport Brundtland.

Aux «utilitaires à outrance» prônant la «mise en valeur des terrains improductifs», Massart demande: «N'y a-t-il donc de valeur que celle qui est monnayée?... Est-ce que la Science ne représente pas une valeur?» Et il rappelle que «ce sont des observations nouvelles qui sont l'origine première de tout progrès». Non sans humour, il poursuit «C'est par un véritable abus de langage qu'on appelle l'industrie et l'agriculture des Sciences appliquées, alors que se sont en somme des applications de la Science pure».

Pour justifier la protection de la nature, Massart défend donc la primauté de la Science pure, suivant en cela son maître Errera, qui, en 1905, déclarait déjà: «Beaucoup de questions biologiques capitales ne peuvent être étudiées que sur des terrains où le développement, la succession, les luttes des animaux et des plantes ne soient pas troublés par l'intervention de l'homme» [18].

Massart accorde une attention spéciale aux relations homme-plante: le but essentiel des collections pédagogiques qu'il préconise pour les institutions d'enseignement et les jardins botaniques [11, 12, 13] est de développer la sensibilité de l'opinion publique au rôle majeur que jouent les plantes dans la vie de l'homme: alimentation, pharmacie, industrie, décoration, en plus de leur rôle fondamental dans les paysages et les espaces verts de loisir.

Les biologistes et l'étude de la nature

Peu après la parution de *Pour la Protection de la Nature en Belgique*, Massart fait à la séance publique de la Classe des sciences de l'Académie royale de Belgique une «lecture» (comme on disait à l'époque) intitulée *«Les naturalistes actuels et l'étude de la nature»* [19]. Dans son discours, Massart défend l'idée de *«l'absolue nécessité qu'il y a pour le biologiste à rester en communion avec la nature»* et fait le procès de ceux qui *«oublient de regarder la nature»*. Il est vrai que depuis la fin du 19ᵉ siècle, l'approche purement descriptive des sciences naturelles a fait place à une approche de plus en plus expérimentale: cultures de plein air ou en serre (Bonnier, Mendel et les autres) et surtout, étude en laboratoire de la structure, du développement et du fonctionnement des organismes.

La sophistication croissante des techniques d'analyse et d'expérimentation exerce une espèce de fascination sur les zoologistes et botanistes qui *«ne consentent à s'extraire du laboratoire que pour se procurer du «matériel»*! Et Massart poursuit: *«Ce n'est pas dans la nature qu'étudient ces naturalistes, mais uniquement devant la table de laboratoire; le rapport de l'organisme avec les innombrables modalités du milieu ne les intéresse pas; leur curiosité scientifique ne s'éveille qu'au moment où l'animal ou la plante sont réduits en tranches assez minces pour être regardées au microscope».*

Toutefois, Massart se défend d'attaquer les hommes de laboratoire (il en est un lui-même!) mais comme il le dit malicieusement: *«Ce que je dis ici ne vise pas ce qu'ils font, mais ce qu'ils ne font pas»* c'est-à-dire prendre en compte les observations éthologiques dans la nature. Car pour Massart, *«il n'est pourtant pas contestable que ces notions là sont aussi nécessaires que les données anatomiques et physiologiques pour la compréhension intégrale de l'être vivant et qu'une théorie synthétique a beaucoup plus de chance d'être féconde si elle est basée sur l'ensemble de la biologie»*. Nous retrouvons une fois de plus l'approche globale de l'étude de la vie qui imprègne toute l'œuvre de Massart.

Dans son discours, Massart n'épargne pas non plus les conservateurs des grandes collections systématiques dans les musées d'histoire naturelle chez qui il constate la *«même insouciance de la nature vivante»*. Pour illustrer son propos, il cite l'exemple d'un botaniste «*fort savant*» qui n'identifiait une plante que lorsqu'«*qu'un séchage et un écrasement approprié l'avaient élevée à la dignité d'échantillon d'herbier*», car il était incapable de la déterminer à l'état frais! Massart ne critique pas la formation des herbiers *«qui sont la condition de tout progrès dans la botanique systématique»* mais rappelle *«qu'avant de faire partie de l'herbier, les plantes étaient vivantes et qu'elles étaient plus intéressantes alors que depuis leur dessiccation»*. Il reconnaît cependant l'utilité des collections qui *«constituent le moyen le plus pratique pour permettre la comparaison des organismes en vue de leur détermination»* car *«il est incontestable que la connaissance précise des espèces est la base nécessaire de tout travail de biologie»*.

Tentant de cerner les causes de «l'indifférence de la génération présente pour les choses de la nature», Massart considère comme responsable principal l'enseignement moyen du degré supérieur qui «tue l'esprit d'initiative pour ne laisser que l'esprit d'imitation» et qui n'incite pas l'élève à l'observation personnelle «car on ne lui a jamais laissé entrevoir qu'on peut apprendre quelque chose d'utile en regardant vivre les bêtes et les plantes, ou en examinant un talus argileux après une pluie d'orage».

Pour remédier à la situation, Massart propose deux stratégies: d'une part, réapprendre aux enfants à aimer la nature par des «promenades à la campagne» pour leur permettre d'observer des accidents géographiques, des roches, des fossiles (dans les carrières), des fleurs, des animaux, des champignons, etc. qui leur donneraient le goût des excursions scientifiques pendant leurs études ultérieures, et d'autre part, l'installation des instituts de biologie à la campagne avec suffisamment de terrains de cultures expérimentales ou d'élevage (pour les animaux) attenants aux laboratoires pour permettre aux chercheurs de suivre minutieusement leurs expériences.

Après 1912

A l'issue de la première guerre mondiale (1914–1918), l'urgence n'est plus de protéger la nature mais de reconstruire les villes dévastées par les incendies et les bombardements et de restaurer les sols inondés artificiellement pour protéger les ports de la mer du Nord pendant les hostilités (Fig. 1). Par ailleurs, l'argent manque pour réaliser les 3e, 4e et 5e volumes de grandes planches photographiques (30 x 40 cm) initialement prévus pour l'ensemble des districts naturels de la Belgique et dont seuls les deux premiers avaient été publiés. C'est précisément dans ces deux volumes que se trouvent les photos de végétations et de paysages avant leur détérioration par l'eau saumâtre des inondations stratégiques. Massart y voit une gigantesque expérience grandeur nature pour étudier les effets des inondations et de la salinité sur le comportement des plantes terrestres et sur la qualité des sols. En 1920, il demande à la Commission royale des Monuments et des Sites de classer le site pour des raisons scientifiques; il y installera un laboratoire de campagne et commence immédiatement des observations qui feront rapidement l'objet d'une publication [28].

En 1921, Massart reprend sa campagne pour la protection de la nature en Belgique: il rédige une circulaire à envoyer à diverses sociétés (notamment le Touring Club de Belgique et la Société royale de Botanique) pour demander à leurs membres de faire des propositions de création de réserves naturelles à la Commission royale des Monuments et des Sites [26]. En même temps, Massart déploie une activité fébrile à l'université: recherches en laboratoire, rédaction du cours donné aux étudiants [27, 29], mission scientifique au Brésil, tournée de conférences aux USA, début de réalisation de son projet de création d'un institut moderne de botanique «à la campagne».

Sa mort prématurée en 1925 interrompit brutalement ses activités.

Après Massart

Les stratégies de protection des sites n'évoluent guère dans les années qui suivent la disparition de Massart. Des initiatives locales sont prises, comme par exemple la création d'une association (*Les Amis de la Fagne*) qui obtient de la part de certaines communes de renoncer à enrésiner des landes et des fagnes en échange de compensations financières payées par les membres de l'association! Mais ailleurs, l'enrésinement des landes et des fagnes continue à progresser!

La deuxième guerre mondiale (1940–1945) interrompt à nouveau la mise en place de sites protégés. Ce n'est qu'après 1950 que la création d'associations de protection de la nature et de réserves naturelles progressera d'une manière significative. La première association nationale, *Les Réserves Ornithologiques de Belgique* a été créée en 1951; elle se développera rapidement et deviendra en 1972 l'association de protection de la nature la plus importante du pays et s'intitulera désormais les *Réserves Naturelles et Ornithologiques de Belgique* (RNOB). En 1986, suite au processus de régionalisation et de fédéralisation de l'Etat belge, l'association nationale se scindera en deux branches distinctes: la branche francophone qui deviendra *Natagora* après sa fusion avec la *Société d'Etudes ornithologiques Aves* (fondée en 1953), et la branche néerlandophone, qui s'intitulera *Natuurpunt.* Les deux associations régionales poursuivront le but initial de l'association nationale: développer un réseau de réserves naturelles représentatives de la diversité des milieux naturels si bien mise en évidence par Massart.

En 1981, le Jardin botanique national de Belgique et les RNOB publient conjointement *«Paysages de Flandre jadis et aujourd'hui»* réalisé par Léo Vanhecke, botaniste du Jardin botanique, Georges Charlier, photographe aux RNOB et Luc Verhelst, également photographe [37]. L'ouvrage permet de comparer 60 paysages photographiés par Massart entre 1904 et 1912 et par les photographes précités en 1980 et en 1981. Chaque photo est accompagnée d'un commentaire rédigé par Vanhecke à partir d'informations fournies par Massart d'une part et, pour les photos récentes, par des enquêtes de terrain auprès d'autochtones âgés ayant encore le souvenir des pratiques agricoles et du mode de vie de la population rurale qui, au début du 20e siècle, était encore largement majoritaire dans le pays.

Cet album illustre magnifiquement l'intérêt historique, scientifique et socioculturel des photos-documents de Massart; son sous-titre: *«De la pauvreté dans la verdure à la surabondance dans la grisaille»* traduit fidèlement le sentiment d'amertume de L. Vanhecke qui juge sévèrement l'évolution des paysages et du mode de vie au cours des soixante-dix années qui séparent les deux séries de photos. Selon cet auteur, contrairement à ce qui s'est passé dans les pays voisins, le «développement anarchique et le chaos» illustrés par les photos récentes résultent de l'absence de planification et de voies légales dans le développement de l'industrie, de l'agriculture intensive, des travaux d'utilité publique et des agglomérations.

Il est permis de penser que les recommandations de Massart n'avaient guère été suivies. Heureusement, les réglementations en matière de protection de la nature et des paysages se sont nettement améliorées et sont aujourd'hui souvent plus contraignantes en Flandre que dans les autres régions de la Belgique.

En 1995, à l'occasion de l'Année européenne de la Conservation de la Nature, le Ministère de la Région wallonne publie *«Le grand Livre de la Nature en Belgique»* [38]. Cet ouvrage fait le point sur les nouvelles stratégies mises en place pour maintenir la biodiversité sur l'ensemble du territoire. 46 auteurs (scientifiques, naturalistes, fonctionnaires, etc.) décrivent les deux grands types de milieux à protéger: les réserves naturelles, principalement déterminées par la nature du sol, et une série de biotopes caractérisant la «nature au quotidien»: friches industrielles, terrils, voies ferrées et autres voies de circulation, milieux souterrains, carrières, ruines et vieux murs.

A quelques exceptions près, d'origine récente, nous retrouvons dans ce choix les types de réserves proposées par Massart en 1912.

3. POUR CONCLURE

Près d'un siècle nous sépare de la parution des ouvrages de Massart consacrés à l'étude et à la protection de la nature en Belgique. Ni le temps, ni l'évolution de la société n'ont estompé l'œuvre avant-gardiste du grand botaniste et naturaliste belge. Sa conception à la fois scientifique et humaniste de la protection de la nature imprègne le vaste programme *Natura 2000* mis sur pied par l'Union européenne en 2002.

Son approche pluridisciplinaire de la géographie botanique peut à juste titre être considérée comme la première synthèse écologique réalisée en Europe occidentale.

Son approche globale de l'étude des paysages est largement présente dans le texte de la *Convention européenne du paysage* entrée en vigueur en mars 2004.

Mais, plus encore que les concepts novateurs de Massart, c'est le caractère prémonitoire de ses préoccupations sociétales qui nous interpelle: la dévalorisation du savoir naturaliste au profit des sciences de laboratoire et la primauté du pouvoir économique sur la science et la protection de la nature.

Qu'en est-t-il aujourd'hui? Les sciences biologiques de terrain (écologie, taxonomie, etc.) sont dominées (au niveau de l'intérêt qu'elles suscitent et des crédits qui leur sont alloués) par la biologie moléculaire dont les découvertes prodigieuses nous aident certes à mieux comprendre les mécanismes de la vie mais dont certaines applications industrielles posent de graves problèmes d'éthique et d'environnement (brevetabilité des gènes et des semences, biopiraterie du patrimoine génétique des espèces naturelles, contamination génétique des plantes sauvages et cultivées par des plantes transgénétiques, etc.).

Quant à la primauté de l'économie sur tout autre considération et son corollaire, la conception utilitariste de la nature, elles montrent aujourd'hui leurs limites. Dénoncés depuis plus d'un quart de siècle par des esprits éclairés, les dégâts avérés et potentiels du productivisme sont confirmés par le récent rapport sur l'état de la planète récemment publié par l'ONU et auquel ont participé 1.360 spécialistes de 93 pays (*Millenium Ecosystem Assessment Synthesis Report*, mars 2005). L'avertissement est sans équivoque: *«L'activité humaine exerce une telle pression sur les fonctions naturelles de la planète que la capacité des écosystèmes à répondre aux demandes des générations futures ne peut plus être considérée comme acquise. Les comptabilités nationales traditionnelles ne mesurent pas la diminution ni la dégradation des ressources naturelles alors que celles-ci représentent la perte d'un patrimoine essentiel»*.

Selon les estimations du rapport, nous ne disposerions plus que de quelques décennies pour repenser fondamentalement notre mode de développement actuel.

Allons-nous enfin admettre que *«Nous ne possédons pas la Terre, nous l'empruntons à nos enfants»* (parole de sage, attribuée au chef indien Seattle ou à Thucydide, selon les auteurs).

BIBLIOGRAPHIE

1. Bommer C., Massart J., 1904 - Projet d'une étude détaillée de la géographie botanique la Belgique. Bull. Soc. Roy. Bot. Belg.: 37.
2. Brien P., 1967 - En souvenir d'un maître naturaliste et botaniste Jean Massart. Revue de l'Université libre de Bruxelles, **3**: 1-11.
3. Héger P., 1926 - Notice sur la vie et les travaux de Jean Massart. Rapport de l'Université libre de Bruxelles sur l'année académique 1924-1925: 36-51.
4. Marchal E., 1927 - Notice sur Jean Massart. Annuaire Acad. Roy. Sc. Belg.: 69-158.
5. Massart C., 2004 - Communication personnelle de la petite-fille de J. Massart.
6. Massart J., 1888 - Sur l'irritabilité des spermatozoïdes de la grenouille. Bull. Acad. Roy. Belg., **25**: 750-754.
7. Massart J., 1888 - Les études de W. Pfeffer sur la sensibilité des végétaux aux substances chimiques. Bull. Soc. Roy. Bot. Belg., **27**: 86-96.
8. Massart J., 1895 - Lettres de Jean Massart à sa tante Melle à l'occasion de son voyage à Java (1894-1895). Bibl. Jard. Bot. Belg., inv. n° 21688: 155.
9. Massart J. 1895 - Un botaniste en Malaisie. Bull. Soc. Bot. Belg., **34**: 151-343.
10. Massart J., 1898-1899 - Un voyage botanique au Sahara. Bull. Soc. Roy. Bot. Belg., **37**: 202-239.
11. Massart J., 1902 - Un jardin botanique pour les écoles moyennes. Bull. Jard. Bot. Etat: 623-694.
12. Massart J., 1904 - Les collections éthologiques au Jardin botanique de l'Etat. Bull. Jard. Bot. Etat: 463-526.
13. Massart J., 1905 - La collection phylogénique au Jardin botanique de l'Etat. Bull. Jard. Bot. Etat: 238-264.
14. Massart J., 1908 - Les districts littoraux et alluviaux. In: Bommer C., Massart J. (eds.), Les aspects de la végétation en Belgique, 1. Jardin botanique de l'Etat, Bruxelles.
15. Massart J., 1910 - Esquisse de la Géographie botanique de la Belgique. Lamertin, Bruxelles.
16. Massart, J., 1911 - Nos Arbres. Lamertin, Bruxelles.
17. Massart J., 1912 - Les districts flandrien et campinien. In: Bommer C., Massart J. (eds.), Les aspects de la végétation en Belgique, 2. Jardin botanique de l'Etat, Bruxelles.
18. Massart J., 1912 - Pour la Protection de la Nature en Belgique. Lamertin, Bruxelles.
19. Massart J., 1912 - Les naturalistes actuels et l'étude de la nature. Bull. Cl. Sc. Acad. Roy. Belg.: 944-965.
20. Massart J., 1914 - Conservation des noms des lieux-dits. Comm. Roy. Monuments et Sites., **53**: 128-130.
21. Massart J., 1916 - Quelques adaptations végétales au climat de la Côte d'Azur. Ann. Géogr. Paris, **26**: 94-105.
22. Massart J., 1916 - Comment les Belges résistent à la domination allemande. In: Contribution au livre des douleurs de la Belgique.
23. Massart J., 1916 - La presse clandestine de la Belgique occupée. Paris – Nancy.
24. Massart J., 1918 - Les intellectuels allemands et la recherche de la vérité. Revue de Paris. Id. Imprimerie médicale et scientifique, Bruxelles (1920): 643-672.

25. Massart J., 1920 - Ce qu'il faut voir sur les champs de bataille et dans les villes détruites de Belgique. Touring Club de Belgique, 2.
26. Massart J., 1921 - Circulaire relative à la création de réserves naturelles en Belgique. Bull. Comm. Royale Monuments et Sites: 85-88.
27. Massart J., 1921 - Eléments de Biologie générale et de Botanique. 1: La Biologie générale et les Protistes. Lamertin, Bruxelles.
28. Massart J., 1922 - La biologie des inondations de l'Yser et la flore des ruines de Nieuport. Recueil de l'Institut botanique L. Errera: 411-429.
29. Massart J., 1923 - Eléments de Biologie végétale et de Botanique. 2: Physiologie et Ethologie-Paléobotanique-Géobotanique. Lamertin, Bruxelles.
30. Massart J., 1923 - La préservation des sites au Brésil. Bull. Comm. royale Monuments et Sites: 325-328.
31. Massart J., Bordet C., 1891 - Le chimiotaxisme des leucocytes et l'infection microbienne. Ann. Inst. Pasteur: 417-444.
32. Massart J., Bouillenne R., Ledoux P., Brien P., Navez A., 1929 - Une mission biologique belge au Brésil. Imprimerie médicale et scientifique, Bruxelles.
33. Massart J., Demoor, J., Vandervelde, E., 1897 - L'évolution régressive en biologie et en sociologie. Paris- Alcan.
34. Massart J., Vandervelde E., 1893 - Parasitisme organique et parasitisme social. Bull. Sc. Fr. et Belg., **25**: 277-294.
35. Schouteden-Wéry J., 1913 - Relations des excursions scientifiques dirigées par Jean Massart. 2, en Brabant. Lamertin, Bruxelles.
36. Stockmans F., 1968 - Jean Massart. Florilège des Sciences. Université libre de Bruxelles: 705-726.
37. Vanhecke L., Charlier G., Verhelst L., 1981 - Paysages de Flandre jadis et aujourd'hui. Jardin botanique national de Belgique et Réserves naturelles et ornithologiques de Belgique.
38. xxx, 1995 - Le grand Livre de la Nature en Wallonie. Ministère de la Région wallonne. Casterman.

THE JOURNAL OF RENZO VIDESOTT
The "Historical Archives" of Nature Protection in Italy in the period 1944-1953

Cecilia VIDESOTT, Elena VIDESOTT*
Via Tonco 21, **10131 Torino (ITALY)**; e-mail: e.videsott@libero.it

Abstract: Professor Renzo Videsott was a pioneer of Nature protection in Italy from 1943 to 1974. His daughters describe his background and early experiences, as rock climber, student of Veterinary Medicine and hunter, which contributed to the development of his future involvement with Parco Nazionale Gran Paradiso (PNGP), the culmination of his interest in mountains and animals, and to which he dedicated his energies until his death. He was active in the field of respect for Nature and conservation in Italy and abroad: in 1947 he took part in the first post-war convention of nature conservationists at Brunnen (CH), in 1948 he founded the Movimento Italiano per la Protezione della Natura (MIPN) and contributed to establishing the Union Internationale pour la Conservation de la Nature (UICN). He achieved many important results for PNGP, particularly for its Alpine *Capra ibex*, carefully taking notes of all the events and personalities concerned. This is why the Journal of Renzo Videsott (1943-54) can be considered the "Historical Archives" of the protection of Nature in Italy during that period, and should be published.

Introduction

Professor Renzo Videsott was one of the great personalities in the field of nature conservation in Italy from 1943 until his death in 1974 (Fig. 1). We shall present pertinent details of his life and personality, as well as the experiences that proved fundamental to the further development of his interest in Nature.

Historical background of his family and birthplace

He was born on 10th September 1904 in Trento (then named Trient, being still under Austrian rule), a lively intellectual city in a border district of the Habsburg Empire, inhabited by Italian-speaking people and about to become part of the Kingdom of Italy after World War I. (Incidentally, Trento is also the birthplace of Professor Franco Pedrotti!).

His ancestors had settled here, coming down from the valleys of the nearby mountains where still today about 30.000 of the inhabitants speak Ladin or Romansch, the language of the local ancient Raeto-Roman population, as witnessed by the family name Vi-de-sott (*Vila-de-supto*).

During his youth, he was influenced by the atmosphere of the Italian nationalist movement, as his family sympathised with Italy; in fact, at the outbreak of war, the Videsott children moved to Tuscany, first to Florence and then to a boarding house in Arezzo. He finished his secondary school studies back in Trento, when the region became Italian.

Rock climbing and veterinary medicine: his main juvenile interests

The Dolomites, the famous calcareous mountains of the Trentino region, had by then become the theatre of activity for the great movement of rock climbers belonging to the schools of Vienna and Munich, which were vying to reach the limit of human possibilities in free climbing. Renzo Videsott and some youths of his age responded to the challenge and accomplished historic first ascents in the nearby Gruppo del Brenta (Fig. 2).

Then, he had to leave Trento for his University studies. It had taken him a whole summer, lying prone on the ground in the park of his house, to speculate what course to choose and where to go. He decided to dedicate his activity “to the Earth and to animals”.

Thus, he selected the Veterinary Medicine course in Turin, in Piemonte at the opposite western side of the Alps, first capital of a united Italy under the Savoy dynasty, famed for its lively intellectual tradition, one that had attracted so many vibrant personalities. There he met Domenico Rudatis, an engineering student from the eastern Dolomites living in Venice, who became his perfect climbing companion and the historian of the "sesto grado" climbing style of that heroic period. Together in 1929 they accomplished the legendary "VI grado" ascent of the Busazza Spigolo Sud-Ovest in the Civetta Group.

After graduation in 1928, he became a member of the teaching staff of the Torino Veterinary Faculty, devoting his efforts to scientific research and to clinical practice, happy to treat animals other than the pets and horses of aristocrats (Fig. 3)!

Fig. 1: Renzo Videsott as a wildlife photographer.

Fig. 2: Renzo Videsott as a Dolomite climber.

In the meantime, he had given up alpinism, as on the Busazza he had reached his target and had attained his own limits in free climbing. Now he hunted for sport in the mountains of his ancestors and led the hunting district of San Vigilio di Marebbe, in Val Badia. One day, an accidental experience as a hunter changed his life and gave his previous interests a new direction, leading him to his future "mission" for Nature: he had shot down the only surviving chamois in a certain area, thus quenching there the life of that noble species. He would never forget the dying animal's last glance and the thumps of its falling body. He gave up hunting altogether.

By now, two fundamental streams of interest in Renzo Videsott's spirit had settled deeply: love for the mountains and love for animals.

Renzo Videsott and the Ibex of Gran Paradiso National Park

Quite by chance, in February 1944, his attention was attracted by an old, 1879, issue of the quarterly publication edited by Club Alpino Italiano on display on an antiquary's bookstand, containing an article by Dr. A. Girtanner, commemorating the death of King Vittorio Emanuele II, the first king of Italy, praised as the defender of the last existing colony of *Capra ibex* in the Alps of Aosta, formerly the Royal Hunting Preserve, and now in danger.

From that moment, Renzo Videsott devoted his mind, his activity, his whole life, first to the mission of "new protector" of the ibex and other animals of Parco Nazionale del Gran Paradiso, and, later, of Nature in general, carefully entering in his notebooks every relevant event and detail [3, 4, 6, 7, 8,].

The Journal of Renzo Videsott

His Journal was now in progress; it also became his confident, to which he expressed his feelings and opinions, thus revealing his complete dedication to the task to which he felt called.

It was begun on 18th August 1944, when he had his first contact with the reality of the Parco Nazionale del Gran Paradiso: riding his bicycle – it was deep in wartime – he went to Aosta to assess the situation there on behalf of the partisan Comitato di Liberazione Nazionale (C.L.N.). The first page of the Journal, containing the inventory of the surviving animals, reveals that he found the Alpine ibex and the National Park (based on the former Royal Hunting Preserve) on the verge of extinction, and thus committed himself to the hard struggle for their survival. Of course, the autonomy of the National Park, the indifference of the authorities, the raising of funds to pay the keepers' salaries, the fight against poachers and so on, were only some of the problems he had to face.

Among the events recorded during the nearly 10 years of activity described in the Journal, here is a selection of some fundamental stages.

In 1947, Renzo Videsott attended the convention held at Brunnen, Switzerland, where, for the first time after World War II, the international representatives of the Nature "diplomatic corps" gathered again.

In 1948, in the Castle of Sarre, Valle d'Aosta – formerly the Royal Hunting Residence – Renzo Videsott founded the Movimento Italiano per la Protezione della Natura (M.I.P.N.), and in the same year, in Fontainebleau, France, he contributed to the establishment of the Union Internationale pour la Protection de la Nature (UIPN), later Union Internationale pour la Conservation de la Nature (UICN) [2, 5].

Glancing through the Journal, page after page, the very life of post-war Italy appears in glimpses and flashes; names of important personalities are mentioned (and sometimes humorously commented upon!).

In fact, the Journal mentions: Gran Paradiso keepers and political personalities that Renzo Videsott contacted for his work and during his journeys; Italian nature conservationists, such as O. de Beaux, G. Castelli, G.G. Gallarati Scotti, his own brother Paolo and many others; also conservationists from different countries in the world, with whom he was in correspondence, namely V.H. Cahalane, W. Vogt and Bump Gardiner (USA), J. Huxley (UK), E. Bourdelle and R. Heim (France), J.P. Harroy and V. van Straelen (Belgium), A. Wodziczko and J. Hryniewiecki (Poland), H. Gams (Austria), V. Puscariu (Romania), Stephan Brunies, J. Buettikofer and J. Baer (Switzerland), to list some at random, and many others (Fig. 4).

Fig. 3: Renzo Videsott as a veterinarian practitioner.

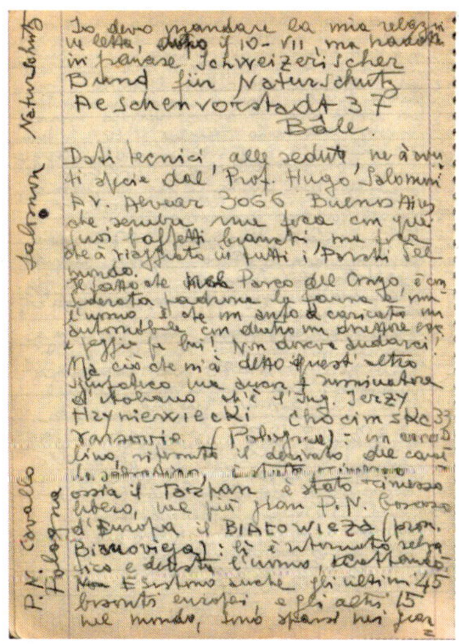

Schweizerischer Bund für Naturschutz
Aeschenvorstadt 37
Bâle.

Fig. 4: A manuscript page from the Renzo Videsott's Journal.

Toil, enthusiasm, a sort of mystic commitment, endurance, an undaunted belief in the cause of Nature, are revealed in short commentaries or personal notes. We ourselves have discovered another man, sometimes with surprise, always with emotion.

We realize that Renzo Videsott belonged to Parco Nazionale Gran Paradiso and to Nature: that is why his Journal shall be published, as "Historical Archives" of Nature Protection in Italy over the years 1944–1953. The complete text of the Journal, edited by C. Rastel Bogin, will soon be published.

Conclusion

You may wonder why Renzo Videsott stopped compiling his Journal. We do not know for certain, but perhaps progress had offered him a new means of expression, as this photo of him might suggest.

When you come to visit the Gran Paradiso National Park, do not forget to pay a visit to "il magico Pian Resello", the place where Renzo Videsott wished his ashes to be scattered [1].

We did not know this at the time of his death, but later on, when a passer-by left the message that was subsequently engraved on a metal plaque to be fixed on a rock at the very spot. Twenty years after the first plaque, on the day of the 100th anniversary of Renzo Videsott's birth, another was added, quoting the words he had written to express his wish.

His desire has not yet been fulfilled.

Perhaps, in the future, the new generation will!

REFERENCES

1. Gabutti A., 2005 - Schegge del Parco. L'Uomo e l'Ambiente, **42**: 1-94.
2. McDowell D., 2000 - Half a century of global nature conservation. L'Uomo e l'Ambiente, **34**: 23-28.
3. Meyer E., 1995 - I pionieri dell'ambiente. Carabà, Milano.
4. Pedrotti F., 1998 - Il fervore dei pochi. Temi Editrice, Trento.
5. Pedrotti F., 2000 - Il 50° anniversario di fondazione del Movimento Italiano per la Protezione della Natura e dell'Union Internationale pour la Protection de la Nature 1948 –1998. L'Uomo e l'Ambiente, **34**: 8-22.
6. Piccioni L., 1999 - Il volto amato della patria. L'Uomo e l'Ambiente, **32**: 1-317.
7. Sievert J., 2000 - The origins of nature conservation in Italy. Peter Lang, Bern.
8. Videsott C., 1983 - Renzo Videsott e Parco Nazionale Gran Paradiso. Scritti scelti e riproposti dalla figlia Cecilia. Lions Club Alto Canavese, Torino.

II

SOCIAL AND LEGAL ASPECTS OF CONSERVATION

NATURE AND CONSERVATION: A NEW, MORE ACTIVE ROLE FOR UNIVERSITIES

Franco de BATTAGLIA
Columnist and Editor-at-Large of the daily newspapers "Trentino", "Alto Adige", "Corriere delle Alpi" - **Trento, Bolzano, Belluno (ITALY)**; e-mail: fdebattaglia@katamail.com

Abstract: Governments and local communities throughout Europe (and the rest of the world) are losing interest in conserving and defending Nature. The economic deregulation of recent years, which has been a characteristic of Western Europe and the USA, is jeopardizing our last natural reserves, national and regional parks also. To overcome this negative trend, a massive new effort is needed by the Scientific Community, institutions and universities in the forefront of conservation. They must set up a network of contacts, relationships and inter-related studies, capable of moving public opinion, to initiate a new system of values – and take action as well.

Left alone to their studies and research, scientists cannot by themselves outline "conceptions and methods" of Nature Conservation, either in Europe or in the various national states in which they operate. Academic researchers in this "Liberal and Market Society", as it is now called, are becoming too dependent upon those who finance them to be able to give their research innovative aims. In too many cases research follows money, not the need of the people or the health of our planet. That's why, to be really free – better, to be just themselves, with the capability to dig deeper into their interests – scholars and researchers need a 'counterface' not only to discriminate their sources of finance, but also to engage a wider audience of public opinion, among ordinary people, not just in specialized circles. Politics now, in the Western world, does not guarantee true representation of cultural interests and ideals; it is necessary, as a consequence, to build a new pattern of relationships among schools, the media, universities, all united by a common view about Man and his destiny in a Nature that can be preserved.

Being a journalist and (even if with the honorary title of "Conservatore Onorario" to the Museo Tridentino di Scienze Naturali in Trento) not a scientist, it is not my intention to enter into specific themes and topics, but I'll try, instead, to share with you some of my experiences and thoughts about 30 years of struggles and battles for the environment and nature protection in Italy, and also in my region, Trentino, from where Professor Franco Pedrotti comes.

Trentino is an interesting case study region in the Alps – the Dolomites – because it faces a very strong impact from the modern world (40% of all traffic through the Alps crosses the Brenner Pass, and follows the Adige Valley where Trento city is located, that is to say about 50,000 cars and trucks every day), while at the same time it is set in a rather well preserved district (half of the land is covered by woods, 80% of them public rather than private, property). On one hand, Trentino is the landscape where hotels, guest houses, ski-slopes, cable-cars have their maximum extent in Europe, if not the world. On the other hand, it is the only area in the Alps where the last free wild bears survive. As for the management of land resources, 12% of the territory is covered – at least officially – by national and regional parks and biotopes.

Trentino could be a wonderful case study for interdisciplinary research in geobotany and ecology, following the patterns that many contributors to this symposium have pointed out, particularly Professor Pop and Professor Fujiwara. In fact, like the North Sea Islands, Trentino has to face impact with industry (traffic), tourism (a 'heavy', not a light industry) and needs to be 'restored' in some ways, according to the guidelines and purposes of true preservation that Professor Fujiwara has pointed out: 1. gene pool; 2. traditional landscapes (not only for tourism but for agriculture, memory, identity); 3. citizens' participation (something more and better than leisure); 4. international co-operation. These two last points deserve more detailed discussion.

Whoever has followed the debate about nature conservation in the last few years has realized that political and public interest on this issue has progressively faded away. In part, this is certainly due to the cultural 'involution' of the concept of nature conservation and national parks. At the same time, it is due to the progressive emergence of a massive 'virtual' way of life (computers, TV, internet, "cloning" of life itself) instead of the old natural one. The trend is accelerated especially by political and economic global interests. Future generations do not appear to be so important; money 'now' is more important. In fact, even in the field of natural sciences, the world of Nature, with its patterns of biodiversity (which means the freedom of following the patterns of creation and evolution: Jacques Monod's "chance and necessity"), gives way, as an objective of study and research, to biotechnologies in which nature can be shaped and used by Man for purposes of power, not life. While we know by scientific means that keeping a natural dimension and balance in the environment is of the utmost importance for mankind (for economic development too), other methods are followed by global investors, and other messages to public opinion implemented, so that nature conservation concepts, and ideas, enter less and less into the global web of communication and perception. Popular and civic participation becomes difficult (a growing 'consumer' attitude leads to a view of nature as just capital to spend, in the best cases a playground to exploit), because international co-operation is not sufficient to match the forces of the 'market' and speculation. In fact, we lack international co-operation in this field.

In this context, recalling Professor Franco Pedrotti's life's work, aimed – as if it was for the old Masters – at putting together scientific research and civic commitment, at the same time, looking at nature as a common treasury for Mankind, I believe that one of the 'methods' of nature conservation should be to create a global web of those who study and protect it, that is to say to make universities the first knots, actors, of this web.

Is it a Utopia? – not necessarily. Universities are the only organized international system that can match and counterbalance the drive toward global exploitation of nature. They form the only global system which can balance globalisation. This is a rather new fact in international organization. The universities' web can set up a source of public opinion and thought about nature. The energy and capability that scientists and scholars have shown by organizing this symposium, putting together so many people from different parts of the world, demonstrate that this can be a path to the future that is possible to follow. In this way, universities must become not only research centres, 'hungry' for finance no matter whence it comes, but communication agencies as well. Experience demonstrates that governments, laws, ecological clubs, protection groups, green parties are not sufficient to show true defiance against the new barbarians who want to destroy biodiversity in life. Universities must show the world how nature problems are inter-related. They can show, as well, how testifying scientific experience can become a way of civil resistance, on behalf of Nature, which means on behalf of Man. The "Babeş-Bolyai" University of Cluj-Napoca, where the 'naturalist' Professor Borza taught and worked so truly for Nature and Creation, in spite of political oppression under the harsh dictatorships which ruled Eastern Europe for so long, can be and become, an example of goodwill, an inspiration for all of us.

Being knots on a web toward freedom (of thought and communication as well) could be a special aim for Eastern Europe universities to achieve, to "shake and shock" a little, their western counterparts, all too tied to 'market' patterns. They can learn from their difficult and heroic history, form their intellectual resistance, an aim which can become a purpose for the future.

We cannot forget, studying Nature – communicating Nature through our studies – the true words that Professor Fujiwara has told us, when she described the beautiful islands of Japan after World War II, shaken and destroyed from top to bottom, without any tree. And now life is returning.

Studying nature, then, means not only preservation, but also restoration as well. Otherwise the world will have no future. When Man destroys, he must also restore and preserve. To this end must science be aimed.

FROM THE PROTECTION OF LANDSCAPE AND "NATURAL BEAUTIES" TO THE DEFENCE OF ECOSYSTEMS IN ITALY

Gianluigi CERUTI
45100 Rovigo (ITALY); e-mail: gianluigi.ceruti@libero.it

Abstract: Between the end of the 19th Century and the first decades of the 20th, in the wake of similar North American and European experience, a movement of public opinion began to develop in Italy, aimed at protecting nature and the natural environment. It was principally launched and supported by small scientific societies, often created for just such purposes.

The dedicated activity of these early groups succeeded in stimulating public bodies (in particular the Italian National Parliament and Government) to preserve some natural monuments and create the first national parks. Some positive outcomes were thus achieved; and fauna, flora and geological formations started to have a destiny other than mere human spoilage, lack of concern, and destructive threats as a result of ever-increasing industrial activities (such as exploitation of the water resources for hydro-electric dams and power plants) and the occupation of coastal areas for tourism and bathing beaches.

At first, nature was merely regarded as deserving protection for "natural beauty" or because it was inherently linked with sites of historical, artistic or literary heritage. Later, nature protection became mostly a means to conserve landscape beauty. At last, starting a new third era, nature conservation began more correctly to be considered as implying the preservation of relevant natural and ecological values.

The present paper aims to explain this progressive evolution of both the Italian cultural scene and the actual regulations, thus showing the basic reasons why conservation of ecosystems and biodiversity in Italy has been so long delayed, and still must face so many obstacles.

Introduction

Under the Italian legal system, nature and landscape protection has undergone a long, slow process that is still far from complete, and as a consequence the outstanding Italian natural heritage is still in great danger.

The causes of this slowness lie in the blind spots of Italian environmental legislation, the ethical and political frailty of those who should enforce it (a situation I have called "eco-liability"), the aggressive abrasiveness of so many businessmen, and that lack of a true naturalistic culture that a great naturalist, Alessandro Chigi, duly deplored over 50 years ago (1954): "Italy, unfortunately, is one of those countries where a culture about the natural sciences is almost lacking, not only among the general public but also among the ruling classes, a lack which is one of the main causes of greatest disasters, such as mountain deforestation and degradation, abandonment of highlands, lowland floods".

In the early 20th century all nature conservation proposals met with open opposition and covert resistance, as can still be experienced at the present day too in an even stronger, more vehement and overt manner, both in Italian society and Parliament, even if a better awareness of environmental problems has admittedly spread greatly in the meantime – and is still increasing.

Between the end of the 19th century and the first two decades of the 20th, a small number of natural scientists (mostly botanists), officials of the Ministry of Education, isolated politicians, tourism enthusiasts, and prominent intellectuals such as the philosopher Benedetto Croce (1866–1952) and writer and journalist Ugo Ojetti, helped transplant into Italy the culture and the experiences of the USA (Yellowstone National Park) and some European countries such as Switzerland.

Scientific and naturalist groups such as the Italian Botanical Society (Florence, 1888), the Ragazzoni Naturalist Group (Brescia, 1898), and the Emilian Society "*Pro Montibus et Sylvis*" (Bologna, 1899), developed profound cultural debate and actively lobbied Parliament and Government. These institutions, however, mirrored the same educational deficiencies as the general public, coupled with the indifference of the socialist movement (occupied as it was in wage claims for the working classes), and were conditioned by the conservatism of the emerging industrial economy.

The protection of landscape "beauties" dependent upon history, arts and literature

The very first Italian law aimed at nature protection was issued in 1905, on the initiative of the Minister of Agriculture, the Hon. Rava, for making "inalienable" the remnants of the famous Ravenna pinewoods.

Apparently in order to justify such a law and facilitate its passage, a link was suggested between these pinewoods and the "dark wood" ("selva oscura") where Italy's national poet, Dante Alighieri (1265–1321), found himself "nel mezzo del cammin di nostra vita", in the opening section of his "Divina Commedia".

Reporting speaker for this law before the Chamber of Deputies was Giovanni Rosadi, a Florentine lawyer who was very active in favour of the country's historical, artistic and natural heritage.

Coinciding with the final vote on the special law for the Ravenna pinewoods, a resolution was also approved which read as follows: "The Chamber invites the Government to present a bill for the conservation of natural beauties which are linked to literature, arts, and history".

At that time (1905), the Government did not heed such an invitation.

The official inspirational reason for the declaration of the "inalienable" status of the Ravenna pinewoods, as well as the tenor of Rosadi's resolution, clearly and significantly indicated that at that time the protection of landscape and of "natural beauties" was necessarily correlated either with some artistic or literary works or with important battlefields. Other examples of this attitude were the protection of Mt. Tabor at Recanati (inspiration for Giacomo Leopardi's famous poem "L'infinito"), the hills around Urbino (the background of some famous paintings by Raphael), and the moraine slopes of Lake Garda (the scene of some military events of the Italian Risorgimento).

If at the time a Governmental representative considered it necessary to invite the most important national poet for passing a minor act aimed at the protection of a limited portion of Italian landscape; if an important member of Parliament, in order to support a bill for the protection of "natural beauties", had to suggest their linkage with national literature, arts and history, one could well wonder as to whether these were the effects of strategic choice or genuine belief.

I would personally tend to think that these linkages with national literature, arts and history chiefly constituted expedients to break through all resistance and overcome disregard. Such an opinion is not an arbitrary and random inference – and it rests on small but significant signs.

In 1906 a royal commission, also proposed by Giovanni Rosadi, presented a bill (which three years later became the Act no. 364/ June 20th, 1909) that limited State authority and responsibility only to those structures of historical, archaeological and artistic merit, without considering "natural beauties". Such an exclusion was due mainly to several difficulties which arose before the Senate, when a separate bill for landscape protection was suggested and a special resolution was voted for, inviting the Government to submit a bill for the protection and conservation of villas, gardens and other areas inherently linked to history and literature or otherwise representing the public interest because of their particular beauty.

The exclusion of "natural beauties" from any protection and the contents of the approved resolution were clearly the result of delaying and obstructive tactics. Proper wording and agendas could not easily be issued since they were not mandatory.

Anyway, relative to the 1905 resolution (Ravenna Pinewoods Act) a difference can be noticed in that the aesthetic protection of landscape was not necessarily dependent upon historical events or artistic and literary works.

A subsequent law (Act no. 388/ June 23rd, 1912), while extending the protection under Act 364/1909 to villas, gardens and parks of historic and artistic interest, would exclude from such protection "natural beauties" located outside the towns and not connected to an area's cultural highlights.

To expand protection outside the town walls was regarded as arousing concern and problems, as landscape protection was sometimes thought to collide with economic activities such as the exploitation of hydroelectric resources in hilly or mountainous areas.

As all solicitations to the Government to issue a protective framework for landscapes had proved ineffective, the Hon. Rosadi officially presented a bill in order to overcome such governmental (and parliamentary) inactivity.

On 13 May 1910 Rosadi presented his bill (no. 496 of March 19th) to the Chamber. Several parts of his speech – which will be largely reproduced below - were then presented in the Italian Parliament: "Honourable Colleagues", Rosadi said, "our law on antiquities and the arts, to which I am honoured to have someway contributed, doesn't contemplate the protection of Italian natural beauties. I've repeated since long ago, however much in vain, that monuments are not just walls, arches, columns and statues, but also landscapes, forests and waters, and all those areas which present some kind of natural beauty or are illustrated by important memories or high artistic or literary works".

In the same way as it was considered injurious to cut down the laurel of Arcetri (which had witnessed the words of Galileo and Milton) and the cypress trees of Villa Ludovisi (immortalized in Goethe's poetry), or to ravage the Ravenna pinewoods (that forest "spessa e viva" which inspired Dante's vision of Eden), it should also be injurious to impoverish the Tivoli or Marmore waterfalls, to in any way disfigure Villa Borghese, or eradicate the Bolgheri cypresses ("che a Bolgheri alti e schietti van da San Guido in duplice filar") which inspired the youthful memories of our last national poet.

The respect due to these natural monuments begins to be shared by our common feelings, and it seems that voices of protest are emerging everywhere, telling us: 'Why do you destroy me? Don't you have any spirit of mercy?' ("Perché mi scerpi? Non hai tu spirto di pietade alcuno?").

Meanwhile, new associations for the protection of landscape are being created, meetings on these matters are being held, and one looks enviously at the example of France, which since 1906 has issued a law for protecting its picturesque landscapes, thanks to that same Hon. Briand who is currently Prime Minister of that Nation.

Our Parliament too has for once expressed concern for these topics.

When in 1905, on the initiative of the then Minister of Agriculture, Hon. Rava, and of my own approved report, the Act was issued which decreed the inalienability of the residual Ravenna pinewoods, this Chamber passed a resolution which read thus: "The Chamber invites the Government to present a bill for the conservation of those natural beauties which are linked with the Italian literature, arts and history".

Well, colleagues, on that day this Chamber took on the commitment of honouring its own auspices. Your Parliamentary Commission, which prepared that fundamental law for antiquities and the arts which was finally passed on June 20th last year, probably supposed to honour such a commitment by simply stating (article 1) that 'Amongst the immovables are included the gardens, forests, landscapes, waters and all those natural areas which show the above mentioned interest' (artistic, historical, archaeological). However, the Central Department of Senate proposed, and the Senate approved, the abolition of such a statement, allegedly for reasons of symmetry and context. And some affirm that there is no love for aesthetics and architecture!

The Senate, in fact, stated that that paragraph was not connected with the subsequent dispositions, and that, while affecting the general symmetry of the bill, was not appropriate for fitting its aims, which would require, because of their peculiarity, special rules. At the same time, however, the Senate passed a resolution which was later approved by this Chamber too.

This resolution read as follows: "The Government is invited to present a bill for the protection and conservation of the villas, gardens and other areas which are linked with history or literature or anyway involve some sort of public interest because of their outstanding natural beauty".

Thus, as you can see, three resolutions have been passed by Parliament, all asking the Government finally to do its duty, i.e. to protect those national beauties which had been excluded from last year's fundamental law. Well, my modest proposal would intend to finally honour these three resolutions. After having determined the boundaries of its action, it establishes that, without authorization from the Ministry of Education, no forests, parks, gardens, waters, villas, and all those places that are of relevant public interest because of their natural beauty or their links to history and literature, can be destroyed or modified. Thus, not just those 'beauties' which can fire the eyes and hearts of all kind beings, but all those which are of relevant public interest.

So, whatever the interest, there is no excessive widening of this law's boundaries. This law decrees that in those places neither new buildings nor renovations can be done that may in any way spoil their appearance. Whoever can feel natural beauty, or has experienced the injury of any contending or offending obstacle (either aiming at sacrilegious publicity or due to poor taste) can well understand the protective reasons for such a disposition."

On behalf of the Government, then came the speech of the Minister of Education, Credaro, who, after the ritual eulogies, expressed a significant reservation: "I cannot do anything but associate myself with the high and noble feelings expressed by the Hon. Rosadi. He has illustrated to the Chamber a bill which is well worthy of being considered with much attention, as the Government asks the Chamber to consider this proposal. However, I have to note that the influence of such a bill on the law about the derivation of public waters of August 1884 has to be duly studied, because this bill could well put restrictions on the development that Italian industry would be able to yield thanks to the exploitation of the richness of our waters. This is a reservation that I feel I have to express".

Such a reservation, in fact, clearly explains why, despite the approval and oral encouragements of the Government, "Rosadi's law" had to wait twelve more years before being passed by the Italian Parliament; and that was only thanks to the influence of a person of the highest cultural prestige – Benedetto Croce, who, having been appointed Minister of Education in the second Giolitti Government (15th of June 1920–4th of July 1921), personally supported Rosadi's text.

The appeal of the Italian Botanical Society (1911)

In 1911 a most significant event occurred. On 12–16 October a meeting of the Italian Botanical Society was held in Rome and the Society's Secretary, Dr. Renato Pampanini, presented a seminal report entitled "For the protection of the Italian flora" [6]. All members concluded the meeting by approving an appeal "to give the law for landscape protection presented by the Hon. Rosadi the widest significance, in order to make it consider also the scientific aspects of natural monuments and thus protect the flora as well".

Such an appeal by the Italian Botanical Society, seen in retrospect, acquires a great historic, cultural and political importance, because it expressed clearly and vigorously the necessity of preserving landscapes – not only because of their links with history, art and literature, not only because of the aesthetic values of their form, but also because of their botanical elements, which (together with the faunal and geological ones) are essential constituents of natural landscapes.

Thus protection was slowly enlarging its scope, but it also had to face more opposition.

The reasons for Giovanni Rosadi's prudent and gentle strategy emerges from his foreword to the above-mentioned report of Pampanini [6], which was published in Florence in 1912 (page viii): "All those who are mindful of the legend of the Horatii and Curiatii, and well know the hostilities which regularly arise in parliamentary joust towards any restrictive bill, may well ask themselves if it would not be more convenient to raise a lesser amount of hostilities and enemies by using the greatest discretion and by facing them one by one. Our Parliament, when confronted with the matter of landscape protection, will probably convince itself that nascent industry will not be harmed if new disfiguring buildings will be forbidden in the places of outstanding beauty; that the owners of wonderful villas and gardens will not be ruined if in case they are selling or allowing them to deteriorate they will be forced to respect them or to sell them to the State; that the working hydraulic forces will not be halted if it will be permitted to draw water from all rivers and streams except from a few beautiful waterfalls; that trading will not fail if unpleasant and tasteless advertising items (unlike discrete and graceful ones) will be prevented from being placed in those localities which are of outstanding beauty; but it probably couldn't be so easily persuaded, during the discussion of the said bill, that it wouldn't be exaggerated to curb the waste of gentians, yarrows, anemones, primroses, saxifrages, narcissus, edelweiss and all those plants which in our various regions are every day becoming more and more uncommon because of vandalistic unacceptable exploitation.

This is a simple doubt of practical nature, which is convenient not to hide. What is certain is that all lovers of the Italian flora, who have recently leapt to its defence, have to make their action more concentrated and pugnacious on the occasion of a bill for the landscape protection. Without their brave support such a doubt may become an unfortunate reality.

"Then it will be a matter of time. When landscape protection will be ratified in the name of aesthetics and history, it will become an act of coherence and indisputable justice to give the flora official protection as well, in the name of nature's variety and beauty. It is a matter of strategy and time. After having overpowered the first of the three Curiatii, who undermined the safety and the national relief of our arts, after having knocked down the second one, who still makes attempts against the beauty of our landscape, it will become easier and unavoidable to beat the third one as well, who makes attempts against our dazzling and fertile flora. Maybe our three enemies have to be faced one by one, as did the victorious Horatius". In Rosadi's phrase "nature's variety", we can already appreciate the modern concept of 'biodiversity'.

During the Chamber session of the 5th of July 1911, the Hon. Rosadi supported the approval of his bill by vigorously affirming: "Could ever this Parliament remain indifferent and inactive, as if it was unaware that the issue of landscape protection is presently felt and agitated in Italy too, perhaps more in Italy than elsewhere? A law for the protection of natural beauties has not to necessarily embrace them all (our proposal does not comprise them all), even if they are lovely and much preferable to a mill or a factory. In contrast, it will protect just those that possess outstanding values of nature and memories".

Rosadi was fully aware that "it is a matter of strategy and time". In order to get dispositions for ensuring the conservation of the Italian landscapes, it was necessary to execute operations of secret and subliminal persuasion, by appealing to the sensitivity of many parliamentarians for certain historical events, the arts and literature, whenever these were linked to localities which were to be placed under public control.

Landscape is protected only if "beautiful"

The first general measures for landscape protection, valid over the whole national territory, were introduced into the Italian legal system only in 1922, twelve years after the unsuccessful presentation of Rosadi's bill.

The Act no. 778/ June 11th, 1922 "For the protection of natural beauties and estates of special historical interest" was certainly a major step towards the protection of natural landscapes, but its contents were still by far too cautious, limited and narrowly defined.

In accordance with such new regulations, landscapes, forests, parks, gardens and waters were considered to deserve protection only when they were of remarkable public interest "because of their natural beauty" or "panoramic scenery"; in contrast, no protection was envisaged for their possible scientific or ecological importance or their unique value. The State had officially to declare such public interest, and the protected natural or monumental item couldn't be disfigured or destroyed without express ministerial authorization.

The subsequent Act no. 1497/ June 29th, 1939 would establish more organic and updated regulations for landscape protection, but basically it would embrace exactly the same, merely aesthetic vision, even if for the first time it mentioned, besides the conspicuous characters of "natural beauty", those of "geological singularity" (article 1, first clause, no. 1).

From landscape protection in Article 9 of the Italian Constitution (1948) to the protection of environment and ecosystems (2001)

The Italian Republic, according to article 9, paragraph 2, of its Constitution (which came into effect on the 1st of January 1948), "protects the landscape and the historical and artistic heritage".

Consultation of the Proceedings of the Constituent Assembly, unfortunately, does not provide any additional novelty elements, as "landscape" was still conceptually considered synonymous with the "natural beauties" mentioned in Act 1497/1939.

However, a progressive interpretation of article 9, by both juridical doctrine and jurisprudence, has slowly enriched the "landscape" concept – thanks also to the stimulus of the cultural association "Italia Nostra" – with those environmental and nature interpretations which at last will be explained by the Act 431/1985 – usually known as "Galasso's law" ("legge Galasso") – after the Hon. Galasso, from Naples, historian and Under-Secretary of State, who was the true 'father' of this law). These contents were expressly and directly recognized by the new reading of article 117 of the Italian Constitution, as modified by the 2001 constitutional reform, which finally reads as follows: "The State protects the environment, ecosystems and cultural structures".

Therefore, the definition of landscape currently comprises its beauty, its history, and furthermore the intrinsic public interest of its flora and fauna, and its geological characters.

This long process towards the legislative protection of the natural environment, however, had been fully concluded and it had reached its culmination with the Act 394/1991 (on National Parks and other protected land and marine areas), one that I am most honoured to have strongly wished for and supported in the Italian Parliament, and which I am still supporting in the University where I have been called to teach, as well as in my country.

The first aim of any protected area, in accordance with article 1 of this law, is "the conservation of animal and plant species, herb and forest associations, geological singularities, paleontological formations, biotopes, scenic and panoramic values, natural processes, water and hydro-geological equilibria, ecological balances".

The entire discipline of the nature protected areas affirms and guarantees the absolute priority of the conservation of the outstanding aesthetic, natural historical, scientific and ecological values, which, in accordance with several decisions of our Constitutional Court, are primary values, not to be subordinated to whatever other interests (including economic).

The challenge of the Third Millennium

The above-mentioned Act 431/1985, while certainly meritorious, did not differ too much from previous regulations, as the protection it enforced was not absolute but dependent upon discretionary decisions of a public authority (changed over the course of time). Such discretionary power, on the whole, did not prove to work well in the last century.

The ruination allowed by the public authorities, during the last and early present century, in many localities which had been declared to be of public interest because of their landscape, natural or environmental values, were frequent, excruciating and devastating enough to persuade me to launch a strong proposition for the Third Millennium, inspired by the sort of wishful thinking which can well come about if only all citizens will play their part.

By complying with the cyclical theory of history ("corsi e ricorsi storici") of philosopher Giambattista Vico (1668–1744), I look forward to a period of widespread and much more focused ethical and political sensitivity that will replace the present-day unlawful and unregulated permissiveness.

If this happens, the immutability (except for management reasons) will have to be established of all those places which are considered to be 'sacred' for their beauty, natural and ecological importance, historical events, artistic or literary memories.

Of these 'sacred' places the first difficult tasks to carry out will be to draw up an open list, to demarcate their borders (based on strictly scientific standards only, non-dependent on side conditioning), and to establish accurate and strict management rules in such a way that all evasive or circumventing interpretations are made impossible.

However, to fulfil such a crucial turning point, widespread, slow and sure absorption will be necessary, as well as a strong popular push which only will force the legislators to implement such a political will.

Thus, if only a place can be bequeathed to future generations, the power of deciding its fate should not have to belong just to the relevant authorities.

This is the lesson which has to be learnt from a century of unsuccessful or unsatisfactory legislative efforts to restrain the egotism, speculation, avarice and greed which are a part of general life.

My personal commitment in the immediate future will be to build up such a perspective, to turn this dream into reality, among and in company with all my country's citizens.

REFERENCES

1. Cassola F., 1973 - La conservazione della natura in Italia: situazione legislativa e aspetti giurisprudenziali. In: Atti III Simposio nazionale sulla Conservazione della Natura, Bari, **1**: 241-261.
2. Cassola F., 1978 - Strade e ambiente naturale: le fonti storiche e normative di un contrasto non risolto. Impresa, Ambiente e Pubblica Amministrazione, **5**: 449-464.
3. Cassola F., 2001 - L'associazionismo ambientale in Italia. Storia, antropologia e scienze del linguaggio, Roma, **16** (3): 69-89.
4. Ceruti G., (ed.) 1996 - Aree naturali protette. Commentario alla legge n. 394/1991. Documenti. Nuova edizione. Editoriale Domus, Rozzano-Milano.
5. Corbetta F., 1986 - I movimenti portatori di interessi ambientali. In: Giornata dell'ambiente. Atti dei Convegni Lincei (Roma), **76**: 115-128.
6. Pampanini R., 1912 - Per la protezione della flora italiana. Stabilimento Pellai, Firenze.
7. Pedrotti F. (ed.), 1988 - La Società Botanica Italiana per la protezione della natura (1888-1990). L'Uomo e l'Ambiente, **14**: 1-181.

8. Pedrotti F., 1998 - Il fervore dei pochi. Il movimento protezionistico italiano dal 1943 al 1971. Temi Editrice, Trento.
9. Piccioni L., 1999 - Il volto amato della Patria. Il primo movimento per la protezione della natura in Italia 1880-1934. L'Uomo e l'Ambiente, **32**: 1-320.
10. Sievert J., 2000 - The Origins of Nature Conservation in Italy. Peter Lang, Bern.

WILD PLANT SPECIES ENDANGERED THROUGH INTENSIVE HARVEST

Anna MAROSSY
Parcul Traian, nr. 29, ap. 11, **410044 Oradea (ROMANIA)**

Abstract: The spontaneous vernal and pre-vernal flora faces a serious threat from massive harvest by tourists and flower-sellers. This has resulted in an inevitable gradual decrease in plant numbers, even endangering the existence of certain populations. The present paper considers 16 pre-vernal species that have been cultivated and now reproduce themselves in private gardens; it is proposed that 36 wild species be introduced into cultivation and even replanted in their original habitats from where they are starting to disappear. An education programme among local people and decision-makers is needed to stop the intensive harvest of certain plants from their natural habitats.

Introduction

Plant species that are relict, endemic, rare or are about to disappear *(Leontopodium alpinum, Nymphaea lotus* var. *thermalis*, *Cypripedium calceolus*, *Taxus baccata*, *Ruscus aculeatus*, *R. hypoglossum*, etc.) are protected by law in Romania as 'monuments of nature'.

Some species have been proclaimed as protected by the decisions of the Bihor county authorities, bearing in mind their rarity and regional importance. This is the case for the following species: *Syringa josikaea*, *Paeonia officinalis* subsp. *banatica*, *Adonis vernalis*, *Gentiana punctata*, *Narcissus angustifolius*, all species of *Iris*, *Pulsatilla* and orchids (*Orchideaceae*), *Nymphaea alba*, *Fritillaria montana*, *Lilium martagon*, *L. jankae* and *Gladiolus imbricatus*.

There are also a large number of species which, for the time being, are not in a critical situation but have been collected *en masse*, mainly in the last 15 years, and are sold as decorative flowers, medicinal plants, for school herbaria, etc. They are not officially protected, but their 'industrial' exploitation affects the local populations.

Material

These decorative plants are mass-harvested and sold in markets without any restriction. Thus, pre-vernal flowering species such as *Galanthus nivalis*, *Leucojum verum*, *Crocus heuffelianus*, *Erythronium dens-canis*, *Primula vulgaris*, *Scilla bifolia* and *Hepatica nobilis*, appear in the markets in large quantities, and when the spring traditional festivities (the 1st and 8th of March) are over, they are thrown out with the garbage. A population of *Galanthus nivalis* has been monitored for 20 years and a gradual decrease, both of its size and vitality, has been noticed.

Later, other species become the object of intensive harvest: *Fritillaria montana*, *Amygdalus nana* (*Prunus tenella*), species of *Paeonia*, *Narcissus angustifolius*, *Doronicum sp.* and *Nymphaea alba* (on the markets in Bucharest); whereas ferns (*Dryopteris* sp., *Phyllitis scolopendrium)*, large branches of *Abies alba*, *Pinus strobus* (for mortuary wreaths), *Juniperus communis*, mosses and *Stipa* sp. are sought out during the entire year. *Ruscus aculeatus* is sold especially in autumn and winter, even with its roots, and is also exported in Hungary.

Medicinal plants are systematically harvested by both pharmaceutical companies and country people (especially gypsies), who sell them in the markets. The companies buy large quantities of *Lycopodium clavatum*, *Helleborus purpurascens* and buds of *Pinus mugo*. Medicinal plants such as mint (*Mentha*), St John's-wort (*Hypericum*), thyme (*Thymus*), chamomile (*Matricaria*), *Lycopodium*, *Equisetum* and *Centaurium* are offered without any control in large quantities in public markets.

The village of Poieni de Jos (Bihor county) has specialized in harvesting medicinal plants as a main source of income since the last century. As well as the well-known species, local people collect the roots of *Bryonia alba*, *B. dioica*, *Tamus communis, Gentiana asclepiadea* and *G. cruciata*, so destroying the entire plant. There has been a conspicuous decrease in numbers of individuals in local populations of these species.

Results and Discussion

What is to be done?

- The authorities should promote and severely apply legislation against the collection of spontaneous plants for commercial purposes.
- The plant species that are the object of intensive collecting should be cultivated, thus saving wild populations (Tab. 1). There are successful examples from other European countries, where cultivated plants of particular wild species are sold in flower-shops.

Table 1: Wild plant species grown in private gardens.

Vinca minor	*Helleborus purpurascens*	*Fritillaria meleagris*
Primula acaulis	*Erythronium dens-canis*	*Scilla bifolia*
Galanthus nivalis	*Anemone nemorosa*	*Leontopodium alpinum*
Leucojum verum	*Muscari comosus*	*Paeonia officinalis* ssp. *banatica*
Hepatica nobilis	*Pulsatilla patens*	*Aquilegia vulgaris*
Crocus heuffelianus	*Adonis vernalis*	

Numerous medicinal plant species have already been introduced into cultivation, including: *Matricaria chamomilla, Digitalis lanata, D. rubra* and *Mentha* species (Tab. 2).

Table 2: Wild plant species introduced as an experiment into cultivation.

Inula helenium	*Iris* sp. pl.	*Lilium martagon*
Digitalis lanata	*Campanula glomerata*	*Veronica spicata*
Trollius europaeus	*C. cervicaria*	*Veronica orchidea*
Gladiolus imbricatus	*Lunaria rediviva*	*Sempervivum tectorum*
Doronicum austriacum	*Carlina acaulis*	*S. marmoreum*
D. columnae	*Physalis alkekengi*	*Sedum* sp. pl.
Aconitum firmum	*Centaurea* sp. pl.	*Dictamnus albus*
Gentiana lutea	*Ajuga reptans*	*Eranthis hiemalis*
G. asclepiadea	*Arabis alpina*	*Syringa josikaea*
Nymphaea sp. pl.	*Colchicum autumnale*	*Sorbus aucuparia*

Conclusion

Local people of all ages, and decision-makers, need to become aware of the danger of loss of biodiversity in natural habitats. To this end, a continuous programme of instruction and education (boards, leaflets, mass-media) needs to be undertaken. Only in this way will people stop collecting wild plants for trade and will start to cultivate the species of interest on their own land.

REFERENCES

1. Carmazinu Cacovschi V., 1978 - Peisajul estetic vitalizant. Ed. Ştiinţifică şi Enciclopedică, Bucureşti.
2. Correvon M.H., 1895 - Les plantes alpines. Paris.
3. Dané T., 1957 - Vadvirágszeliditők. Ed. Tineretului, Bucureşti.
4. Domokos J., 1973 - Páfrányok a lakásba és a kertbe. Mezőgazdasági Kiadó, Budapest.
5. Galantai M., 1981 - A kivesző növények szaporíthatók! Búvár, 3, Budapest.
6. Marossy A., 1983 - Introducerea în cultură a speciilor de plante spontane periclitate. Folia naturae Bihariae (Nymphaea), **10**: 221 – 226.
7. Paun M., Palade L., 1976 Flora spontană sursă de plante pentru spaţiile verzi. Ed. Scrisul românesc, Craiova.
8. Priszter S., 1974 - Hagymás kerti virágok. Mezőgazdasági Kiadó, Budapest.
9. Sigarteu Petrina L., 1977 - Arte tradiţionale japoneze. Ed. Albatros, Bucureşti.
10. Sulyok M.., Timár Z., 1972 - Képes virág abc. Mezőgazdasági Kiadó, Budapest.
11. Tauber F., 1977 - Conservarea florei vernale maramureşene. Ocrotirea naturii maramureşene. Acad. R.S.R., Subcomisia Monumentelor Naturii, Cluj-Napoca.
12. Tergit G., 1969 - A virágok regénye. Gondolat Kiadó, Budapest.

PEATLANDS OF NORTH-EASTERN EUROPE AND THEIR CONSERVATION

Tatiana YURKOVSKAYA
Department of Vegetation Geography and Cartography, Komarov Botanical Institute, Prof. Popov Street 2, **197376 St.-Petersburg (RUSSIA)**; e-mail: Yurkovskaya@hotmail.ru

Abstract: The north-east of Europe is a territory where latitudinal zonation in distribution of mires and its relation to bioclimatic zones is distinctly pronounced. From tundra to the limit of taiga, polygonal, palsa, aapa mires and raised bogs follow each other in succession. The preservation of mires depends upon the personal initiative of researchers, local administrations and the support of international opinion.

Study area

The north-east of Europe includes two large administrative regions of Russia – the Arkhangelsk region with Nenets Autonomous District and the Republic of Komi. They are shown on the map (Fig. 1).

In the north the territory includes the eastern coast of the White Sea and the Barents Sea coast, together with adjacent islands, the largest being Kolguev (in the north-west), Novaya Zemlya and Vaigach (in the north-east). In the west it borders Karelia and in the south the border approaches 60^0 N. The eastern border is with the Ural Mountains.

Geography of mires and their features

In the vast expanses of north-eastern Europe two bioclimatic zones are distinguished: arctic (tundra) and boreal (taiga). The arctic zone is divided into four subzones: high arctic tundra (N. Zemlya Island), arctic tundra (south of N. Zemlya and north of Vaigach Isles), north tundra and south tundra. The boreal zone is represented by three subzones: forest-tundra, north and mid taiga. It is curious that the administrative borders between mid and south taiga almost coincide with the natural ones. The distribution of mires is clearly connected with bioclimatic zones and subzones. From north to south the following types of mires can be found: polygonal, palsa, ribbed fens (aapa) and raised bogs. The most widely distributed are the herb and herb-moss fens; they spread from high-arctic tundra to the southern limit of the territory, but their floristic composition, syntaxonomy and types change along the latitudinal gradient.

Mires constitute a significant portion of vegetation cover in the tundra. The most remarkable type of Arctic mire is the polygonal mire. The polygonal structure arises through a net of deep cracks breaking the mire surface into polygons. In the north-east of Europe polygonal mires are confined to northern (typical) tundras. The westernmost localities of polygonal mires have been recorded in the east of the Malozemelskaya tundra in the region of Nenetskaya Ridge. Here they are especially numerous in the Neruta River Valley. The southernmost localities of polygonal mires are recorded in the upper reaches of the Kolva River. Data on the spread of polygonal mires in north-eastern Europe has recently been published [3]. These data have changed the traditional idea that polygonal mires are characteristic of the north of Asian Russia and North America.

Herb and herb-moss fens are distributed throughout the whole Arctic. Herb-moss fens only occur in the high Arctic. The most typical fens of the coastal part of the Arctic are grass-sedge fens with *Dupontia fischeri* and *Carex stans*. These fens are characterised by high water saturation and often merge with saltmarshes, forming large wetlands. Sedge-cottongrass-moss fens occur throughout the whole tundra (except the High Arctic tundra subzone) in river and stream valleys and in lake depressions, as well as in coastal lowlands at some distance from the sea. These fens include *Eriophorum medium*, *E. russeolum*, *E. polystachion*, *Carex chordorrhiza*, *C. concolor*, etc.

Another peculiar type of mire is the palsa mire which is connected with sporadic distribution of permafrost. Palsa have various shapes and heights. Most characteristic of palsa is dwarf birch (*Betula nana*), which is especially abundant in the low parts of a palsa. European palsa are treeless. In European Russia palsa mires appear in the south Arctic but they are at their most typical outside it, in the forest-tundra subzone. Their northernmost location is Kolguev Island. Thus their geographical range may be termed arctic-boreal.

The next type of mire along the north-south direction is the ribbed fen (aapa mire). The main features of ribbed fens are: concave surface shape, high water saturation in the centre and a heterotrophic string-flark complex. The centre of this mire is occupied by communities with higher mineral nutrition demands. Usually these are herb and herb-moss flarks and herb-sphagnum strings, whereas the edges are meso-oligotrophic, low shrub-peatmoss with pine and *Sphagnum fuscum*.

Three dominating synusia are characteristic of the ribbed fens – herb, peatmoss and brown moss (*Amblystegiaceae*). Their ratio and role in the composition of vegetation cover clearly varies. The dominance of peatmoss synusia is indisputable on the positive elements of micro-relief and on the marginal parts of mire massifs. Herb and brown moss synusia are dominant in the negative forms of micro-relief, mainly in central parts of mire massifs. The presence of herb flarks that lack moss cover is one of characteristic features of ribbed fens. They are clearly visible from the air as well as on aerial and space photographs, owing to the peculiar structure of string-flark complexes: the thin winding light strings are seen against the dark water of flarks. They are restricted to forest-tundra, northern and mid taiga. Thus the north part of their range is overlapped by palsa mires, while the south part is overlapped by raised bogs. The ribbed fens are restricted to forest-tundra, northern and middle taiga. Up to the present day many authors still rely on old data from well-known publications [2, 4] and write about the concentration of ribbed fens in north-eastern Europe in the north taiga in the Pechora basin. The latter is actually not the case. In fact ribbed fens are spread throughout the whole of the north-eastern European taiga in the subzones of forest tundra, north and mid taiga. They also occur in south taiga just outside the limits of the region. They can be well deciphered on aerial and space photographs.

Raised bogs dominate over all other types of mire in the north and mid taiga subzones. Two types of raised bogs are distinguished: coastal (Southern White Sea raised bogs) and continental (Pechora-Onega raised bogs). The most typical coastal bogs are situated in the West, in Karelia.

The continental raised bogs are distributed beyond the narrow coastal area, setting them apart from the Pechora-Onega continental raised bogs across the border. We have carried out a comparison of these two types [9]. Here we will only mention their essential differences.

The flora of continental raised bogs lacks *Calluna vulgaris*, the essential element of ridges and heaths of the coastal bogs. *Chamaedaphne calyculata* becomes the main dominant species on the ridges of continental bogs. *Carex rariflora*, which is so characteristic of hollows of coastal bogs, is replaced in continental bogs by *C. limosa*. The phytocoenotic role of some species changes. Thus the role of pine substantially increases on continental bogs. The prevailing ridge communities are significantly different. On the other hand, phytocoenoses of hollows are uniform except for *Caricetum limosae* Osvald 23 (continental) and *Caricetum rariflorae* Fries 13 (coastal). Massifs of continental and coastal bogs differ in the pattern of their complexes, surface morphology and composition of peat deposit. The role of bogs themselves in landscape is different. Coastal bogs dominate the landscape of the White Sea coast while continental bogs give way largely to spruce forests and their derived communities, though they play a significant role in the structure of taiga vegetation cover.

Among all Sphagnum mires a special place belongs to transitional mire. This type of Sphagnum mire occurs further north up to the northern forest-tundra. It is by these mires that many boreal species advance northward.

Unpatterned herb and herb-moss fens occur practically in all zones and regions. We have already noted their characteristic features in the Arctic. In the boreal zone the small-sedge and large-sedge fens prevail, with *Carex nigra, C. lasiocarpa, C. rostrata, C. vesicaria*, *Equisetum fluviatile*, etc.

A special place is occupied by spring fens. They are small in size but their flora and vegetation are very peculiar because of the influence of calcium-enriched waters. They occur sporadically throughout the whole taiga of the region.

And finally, forest swamps, typical of the boreal region. In the north-east of the European boreal region these are predominantly birch and spruce-birch swamps. They are especially widespread in the Pre-Ural territory.

Conservation of mires

Until very recently almost nothing threatened the mires in the north-eastern Europe. In 1960s, however, the amelioration boom began, oil was discovered and the oil industry boosted. The tundra and taiga mires became subject to intensive destruction or transformation. Hence it has become necessary to conserve tundra and taiga mires. The first perceptible measures towards the preservation of mires were developed in the Republic of Komi. They were initiated by activists of the TELMA group under the leadership of Prof. Hugo Sjörs (Sweden) and V. Mazing and M. Boch in Russia.

In the Republic of Komi the conservation of mires was carried out very successfully owing to the activity of Alekseeva [1] with the support of the regional administration. In the Arkhangelsk region there are currently no professional geobotanists and therefore no initiatives on mire conservation. One look at a map is enough to estimate the situation with mire conservation.

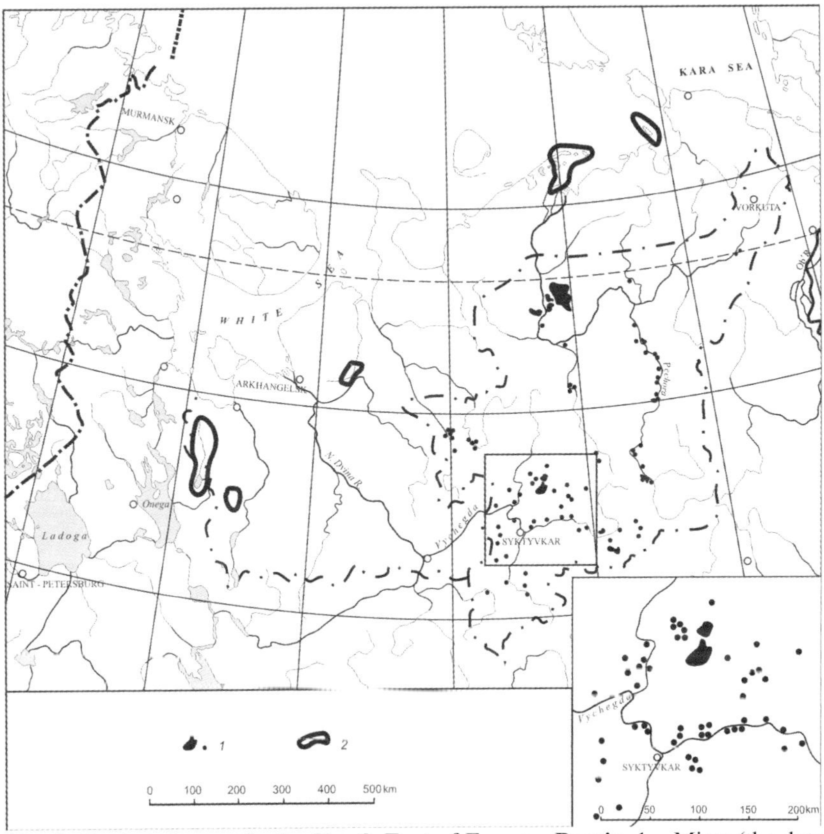

Fig. 1: Protected areas in the North-East of Europe, Russia. 1 – Mires (the dots correspond to mires with an area <25 ha); 2 – State strict nature reserves and national parks.

The map of the territory of Komi is spotted with black marks which show protected mires (Fig. 1/1). The scale does not permit all of them to be shown. For comparison the area near Syktyvkar, singled out as a separate square, is shown in a larger scale to give the real impression of the number of mires under conservation.

The protected mires of Komi have various types of status (nature monuments, natural reserve, national parks etc.) and different causes determining their conservation category (berry lands, regional and zonal standards of mire massifs, mires with high species diversity or with rare or disappearing species, etc.).

In the Arkhangelsk region or the Nenets National District there are no specially protected mires. Several areas shown on the map are the recently established national parks and strict state natural reserves (Fig. 1/2). These can also include mires: for instance mires prevail in the landscape of Dolgy Island, which is part of the state strict nature reserve “Nenetsky” (oral communication, N. Matveeva). It must be noted that we even did not show national parks and reserves (possibly including mires) on the map of Komi in order not to distort the overall picture. Not included is the oldest Pechora-Ilych state strict natural reserve, established 70 years ago, the large national park of Yugyd-Va. In order to estimate the present level of nature conservation activity in the north-east of Europe I shall present some figures. Protected areas cover 14.6% of the total area of the Republic of Komi, 4% of the Nenets District, and 1.32% of the Arkhangelsk Region.

At present the nature conservation activity has intensified in all these regions, but that in the Republic of Komi is the most intensive. The great number of the research and nature conservation projects, including international, is most striking [8]. The non-commercial organisations also support the nature conservation initiative [7]. Nature conservation activity is also being intensified in Nenets District [5]. Finally, in the Arkhangelsk region, a conservation movement has began, supported by the regional administration, WWF and other national or international organizations [6]. Among these initiatives the most interesting is that connected with the study and conservation of primeval forests. The future will show how this project will be fulfilled in practice, for the forest of this region is on sale to Sweden.

Conclusion

Russia has joined the Ramsar Convention on wetlands, and the strategy for Russian mire conservation has been established. If one analyses the real state of wetland conservation, however, it becomes obvious that protection of mires depends upon the personal initiative of researchers, local administrations and the support of international opinion.

Acknowledgements: I appreciate the financial support of the Russian Fund of Fundamental Researches, Grant 03-04-48791.

REFERENCES

1. Alekseeva R.N., 1988 - Mires of the Pechora region. Nauka, Leningrad (in Russian).
2. Boch M.S., Masing V.V., 1979 - Ecosystems of mires of the USSR. Nauka, Leningrad. (in Russian).

3. Gribova S.A., Yurkovskaya T.K., 1984 - On the geography of polygonal mires in European part of the USSR. Geography and natural resources **2**: 41-46. (in Russian).
4. Kats N.Y., 1971 - Swamps of the Earth. Nauka, Moscow. (in Russian).
5. Kozlov S., Glotov A., 2004 - Creation and Development of Protected Areas in the Nenets Autonomous District. The Finnish Environment, **671**: 96-97.
6. Kuznetsov V., 2004 - Protected Areas in the Arkhangelsk Region and Old-Growth Forests in the Organisation of New Protected Areas. The Finnish Environment. **671**: 106-107.
7. Mariev A., 2004 - Boreal Forest Conservation and Management: the Komi model project. The Finnish Environment, **671**: 112-113.
8. Ponomarev V., 2004 - Protected Areas Network of the Komi Republic: environmental projects and initiatives. The Finnish Environment, **671**: 122-123.
9. Yurkovskaya T., 2004 - Raised bogs on the north-east of Europe. Annali di Botanica, **4**: 61-69.

III

BIODIVERSITY ASSESSMENT

WOODY FORMATIONS IN A MESOTHERMIC VALLEY OF TARIJA PROVINCE, BOLIVIA

Maximo LIBERMAN CRUZ[1], **Franco PEDROTTI**[2]
[1] Bolivian National Service of Protected Areas, **La Paz (BOLIVIA)**;
e-mail: mliberman@sernap.gov.bo
[2] Department of Botany and Ecology, University of Camerino, Via Pontoni 5,
62032 Camerino (ITALY); e-mail: franco.pedrotti@unicam.it

Abstract: In the Bolivian Andean mountain range there is a series of mesothermic valleys with a particular arboreal flora. This study describes the arboreal and scrub flora of the Camacho river basin, and its distribution dependent upon numerous ecological factors including anthropogenic influence.

Introduction

The transformation of natural vegetation in the Tarija Central Valley dates back mainly to the time of arrival of the Spaniards in the Americas. In 1794 Father Mingo reports that "... from the lowlands of the Calama one could see the deep forests of Yellow flowered Hardwood trees, Quebracho barks, Cedars, Pines, Walnut trees, Carob trees, Silk-cotton trees, Tacos, Willow trees, Pepper trees and Wood Sorrel trees that beautified these valleys..." [17] in the recent past.

Almost 100 years later in 1884, Fr. Corrado describes a different landscape that resembles the current situation, when he states: "... man's greed and animal voraciousness destroyed nature's magnificence little by little; now, the arid soil, torn by ugly and deep chasms tries to partly cover its nudity with thorny bushes and melancholy shrubs, and as a testimony of its former wealth scarcely presents beautiful pepper trees that stand tall among midget wood sorrel trees and rachitic carob trees" [8].

Descriptions by Franciscan chroniclers from the 18th–19th centuries, and recently by Gerold [12, 13], Coro [7] and Bastian [2], show us that transformations of vegetation and landscape occurred over a relatively quick period. The few forests in acceptable condition are found in zones that are almost inaccessible to man such as humid cañons or streams, hillsides and rocky zones with steep slopes and riverbeds. The original forests have been, in many cases, replaced by secondary formations of chaparral and scrub-like bushes. In extreme cases, isolated trees appear, standing witnesses to the original vegetation.

Location and description of the study area

The Camacho river basin is located at the extreme south of Bolivia in Tarija province within an inter-Andean valley. Politically, it belongs to Avilez, Arce and Cercado provinces. The Camacho river is one of the five great rivers occuring in the Tarija Central Valley, located in the Valley's southern third. The approximate land surface area covers 1050 km^2 and varies in height from 1600 m a.s.l., at the juncture of the Camacho and Guadalquivir rivers, to 4500 m a.s.l., around the Taxara mountain range. The zone of the Tarija central valley basin corresponds to a mesothermic valley in which the distribution of rain is seasonal, concentrated between the months of December and March. In the Camacho river basin, the rainfall range is great, from 300 mm per annum on average in the Angostura zone to 1400 mm in the basin's south-western mountainous zone. The yearly average temperature according to registers of the Tarija and Padcaya season reaches 17.9° and 17.7°, respectively.

The Tarija valley including the Camacho river basin is located in the sub-Andean region of the Eastern Royal Range of Los Andes. In view of its geographical location and dominant vegetation type, the zone belongs, according to Cabrera [5] to the Amazon, Andean and Chaqueño mountainous phyto-geographical domains. The Tucumano–Bolivian provinces represent the Amazon domain, with two districts of mountane to subalpine forests, with Alder and Mountain Pine, and Transition Forests with Yellow-flowered Hardwood. The Andean domain is covered with Keñua forests, its distribution in the basin limited to mountainous zones of northern aspect. The Chaqueño mountainous domain, which was represented by high timber woods, has practically disappeared in the Camacho riverbed; currently there remain secondary shrubby species characteristic of the Chaqueño Province.

Description of the woody plant formations

White alder forests

Alder forests occupy the higher portion of the humid mountainous forests in the sub-Andean strip of the Andes. The dominant species is Alder (*Alnus acuminata* subsp. *acuminata*), a deciduous tree that reaches an average height of 10 m. It grows in almost pure forests with an important element of shrubby and herbaceous layers, and a large quantity of ferns. In some cases, under natural development, it reaches 100% cover, forming continuous strips of woodland; in general, cover varies between 60 and 90%, presenting itself as forests isolated in streams and on slopes of the mountains oriented towards the South. They are distributed in the mountainous belt towards the subalpine level, with slopes of southern aspect, between 2300 and 3000 m a.s.l., in zones where humid winds and clouds that originated in Argentinean territory lay stationary to leeward due to an orographic effect.

The alder forests, when not under a strong anthropogenic influence, show a distribution zone above the forests of Mountain Pine (*Podocarpus parlatorei).* Alder forests develop in mountainous areas with steep to steeper slopes varying between 25° to 45° in zones of south to south-east aspect. Their distribution coincides approximately with the isohyets of 600–1,200 mm of average yearly rains. Frequent southern winds bring humid and fresh air to the zone. A dry layer superimposes the humid air layer, and both are separated by a vertical temperature inversion. In this zone, especially between the months of December and March, the phenomenon of water vapour condensation takes place, and a broad stationary cloud layer is developed; it coincides with the distribution of alder woods. These cloud layers that are in contact with the woods capture additionally quantities of water due to fog precipitation. This is important as a hydric resource source, aside from the value of the forestry masses, which are the main obstacles and act as water recipients of the water condensed by the fogs. On the other hand, the alder woods develop on slopes and mountainous environments where there are humid soils, not being capable of standing long periods of stagnant waters. In some cases there may be alders on rocky soils. These soils have abundant accumulated organic matter, and this determines a significant increase in porosity and water infiltration capacity in the soil.

The tree layer is composed mainly of Alder, but subdominant trees are also present, such as Hawthorn (*Duranta serratifolia*), Keñua (*Polylepis hieronymi* and/or *P. besseri*) and tall bushes of the following species: Jalancacho (*Eupatorium saltense*) and Mountain Pine (*Podocarpus parlatorei*). The shrub-like layer is characterized by showing cover lower than 20–60%, with the following species: *Brachyotum microdon*, *Minthostachys* sp., *Lepechinia graveolens* and *Berberis* aff. *fiebrigii*. The main forest is characterized by a series of species with a variable cover that ranges between 30 and 90%, with: Santa Lucia (*Commelina oblicua*), *Oxalis spiralis*, *Selaginella sulcata*, *Festuca dolichophylla*, *Rhynchospora hieronymi*, *Luzula excelsa*, *Bidens tenera* and *Begonia* aff. *veitchii*, and ferns such as *Anemia tomentosa* var. *anthriscifolia*, *Woodsia montevidensis* and *Cheilanthes marginata*. In very open degraded zones a secondary community occurs dominated by the gramineous species *Lamprothyrsus hieronymi*.

Keñua forests

Dominant species: *Polylepis besseri* Hieronymus (synonyms according to Simpson [23]: *P. crista-galli* Bitter, *P. crista-galli* var. *longiracemosa* Bitter, *P. incana* subsp. *subtusalbida* Bitter).

Keñua forests (*Polylepis besseri*) correspond to the semi-humid to semi-dry forest district of the Andean mountainous region. Very open, monospecific formations develop preferentially on slopes of northern aspect. Physiognomically, they resemble a hay-field by the presence of numerous trees whose crowns do not touch. In their natural state, distribution of individuals is homogeneous, with 25–40% cover. The size of trees ranges between 2 and 6 m; exceptionally, 10 m-tall trees may be found. They are distributed on mountainous slopes of northern aspect, at an altitude of 2,300–3,000 m. The Keñua woods generally appear on slopes opposite the White Alder forest, thus constituting an asymmetry in vegetation distribution due to the effect of aspect. Despite having located all the most representative forests in the Camacho River basin, it is difficult to come up with an assessment of its potential distribution, although we know that it was once wider than at present. In some zones, such as the Keñuahuani river streams, there are no Keñua specimens; however, that area is named after them.

The Keñua woods are distributed along mountainous regions with slopes of 30–40°. According to the isohyets map of the Camacho river basin [16], the woods would be located in the zone where the average yearly rains range between 900 and 1,200 mm. The fact that they are almost always exposed to the north of the mountains means that moisture is lower; therefore the Keñua woods are xeric formations as compared with the alder woods. It is important to stress that on the southern extreme of the basin, where there are valleys with northern aspect as well as mountains, the Keñua woods reach their highest elevations due to the effect of mountain masses.

The main accompanying species in the *Polylepis* woods are: Sacha (*Tillandsia usneoides*), especially in zones exposed to winds, Thola (*Baccharis dracunculifolia*), *Brachyotum microdon*, *Lepechinia graveolens*, *Satureja* sp., *Elionurus muticus, Stipa ichu*, *Begonia* aff. *veitchii*, *B. boliviensis*, and several ferns – *Asplenium lorentzii*, *Anemia tomentosa* var. *australis*, *Notholaena squamosa*. On steep rocky slopes, mainly of southern aspect, there are Keñua woods of *Polylepis hieronymi* Pilger (syn. *P. racemosa* var. *hypoleuca* Wedd., *P. australis* var. *bijuga* Bitter), forming reduced and non-continuous bushes, with typically twisted trunks.

Mountain pine forests (tarija pine, mountain pine)

Mountain Pine forests (*Podocarpus parlatorei*) belong to the zone of the humid high-mountain forests distributed in a vegetation belt below the alder woods. Despite the name Mountain Pine, it does not belong to *Pinaceae* but to *Podocarpaceae*. These trees form pure or almost pure woods in the river valleys or on very humid slopes, preferably on southern slopes. Cover of the woods ranges from 70 to 100% with a shrub-like herbaceous stratum very rich in broad-leaved species, gramineous plants and mosses. Trees reach variable heights ranging from 4 to 12 m and 80 cm in diameter.

Pine trees appear distributed in diverse environments: mountainous areas with slopes of southern aspect: southwest- and east-facing slopes, riverbeds and outwash cones. It is possible to find continuous strips of woods in a range of environments, from riverbeds with abundant rocky material of alluvial and colluvial origin, to the mountainous zones, especially in places with low anthropogenic pressure or inaccessible to man. In other cases, they are found as little woods isolated on the slopes. Altitude variation ranges from 2000 to 2,500 m.

The mountain pine forests develop in habitats of variable slope, from level to inclined (10–15°) to steep (30–40°). Their distribution coincides with the isohyets of 600–1,200 mm of average annual rainfall. In this case, the humidity contribution of fogs is not as important as it is in the alder woods, as they grow in a lower vegetation belt and altitude. According to Bass-Werner [1], the Mountain Pine in pure stands may reach 108 trees per hectare. The mountain pine, when forming forests, is a dominant tree, and constitutes the majority of the biomass of the higher tree layer.

In some cases floristic variations of structure and density are correlated with the variation of the bedrock and relative topographical position. Near its higher limit, almost more humid, it shares space with Alder (*Alnus acuminata* subsp. *acuminata*); occurs with *Weinmannia boliviana* and *Polylepis hieronymi* in encased zones close to the riverbeds such as the Cabildo river stream; toward its lower limit, it is more xeric with Jacaranda or Rosewood (*Jacaranda mimosifolia*); in open or exposed areas, it alternates with Red Alder *(Myrica pubescens*), Jalancacho (*Eupatorium saltense*) and Wood Sorrel (*Acacia caven*). Other tree species present in the mountain pine formation are Sauco (*Fagara coco*) and *Duranta serratifolia.* The shrub-like layer presents a variable cover of 20–60%, the most frequent species being: *Myrica pubescens*, *Lepechinia graveolens*, *Eupatorium bupleurifolium*, *Satureja* sp., *Brachyotum microdon.* The herbaceous layer is very rich, with ferns, fungi and grasses. Its cover fluctuates between 60 and 95%. Those species of higher occurrence are: *Selaginella sulfata, Rhynchospora hieronymi*, *Oplismenus hirtellus*, *Panicum* cf. *demissum, Melica sarmentosa*, *Adiantum poiretti*, *Anemia tomentosa* var. *australis*, *Asplenium lorentzii* and *Woodsia montevidensis*. It is necessary to stress the presence of an important coverage of Sacha (*Tillandsia usneoides*), epiphytic in the tree layer, in almost all the woods where relévés were recorded, forming veritable curtains alongside *Usnea* sp. and *Tillandsia capillaris.*

Red alder bushes

Red Alder (*Myrica* cf. *pubescens*) bushes correspond to a stage in the Mountain Pine (*Podocarpus parlatorei*) forests substitution series and, to a higher degree, to that of White Alder (*Alnus acuminata* subsp. *acuminata*). Red Alder is found in the shrubby layer of the Mountain Pine and White Alder forests, with an insignificant coverage. When the forests are cut, devastated by overstocking or uncontrolled fires, red alder develops secondarily, covering areas above the remains of the original forest. These trees grow on southern slopes, forming pure or almost pure thickets that may reach a height of 3 m. In other cases they form small isolated shrubby clusters. The replacement is more evident in the lower limits of the forest distribution cited, and they grow at an altitude of 2000-2500 m a.s.l. The Red Alder forests are present in zones with inclinations of 15-45°, of southern to south-western aspect, with important cover varying between 95 and 100%.

The secondary accompanying species, for the most part, are the same that appear in the mountain pine forests, but with very low cover. These are: *Baccharis* sp., *Eupatorium saltense*, *Lepechinia graveolens*, *Brachyotum microdon*, *Gaylussacia hieronymi*, *Panicum* cf. *demissum, Oxalis spiralis.* Their distribution coincides with the 600–1,000 mm isohyets of average yearly rainfall.

Yellow-flowered Hardwood tree forests

The Yellow-flowered Hardwood Tree (*Tipuana tipu*) forests correspond to the district of the transition forests. It is a climatic community that has practically been destroyed, as its distribution area coincides with zones where man has intensively developed agricultural activities, especially in wetlands – lake sediment of the Camacho river basin. Currently only some groups of trees are present on slopes where moisture accumulates. The Yellow-flowered Hardwood Trees are the most conspicuous of the entire basin. They are very tall, reaching up to 20 m and over 120 cm in diameter. Their foliage usually branches at great heights. The Yellow-flowered Hardwood Tree groups constitute remnants of the old splendour of forests that existed in the past [15]. In altitude it is distributed between 1900 and 2100 m. The distribution of Yellow-flowered Hardwood Tree samples located in the basin coincides approximately with the isohyets ranging from 800 to 1,000 mm of average yearly rainfall. It is possible to find them in zones with level or almost level slopes (0–5°), or on 10–15° of slope.

The Yellow-flowered Hardwood Tree is of great height, medium foliage, flexuous branches and fresh aspect. It constitutes an excellent substrate for epiphytic plants (ferns and also cactaceous species) that cover it from the base of the trunk to the highest branches. Frequently it is accompanied by sub-dominant species such as Jantarqui (*Escallonia millegrana*), Jarca (*Acacia visco*), Carob Tree (*Prosopis laevigata* var. *andicola*) and Wood Sorrel Tree (*Acacia caven*).

Pepper tree forests

The Pepper Tree (*Schinus molle*) forests are present as very open formations with a variable cover ranging between 30 to 60% in zones with enough edaphic humidity, like the alluvial plains, riverbeds and river and stream terraces. Trees grow in isolation or form groups; their height varies between 5 and 10 m, but under favourable conditions, it is possible to find trees up to 15 m high and 80 cm in diameter. The forest belongs to the woody dry formations of low mountains. It is mainly distributed in the old riverbeds or river courses at 1700–2400 m a.s.a. It is possible to find isolated Pepper Trees on some mountain slopes and hills of southern aspect. In many zones, especially on the Camacho riverbank, man favours their distribution, and uses them to protect crop areas from periodic floods due to river overflow. Their distribution coincides with the isohyets of 300–500 mm average yearly rainfall. According to Pretell Chiclote *et al.* [19], the Pepper Tree is likely to be found in inter-Andean valleys where the amount of precipitation reaches up to 200 mm a year, producing a deep, abundant root system, three or more times the height of the tree.

In many cases the Pepper Tree forests appear accompanied by other trees: Willow (*Salix humboldtiana*), Carob (several species, e.g. *Prosopis laevigata* var. *andicola*, *P. alba*, *P. nigra*), Jarca (*Acacia visco*), Wood Sorrel (*Acacia caven*), Jantarqui (*Escallonia millegrana*), Chilca (*Baccharis* cf. *dracunculifolia*) and Elder (*Fagara coco*).

Carob tree forests

Dominant species: *Prosopis laevigata* var. *andicola*, *P. alba*, *P. nigra*.

The Taco or Carob Tree formations are characteristic of the Chaco Province. Composed of microfoliate, xerophilous woody species they are present in low-lying places, streams and terraces where there is more humidity. Like the Pepper Tree forests, they are found in isolation or in small groups with cover varying from 20 to 60%. The tree reaches 4–8 meters in height, and 60 cm diameter. The Carob Tree forests include several *Prosopis* species, which are mixed and receive the same vernacular denomination. The collected samples were used in the works by Burkart and Simpson [4].

They are distributed in streams, riverbeds and alluvial terraces. Their altitudinal variation ranges from 1700 to 2100 m a.s.l. According to Cardenas [6] "… in a recent past, the Taco or Carob Tree fruits would be eaten raw. They would be cooked to prepare chichi". Its fruits are much consumed, especially by animals, during droughts. In view of this, man has favoured their distribution to some extent [18]. Their presence generally coincides with population centres or housing construction. In the Camacho basin there is no place where the Carob Tree is considered to form a true natural forest.

According to FAO/PNUMA [9] and Garcia Bes *et al.* [10], the Carob Tree or Taco is an accompanying species of the Soto climax forests (*Schinopsis haenkeana*) and of the White Quebracho Bark or Cacha Cacha (*Aspidosperma quebracho-blanco*) and, according to Pedrotti *et al.* [18], the Carob Tree constitutes a substitution xeric formation of the mountainous chaqueño forests. At present the Quebracho Bark has disappeared from the Tarija central valley, and only a small forest remains in the proximities of Santa Ana, with trees up to 20 m high and abundant regeneration. In the Camacho riverbed, Carob trees are found over the river and lake sediments. According to Hueck [14], Tacos develop very well in zones with soils that have some degree of salinity.

Chañar groves

The Chañar Grove Oak (*Geoffroea decorticans*) corresponds to a formation from the Chaco Province where open almost pure shrub-like "Charañales" grow. The dwarf oaks reach 2–4 m in height, and 30 cm in diameter. They develop in relatively dry zones (average annual rains, 500 mm) in areas that are not larger than 0.5 ha. This formation represents a substitution stage of the original forests that once populated the zone. They are found in the low zones of the basin, on alluvial terraces, riverbeds and undulating hills between 1600 and 1850 m. According to Hueck [14], *Geoffroea decorticans* is a characteristic tree from the Chaco, on saline soils.

The Chañar groves are distributed in places close to rivers, over river-lake sediments in plains or zones almost in the plains. The groups they form are favoured by their vegetative propagation capacity with bud production on the horizontal roots; they constitute almost pure Chañar groups [24]. According to Ruiz and Flores [22] the Chañar presents densities of up to 206 trees per hectare in certain zones of the Camacho river basin.

In other cases, the Chañar grows in the Acacia (*Acacia aroma*), Wood Sorrel Tree (*Acacia caven*), Pepper Tree (*Schinus molle*) and Carob Tree (*Prosopis laevigata* var. *andicola*, *P. alba* and *P. nigra*) formations. Chañar trees may reach heights of up to 10 m.

Acacia groves

The Acacia (*Acacia aroma*) groves correspond to a formation of the phytogeographical chaqueño province. The Acacia is one of the secondary species in the White Quebracho Bark forests (*Aspidosperma quebracho-blanco*) and Red Bark (*Schinopsis lorentzii*).

The Acacia groves appear as xeric groups of dwarf trees that reach a height of 3–4 m and 15 cm in diameter. Their crowns touch, making a closed formation. The Acacia has the appearance of a very tall thorny bush. The Acacia Dwarf Groves are considered a substitute formation. Isolated Acacia trees are of broad distribution along the basin, accompanied by Wood Sorrel Tree, Carob Tree, Chañar and, in some cases by Pepper Tree, but their cover is very low. With regard to height, they are distributed from 1700 m up to 1850 m. The Acacia Groves are located in zones of variable inclination: from level or almost level zones (0–5°), to undulating areas (10–15° of slope); there is no preferential orientation in their distribution. Average annual rainfall ranges from 600 to 700 mm, and coincides with the Acacia Grove distribution.

Physiognomically, they look like large bushes with a variable coverage that ranges between 80 and 100%. Among the accompanying species are Chañar (*Geoffroea decorticans*) and Wood Sorrel tree (*Acacia caven*). There is also a shrub-like layer under a variable cover ranging from 15 to 40% with the following species: *Abutilon virgatum*, Pinco Pinco (*Ephedra triandra*), Taquillo (*Prosopis alpataco* and *P. flexuosa*). The herb layer covers some 70–90%. The most frequent species are *Boerhaavia coccinea*, *Heterosperma tenuisecta*, Merry Widows (*Zinnia peruviana*) and *Evolvulus sericeus*.

Wood sorrel tree dwarf groves

As the Wood Sorrel Tree (*Acacia caven*) is one of the typical elements accompanying chaqueño formations, its presence and current distribution in the Camacho river basin is related solely to anthropogenic influence. The Wood Sorrel Tree is a species of great plasticity; this has permitted its colonization of diverse environments, arid as well as humid. These attributes are matched by its high reproductive capacity and rapid growth. The Wood Sorrel Tree is a thorny dwarf tree, a heliophyte with hemispherical crowns 2-6 m high and 20 cm in diameter. It frequently presents a shrub-like appearance 1.5–2 m in height. These shrubs group together and provide real dwarf groves. The Wood Sorrel groves in the Camacho river basin constitute the unit that presents the largest geographic distribution, colonizing a diversity of ecosystems due to their great adaptive capacity.

In the zones corresponding to the river-lake sediments, mostly with clayey and sandy-clayey soils, its distribution is wide; in the dry zones it has been favoured by goats, which eat the legumes produced by the plant, so disseminating its seeds. In many cases it is accompanied by Taquillo (*Prosopis alpataco*), Atamisqui (*Atamisquea emarginata*), Sisico (*Lycium cestroides*). It grows from 1600 to 2300 m a.s.l.

The Wood Sorrel bushes correspond to a successful formation in the substitution series of the original forests of Yellow-flowered Hardwood Tree, White Quebracho Bark, Soto and Jacaranda, that existed before the arrival of man in the Tarija Central Valley. Vegetation accompanying the Wood Sorrel is herbaceous and grassy, which in one way or another have been favoured by agricultural activities. Such is the case of *Paspalum* sp., a gramineous species that always appears with significant cover.

After field observation, it is concluded that the Wood Sorrel tree has not yet formed its own community, as its accompanying flora is varied and grows in numerous environments (riverbeds, fallow areas, steep slopes, undulated hills, plains, etc.). Thus it corresponds to a secondary origin physiognomy; therefore, the presence of Wood Sorrel only characterizes a vegetation landscape. The only factors limiting its development are low temperatures and high altitudes. According to Ruiz and Flores [22], Wood Sorrel tree densities in river-lake sediments in the Camacho river basin reach 1100 individuals per hectare. It shows its widest distribution in the xeric bioclimatic zone of the basin (300–600 mm average annual rainfall).

Taquillo bushes

Dominant species: *Prosopis alpataco*, *P. flexuosa*

Taquillo bushes (*Prosopis alpataco*) correspond to a formation of the Inter-Andean Chaqueño Province. They are formed by individuals grouped in sub-circular spots, where this species presents an important vegetation coverage accompanied by a series of shrub-like and herbaceous species. The Taquillo has a dim aspect but may reach a 150 cm height and a 10-m diameter as a group of bushes. The Taquillo bush is the only formation that almost is exclusively distributed over the quaternary sediments of river-lake origin; this zone currently presents the most serious effects of hydric (meteoric) erosion throughout the basin. In altitude it grows from 1700 to 1900 m a.s.l. The Taquillo bush is distributed throughout the most xeric zone of the Camacho river basin, where average annual rainfall varies between 300 and 4000 mm. According to Roig [21] *Prosopis alpataco* prefers saline and alkaline soils and behaves as a typically freatophytic bush. Soils are clayey, or a combination of sand and clay, shallow, with level and almost level variable inclinations (0–5°), and slightly inclined (5–15°).

In the Ventolera zone, where strong wind gusts are frequent, the Taquillo shrubs accumulate wind-transported soil particles, forming elevated mounds 1–1.5 m high.

The shrub-like layer has a variable coverage of 70–80%; species accompanying *Prosopis alpataco* are Pinco Pinco (*Ephedra triandra*), *Cleistocactus tarijensis*, Higuerilla (*Carica quercifolia*) in the Higueras zone, *Platyopuntia sulfurea*, Atamisqui (*Atamisquea emarginata*), *Capsicum baccatum*, Sisico (*Lycium cestroides*) and Bolivian Leppia. The herbaceous layer has a variable coverage of 20–50%. The most significant and frequent species are *Setaria fiebrigii*, *Bidens pseudocosmos*, *Boerhaavia coccinea*, *Pappophorum mucronatulum*, *Heterosperma tenuisecta*, *Allionia incarnata* and *Abutilon virgatum*.

Guaranguay and/or tolilla and/or mountain cedar bushes

Dominant species: *Tecoma* cf. *tenuiflora*, *Eupatorium buniifolium*, *Aloysia gratissima*.

Guaranguay (*Tecoma tenuiflora*), Tolilla (*Eupatorium buniifolium*) and Mountain Cedar (*Aloysia gratissima*) bushes are extensive shrub-like formations 1–1.5 m high, with cover varying from 80 to 100%, mostly found on southern slopes. Cover, presence and dominance of each of the three species is variable. A determining factor for propagation has been the fact they are not palatable to cattle and they have a high regeneration capacity. These bushes are distributed in zones between 1900 and 2000 m altitude, mostly on mountainous slopes of southern aspect, although the dominating *Tecoma tenuiflora* bushes are found in diverse expositions. Their distribution on slopes of mostly southern aspect indicates that these species develop in relatively fresh and humid environments. Slope inclination where they develop varies between 10° and 40°. Their distribution coincides with 300–500 mm average annual rain isohyets.

Accompanying shrub-like species for each of the three shrub types are practically the same: *Verbesina* sp., Pinco Pinco (*Ephedra triandra*) and *Abutilon virgatum*. Herbaceous species have a variable cover that oscillates between 20% and 30% with the presence of *Heterosperma tenuisecta*, *Bouteloua curtipendula*, *Eragrostis* sp., *Setaria fiebrigii* and *Chamaesyce* sp.

Discussion

From the analysis of ecological characteristics of natural woody plant formations that determine their spatial distribution in the Camacho river basin, it is possible to summarize the following considerations in regard to the current status of the native forests.

The primary forest vegetation is now strongly fragmented and covers only 4.3% of the basin (45 km^2) with 2.9% of the surface covered with *Tipuana tipu*, 12.4% with *Podocarpus parlatorei*, 16.9% with *Polylepis* sp., 20% with *Schinus molle*, and 45.8% with *Alnus acuminata* subsp *acuminata* (see the attached vegetation map). Deforestation has been resulting in a desertification process, originated by destruction and soil transportation.

The zones located in the north-eastern part of the basin show low rainfall due to their orographic conditions; therefore, the potential vegetation corresponds to dry forests, which in the near past were at the limit of their survival as rains were scarce. When man began his agricultural activities, the consequences turned doubly negative, accelerating their destruction.

The original forests that populated the basin in the past, were characteristically of mesomorphic structure and were eventually replaced by bush-like dwarf groves of xeromorphic structure from species of highly reproductive capacity.

The Wood Sorrel, Chañar, Acacia, and Carob Tree formations as well as the Taquillo and Guaranguay bushes, whose distribution are broadly extended over river-lake sediments, correspond to secondary formations substituting originary forests with a distribution not following a regular pattern throughout the study area. Colonization of the different ecosystems by the tree groups mentioned has been favoured by the intensive exploitation of forests since the arrival of man in the Tarija Central Valley, creating conditions for a vigorous xeric regeneration.

Natural formations of White Alder, Mountain Pine and Keñua are preserved as they are found in zones almost inaccessible to man and cattle, aside from demanding particular ecological characteristics such as aspect, topographical position, rainfall quantity and soil type that favour their distribution.

Isolated Yellow-flowered Hardwood Tree, White Alder, Keñua and White Quebracho or Cacha Cacha (the latter out of the basin) distributed in different zones, constitute remnants of the original vegetation. This allows us to determine the potential as well as the real forest suitability of the degraded zones in the basin.

Experiences in other Andean zones [3] have shown that it is possible to establish forests with native woody species in zones where introduced species have not obtained good results as they had not adapted to the particular ecological conditions of the Andes such as: frost, prolonged dry periods, diseases, etc.

To perform a zonation of the Camacho River basin with the objective of aforestation with native species, it will be necessary to differentiate two large territories or geographical areas that are ecologically homogeneous, according to bioclimatic characteristics: a humid zone with variable annual rains oscillating between 600 and 1400 mm, with the presence of mountainous forests and transition woods, and another xeric zone with variable annual rains ranging from 300 to 600 mm, with chaqueño mountainous elements.

REFERENCES

1. Bass–Werner M., 1975 - Porcentaje de germinación, capacidad y energía germinativa y sobrevivencia del Podocarpus parlatorei en cuatro suelos diferentes. Tesis no publicada, Ing. Forestal UJMS Tarija.
2. Bastian E., 1986 - Grundzüge der Vegetationsdegradation in Südost – Bolivien. Jahrb. Geograph. Gesell., Hannover, 1985: 23-67.
3. Brandbyge J., Holm-Nielsen L.B., 1987 - Reforestation of the high Andes with local species. Reports from the Botanical Institute (University of Aarhus), **13**: 1-118.
4. Burkart A., Simpson B.B., 1977 - Apendix: The genus *Prosopis*. An annotated key to the species of the world. In: Simpson B.B. (ed.), Mesquite. Its biology in two deserts ecosystems. US-IBP Syntesis Series **1**: 201-215.
5. Cabrera A.L., 1971 - Fitogeografía de la República Argentina. Bol. Soc. Arg. Bot., **14** (1-2): 1-42.
6. Cardenas M., 1955 - Plantas económicas de Bolivia. Ichtus, La Paz.
7. Coro M., 1982 - El Carob Tree. La vegetación del valle de Tarija. Rev. Ciencia Técnica, **3** (4): 29-107.

8. Corrado M., 1884 - El colegio franciscano de Tarija y sus Misiones. Noticias históricas. Quarachi cerca de Florenzia. Tipografía del Colegio San Buena Aventura.
9. FAO-PNUMA, 1985 - Un sistema de áreas silvestres protegidas para el Gran Chaco. Proyecto FAO/PNUMA 615-85-01. Documento Técnico, **1**: 1-159.
10. Garcia Bes J., Ratier de Colina A., Colina, S., Moreno C.A., 1985 - Fitogeografía de las provincias de Salta y Jujuy. Salta. Universidad Nacional de Salta.
11. Gerold G., 1979 - Untersuchungen zur Klima-, Vegetations-Hohenstufung und Bodensequenz in SE-Bolivien. Aachener Geogr. Arb., **19**: 1-70.
12. Gerold G., 1981 - "Desertification" in Südbolivien – Untersuchungen im Badlandbereich des andinen Beckens von Tarija. Würzburger Geogr. Arb., **53**: 73-109.
13. Gerold G., 1985 - Untersuchungen zur Badlandentwicklung in den wechselfeuchten Waldgebieten Südboliviens. Geoökodynamik, **6** (1-2): 35-70.
14. Hueck K., 1978 - Los bosques de Sudamérica. Eschborn, GTZ.
15. Liberman M., 1988 - Una reflexión sobre un caso dramático de conservación de la naturaleza. LIDEMA (La Paz), **2** (6): 5.
16. Martinez R., 1985 - Estudio de control y protección de la cuenca del río Camacho. Tesis no publicada Ing. Hidráulica. UMSA La Paz.
17. Mingo M., 1981 - Historia de las Misiones Franciscanas de Tarija entre los Chiriguanos. I. Tarija, Universidad Juan Misael Caracho.
18. Pedrotti F., Venanzoni R., Suarez Tapia E., 1988 - Comunidades vegetales del Valle de Capinota (Cochabamba – Bolivia). Ecología en Bolivia, **11**: 25-45.
19. Pretell Chiclote J., Ocaña Vidal D., Jon Jap R., Barahona Chura E., 1985 - Apuntes sobre algunas especies forestales nativas de la Sierra peruana. Proyecto FAO/Holanda/Infor. 120p.
20. Roig F., 1986 - Cartilla del Algarrobo. Comité ecológico IADIZA, Ministerio de Economia. Mendoza.
21. Roig F., 1987 - Árboles y arbustos en Prosopis flexuosa y Prosopis alpataco Leg. Parodiana, **5** (1): 5-22.
22. Ruiz M., Flores M., 1988 - Estudio dasométrico de cuatro especies nativas: *Acacia caven*, *Geoffroea decorticans*, *Prosopis alba* y *Acacia aroma*. Informe interno no publicado PERT-GTZ, Tarija.
23. Simpson B., 1979 - A revisión of the genus *Polylepis* (*Rosaceae: Sanguisorbeae*). Smithsonian Contributions to Botany, **43**: 1-59.
24. Talmon L., Mujica B., 1986 - Aspectos morfológicos y anatómicos de la formación de yemas de raíces de *Geoffroea decorticans* (*Leguminosae*). IV Congreso Latinoamericano de Botánica (Medellín): 108.

THE PROTECTION OF BIODIVERSITY IN TUSCANY

Paolo Emilio TOMEI[*], **Andrea BERTACCHI**
Department of Agronomy and Agroecosystem Management
Università degli Studi di Pisa, Via San Michele degli Scalzi 2, **56124 PISA (ITALY)**;
e-mail: petomei@agr.unipi.it

Abstract: The work focuses on the RE.NA.TO Project that proposes guidelines for the preservation of both faunal and floral elements as well as plant communities at high risk of extinction in the Tuscany region. The main goal of the project is to create a Catalogue in which all the information on current and future natural history data will be collected by means of descriptive files and distribution maps including geo-referred GIS data.

472 plants and 84 plant community types were identified as at the greatest danger. In addition, 88 habitats in which these plants and communities are likely to occur were located.

The data collected, thanks to this project, will allow a detailed survey of biodiversity in the area and better management of the regional territory.

Introduction

Tuscany is situated on the west coast of central Italy, between two different phytogeographical areas of which the biodiversity is very rich: the Mediterranean region and the Central European region [5]. The public institution has set up a project called RE.NA.TO. to improve biodiversity protection [1]. The principal aim of the project is to create a natural history catalogue, obtained by means of the collection, in-depth study and reorganization of the current knowledge of rarity and threat for fauna, flora and vegetation throughout the Tuscan territory, and to rework them.

Methods

In greater detail, the main specific aims, corresponding with the development phases, are the following:

- Identification of the elements of fauna, flora and vegetation which need to be especially protected.
- Compilation of standard files by collecting and reorganizing the current knowledge of these elements.
- Execution of surveys, especially concerning the elements and the areas not well-known, to expand existing knowledge.
- Compilation of an updated, user-friendly catalogue, including geo-referred data through GIS.
- Processing of a "Red List" of flora and fauna.
- Processing of synthetic, descriptive files and distributive maps for every important datum and community type, describing their status, the

knowledge of them, causes of threat and measures necessary for their preservation.
- Identification of the most important areas for biodiversity conservation, showing the guidelines to preserve and develop them.
- Identification of the lack of knowledge concerning particular lands and specific groups or elements of attention.
- Divulgence of the results.

The development of the above-mentioned phases has led to several identifiable modules that, as a whole, form the Natural History Catalogue of Tuscany:
- Red – Geo-referred database with all the indications concerning each element included on the Red List.
- Synthetic files and distributive maps of each element.
- Synthetic maps concerning the distribution and the amount of knowledge of the elements.
- Files and maps concerning the more relevant areas for biodiversity preservation.
- Report on the state of biodiversity of animal and plant kingdom in Tuscany.

Results

It is now possible to locate up-to-date data in the Catalogue, right up to the end of 2000. It constitutes the 'starting point' for the gathering of information on the natural history elements at a higher risk of extinction in the territory of Tuscany. Nevertheless, the catalogue is not only an updated information database, but also it is a sort of structured and organized container to enrich all the future results reached by means of analysis and monitoring the territory.

Attention lists of flora, habitat and plant communities have been compiled concerning the floristic-vegetation aspects.

With regard to flora, attention has been focused on endemic plants, exclusive to the region, and generally on species considered rare and at risk everywhere and with reference to Tuscany. In particular, the species represented in no more than three stations within the Tuscan territory (five stations for species living in high-risk areas, like wetland sites) have been considered. The preliminary Catalogue has been written principally according to the works by Caruel [3], Baroni [2], Fiori [4] and Pignatti [6].

The provisional list of the habitats, with an interest in preservation, has been compiled from the Enclosure I, directive 92/43/CEE, with additional enclosures suggested by the working group of the Italian Botanical Society, during the work of the Bioitaly – Natura 2000 Project.

The plant communities considered on the attention list are those which have at least one of the following characteristics:
- being rare in Tuscany Region;
- being a biological shelter station for rare species;

- coming from the habitat according to the directive 92/43/CEE;
- being ecologically important because they are very little affected by anthropogenic action;
- having bio-geographical value (i.e. heterotopic communities).

Some cases concerning sandy coasts, wetland and highland areas will be illustrated as explanatory examples.

In three sites on the sandy coasts of Tuscany, the habitat "Shifting dunes along the shoreline with *Ammophila arenaria* (white dunes)" has been identified. In this habitat, there are also plant communities of *Agropyrum junceum* (L.) Beauv., and it is possible to see *Solidago virgaurea* L. subsp. *litoralis* (Savi) Burnat, a species that appears on the attention list.

In the bogs in the northern part of the region, the habitat "Middle-course mires floating in planitial areas" is located at two sites, where *Sphagnum* sp.pl. and *Drosera rotundifolia* L. plant communities are found (*Sphagno-Droseretum rotundifoliae* Tomei, Guazzi *et* Barsanti 1997) [7]; several species which appear on the attention list occur in these sites, such as *Drosera rotundifolia* L., *Anagallis tenella* (L.) L., *Rhynchospora alba* (L.) Vahl.

In the Tuscan-Emilian Apennines and the Apuan Alps, generally at an altitude of more than 1500 m, the habitat "Chasmophytic vegetation of siliceous rocks" has been observed; there, no particular plant associations have been identified, but there are species belonging to the attention list: *Primula appennina* Widmer, *Soldanella alpina* L., *Silene acaulis* (L.) Jacq.

Conclusion

The whole research pinpointed 88 habitats, 84 plant communities and 472 species that need particular attention. The natural history and zoological rare entities have been mapped, which give a satisfactory evaluation of the condition of the environment. The information gathered from the RE.NA.TO. Project has helped to set up a correct policy for the biodiversity protection in Tuscany. The measures adopted will gradually improve as the information enhances the project itself.

REFERENCES

1. AA.VV., 2005 - Il Progetto RE.NA.TO. Regione Toscana, Firenze.
2. Baroni E., 1987-1908 - Supplemento generale al "Prodromo della Flora toscana di T. Caruel". Firenze.
3. Caruel T., 1860 - Prodromo della Flora toscana. Firenze.
4. Fiori A., 1923-1929 - Nuova Flora analitica d'Italia. Tipografia M. Ricci, Firenze
5. Pignatti S., 1979 - I piani di vegetazione in Italia. Giorn. Bot. Ital., **113**: 411-428.
6. Pignatti S., 1982 - Flora d'Italia. Edagricole, Bologna.
7. Tomei P.E., Guazzi E., Barsanti A., 1997 - La carta della vegetazione delle paludi e del Lago di Massaciuccoli. Lago di Massaciuccoli – 13 ricerche finalizzate al risanamento: 275-288. Ente Parco Regionale, S. Rossore, Masaciuccoli.

THE REASONS IN FAVOUR OF SETTING UP A NEW NATURAL RESERVE IN THE BLACK SEA SHORE AREA BETWEEN NORTH AND SOUTH EFORIE (CONSTANȚA COUNTY)

Marius FĂGĂRAȘ[*], **Rodica BERCU, Loreley JIANU**
"OVIDIUS" University of Constanța, Natural Science Faculty, Mamaia Avenue 24,
900670 Constanța (ROMANIA); e-mail: fagarasm@yahoo.com

Abstract: In the Romanian Black Sea shore area between North and South Eforie, c. 15 km south from Constanța city, lies a floristically interesting area. Here, on mobile or fixed sand dunes, large populations of many rare psammophilous species occur (*Cakile maritima* ssp. *euxina, Elymus farctus ssp. bessarabicus, Silene thymifolia, Eryngium maritimum, Glaucium flavum, Polygonum maritimum, Gypsophyla perfoliata*). Some have sporadically spread in the rest of littoral area. These species are included in different IUCN categories in the Romanian Red Lists. The high percentage of rare and threatened species (18.29%), higher than in other southern protected seashore areas (Agigea, 2 Mai - Vama Veche), demonstrates the floristic importance of this site. In the area studied are also some dune habitat types of European conservation interest, which are included in the Habitats Directive and Bern Convention. The beginning of some building works, even in the dune habitats, threatens with extinction the specific psammophilous flora and plant communities. On the base of this evidence we propose the setting up of a natural reserve in this area, preferably as part of the Natura 2000 protected areas network.

Introduction

The diversity of natural habitats in the seashore area, the specific climate and soil types allow the development of numerous plant species, many of them rare in the Romanian flora. The conservation of this floristic richness is achieved in acceptable conditions only within protected areas from the northern side of Romanian Black Sea shore (Chituc sandbank and Danube Delta). As regards the southern coast (Cape Midia –Vama Veche), it is important to mention that the last 'oases' of psammophilous vegetation are seriously affected by human activities. We observed such a situation between North and South Eforie (Fig. 1), about 15 km south from the city of Constanța. Here, on mobile or fixed sand dunes, large populations of many rare psammophilous species occur. The beginning of some building works, even in the dune habitats (Fig. 2), threatens the special flora and plant communities with extinction. In the area studied, some dune habitat types of European interest for conservation included in the Habitats Directive and Bern Convention are also endangered.

This paper aims to stress the floristic importance of this area and to present some scientific data in support of setting up of a new natural reserve, preferably as part of the Natura 2000 protected areas network. Talking into account the above observations, the dune habitats would be preserved – and at the same time certain numerous species that are included on the Romanian Red Lists.

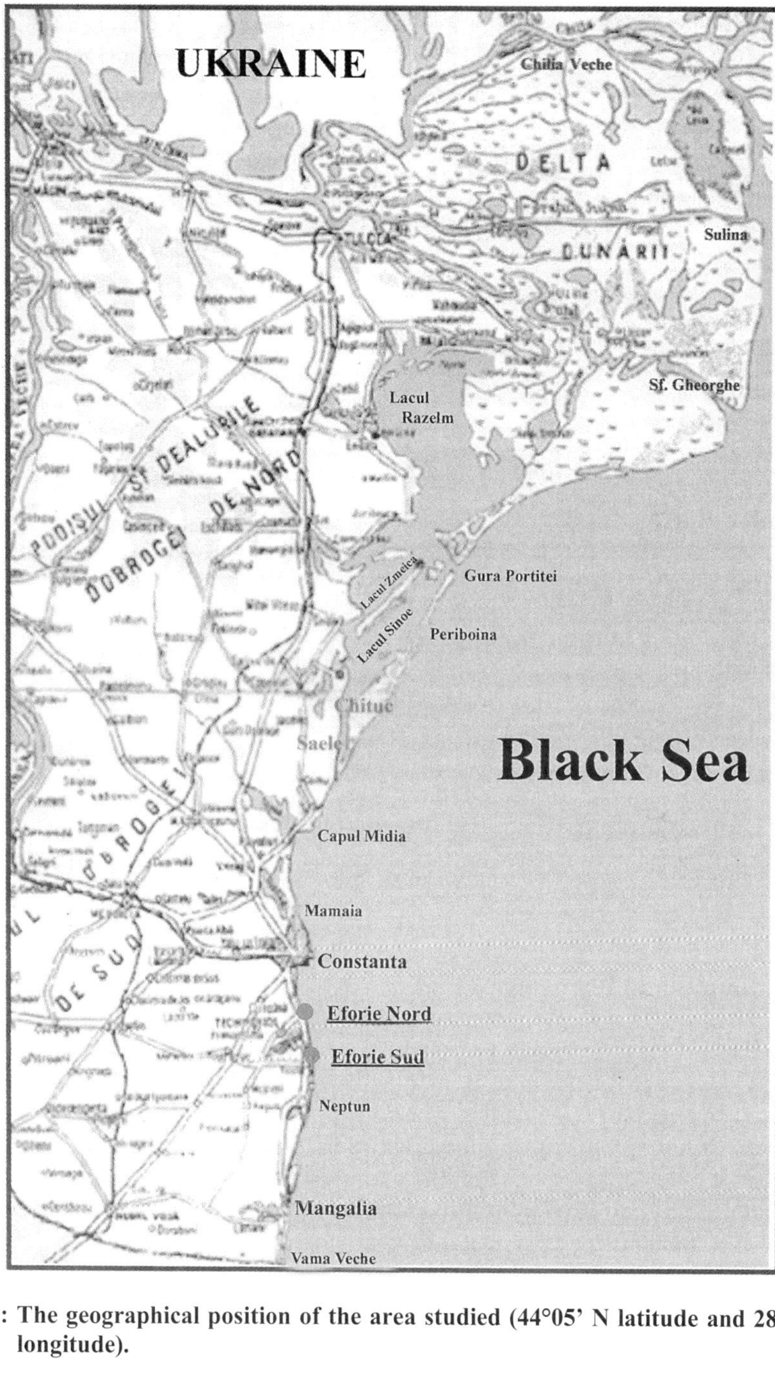

Fig. 1: The geographical position of the area studied (44°05' N latitude and 28°40' E longitude).

Fig. 2: The building works in the proximity of the sand dunes area between North and South Eforie.

Materials and Methods

The floristic inventory was carried out on the basis of our observations over the last five years in the seashore area between North and South Eforie. Amongst all the taxa identified, the rare and threatened plants have been selected (in accordance with Romanian Red Lists) and their percentages compared with those of other seacoast protected areas (Agigea, Vama Veche) [3, 8, 11, 12, 13, 14].

Results and Discussion

The floristic inventory contains 142 taxa (130 species and 12 subspecies). Among these, 7 taxa are Endangered (4.92%), 14 are Vulnerable (9.85%) and 5 are rare (3.52%), in accordance with Romanian Red Lists [2, 5, 9, 10]. Compared with the general situation of Romanian seashores, where the percentage of rare and threatened species is 20.22% [6], the proportion of these plants categories is lower (18.29%) in the study area. Pointing to the percentage of Endangered and Vulnerable plants, we have to mention that these are higher (Tab. 1), and it is just these IUCN plant categories that are the most threatened by the disturbance influence of human activities.

In the area studied, the rate of Vulnerable and Endangered species is higher than in some protected areas such as the Agigea sand dunes reserve and the 2 Mai-Vama Veche seashore area (Fig. 3). This evidence emphasizes the floristic importance of this area and is a significant reason for setting up a natural reserve between North and South Eforie.

Table 1: The rate of rare and threatened taxa from some Romanian Black Sea coast areas (E – endangered; V – vulnerable; R – rare).

Compared southern Romanian sea coast areas	The number of taxa	Rare and threatened taxa (%)	IUCN plant categories (E) (%)	(V) (%)	(R) (%)
North Eforie-South Eforie	142	18.29	4.92	9.85	3.52
Agigea sand dunes reserve (Mititelu *et al.*, 1992)	458	11.79	3.05	4.36	3.93
Vama Veche seashore area (Pop I., 1969, 1970, 1985)	137	10.21	4.37	4.37	1.45
All Romanian sea coast area (Făgăraş, 2004)	702	20.22	4.41	6.83	8.54

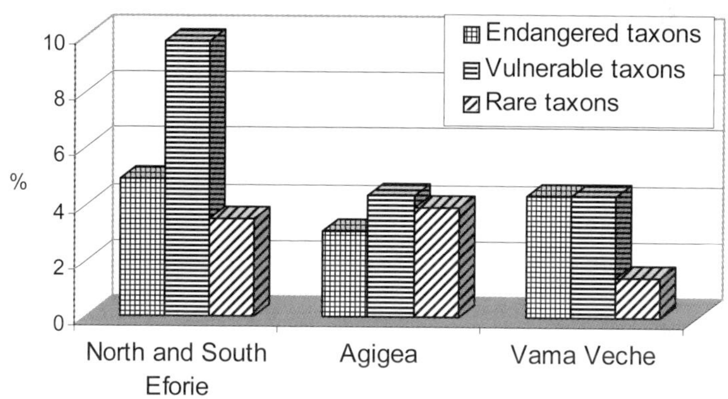

Fig. 3: The proportion of the main IUCN plant categories in the study area, Agigea sand dunes reservation and Vama Veche sea coast zone.

The plant taxa identified in the study area are as follows (with Rare, Vulnerable and Endangered taxa indicated by bold): *Achillea setacea* Waldst. et Kit., *Althaea officinalis* L., ***Alyssum hirsutum*** Bieb. **[R]**, *Alyssum desertorum* Stapf, *Amaranthus retroflexus L.*, ***Anchusa thessala*** Boiss. et Spruner **[E]**, *Apera spica-venti* (L.) Beauv. ssp. *maritima* (Klokov) Tzvelev, *Artemisia austriaca* Jacq., *Artemisia santonica* L. ssp. *monogyna* (Waldst. et Kit.) Leonova, ***Artemisia tschernieviana*** Besser **[V]**, *Arctium lappa* L., ***Argusia sibirica*** (L.) Dandy **[V]**, *Asperugo procumbens* L., ***Astrodaucus littoralis*** (Bieb.) Drude **[E]**, *Atriplex tatarica* L., *Atriplex sagittata* Borkh., *Atriplex oblongifolia* Waldst. et Kit., *Ballota nigra* L., ***Bassia hirsuta*** (L.) Ascherson **[V]**, *Bassia prostrata* (L.) G. Beck, *Bassia scoparia* (L.) Voss, *Berteroa incana* (L.) DC., *Bromus hordeaceus* L., *Bromus* *sterilis* L., *Bromus squarrosus* L., *Bromus tectorum* L., ***Cakile maritima*** Scop. **ssp. *euxina*** (Pobed.) E.I.Nyárády **[V]**, *Calepina iregularis* (Asso.) Thell., *Capsella bursa-pastoris* (L.) Medik., *Carduus acanthoides* L., *Carex colchica* Gay ssp. *colchica, Centaurea arenaria* Bieb. ssp. *borystenica* (Gruner) Dostal, *Centaurea diffusa* Lam., *Cerastium semidecandrum* L., ***Chamaesycae peplis*** **(**L.) Prokh. **[V]**, *Chenopodium album* L., *Chondrilla juncea* L., *Cicorium intybus* L., *Conyza canadensis* (L.) Cronq., *Convolvulus arvensis* L., ***Corispermum nitidum*** Kit. in Schultes **[V]**, *Coronilla varia* L., ***Crambe maritima*** L. **[V]**, *Crepis foetida* L. *ssp. rhoeadifolia* (Bieb.) Celak., *Crypsis aculeata* (L.) Aiton, *Cynanchum acutum* L., *Cynodon dactylon* (L.) Pers., *Daucus carota* ssp. *carota* L., ***Daucus guttatus*** Sibth. et Sm. **ssp. *zahariadi*** Heywood **[R]**, *Descurainia sophia* (L.) Webb ex Prantl, ***Dianthus bessarabicus*** (Kleopov) Klokov **[V]**, *Diplotaxis muralis* (L.) DC., *Diplotaxis tenuifolia* (L.) DC., *Elaeagnus angustifolia* L., *Elymus elongatus* (Host) Runemark, ***Elymus farctus*** (Viv.) Runemark ex Melderis **ssp. *bessarabicus*** (Săvul. et Rayss) Melderis **[E]**, *Elymus repens* (L.) Gould, *Erophyla verna* (L.) Chevall., ***Eryngium maritimum*** L. **[E]**, *Eryssimum diffusum* Ehrh., *Erodium cicutarium* (L.) L'Hérit., *Erodium ciconium* (L.) L'Hérit., ***Chamaesycae peplis*** L. **[V]**, *Euphorbia helioscopia* L., *Euphorbia sequieriana* Necker, *Fumaria vaillanti* Loisel., *Gallium humifusum* Bieb., ***Glaucium flavum*** Crantz **[V]**, ***Gypsophyla perfoliata*** L. **[R]**, *Heliotropium europaeum* L., *Holosteum umbellatum* L., *Hordeum geniculatum* All., *Hordeum murinum* L., *Lactuca tatarica* (L.) C.A. Meyer, *Lamium amplexicaule* L., *Lamium purpureum* L., *Lappula squarrosa* (Retz.) Dumort., *Lathyrus tuberosus* L., *Lepidium perfoliatum* L., *Linaria genistifolia* (L.) Miller, *Lotus corniculatus* L., ***Leymus racemosus*** (Lam.) Tzvelev. **ssp. *sabulosus*** (Bieb.) Tzvelev. **[V]**, *Lycopsis arvensis* L. ssp. *orientalis* (L.) Kuntze, *Malva sylvestris* L., *Marrubium peregrinum* L., *Matricaria recutita* L., *Melilotus albus* Medik., *Medicago falcata* L., *Medicago lupulina* L., *Medicago minima* L., *Medicago sativa* L., *Nonea pulla* DC. in Lam. et DC. ssp. *atra* (Griseb) Ciocârlan, *Onopordum tauricum* Willd., *Ornithogalum refractum* Kit., *Papaver rhoeas* L., *Picris hieracioides* L., *Plantago lanceolata* L., *Plantago media* L., *Plantago major* L., *Plantago maritima* L., *Plantago scabra* Moench, *Poa annua* L., *Poa pratensis* L., *Polygonum arenarium* Waldst. et Kit., *Polygonum aviculare* L., ***Polygonum maritimum*** L. **[E]**, ***Polygonum mesembricum*** Chrtek **[V]**, *Portulaca oleracea* L., *Reseda lutea* L., *Rubus caesius* L., *Rumex crispus* L., *Salsola kali* L. ssp. *ruthenica* (Iljin) Soó, ***Scolymus hispanicus*** L. **[R]**, *Scorzonera hispanica* L., *Sclerochloa dura* (L.) Beauv., ***Secale sylvestre*** Host. **[V]**, *Senecio jacobaea* L., *Senecio vernalis* Waldst. et Kit., ***Silene borystenica*** (Gruner) Walters **[E]**, *Silene conica* L., ***Silene exaltata*** Friv. **[R]**, ***Silene thymifolia*** Sibth. et Sm. **[E]**, *Sisymbrium orientale* L., *Solanum nigrum* L., *Sonchus arvensis* L., *Stachys atherocalyx* C. Koch, *Stellaria media* (L.) Vill., ***Syrenia montana*** (Pallas) Klokov **[V]**, *Tamarix ramosissima* Ledeb., *Taraxacum officinale* Weber ex Wiggers, *Thlaspi perfoliatum* L., *Torilis arvensis* (Hudson) Link, ***Tragopogon floccosus*** Waldst. et Kit. **[V]**, *Tragopogon pratensis* L. ssp. *orientalis* (L.) Celak, *Tragus racemosus* (L.) All., *Tribulus terestris* L., *Trifolium fragiferum* L., *Trifolium repens* L., *Valerianella locusta* (L.)

Laterrade, *Veronica polita* Fries, *Xanthium italicum* Moretti, *Xanthium spinosum* L., *Xeranthemum annuum* L.

The setting up of a new natural reserve between North and South Eforie is also supported by the presence of large local populations of some rare and threatened psammophilous taxa (*Cakile maritima* ssp. *euxina*, *Elymus farctus* ssp. *bessarabicus*, *Silene thymifolia*, *Eryngium maritimum*, *Glaucium flavum*, *Gypsophyla perfoliata*, *Polygonum maritimum*). Among these plants, *Cakile maritima* ssp. *euxina* and *Elymus farctus* ssp. *bessarabicus* are critical taxa, included in the List of subendemics and endangered species from Romania (Aiv) [17]. These plant species have only a sporadic distribution in the rest of the southern Romanian Black Sea coast and they are threatened with extinction by anthropogenic factors.

In the study area there are also some dune habitat types of European interest for conservation, which are included in the Habitats Directive and Bern Convention [17]. Such habitat types are the embryonic mobile dunes (code 2110) and the seacoast fixed dunes with herbaceous vegetation (priority habitat-code 2130). Among the new habitat types included in Habitats Directive we mention from the researched area the Euxinic coastal salty sand communities with *Cakile maritima* ssp. *euxina*, *Crambe maritima*, *Argusia sibirica*, *Lactuca tatarica*, *Glaucium flavum*, *Euphorbia peplis* and *Scolymus hispanicus* (code B1, code EUNIS level 2). All these habitat types are endangered by building and other commercial activities within dune habitats.

Human activities adjacent to dune habitats have favoured the development of some weed and ruderal plant communities. As conservation measures are absent, the number of anthropophytic plants will increase, modifying the structure and floristic composition of native psammophilous plant communities.

In the investigated area, the following plant associations have been identified: *Atripliceto hastatae-Cakiletum euxinae* Sanda, Popescu 1999 (Syn. *Cakiletum frisicum* Tx. 50), *Lactuco tataricae-Glaucietum flavae* Dihoru et Negrean 1976, *Salsolo ruthaenicae-Xanthietum strumarii* Oberd. et Tx. 1950, *Atriplicetum hastatae* Poli et J. Tx.1960, *Agropyretum juncei* (Br. Bl. et De L. 1936) Tx. 1952, *Elymetum gigantei* Morariu 1957, *Secali sylvestris-Brometum tectorum* Hargitai 1940, *Aperetum maritimae* Popescu *et al.* 1978, *Bromo-Cynodontetum* I. Pop 1970, *Tribulo-Tragetum racemosi* Soó et Timar 1954, *Descurainetum sophiae* Krech 1933 corr. Oberd. 1970, *Hordeetum murini* Libbert 1932 em. Pass. 1964, *Balloto-Malvetum sylvestris* Gutte 1966, *Lepidietum drabae* Timar 1950.

Conclusions

The main reasons in favour of setting up a new natural reserve between North and South Eforie are as follows:

The percentage of Rare, Endangered and Vulnerable species (18.29%) is higher than that observed within the Agigea dunes reservation (11.79%) or the seashore area between 2 Mai-Vama Veche villages (10.21%), two protected areas not very far from Eforie.

In the area studied, large populations of many rare and threatened psammophilous species occur, some of sporadic distribution in the rest of the southern Romanian seashore area.

The beginning of some building works, even in the dune biotopes, threatens with extinction the specific flora and plant communities between North and South Eforie.

The presence in the study area of some sand dune habitat types of European interest for conservation (concordant with Habitats Directive and Bern Convention) sustain our argument concerning the importance of this area for the conservation of psammophilous flora.

In our opinion, the conservation of the rare psammophilous flora and typical dune biotopes is not possible without setting up this area as a natural reserve, preferably as part of the Natura 2000 protected areas network.

REFERENCES

1. Beldie A., 1977-1979 - Flora României. Determinator ilustrat al plantelor vasculare. Ed. Acad. RSR, I, II, Bucureşti.
2. Boşcaiu N., Coldea G., Horeanu C., 1994 - Lista roşie a plantelor vasculare dispărute, periclitate, vulnerabile şi rare din flora Romaniei. Ocrot. Nat. Med. Înconj., **38** (1): 45-56.
3. Boşcaiu N., 1975 - Aspecte de vegetaţie din rezervaţia dunelor maritime de la Agigea. Studii şi Comunicări, Muzeul Bruckenthal, Ştiinţele Naturii, Sibiu, **19**: 181-184.
4. Ciocârlan V., 2000 - Flora ilustrată a României (*Pteridophyta et Spermatophyta*). Ed. Ceres, Bucureşti.
5. Dihoru G., Dihoru A., 1994 - Plante rare, periclitate şi endemice în flora României – lista roşie. Acta Botanica Horti Bucurestiensis, (1993-1994): 173-197.
6. Făgăraş M., Gomoiu M.T., 2004 - The conspectus of vascular flora from Romanian Black Sea shore. "Ovidius" University Annals of Natural Sciences, Biology-Ecology Series, Constantza, **6**: 1-50.
7. Făgăraş M., 2003 - The conspectus of plant associations from Romanian Black Sea shore. "Ovidius" University Annals of Natural Sciences, Biology-Ecology Series, Constantza, **7**: 133-138.
8. Mititelu D., Costică M., Parincu M., 1992 - La flore et la végétation de la réserve botanique de dunes littorales d'Agigea – Constanţa. Analele Ştiinţifice ale Univ. "Al. I.Cuza" Iaşi, Biologie vegetală, **38** (2): 35-40.
9. Negrean G., 2001 - Lista Roşie a plantelor din România existente în pajişti, inclusiv endemite şi subendemite (*Tracheophyta*). In: Sârbu A., Coldea G., Sârbu I., Negrean G., (eds.), Ghid pentru identificarea şi inventarierea pajiştilor seminaturale din România, pp. 30-57. Ed. Alo Bucureşti!, Bucureşti.
10. Oltean M., Negrean G., Popescu A., Roman N., Dihoru G., Sanda V., Mihăilescu S., 1994 - Lista roşie a plantelor superioare din România, Bucureşti, Studii, Sinteze, Documente de Ecologie, **1**: 1-52.

11. Pascal P., 1962 - Rezervaţia naturală de dune de la Agigea. Ocrotirea naturii (Bucureşti), **6**: 122-126.
12. Pop I., 1969 - Contribuţii la cunoaşterea vegetaţiei litoralului românesc al Mării Negre din împrejurimile localităţii Vama Veche (Dobrogea). Studia Univ."Babeş-Bolyai" Cluj, Seria Biologie, **14** (1): 9-19.
13. Pop I., 1970 - Adnotaţii şi studii comparative asupra vegetaţiei Mării Negre din împrejurimile localităţii Vama Veche (Dobrogea). Contr. Bot.: 99-110.
14. Pop I., 1985 - Contribuţii la cunoaşterea vegetaţiei arenicole şi ruderale de la Vama Veche – Dobrogea (judeţul Constanţa). Contr. Bot., Cluj-Napoca: 131-139.
15. Sanda V., Popescu A., Barabaş N., 1998 - Cenotaxonomia şi caracterizarea grupărilor vegetale din România., Studii şi Comunicări, Biologie vegetală, Complexul Muzeal de Ştiinţe Naturale Bacău, **14**: 5-366.
16. Sanda V., 2002 - Vademecum ceno-structural privind covorul vegetal din România, Ed. Vergiliu, Bucureşti.
17. Sârbu A., Coldea G., Cristea V., Negrean G., Cristurean I., Sârbu I., Oprea A., Popescu G., 2003 - Ghid pentru identificarea importantelor arii de protecţie şi conservarea plantelor din România. Ed. Alo Bucureşti!, Bucureşti.
18. Săvulescu T. (ed.), 1952-1976 - Flora R.P.R – R.S.R., Vol. 1-13. Ed. Academiei Române, Bucureşti.
19. Tutin T.G., Heywood V.H., Burges N.A., Moore D.M., Valentine D.H., Walters S.M., Webb D.A. (eds.), 1964-1980 - Flora Europaea, Vols. 1-5, Cambridge, Cambridge University Press.
20. Tutin T.G., Burges N.A., Chater A.O., Edmonson J.R., Heywood V.H., Moore D.M., Valentine D.H., Walters S.M., Webb D.A. (eds., assist. by Akeroyd J.R. & Newton M.E.; appendices by Mill R.R.), 1993 - Flora Europaea, 2nd ed., Vol. 1, Cambridge, Cambridge University Press.
21. ***, 1997 - Globally threatened plants in Europe. A subset from the 1997 IUCN Red Lists of Threatened Plants, World Conservation Monitoring Centre, Draft Version – July 1997: 1-68.

ASSESSMENT OF THE VASCULAR FLORA CONSERVATION THROUGH SPECIFIC INDICES – A COMPARISON STUDY IN CENTRAL EUROPE

Arkadiusz NOWAK
Department of Biosystematics, Division of Phytosociology, University of Opole, Oleska str. 22, **45-052 Opole (POLAND)**; e-mail: anowak@uni.opole.pl

Abstract: In almost all countries with a well-founded law system, species-targeted conservation has legal support and plays a fundamental role in the protection of the most threatened plants. However, even in countries with a long tradition of plant conservation, there is little knowledge about its effectiveness and there is little or no possibility to compare the results of this kind of protection in different countries. Conservation methods using numerical indicators seem especially not to have been developed enough. To make the strategies of plant protection more effective and efficient, the author attempts to propose some measures for evaluating the status of flora conservation in a given area. The proposed indicators could also be useful in analyses of the results of protection, seen as the synanthropo-dynamic state of the vegetation (*e.g.* the amount of Red-listed or threatened species). For example, sozophytes as a group of sensitive plants could play the role of indicators of environmental quality. The proposed indicators are as follows: conservation indicator (C), threat conservation indicator (C_K), tanatophytisation indicator (R), sozophytisation indicator (W) and conservation effectiveness indicator (E).

Introduction

Species-targeted conservation is widely implemented all over the world to protect the most threatened plants [e.g. 3, 4, 6, 15]. However, it is remarkable that, even in countries with a long tradition of plant conservation, there is little knowledge of the effectiveness and results of this kind of protection [16]. Conservation methods using numerical indices seem especially not to have been developed enough. Only a few efforts worth noticing have been done recently concerning biodiversity risk assessment [21, 22], prioritisation in conservation [23] or environmental vulnerability and sustainable use of natural resourses [11, 24]. Nevertheless, there is an apparent scarcity of comparative studies, which aim to assess the methods, results and potential effectiveness of flora conservation in different regions. To make the strategies of plant protection more effective and efficient, it is indispensable to implement adequate measures for evaluating the status of flora conservation in a given area and also the results of protection, seen as the synanthropo-dynamic state of the vegetation (e.g. the amount of Red- listed or threatened species). Especially helpful would be the development of methods of comparison of the means of conservation in different countries, which are to be unified, as in the case of new member states of European Union. Those countries would be embraced by the EU plant conservation strategies and included in common planning and legal processes regarding financial, organisational and methodological means. It would also be useful to compare conservation practices from other parts of the world with the aim of implementing the most effective methods of protection. Conservation interest is still growing in ways in which we can monitor and evaluate the effectiveness and progress in species conservation and improve ongoing management. Introducing plant conservation indices seems also to be useful in environmental monitoring. Sozophytes as a group of sensitive plants could be used as indicators of environmental quality. The Trent Biotic Index [25] could serve as a good example of such biotic monitoring. Easy to calculate, this cost-effective index is based on the presence or absence of a certain group of indicator organisms, in this case aquatic species.

An example of the typical history of plant conservation in Central Europe can be shown by Poland.

The tradition of Polish plant conservation goes back to the early 1920s, when in 1919, the Minister of Religions and Public Education issued a Natural Monuments Conservation Ordinance. Numerous rare plant localities and seven plant species were taken under legal protection. In the following years (1938, 1957, 1983, 1995) the list of protected plant species was changed, becoming longer [17]. The last ordinance about plant species conservation was released on 11th September 2001, enumerating 268 taxa of vascular plants considered protected by state law [18]. The main goal of species protection is to preserve and safeguard every plant species, which are valuable for many reasons and generally endangered by human activities. As well as the formal and lawful context of flora preservation, for some decades the specific, scientific methodology of threat assessment has been developing, resulting in the compilation of several Red Data Books and Red Lists. In these, the threat factors for plant populations were analysed, and the level of endangerment and extinction risk were estimated. In Poland, the first Red List of vascular plants was compiled by Zarzycki [27] and improved and supplemented afterwards by Zarzycki and Szeląg [29]. In 1993 and 2001 Red Data Books were issued [7, 28].

At the end of every stage of the development of given knowledge, reasonable assessment with appropriate methods has to come. This paper attempts to find useful measurement techniques of plant conservation in the region considered and tries to propose a simple method of comparisons of plant conservation means, methods and needs.

Methods

Study Area

We have compared indigenous flora conservation in four Central European countries: Czech Republic, Germany, Poland and Slovakia. These states are located in a mesothermal climate of the temperate Zone. The average daily temperatures are -2,5°C in January and 18°C in July. The average annual precipitation is about 600mm [1]. The prevailing natural vegetation is composed of broad-leaved forests (*Querco-Fagetea*). In uplands and mountainous areas (above 500 m a.s.l.) beech forests (*Fagion sylvaticae*) and coniferous forests

(*Vaccinio-Abietenion*) are dominant. Humans have inhabited the region for thousands of years and the majority of the landscape is intensively exploited for agriculture. Nowadays more than 136 million people live in the area, about 170 inhabitants per square kilometre, with the highest population density in Germany and lowest in Slovakia.

The domestic flora of the area (indigenophytes and archeophytes) comprises about 4000 species of vascular plants, of which some 40% are considered to be of importance to conservation [5, 9, 12, 29]. The region is characterised by agricultural lands, which cover over 60 % of the area, forests (about 25%), communication areas (about 5%), open waters (3%) and urbanised, housing and industrial areas (about 5%).

Indicators

The vascular flora of the investigated area comprises several groups of plant species divided according to their history of habitation, conservation importance, rarity and threat category. The study focuses on so-called "domestic plants" consisting of plants indigenous to the area (spontaneophytes *sensu* Mirek [13]) and plants introduced before 1500 A.D. (archeophytes *sensu* Kornaś [8]). Native plants are regarded as an important group for natural biodiversity conservation. Species introduced into Central Europe after 1500 (kenophytes) are not considered as requiring conservation.

For the selected group of species within the domestic flora, a special nomenclature was proposed. This should ease communication and eliminate misunderstandings (Tab.1).

Table 1: Explanation of terms with Greek or Latin roots used in the text.

Term	Meaning of Greek/Latin root	Sense
Domesticophytes (D)	*domesticus-* = domestic	Domestic flora (Arch+I)
Archeophytes (Arch)	*archaios-* = old	Plants introduced in the area before 1500 A.D.
Sozophytes (S)	*sodzein-* = rescue,	Plants listed on a "Red list" with EX, CR, EN, VU and DD category
Kindynophytes (K)	*kindyno-* = threat, to be threatened	Threatened plants (CR, EN and VU category)
Tanatophytes (T)	*thanato-* = death	Extinct plants (Ex category)
Nomosozophytes (NS)	*nomo-* = law, legal	Plant legally protected
Indigenophytes (I)	*indigena-* = born in a place, native	Native plants

Five indicators were proposed as a tool for the plant conservation comparisons between different countries.

Conservation indicator (C)

$$C = \frac{NS}{D} \times 100 \ (\%)$$

where:
NS – number of nomosozophytes;
D – number of domestic taxa (indigenophytes and archeophytes).

Threat conservation indicator (C_K)

$$C_K = \frac{NS_K}{K} \times 100 \ (\%)$$

where:
NS_K – number of legally protected kindynophytes;
K – number of kindynophytes (CR, EN, VU).

Tanatophytisation indicator (R)

$$R = \frac{T}{D} 100 \ (\%)$$

where:
T – number of extinct species (tanatophytes);
D- number of domestic taxa (indigenophytes and archeophytes).

Sozophytisation indicator (W)

$$W = \frac{S}{D} \times 100\% \ (\%)$$

where:
S – number of sozophytes;
D- number of domestic taxa (indigenophytes and archeophytes).
As sozophytes we consider every EX, CR, EN, VU, NT and DD species occurring naturally in the studied area, chosen from domestic flora.

Conservation effectiveness indicator (E)

$$E = \frac{NS}{S} \times 100\% \ (\%)$$

where:
S – number of sozophytes;
NS - number of nomosozophytes.

To calculate the indicators value, the relevant literature data were collected for each country (Tab. 2).

Table 2: Number of sozophytes (*S*), indigenophytes (*I*), archeophytes (*Arch*), kindynophytes (*K*), tanatophytes (*T*), nomosozophytes (*NS*), threatened nomosozophytes (NS_K) and richness of domestic flora (*D*) in Poland, Czech Republic, Slovak Republic and Germany.

Country	*D*	*S*	*I*	*Arch*	*K*	*T*	NS_K	*NS*
Czech Republic[1]	2036	1547	1766	270	1459	118	468	481
Germany[2]	3001	943	2805	196	804	47	238	434
Poland[3]	2400	418	2260	140	378	40	111	263
Slovakia[4]	2560	1054	2418	142	1017	37	632	660

[1] Indigenophytes from [10], archeophytes from [19], sozophytes, kindynophytes and tanatophytes from [5] and nomosozophytes (legally protected sozophytes) from [20].
[2] All data taken from [9].
[3] Indigenophytes from [14], archeophytes from [26], sozophytes, kindynophytes and tanatophytes from [29] and nomosozophytes from [18].
[4] All data taken from [12].

Results

The conservation indicator (*C*), threat conservation indicator (C_K), tanatophytisation indicator (*R*), sozophytisation indicator (*W*) and conservation effectiveness indicator (*E*) were calculated for each country (Tab. 3.).

Table 3. The indicators value associated with the four compared countries.

Country	Indicator values				
	C (%)	C_K (%)	*R* (%)	*W* (%)	*E* (%)
Czech Republic	23.6	32.07	5.79	75.9	31.09
Germany	14.46	29.6	1.56	31.4	46.02
Poland	10.9	29.36	1.6	17.4	62.9
Slovakia	25.78	62.14	1.44	41.1	62.6

Discussion

The conservation indicator shows directly the scope of legal plant protection within the considered area. It indicates the proportion of domestic, conservation-important flora that is legally protected. It also assists in the consideration of the percentage of domestic flora, which could be maintained by the implementation of legal conservation.

The conservation indicator is not useful and could also be misleading when species other than indigenophytes or archeophytes are under legal protection. But, such cases should be excluded as being contradictory to rational conservation of domestic biodiversity, because nature conservation is not simply a biological diversity protection. For example kenophytes as newly arrived anthropophytes should not be the subject of conservation, the only exception being when controlling plant invasions. However, there are recent signs that some new species occur naturally within areas they did not occupy before, because of the expansion of their geographical range. But this is the only exception from the above mentioned principle. Such cases could be ignored because of their rarity within the time period in which plant conservation strategies have been implemented and assessed.

There is obviously no need to place the whole domestic flora under legal protection. Plants with numerous, stable populations and a large distribution do not require species-targeted legal action. It would be a waste of conservation means and funds, and could decrease the conservation effectiveness of floral diversity. So, there is no demand to increase the conservation indicator up to 100%. In different regions, with particular floras, an adequate number of protected species chosen from the group of national flora should undergo legal preservation. The conservation indicator gives the relevant information, important for the elaboration and implementation of plant conservation strategies, especially in the budget section.

Comparing the conservation indicator for the chosen countries, we stated that almost a quarter of the domestic flora of Slovakia and Czech Republic is under legal protection. Poland has a lower level, with only about 10% of plants considered of conservation importance the subject of legal action. The indicator value is not proportional to the total number of species or to the sozophytes number in chosen countries. The results seem to be proportional to the sozophytisation indicator, although in the case of Czech Republic there is apparent inequality. Over 75% of the domestic flora is Red-listed and only about 23% is under legal protection. Such a situation is not suitable and needs to be clarified and possibly corrected.

Better, but still not good, is the state of threatened species preservation. For the Czech Republic but also for Poland and Germany values fluctuate around 30%. Substantially better is the situation in Slovakia where over 60% of threatened plants are legally protected.

The threat conservation indicator shows how many species from the threatened group is under legal conservation. It indicates the state's activity in threatened flora conservation – the group of the most important plants, which undergo several negative impacts especially from human economy. The threat conservation indicator is also an indirect test of conservation effectiveness for visibly declining species. It seems to be valid, that this indicator should tend to be 100% or at least close to it. So, it is suggested that it should be obligatory for every threatened plant species, whether attractive, alluring and sumptuous or inconspicuous, small and colourless, to be preserved by placing it under legal protection. The threat conservation indicator points out the current needs regarding the most important species. So, it seems crucial in the case of Poland, the Czech Republic and Germany to expand the state list of protected plants.

The tanatophytisation indicator shows directly the percentage of extinct species from the group of national flora. Its value also informs us about the anthropogenic pressure on the given area and the scope of necessary recovery undertakings. The worst is the situation again in the Czech Republic, where 6% of the national flora has been exterminated. Is it a consequence of a relatively low level of legal, species-oriented conservation? Undoubtedly the success of any rescue action for extremely endangered species also depends on the legal measures taken by the state.

To find what the proportion between the numbers of the native flora and number of sozophytes i.e. plants gaining EX, CR, EN, VU, NT or DD category is, the sozophytisation indicator was introduced. It shows the level of conservation demand in the studied area. The sozophytisation indicator points out the conservation priority areas, where for example despite the comparatively lower number of species the conservation demand is high because of the substantial percentage of threatened, extinct and near to threatened plants. This index is quite differentiated within the chosen countries. The worst is the situation in the Czech Republic. Over three-quarters of the domestic flora have been considered to be sozophytes. Poland has the shortest Red List. The results of sozophytisation indicator comparison shows surprisingly, that the number of sozophytes is loosely related to common thinking about the anthropogenic pressure and industrialisation level in given countries. Germany has less than half the percentage of sozophytes in comparison to the Czech Republic.

The conservation effectiveness indicator answers directly the question whether the legal protection of the conservation important flora is appropriate or not. This indicator assumes that all sozophytes should undergo legal protection irrespective of their attractiveness, life form, access to their biotopes, etc. Legal protection should also be given to extinct species, because in a case of spontaneous reoccurrence or as a result of recovery actions the probably small populations would certainly require legal assistance. Also, other groups of sozophytes, such as threatened, near-threatened and data-deficient species should obviously be the subject of legal conservation to allow the implementation of the restrictive administration procedures within human activities such as agriculture, forestry or fishery. The indicator formula means that there is no sense in providing legal protection for those plant species that are not sozophytes. Implementation of legal conservation for the plant species with numerous, stable, increasing, expanding populations seems to be irrational and against the idea of concentrating means and funds on the most crucial and priority conservation tasks. In the author's opinion, the conservation effectiveness indicator should tend towards a value of 100%. If a greater amount of sozophytes is under legal preservation, it could be assumed that the conservation of the floral diversity of a given area brings better results. The indicator shows directly what is the distance from achieving the most desirable level of legal protection of the important group of species. In comparison to the threat conservation indicator it gives more general and ultimate view.

Analysing the examples here, the difference between less developed countries and those with higher level of industrialisation is clearly seen. Poland and Slovakia achieve almost the same, quite high value while Germany and the Czech Republic, countries with the highest industry potential in the region, have less than half of their sozophytes under legal preservation. Looking for the causes of such results, the strength of lobby groups should be considered. It is a common issue that the compilation of longer lists of legally protected plants is blocked by powerful players and also administration representatives like Water Management Services, Agricultural Authorities and others. The significant example is that until now there has been no legal conservation of weeds in countries considered.

Acknowledgements: The author would like to thank Prof. Halina Rostropowicz from Opole University and Mrs. Donka Koch from Wrocław University for advice and assistance in Greek and Latin translations. Many thanks also to Aleš Pečinka from the Department of Cytogenetics, Institute for Plant Genetics and Crop Plant Research in Gatersleben, Germany, for advice and sending indispensable data.

REFERENCES

1. Bateman G., Egan V. (eds.)., 1993 - The encyclopedia of World geography. Andromeda, Oxford.
2. Davis S.D., Droop S.J.M, Gregerson P., Henson L., Leon C.J., Villa-Lobos J., Synge H., Zantovska J., 1986 - Plants in danger. What do we know? Int. Union Conserv. Nat. Resources, Gland, Switzerland & Cambridge, UK.
3. Given D.R., 1994 - Principles and practise of plant conservation. Chapman and Hall, London.
4. Heywood V.H., Iriondo J.M., 2003 - Plant conservation: old problems, new perspectives. Biological conservation, **113**: 321-335.
5. Holub J., Procházka F., 2000 - Red list of vascular plants of the Czech Republic. Preslia, Praha **72**: 187-230.
6. Ioras F., 2003 - Trends in Romanian biodiversity conservation policy. Biodiversity and Conservation, **12**: 9-23.
7. Kaźmierczkowa R., Zarzycki K. (eds.), 2001 - Polish Red Data Book of Plants. Pteridophytes and Flowering Plants. W. Szafer Institute of Botany, Institute of Nature Conservation. Polish Academy of Sciences, Cracow.
8. Kornaś J., 1968 - Geograficzno-historyczna klasyfikacja roślin synantropijnych. Mater. Zakł. Fitosocjol. Stos. UW, **25**: 33-41.
9. Korneck D., Schnittler M., Vollmer I., 1996 - Rote Liste der Farn- Und Blütenpflanzen (*Pteridophyta* et *Spermatophyta*) Deutschlands. In: Ludwig G., Schnittler M. (eds.), Rote Liste gefährdeter Pflanzen Deutschlands. Bundesamt für Naturschutz. Bonn - Bad Godesberg. ss. 21-187.
10. Kubát K. (ed.), 2002 - Key to the flora of Czech Republic. Academia, Praha.
11. Manoliadis O.G., 2001 - Environmental indices in irrigation management. Environmental Management, **28** (4): 497-504.
12. Marhold K., Hindák F. (eds.), 1998 - Checklist of non-vascular and vascular plants of Slovakia. Veda, Bratislava.
13. Mirek Z., 1981 - Problemy klasyfikacji roślin synantropijnych. Wiad. Bot., **25** (1): 45-54.
14. Mirek Z., Piękoś-Mirkowa H., Zając A., Zając M., 1995 - Vascular Plants of Poland - a checklist. Polish Bot. Stud., Guideb. Ser., **15**: 3-303.
15. Morillo C., Gómez-Campo C., 2000 - Conservation in Spain, 1980-2000. Biological Conservation, **95**: 165-174.

16. Nowak A., Nowak S., 2004 - The effectiveness of plant conservation: a case study of Opole Province, Southwest Poland. Environmental Management, **34** (3): 363-371.
17. Piękoś-Mirkowa, H., 1990 - The functioning of the plant species protection in Poland. In: Klimek K. (ed.), Protected areas and species conservation in Southern Poland – functioning, evaluation, perspectives, pp. 141-167. Zakład Ochrony Przyrody i Zasobów Naturalnych PAN, Kraków.
18. Polish Species Conservation Ordinance, 2001 - Vol. no. 106, pos. 1176 from 29th September 2001.
19. Pyšek P., Sádlo J. Mandák B., 2002 - Catalogue of alien plants of the Czech Republic. Preslia, Praha, **74**: 97-186.
20. Regulation, 1992 - Regulation of the Ministry of the environment of the Czech Republic no. 395/1992b., Appendix 2 – The list of the protected plant species.
21. Reyers B., James A.N., 1999 - An upgraded national biodiversity risk assessment index. Biodiversity and Conservation 8: 1555-1560.
22. Reyers B., James A.N., van Jaarsveld A.S., McGeoch M.A., 1998 - National biodiversity risk assessment: a composite multivariate and index approach. Biodiversity and Conservation, **7**: 945-965.
23. Schnittler M., Günther K.-F., 1999 - Central European vascular plants requiring priority conservation measures – an analysis from national Red List and distribution maps. Biodiversity and Conservation, **8**: 891-925.
24. Villa F., McLeod H., 2002 - Environmental vulnerability indicators for environmental planning and decision-making: guidelines and applications. Environmental Management, **29** (3): 335-348.
25. Woodiwiss F.S., 1964 - The biological system of stream classification used by the Trend River Board. Chemistry and Industry: 443-447.
26. Zając A., 1979 - Pochodzenie archeofitów występujących w Polsce (The origin of archeophytes occurring in Poland). Prace habilitacyjne nr. 29. Uniwersytet Jagielloński, Kraków.
27. Zarzycki K., 1986 - Lista wymierających i zagrożonych roślin naczyniowych Polski [Red list of threatened vascular plants in Poland]. In: Zarzycki, K., Wojewoda, W. (eds.), Lista roślin wymierających i zagrożonych w Polsce [List of threatened plants in Poland], pp. 11-28. PWN, Warszawa.
28. Zarzycki K., Kaźmierczakowa R. (eds)., 1993 - Polish plant red data book. Pteridophyta and Spermatophyte. Polish Academy of Science, Kraków.
29. Zarzycki, K., Szeląg Z., 1992 - Czerwona lista roślin naczyniowych zagrożonych w Polsce [Red list of threatened vascular plants in Poland]. In: Zarzycki K., Wojewoda W., Heinrich Z. (eds.), Lista roślin zagrożonych w Polsce [List of threatened plants in Poland], pp. 87-98. 2nd edition. Instytut Botaniki im. W. Szafera PAN, Kraków.

ANTHROPOGENIC HABITATS CAN SHELTER THREATENED PLANTS

Arkadiusz NOWAK*, Sylwia NOWAK
Department of Biosystematics, Division of Phytosociology and Plant Geography, University of Opole, Oleska str. 22, **45-052 Opole (POLAND)**;
e-mail: anowak@uni.opole.pl

Abstract: This paper describes possible different applications of phytosociology in the conservation of floristic biodiversity. It analyses in detail the occurrence of Red-listed species, so-called sozophytes, in plant communities at different stages of naturalness and degeneration. Based on these analyses, the authors conclude that as a result of distinct transformations of plant cover, endangered species evolve adaptive mechanisms and colonise ecosystem types new for them. Plants can remain in habitats that are not natural for them and can form new syntaxonomic noda. This phenomenon, which could be regarded as a kind of exodus of sozophytes, is very important from both scientific and conservation points of view.

Introduction

Basic knowledge of plant communities, their floristic composition, structure, spatial and temporal relations, diversity and naturalness is of fundamental importance to nature conservation. With reference to the conservation of plant diversity at a species level, it is especially helpful in restoration, translocation, re-establishment, rehabilitation and other active management techniques [8, 9]. At a vegetation level the phytosociological approach is useful in the assessment of the types and degree of alteration of plant communities [18, 6], evaluation of vegetation naturalness, estimation of restoration possibilities, establishment of the goals of conservation in protected areas; and also to show the potential effect of strict protection (the end of a succession sequence), ascertaining legally protected and rare plant communities, estimation of vegetation diversity, establishment of effective protection networks with all important vegetation types, assessment of conservation (e.g. to create priority conservation sites), agro-environment schemes and programmes co-financed by the EU [20]. Phytosociological surveys are indispensable in analysing the environmental conditions of threatened species occurrence, estimation of the synanthropo-dynamic state of sozophytes and its adaptation potential, and an assessment of the importance of the species in conservation [15].

At species level, the phytosociological approach can be specifically used in the following applications:

- To describe the habitat of the population of scarce or declining species, aiming to detect the ecological amplitude of a rare species, its habitat preferences and phytosociological fidelity,

- To find out the abundance of the given taxon (e.g., methods of assignment of harvesting limits for medicinal plants),
- To assess habitats for restitution, translocation or re-establishment of the threatened or extinct species,
- To foresee the succession sequence and future habitat conditions for population of the taxa under consideration,
- To answer the question whether the considered population would be able to survive for a very long time requires finding the climax vegetation for the area in which it occurs,
- To gain knowledge of vegetation cover and the diversity and composition of a plant community is substantially helpful in searching inventories for locations of the rare species,
- To assess the quality and final effects of management in protected areas,
- To evaluate the vegetation for site selection and delimitation of protected areas for the species considered,
- To evaluate the vegetation, aiming to find "hot spots" of floristic diversity,
- To assess the adaptation potential of the taxa of interest,
- To assess the degeneration phases of communities.

In a natural environment changing under the influence of man, gradually larger areas are covered by transformed or completely destroyed vegetation. They are sites both used by man, such as roadsides, drainage ditches, walls, ponds; and abandoned sites, such as worked-out quarries, gravel- and sand-pits, etc. The result of various economic and management activities are remarkable changes in vegetation cover – plant communities undergo degeneration, stenotopic floristic elements, being in most cases characteristic species of given syntaxa, disappear. The anthropophytes and expansive species take advantage, causing disturbances in the floristic composition of communities. Habitat conditions undergo transformation as well, including water conditions and soil profiles crucial for functioning of phytocoenoses. At the final stage, the geology and landscape of the area are exposed to transformation, which is most visible in the case of opencast excavations, industrial heaps, etc. These changes most often lead to a reduction of the natural floristic diversity [8, 19].

However, the analysis of the floristic diversity of the land area of Poland apparently shows that anthropogenic habitats could play an important role in maintaining the habitat for sozophytes, i.e. rare and endangered species. Despite the fact, that the ecosystem is changed and phytocoenoses are deprived of characteristic elements, they could still function as an important site for many threatened species. In many cases the substantially but not completely changed ecosystem conditions give the threatened populations a chance to take hold under the most unfavourable conditions. Sometimes this allows the liberation or acceleration of the adaptation and evolution processes and finally gives conservationists the opportunity to research these cases and find out how to help the endangered survivors in a changed world. As an example from Central Europe, the Polish case could be mentioned. In the Red Data Book of Vascular Plants of Poland [12] as many as 59 taxa among the described 296 occurred in anthropogenic habitats. Among these are such rare and endangered species as: *Aldrovanda vesiculosa*, *Cyperus flavescens*, *Elatine alsinastrum*, *Euphorbia epithymoides*, *Lathyrus latifolius*, *Lindernia procumbens*, *Nasturcium microphyllum*, *Pilularia globulifera*, *Trapa natans*. Many species listed in the Red Data Book have their only localities in eu- or polihemerobic ecosystems, e.g. *Apium nodiflorum*, *Carex stenophylla*, *Crassula aquatica*, *Dichostylis micheliana*, *Marsilea quadrifolia*. Similar phenomena are known, for example, from the area of Opole province [17], Lower Silesia [13] and the Czech Republic [4].

Considering the increase in the surface strongly transformed by human activity and, at the same time, more and more frequent records of occurrence of sozophytes in such sites the authors made an attempt to present a phytosociological approach to analyse phytocoenosis conditions (naturalness and degeneration) of red-listed species populations.

Methods

Geobotanical studies were conducted, especially on anthropogenic habitats, in Opole Silesia in the years 1997–2003 (Fig. 1.). During the field work, 1516 relevés were recorded and another 1242 were taken from literature. The subject of the study was the occurrence of taxa that are endangered, rare and protected by law in areas strongly transformed by man, including meadows, pastures, fields, quarries, gypsum mines, clay-, gravel- and sand-pits, large dam reservoirs, fish-ponds, small anthropogenic ponds, roadsides, railway tracks, channels, walls, boundary strips, harbours, parks, drainage ditches and the remaining urban areas.

Species nomenclature follows Mirek *et al.* [14]. The relevés were recorded using the Braun-Blanquet method [3]. The analyzed group of species (sozophytes) was taken from the regional Red List of threatened vascular plants [15].

The degeneration phases were proposed using the simplified approach of Faliński [6]. Three phases of community degeneration were differentiated: (I) disappearance of species characteristic for association and alliance, (II) disappearance of species characteristic for order, (III) disappearance of species characteristic for class. The classification of plant association responding to human pressure was taken from Faliński [7].

Fig. 1: Location of the Opole voivodship in Poland.

Results

A total of 6367 locations of 529 species from the chosen group were registered. For each population of sozophyte, the naturalness and phase of degeneration of the plant community were stated (Tabs.1 and 2).

Additionally, as a result of the study, the occurrence in anthropogenic habitats of 199 species from the selected group of 529 plants was stated, including species critically endangered, endangered, vulnerable, near threatened, rare and protected by law. In total, 688 sites of selected plants were documented in strongly degraded habitats (eu- and polyhemerobic), which is about 11.5% of all sites of the chosen species (Fig. 2).

Table 1: Naturalness of plant assemblages where red-listed plants occur (according to Faliński's degeneration phases [7]).

<table>
<tr><th colspan="3" rowspan="2">Type of plant community</th><th colspan="7">Number of localities with respect to threat category</th><th rowspan="2">Σ</th><th rowspan="2">%</th></tr>
<tr><th>RE</th><th>CR</th><th>EN</th><th>VU</th><th>NT</th><th>LC</th><th>DD</th></tr>
<tr><td colspan="3">Primeval</td><td>-</td><td>-</td><td>-</td><td>-</td><td>-</td><td>-</td><td>-</td><td>-</td><td>-</td></tr>
<tr><td colspan="3">Natural</td><td>-</td><td>398</td><td>423</td><td>517</td><td>386</td><td>242</td><td>251</td><td>2217</td><td>34.8</td></tr>
<tr><td rowspan="4">Anthropogenic</td><td colspan="2">Seminatural</td><td>7</td><td>659</td><td>685</td><td>720</td><td>495</td><td>245</td><td>311</td><td>3122</td><td>49</td></tr>
<tr><td colspan="2">Xenospontaneous</td><td>-</td><td>-</td><td>10</td><td>24</td><td>50</td><td>47</td><td>25</td><td>156</td><td>2.5</td></tr>
<tr><td rowspan="2">Syn-anthro-pic</td><td>Segetal</td><td>4</td><td>103</td><td>127</td><td>167</td><td>96</td><td>87</td><td>53</td><td>637</td><td>10</td></tr>
<tr><td>Ruderal</td><td>1</td><td>8</td><td>27</td><td>36</td><td>52</td><td>58</td><td>53</td><td>235</td><td>3.7</td></tr>
</table>

Table 2: Degeneration phases of plant assemblages where threatened plants occur (see 'Methods' for codes of degeneration phases).

Degeneration phase	Number of localities with respect to threat category							Total	%
	RE	CR	EN	VU	NT	LC	DD		
(I)	9	713	784	756	565	324	399	3548	55.8
(II)	3	375	400	576	405	222	195	2176	34.1
(III)	0	80	88	132	109	133	101	643	10.1

Explanations: RE – regionally extinct, CR – critically endangered, EN – endangered, VU – vulnerable, NT – near threatened, LC – least concerned, DD – data deficient.

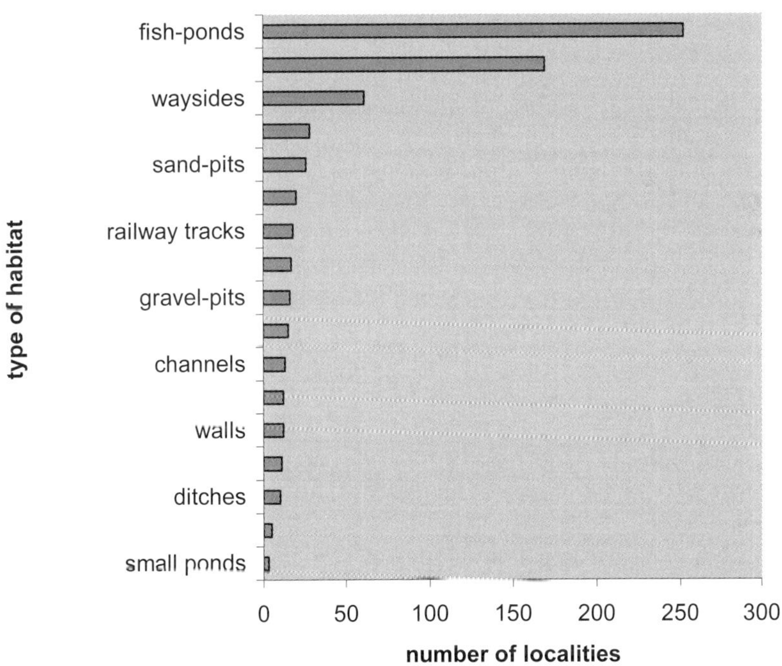

Fig. 2: Number of localities of taxa occurrence by different types of anthropogenic habitats.

Discussion

Opole Silesia is a region that has a relatively strongly transformed vegetation cover, without any preserved fragments of natural vegetation. Natural or almost natural plant associations (*e.g.* forests, aquatic or rush communities) develop in the most valuable wildlife areas. A remarkable percentage of sozophytes (c. 35%) of different threat categories occur there, the dominant types being those which are assigned to category VU. Fragments of natural vegetation are constantly exposed to anthropogenic pressure. Despite the implementation of various nature protection procedures, due to the low effectiveness of such protection many species associated with this type of vegetation often abandon occupied habitats. Semi-natural communities, such as meadows or swards, undergo even greater human pressure. However, they still present relatively high biodiversity indices. Almost half of the known locations of sozophytes are associated with semi-natural vegetation. Maintaining extensive methods of meadow and pasture management is of high importance for the preservation of habitats important for the floristic richness of the region. The greatest challenge, both in conservation and scientific terms, is the occurrence of sozophytes in extreme anthropogenic habitats, such as xenospontaneous communities, formed mainly by alien species (c. 2.5% of localities) or segetal (10% of localities) and ruderal communities (4% of localities). In these types of habitat, taxa of the highest threat category, with small populations that do not have the required adaptive potential, occur rather rarely. However, other threat categories are represented in these habitats, which is evidence of the development of adaptive mechanisms in populations endangered in their natural habitats. The visible exodus of threatened species to ruderal localities, to communities with a high percentage of taxa of allochthonous origin, should be considered while assessing the threat status of a given taxon. From the point of view of the conservation biology, this is an important phenomenon, probably in its initial phase, and its role will develop thanks to the rapid adaptation of plants.

The analysis of the degree of degeneration of plant communities in which sozophytes occur indicate that, under the conditions in Opole Silesia, sozophytes preferred well-developed phytocoenoses, where half their localities were situated. However, the percentage of localities of sozophytes in degraded communities was also remarkable. Even in associations of the IIIrd degeneration level, where the phytosociological class was hard to define, there was still c. 10% of localities of the species under analysis. A decrease in the percentage of taxa assigned to the high threat category with an increase in the level of community degeneration was characteristic.

Such a great participation of sites of sozophytes in anthropogenic habitats prompts a very serious attitude to the problem of occurrence of rare and endangered species in non-natural ecological systems. Nature conservation should develop a proper attitude to sites of valuable species in strongly transformed areas. In the same way that, in the past, semi-natural biocoenoses, for example pastures, meadows and managed forests, have been recognised as an object of interest for nature conservation, the need for the protection of sites of valuable taxa in habitats strongly transformed by man should be taken into consideration at the present time. Obviously, protection of such sites is often very difficult in practice, e.g. in the case of communication tracks or roadsides. However, in many cases, where there are no problems of an administrative or legal origin, it is possible effectively to safeguard valuable sites, for example closed quarries or opencast excavations. Positive attempts are already being undertaken to increase the rate of recolonisation by vegetation of degraded areas and then to cover them with special protection [5] and to analyse the biodiversity of transformed areas in the context of preservation of vegetation [21].

Anthropogenic habitats are often temporary shelters rather than mainstays, from which the plants can expand further. Frequently they play the role of corridors for expansion of taxa and links between natural populations.

Reasons for the presence of sozophytes in ecosystems strongly transformed by man can be traced in several phenomena. Undoubtedly, among the most important direct causes which should be mentioned, are the creation by man of analogous habitats, i.e. those similar to natural ones with respect to the structure, biogeochemical composition and level of moisture. An example can be silt habitats in fishponds, which are analogues of alluvia of rivers that disappear as a result of regulation works. Another reason, much more difficult to observe, is an expansion of the geographical range of taxa caused by the removal of barriers, i.e. climatic or geological barriers, competition or parasitism of other species and in consequence – inhabitancy of new areas by a species. Particularly numerous are examples of the appearance of rare and endangered species in areas of an inhibited natural succession due to human activity. These are namely mowed roadsides, where, in spite of drastically changed edaphic and water conditions, specialised species such as e.g. xerothermophiles remain, thanks to a reduction of the competitiveness of other plants. Another, difficult to notice but undoubtedly the reason for the occurrence of plants in anthropogenic habitats are adaptive changes and appearance of new biological characters in plants. The problem of biological and ecological differences between closely related rare and common species was discussed in different papers [1, 2, 10, 11]. Their investigations proved that emerging new features referring to for example the number of fruits and seeds, year of first flowering, flower structure, surface of leaves, biomass of plants, living form and different genetic modifications such as polyploidy can enable a given taxon to expand into new biotopes, including ecosystems transformed by man.

Another reason for the maintenance of sozophytes in anthropogenic habitats is their limited accessibility. Sites such as quarries, railway areas, gravel- or sand-pits and harbours are not widely open to the public. This is a similar situation to that of fishponds or other ponds that are most often private. The reduction of, or exclusion from, human penetration of such areas allows for the possibility of existence of taxa which are susceptible to a direct, negative influence.

Considering the number of endangered, rare and protected taxa that occur in habitats strongly transformed by man and the total number of their locations in anthropogenic habitats, it seems necessary to recognise thoroughly the occurrence of sozophytes in habitats at the advanced stage of hemeroby. Moreover, geobotanical studies aimed at characterising the process of synanthropization and apophytization of sozophytes would be necessary.

REFERENCES

1. Baskin J.M., Snyder K.M., Walck J.L., Baskin C.C., 1997 - The comparative autecology of endemic, globally-rare and geographically-widespread common plant species: three case studies. The Southwestern Naturalist, **42**: 384-399.
2. Bevill R.L., Louda S.M., 1999 - Comparisons of related rare and common species in the study of plant rarity. Cons. Biol., **13** (3): 493-498.
3. Braun - Blanquet J., 1964 - Pflanzensoziologie, Gründzüge der Vegetationskunde. Dritte Auflage. Springer Verlag, Wien-New York.
4. Čerovský J., Feráková V., Holub J., Maglocký Š., Procházka F., 1999 - Červená Kniha ohrožených a vzácných druhu rostlin a živočichu ČR a SR. Vol. 5 Vyšši rostliny. Priroda a. s., Bratislava. ss. 456.
5. Cullen W.R., Wheater C.P., Dunleavy P.J., 1998 - Establishment of species-rich vegetation on reclaimed limestone quarry faces in Derbyshire, UK. Biological Conservation, **84**: 25-33.
6. Faliński J.B., 1966 - Próba określenia zniekształceń fitocenozy. System faz degeneracji zbiorowisk roślinnych. Dyskusje Fitosocjologiczne (3). Ekologia Polska, Ser. B, tom **XII**, Zesz. 1. ss.: 31-42.
7. Faliński J.B., 1969 - Zbiorowiska autogeniczne i antropogeniczne. Próba określenia i klasyfikacji. Dyskusje fitosocjologiczne (4). Ekologia Polska, Ser. B, tom **XV**, Zesz. 2. ss.: 173-179.
8. Given D.R., 1994 - Principles and practice of plant conservation. Chapman & Hall.
9. Gordon D.R., 1996 - Experimental translocation of the endangered shrub Apalachicola Rosemary *Conradina glabra* to the Apalachicola bluffs and ravines preserve, Florida. Biological Conservation, **77**: 19-26.
10. Hamilton C.W., 1990 - Variations on a distylous theme in mesoamerican *Psychotria* subgenus *Psychotria* (Rubiaceae). Memoirs of the New York Botanic Garden, **55**: 62-75.
11. Karron J.D., Linhart Y.B., Chaulk C.A., Robertson C.A., 1988 - Genetic structure of populations of geographically restricted and widespread species of *Astragalus* (*Fabaceae*). American Journal of Botany, **75**: 1114-1119.
12. Kaźmierczakowa R., Zarzycki K. (eds.), 2001 - Polish red data book of plants. Pteridophytes and flowering plants. Cracow.
13. Kącki Z. (ed.), 2003 - Czerwona Lista Roślin Naczyniowych Dolnego Śląska. Uniwersytet Wrocławski.
14. Mirek Z., Piękoś-Mirkowa H., Zając A., Zając M., 2002 - Vascular Plants of Poland - a checklist. Polish Bot. Stud. Guideb. Ser. **15**: 3-303.
15. Nowak A., Nowak S., 2004 - Changes of the sozofloristic value of Opole Province in Southwestern Poland. Polish Journal of Environmental Studies, **13** (3): 341-346.
16. Nowak A., Nowak S., Spałek K., 2003 - Red list of vascular plants of Opole Province. Nature Journal, **36**: 5-20.
17. Nowak A., Spałek K. (eds.), 2002 - Red data book of vascular plants of Opole Province. OTPN, Opole. (in Polish).

18. Olaczek R., 1972 - Formy antropogenicznej degeneracji leśnych zbiorowisk roślinnych w krajobrazie rolniczym Polski Niżowej. Uniwersytet Łódzki, Łódź.
19. Sutherland W.J. (ed.), 1998 - Conservation science and action. Blackwell Publishing.
20. Yeo M.J.M., Blackstock T.H., Stevens D.P., 1998 - The use of phytosociological data in conservation assessment: a case study of lowland grasslands in mid Wales. Biological Conservation, **86**: 125-138.
21. Zerbe S., Maurer U., Schmitz S., Sukopp H., 2003 - Biodiversity in Berlin and its potential for nature conservation. Landscape and Urban Planning, **62**: 139-148.

GLOBALLY AND EUROPEAN THREATENED PLANTS PRESENT IN DOBROGEA (SOUTH-EASTERN ROMANIA)

Anca SÂRBU, Gavril NEGREAN, Gabriela PASCALE*, Paulina ANASTASIU
University of Bucharest, Department of Botany, Intrarea Portocalelor 1-3, Sector 6
060101 Bucharest (ROMANIA); e-mail:pasgabi@yahoo.com

Abstract: The present work is the result of research carried out within the framework of IPAs project, a practical application of the Global Strategy for Plant Conservation (2002 Hague) in Romania. The data provided constitute a scientific basis support for enlargement of the national network of protected areas and for selection of Natura 2000 sites in Romania. A significant part is occupied by studies performed in Dobrogea. New information regarding 16 threatened taxa at Global (4 species) or European (12 species) level, that still occur in Dobrogea, are provided.

Introduction

The project "Important Plant Areas (IPAs) in Romania" is an integrated part of the European IPAs Programme, co-ordinated by Plantlife International and financed by the Ministry of Agriculture, Nature Management and Fisheries of the Netherlands. This European programme has been deployed in seven Central and East European countries. The programme objective is to identify a network of the best sites throughout Europe using consistent criteria: threatened plants, threatened habitats and botanical richness.

Some of the major results of the Romania IPAs project can be summarized as follows: updating the information on threatened flora and vegetation according to the international legislation (EUNIS, Bern Convention, Habitat Directive, etc.); identification of 276 IPAs as a key element for threatened plants and habitats protection and conservation; and development of a data base including information, maps and images [24].

The Dobrogea province is part of the Balkan Peninsula, located in the eastern sub-Mediterranean zone. Its rather small area, geological past, frequent climatic changes and present-day physical-geographical conditions have determined a high concentration of endemic plant taxa. The characteristic plant taxa are termed "Dobrogean elements".

Material and Methods

In order to reach a good evaluation of the real situation of endangered plants (threatened at Global and European level) still present in Dobrogea, a series of field investigations, collecting and preparing for herbarium use have been carried out between 2002 and 2004, and a vast array of specialized bibliographical references has been consulted. Botanical nomenclature is in accordance with the

Flora of Romania [23] and *Flora Europaea* [28, 29]. The international abbreviation for herbaria follows Index Herbariorum [9].

Results and Discussions

In Dobrogea 16 endangered plant species at Global and European level have been found, of which 6 are endemic and near-endemic: *Aldrovanda vesiculosa* L., *Alyssum borzaeanum* Nyár., *Campanula romanica* Săvul., *Centaurea jankae* Brandza, *Centaurea pontica* Prodan & Nyár., *Echium russicum* J.F. Gmelin, *Fritillaria orientalis* Adams, *Liparis loeselii* (L.) L.C.M. Richard, *Marsilea quadrifolia* L., *Moehringia jankae* Griseb. ex Janka, *Ornithogalum ortophyllum* Ten. subsp. *psammophilum* (Zahar.) Zahar., *Paeonia tenuifolia* L., *Potentilla emilii-popii* Nyár., *Salvinia natans* L., *Trapa natans* L., *Zostera marina* L.

Campanula romanica Săvul. (Globally threatened species – IUCN Red List, Habitats Directive and Bern Convention) is an endemic taxon from Dobrogea, occurring only in the Romanian part of this region. It is probably a paleoendemic species, evolved in the early Pleistocene, when the continental Dobrogea was connected northwards with the Carpathian chain, where other taxa of section *Linophylloides* also occur. It is a saxicolous plant, indifferent with respect to the type of substrate (Jurassic limestones, green shales, sandstone, granite). It thrives in saxicolous associations of the *Pimpinello - Thymion* alliance. Chorology: This species is present from the Munții Măcin as far as Adamclisi, with a higher density north of Valea Cara-Su [17]. Information on its occurrence at Adamclisi has not been confirmed for over 50 years. Likewise the citation in the Flora of Romania [17] of this plant as being present at Hagieni is erroneous. According to our observations this species seems to thrive solely on ancient highlands, but not on loess or sarmatic or other types of rock, south of Cara-Su. From the number of sites at which this species occurs, it could be considered to be rather widespread, but actually it forms small populations in rock crevices. It could become endangered by digging of new quarries for the exploitation of the stone on which the plant is growing.

Centaurea jankae Brandza (Globally threatened species – IUCN Red List, Habitats Directive and Bern Convention) is a Dobrogean paleoendemic taxon, evolved probably at the end of the Tertiary period. It belongs to section *Hyalolema* as a unique representative in Europe; there are close-related taxa beyond the Ural Mountains. This species thrives on stony slopes, preferentially calcareous, from the Platoul Babadag to the border with Bulgaria. Chorology: On the Platoul Babadag it is found at 4–5 sites, beginning with the classic spot from the south of Pădurea Babadag towards Caugagia (where have been no reports since1960). The densest population exists at Capul Doloşman, where individuals grow in optimal conditions; here we have counted almost 300 plants, mostly in bloom. In 2003 we detected a small population about 2 km NW of Capul Doloşman. Other localities to the south were found after the publication of the Flora of Romania: Palazul Mic [11], Pădurea Dumbrăveni [21] and Cotul Văii [BUC] [18].

Centaurea pontica Prodan & Nyár. (Globally threatened species – IUCN Red List, Habitats Directive and Bern Convention) is an arenicolous endemic species from the Danube Delta. Along with *C. calcitrapa* and *C. iberica* it belongs to the section *Jacea*. Its taxonomic position is rather controversial, the present status being probably the most adequate. Extant populations are extremely reduced in number, the one in Sulina town being threatened by extinction. Appropriate measures for its rescue are imperative. Chorology: The area of distribution is concentrated around Sulina, on the streets and in the peripheral zones. This taxon was also found in the south of Braţul Sf. Gheorghe on the sands of Ciotic, leg. G. Negrean [BUCA] [6].

Ornithogalum ortophyllum Ten. subsp. *psammophilum* (Zahar.) Zahar. (Globally threatened species – IUCN Red List, Habitats Directive and Bern Convention) is an arenicolous endemic taxon, originally described at Platoneşti, Ialomiţa County [30, 31]. Afterwards it was found also at Hanul Conachi and Lieşti [15]. Owing to its morphological peculiarities and special ecology, this taxon should deserve the rank of species. Chorology: It was found on Grindul Lupilor [5]. We lack information regarding the state of the population from this site. Nevertheless its presence on saline sands of the lagoon complex Razim-Sinoe is very strange. It could be another taxon, maybe a completely unknown one. Cultivation of this plant is necessary to allow more detailed morphological, genetic and biochemical studies.

Moehringia jankae Griseb. ex Janka (European threatened species – Habitats Directive and Bern Convention) is a Dobrogean sub-endemic plant, taxonomically close to *M. grisebachii*, but much rarer. It is a basophilous plant [28]. Chorology: In Tulcea County it has been found on the mountains between Greci and Măcin, on Dealul Consul, at Topolog and on Dealul Tuşan-Măgurele. It was also reported in Hârşova, Dealul Moşul, Stâncile Călugăreni – Colţanii Mari and Cheia (Constanţa County). Populations comprise always a very small number of plants.

Potentilla emilii-popii Nyár. (European threatened species – Habitats Directive and Bern Convention) is a Dobrogean endemic plant [Bu Rm], a member of section *Rectae* of the genus. Its taxonomic status is uncertain, as in *Flora Europaea* it is included in the *Taurica* group. This species is present only in the south of Dobrogea, within plant associations of the *Pimpinello-Thymion* alliance. Populations are extremely poor in number of individuals, because of reproduction difficulties. Establishment of new quarries or pastures will endanger this species. Chorology: The plant has been reported in Topolog, Dealul Tuşan-Măgurele, Valul lui Traian, Adamclisi, Coroana, Pădurea Dumbrăveni Reserve, Pădurea Canaraua Fetii Reserve, Pădurea Esechioi Reserve, Independenţa, Dealul Alah-Bair Reserve, Pădurea Hagieni Reserve [8, 11, 1, 16, 2, 3, 4].

Alyssum borzaeanum Nyár. (European threatened species – Habitats Directive and Bern Convention) is a Critically Endangered sub-endemic species; it is a littoral, east sub-Mediterranean element. This species is conserved within the Danube Delta Biosphere Reserve and in the Agigea Reserve, being threatened by habitat changes caused by the intense tourism. Chorology: It occurs on Grindul Lupilor, at Cetatea Histria on the shore of Lacul Sinoe, on Grindul Saele-Istria, at Gargalâc, Mamaia, Capul Midia, between Tuzla and the sea-shore, at Techirghiol, Agigea, Eforie Sud and Mangalia [20, 7]. In spite of these numerous reports this plant has no longer been found at some of the sites (Mangalia, Mamaia, Eforie Sud), most probably because of the intense management of beaches for tourism.

Echium russicum J.F. Gmelin (European threatened species – Habitats Directive and Bern Convention) is a continental element extending westwards to eastern Austria. This plant is growing on certain steppic grasslands in almost all Romanian provinces. It was not registered yet in any national Red Lists, but is conserved in Măcin National Park. Chorology: In Dobrogea it is extremely rare. It has been found in the following sites: Măcin, Greci, Luncavița, Telița, Malcoci, Teche, Platoul Babadag, Dobromir, Canaraua Fetii, Esechioi, Hagieni and Pădurea Dumbrăveni [22, 1]. However, we have not found it at the previously mentioned sites.

Fritillaria orientalis Adams (European threatened species – Habitats Directive and Bern Convention) is a sub-Mediterranean geophyte, reported from all regions of Romania except Maramureş. National Red Lists mention it as being Vulnerable and Rare. It is endangered mainly because of its decorative aspect, being collected for commerce. Chorology: This species is extremely rare in Dobrogea, reported solely from Platoul Babadag [23]. It is imperative to include its populations into a natural reserve.

Paeonia tenuifolia L. (European threatened species – Habitats Directive and Bern Convention) is a steppic species that is very rare in Romania. Due to the beauty of its large flowers, it is intensively collected for commerce. The national Red Lists assess it as a Vulnerable taxon. It is protected in the natural reserves of Fântânița-Murfatlar, Cheia, Hagieni and in the Măcin National Park. Chorology: This species is indicated as occurring in Greci, Cheia, Palazul Mic, Basarabi, Pădurea Dumbrăveni, Cotul Văii, Canaraua Fetii and Pădurea Hagieni [19, 10, 11, 12, 21, 18, 1]. We have not confirmed its presence in some sites considered as important protection areas for plants, *e.g.* Măcin National Park.

Marsilea quadrifolia L. (European threatened species – Habitats Directive and Bern Convention) is a rather rare plant, being reported from Crişana, Banat, Oltenia and Muntenia provinces. This species has become vulnerable owing to extended drainage carried out to yield new land for crops. Chorology: In Dobrogea this plant occurs extremely rarely. It has been reported at "Măcin, Iglița" [25] and in the Danube Delta from Sulina and Gârla Madgearu [6], the latter being under protection. Populations of this species are very poor in number of individuals.

Salvinia natans L. (European threatened species – Habitats Directive and Bern Convention) is an aquatic fern recorded from several localities, especially in the south and west of Romania. It is Endangered by the reclamation of ponds.

Chorology: In the Danube Delta it is rather frequent. Further south it has been reported only in Lacul Mamaia and Lacul Gârliţa [26].

Trapa natans L. (European threatened species – Habitats Directive and Bern Convention) is considered a Vulnerable species; its presence is reported in all Romanian provinces, except Maramureş [23]. Chorology: It is distributed solely from northern Dobrogea at Măcin and Crapina-Jijila, and is known to be rather frequent in some ponds within the Danube Delta.

Aldrovanda vesiculosa L. (European threatened species – Habitats Directive and Bern Convention) is a submerged carnivorous plant, considered to be Critically Endangered. It has been reported from several localities in Crişana, Oltenia and Muntenia. In Dobrogea, the brothers Sintenis collected it the first time, in 1873 at Caraorman [13]. Chorology: It was observed in the Lacul Babadag, in the Danube Delta and in several other localities, with fluctuating appearances: Gârla Porcului near Sulina, Canalul Madgearu, Mila 23, Pardina, Perişor, Heracle, Gârla Împuţita, Canalul Litcov, Carasuhat, Dranov, Obretinul Mare, Lacul Roşu and Balta Somova [6, 14].

Zostera marina L. (European threatened species – Habitats Directive and Bern Convention) is a circumboreal sub-littoral element. This plant is extremely rare and vulnerable because of seawater pollution near the coast, and as a consequence of human disturbance of beaches. Chorology: The plant is present in the Lacul Sinoe, near Cetatea Histria, and between Mamaia and Eforie Sud [27]. It was also reported in the Danube Delta, but reliable confirmation is needed.

Conclusions

Dobrogea represents a floristically interesting region, as it shelters a relatively large number of endangered species, and offers a mosaic of unique habitats, some of which are little-known at European and Global level.

Among the plant species mentioned, some have been found in new localities during our field investigations (*Campanula romanica*, *Centaurea jankae*, *Potentilla emilii-popii*, *Echium russicum*) [18]. In some instances we could not reconfirm the presence of some taxa in Dobrogea, as in the case of *Liparis loeselii*, reported from the Danube Delta [13]. There is a stringent need for effective protection and conservation of these species and their habitats, as they are exposed to growing human impact: intensification of tourism (beach zone), opening of new quarries for exploitation of building materials (limestone, granite), expansion of agricultural land (cleaning by fire around settlements), and extensive cutting of wooded land strips.

All taxa presented in this paper are already included in the IPAs network. Our data will be available to the Ministry for Environment and Water Management – Commission for Nature Monuments, and will constitute the scientific background for requesting the statute of national protection for those sites that are not already included in the National System of Protected Areas.

REFERENCES

1. Arcuş M., 1999 - Flora vasculară şi vegetaţia rezervaţiilor forestiere din sudul Dobrogei: Esechioi, Canaraua Fetii, Dumbrăveni şi Hagieni. Ph. D. thesis, Univ. A.I. Cuza, Iaşi.
2. Bavaru A., Sălăgeanu G., Turcu G., Parincu M., 1996 - Aspecte din flora şi vegetaţia rezervaţiei naturale Dumbrăveni (jud. Constanţa). Acta Bot. Horti Bucurest. /**1994-1995**/: 23-36.
3. Bavaru A., Sălăgeanu G., Turcu G., Parincu M., 1997 - Aspects de la flore et de la végétation de la Reserve naturelle Dumbrăveni, distr. de Constanţa, II. Acta Bot. Horti. Bucurest. /**1997**/: 133-140.
4. Ciocârlan V., Costea M., 1997 - Flora Rezervaţiei botanice Dealul Alah Bair (jud. Constanţa). Acta Bot. Horti Bucurest. /**1995-1996**/: 97-104.
5. Ciocârlan V., Sârbu I., 1999 - Flora Rezervaţiei Biosferei Delta Dunării. Addenda et corrigenda. Bul. Grăd. Bot. Iaşi, **7**: 97-100.
6. Dihoru G., Negrean G., 1976 - Flora of the Danube Delta. Peuce (Bot.), **5**: 217-251.
7. Făgăraş M., 2002 - Contribuţii la cunoaşterea ecologică a florei şi vegetaţiei de la litoralul românesc al Mării Negre. Ph.D. thesis, Univ. "Ovidius", Constanţa.
8. Guşuleac M., 1956 - *Potentilla*. In: T. Săvulescu (ed.), Flora României 4th volume, pp. 596-660. Ed. Academiei Române, Bucureşti.
9. Holmgren P.K., Holmgren N.H., Barnett L.C., 1990 - Index Herbariorum, Part I: The Herbaria of the World. 8th Ed. Regnum Veg., **120**: 1-693.
10. Horeanu C., 1973 - Aspecte floristice de la Cheia (Dobrogea). Ocrot. Nat., **17** (1): 83-88.
11. Horeanu C., 1976 - Propuneri pentru înfiinţarea unor rezervaţii în Podişiul Casimcea. In: Ocrotirea naturii dobrogene, pp. 158-166. Cluj-Napoca.
12. Ionescu-Ţeculescu V., Cristurean I., 1967 - Cercetări floristice în Rezervaţia Naturală Pădurea Hagieni. Ocrot. Nat., **11** (1): 25-36.
13. Kanitz A., 1879-1881 - Plantas Romaniae hucusque cognitas. (Ephemeridi ad "Magyar Növövénytani Lapok" anni iii-v). i-xxiii, 1-268. Claudiopoli: E. Demjén; Londini: Dulau & Co.; Vindobonae: W. Braumüller et fil. [Pp. 1-76: decembre 1879; pp. 77-204: 1880; pp. 205-268 + i-xxiii: 1881].
14. Mihăilescu S., Ştefănuţ S., Popescu A., Sanda V., Biţă C., Jalbă A., 2004 - Chorology of the threatened high plant species from Romania (I). Acta Bot. Horti. Bucurest., **31**: 125-130.
15. Mititelu D., Barabas N., Ştefan N., 1987 - Contribuţii la corologia unor plante rare în Moldova şi Muntenia. Analele Şt. Univ. Iaşi, Biol., **33**: 20-24.
16. Mititelu D., Parincu M., Gheorghiţă C., 1993 - Flora rezervaţiilor forestiere "Canaraua Fetei" şi "Esechioi" (din Dobrogea de sud - vest). Stud. Cercet., Biol.-Muzeol., Muzeul Piatra Neamţ, **7**: 31-36.
17. Morariu I., 1964 - *Campanula*, Secţia *Linophylloides*. In: T. Săvulescu (ed.), Flora României 9th volume, pp. 86-111. Ed. Academiei Române, Bucureşti.
18. Negrean G., Anastasiu P., 2003 - Rare plants concentrations in the SE of Dobrogea. Analele Şt. Univ. Iaşi, Secţ. II-a. Biol. Veget., **48**: 85-94.
19. Nyárády A., 1953 - *Paeonia*. In: T. Săvulescu (ed.), Flora României 2nd volume, pp. 400-411. Ed. Academiei Române, Bucureşti.

20. Nyárády E.I., 1955 - *Alyssum*. In: T. Săvulescu (ed.), Flora României 3rd volume, pp. 318-355. Ed. Academiei Române, Bucureşti.
21. Parincu M., Mititelu D., Aniţei L., 1998 - Flora vasculară din rezervaţia botanică Pădurea Dumbrăveni (jud. Constanţa). Bul. Grăd. Bot. Iaşi., **6** (2): 353-358.
22. Prodan I., 1939 - Conspectul Florei Dobrogei – 3rd part. (Reprint from Bul. Acad. Înalte Stud. Agron. Cluj. 1938, **7**: 1-96). Tipografia Naţională S.A., Cluj.
23. Săvulescu T. (ed.), 1952-1976 - Flora României, Volumes **1-13**. Ed. Academiei Române. Bucureşti.
24. Sârbu A. (ed.), 2003 - Ghid de identificare a importantelor arii de protecţie a plantelor din România. Ed. Alo, Bucureşti!, Bucureşti.
25. Ţopa E., 1952 - *Marsilea* L. /" *Marsilia* "/. In: T. Săvulescu (ed.), Flora României 1st volume, pp. 150-151. Ed. Academiei Române, Bucureşti.
26. Ţopa E., 1952 - *Salvinia*. In: T. Săvulescu (ed.), Flora României 1st volume, pp. 152-153. Ed. Academiei Române, Bucureşti.
27. Ţopa E., 1966 - *Zostera*. In: T. Săvulescu (ed.), Flora României 11th volume, pp. 87-88. Ed. Academiei Române, Bucureşti.
28. Tutin T.G., Burges N.A., Chater A.O., Edmonson J.R., Heywood V.H., Moore D.M., Valentine D.H., Walters S.M., Webb D.A. (eds), 1993 - Flora Europaea, Volume **1**, 2nd edition. *Psilotaceae* to *Platanaceae*. Cambridge University Press, Cambridge.
29. Tutin T.G., Burges N.A., Heywood V.H., Moore D.M., Valentine D.H., Walters S.M., Webb D.A. (eds), 1964-1980 - Flora Europaea. Volumes **1-5**. Cambridge University Press, Cambridge.
30. Zahariadi C., 1962 - Caracteres morphologiques, anatomiques et biologiques dans la taxonomie du genre *Ornithogalum*. Rev. Biol. (Bucureşti), **7** (1): 5-41.
31. Zahariadi C., 1966 - *Ornithogalum*. In: T. Săvulescu (ed.), Flora României 6th volume, pp. 317-349, 850-851. Ed. Academiei Române, Bucureşti.

BIOGEOGRAPHICAL CHARACTERIZATION OF THE BRYOLOGICAL FLORA OF THE "MONTAGNA DI TORRICCHIO" NATURE RESERVE (CENTRAL ITALY)

Michele ALEFFI, Carmela CORTINI PEDROTTI, Roberta TACCHI
Department of Botany and Ecology, Camerino University, Via Pontoni 5
62032 Camerino MC (ITALY); e-mail: michele.aleffi@unicam.it

Abstract: The present work examines the bryological flora of the "Montagna di Torricchio" Nature Reserve in order to increase knowledge about the biodiversity of the entire area. Specimen collection was conducted over the years 1975–2003 by various authors. The survey revealed 181 bryophyte taxa of which 19 were liverworts and 162 were mosses. In addition, further analysis assessed which families occur in the same proposed categories, and to what degree, in order to obtain more information on their distribution and ecology. The distribution of species frequency by chorological category – cosmopolitan, circumpolar and Eurasiatic – shows that the families richest in species are the most represented in the Torricchio Reserve. The general rate of the percentages of the chorological elements in the reserve's bryological flora confirms that the temperate element dominates, closely followed by the boreal and oceanic-Mediterranean elements. The factor that promotes the occurrence of the other elements, resulting in a generally higher level of diversity, is the presence of niches with special microclimatic conditions that differ from the surrounding context.

Introduction

The "Montagna di Torricchio" Nature Reserve was established in 1970 after its donation to the University of Camerino by the Marquis Mario Incisa della Rocchetta.

Located on the south-central ridge of the Apennines of the Marches Region, in the Tazza Valley, this reserve covers about 317 ha, at an altitude of 820-1491 m [10]. It is delimited by the slopes of Mts. Fema (1575 m), Cetrognola (1491 m) and Torricchio (1444 m) (Fig. 1).

In terms of its geology, the territory is formed by calcareous, calcareous-marly, and marly formations from the Lower Jurassic to Quaternary [1, 9].

While the area has no watercourses, not even temporary ones, it does hold three springs, the "Fonte della Romita" at 1123 m, half-way up the slopes of Mt. Cetrognola, the "Fonte di Carafiume" on the other side of the valley, on the slopes of Mt. Fema (both springs are intercepted into aqueducts for human use), and the "Fontanelle," which flows in the valley bottom, along the mule track leading to "Casale Piscini".

Finally, the area contains a series of special environments such as rock walls, detritus colluvia, nitrophilous areas, ruins, springs, rise pits, etc.

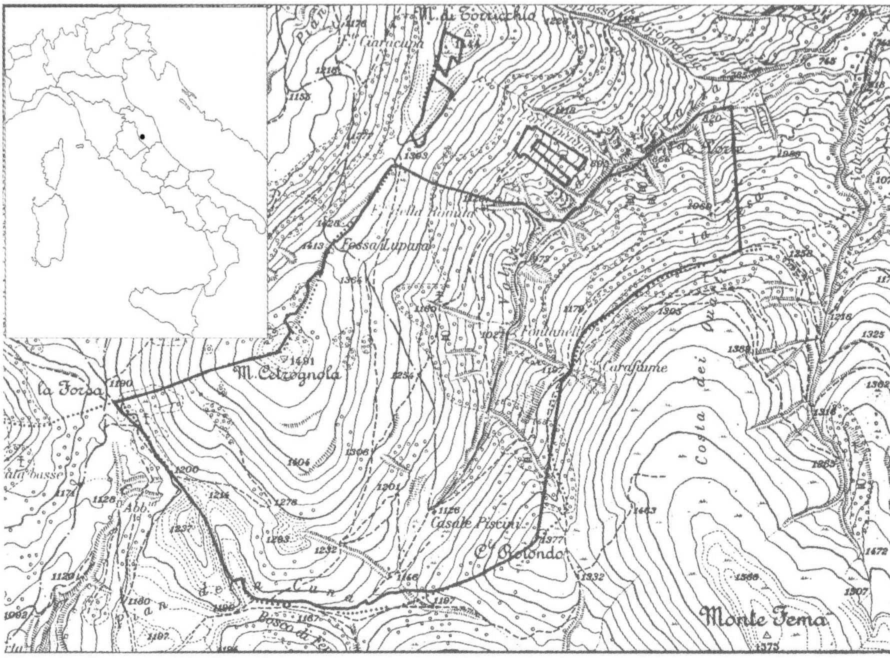

Fig. 1: Geographic location and topographic map of the "Montagna di Torricchio" Natural Reserve.

The reserve territory comprises two vegetation series [15], *Scutellario-Ostryeto carpinifoliae* sigmetum in the hilly belt, and *Polysticho-Fageto* sigmetum in the montane belt.

The hilly belt series occupies the base of the Tazza Valley along the section between the zones of "Le Porte" and "Le Fontanelle", rising up to fairly high elevations on the south-eastern slopes of Mt. Cetrognola. On the more precipitous or altered slopes, *Scutellario-Ostryeto carpinifoliae* is open, with more isolated arboreal species, and the presence of Downy Oak (*Quercus pubescens*).

The montane belt series, extending from 800–900 m to an upper altitude of 1491 m, the reserve's highest point, almost totally covers the north-facing slopes of the Tazza Valley, while it becomes more fragmentary in other zones subjected to greater deforestation in the past to benefit grazing. These are mostly coppiced; many areas are cut, foresters choosing and sparing only younger shoot-bearing trees, while other areas have become high forest.

According to the climatic classification of Rivas-Martinez [11], the Torricchio Mountain district is located in the temperate oceanic macro-bioclimate, eu-oceanic subtype, and supra-temperate thermotype [14].

Results

Specimen collection was conducted over the years 1975–2003 by various authors. Samples were collected throughout the entire area of the reserve, giving particular consideration to the different ecological spheres induced by the altitudinal gradient, the geomorphological structure of the valley in which the area is located, and past anthropogenic activity.

The survey revealed 181 bryophyte taxa, 19 of which were liverworts and 162 mosses.

Fig. 2 shows the chorological spectrum, summarizing the distribution of the bryophytes found in the reserve by their chorological category according to Dierßen [5].

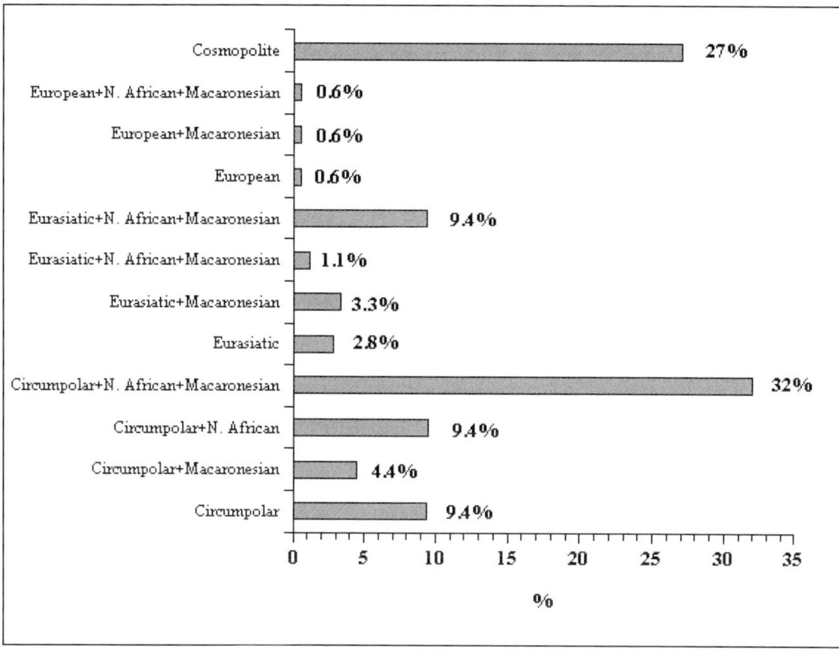

Fig. 2: The chorological spectrum summarizing the distribution of the bryophytes found in the reserve by their chorological category, according to Dierßen [5].

The extent to which various families occur in each chorological category was assessed in order to obtain greater information on distribution and ecology.

The cosmopolitan species, for example *Bryum argenteum, Ceratodon purpureus* and *Hypnum cupressiforme*, occur in a proportion of 27%. The bryophyte genera that are particularly rich in species tend to have a wide distribution within the same climatic zone and also in different climates [3, 12].

In the particular case of the Torricchio Reserve, the family most represented in the cosmopolitan category is *Pottiaceae* with 15 species (31.25%), followed by *Bryaceae* with 12 species (25%), *Dicranaceae* and *Amblystegiaceae*, both with three species (6.25%), *Hypnaceae* and *Grimmiaceae* with two species (4.18%), followed by other families, all with one species (2.08%).

The other species found in the reserve belong to the Holarctic species groups. In this study, the Holarctic region was divided into two components, according to Dierßen [5]: a latitudinal component defined as circumpolar, with some extensions as far as the Mediterranean and Macaronesian regions, and another component defined as Euroasiatic, which extends along a gradient of continentality.

The family most represented in the circumpolar category is *Pottiaceae* with 18 species (18%) (for example, *Didymodon acutus* and *Syntrichia intermedia*), followed by *Brachytheciaceae* with 14 species (14%) (for example, *Brachythecium rutabulum, Eurhynchium pulchellum*), *Orthotrichaceae* with 11 species (11%) (*Orthotrichum affine, Zygodon rupestris*) and *Bryaceae* with 7 species (7%) (*Bryum elegans, Bryum subelegans*). Worth noting is the increased number of families belonging to this category, compared to the cosmopolitan group, and in particular, the increased number of liverwort families.

The category of the Eurasiatics (18.4%) includes those species that extend eastward as far as the continental areas of central Asia. In this category, *Brachytheciaceae* with 10 species (31.25%) (e.g. *Eurhynchium pumilum, Eurhynchium schleicheri, Cirriphyllum tommasinii*) predominated over all the other families that contributed to the species list with one or at most two species.

The distribution of species frequency by chorological category – cosmopolitan, circumpolar and Eurasiatic – suggests that the richest families are the most common in the Torricchio Reserve.

Finally, the chorological elements of Düll [6, 7, 8], which were later united in categories according to Sérgio *et al.* [13], were applied to the previously identified bryological flora categories.

The cosmopolitan component comprises 48 species, representing 27% of the reserve's bryological flora.

Analysis of the chorological spectrum in Fig. 3 shows that about 60% of the reserve's cosmopolitan species belong to the temperate element (temp-mont), followed by the submediterranean species (submed-mont) with 12.5%, then boreal species (bor-mont, sub-bor) with 10.42%, and oceanic-Mediterranean species (suboc-med, oc-submed) with 8.32%.

The circumpolar component encompasses 100 species, equivalent to 55.5% of the reserve's bryological flora.

The analysis of the chorological spectrum in Figure 4 reveals that within the circumpolar component, the temperate species (temp-mont) are the most represented (40%). The boreal element ranks second (subbor, subbor-mont, bor-mont, bor-mont/dealp) (33%), followed by oceanic-Mediterranean elements (suboc-submed, oc-submed, suboc-med) and suboceanic (suboc-mont) with the same percentages (7%). In addition, the subarctic-subalpine element is represented with 1%.

The temperate element, while remaining the best represented, shows a reduction compared to the cosmopolitan group, whereas the "boreal" element, in the broadest sense of the term, increases, considering as well the occurrence of a subarctic-subalpine component.

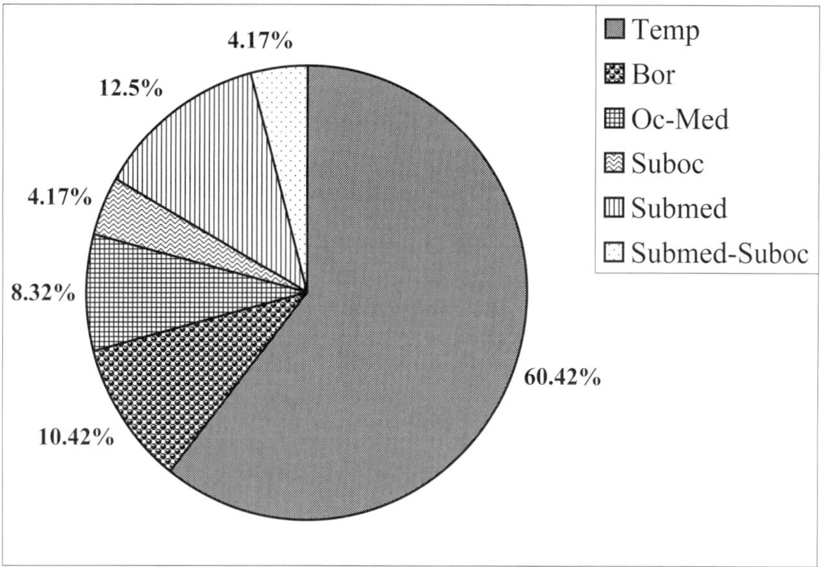

Fig. 3: Chorological spectrum of the cosmopolite species.

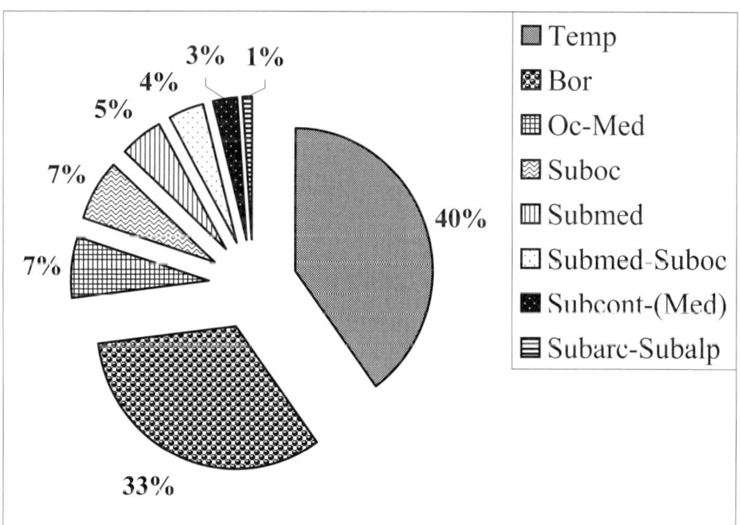

Fig. 4: Chorological spectrum of the circumpolar component.

The Eurasiatic component is composed of 33 species, equivalent to 18.23% of the reserve's bryophyte flora. The chorological spectrum in Figure 5 shows that the temperate component still predominates over the other elements (30.3%), followed by the oceanic-Mediterranean (oc-submed, suboc-submed), and suboceanic elements (oc, suboc-mont) in the same proportion (18.18%), and by the submediterranean element (submed-mont) (12.13%). It is worth noting the increase of the subcontinental species (subcont-mont, subcont-med-mont/dealp) within the Eurasiatic component.

The observations made for the temperate species apply as well to this category; in addition, there is a notable decrease in the boreal component, to the detriment of the other elements.

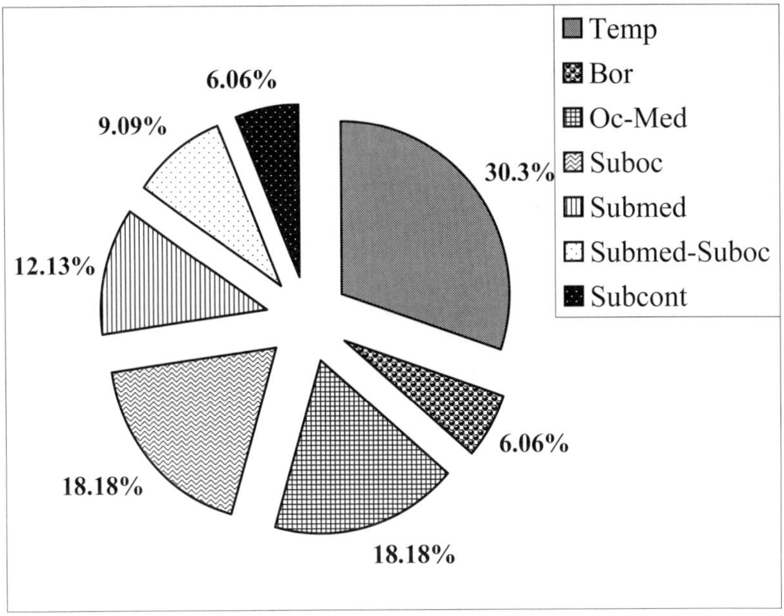

Fig. 5: Chorological spectrum of the Eurasiatic component.

Conclusions

The distribution of the chorological elements of the bryological flora of the Torricchio Reserve confirms observations by Cortini Pedrotti [2] and Aleffi & Cortini Pedrotti [1] for the Marches Region, inasmuch as the temperate element dominates over the others, closely followed by the boreal and oceanic-Mediterranean elements. The factor that promotes the occurrence of the other elements, resulting in generally higher biodiversity, is the presence of niches with special microclimatic conditions that differ from the surrounding context.

Throughout the territory of the reserve, the species belonging to the circumpolar group are found more frequently in cool environments like the Tazza Valley ravine, under the forest canopy, or in proximity to environments characterized by continuous water flow, such as flushes (exsurgences) and springs, and on all types of substrate, but mostly on tree-trunks.

The species of the Eurasiatic group also show a greater preference for the reserve's cool and moist habitats, such as shady zones near the little shrine of the Tazza Valley and under-storeys of Beech (*Fagus sylvatica*) or Hop-hornbeam (*Ostrya carpinifolia*) woods, mostly on rocky substrates.

In the cosmopolitan category, instead, one does not observe a tendency to occupy prevalently one habitat over another, but the species are distributed more or less homogeneously in semi-arid habitats present along the slopes of Mount Cetrognola, in shaded environments in the woods, in areas affected by water dripping, on walls, in grazing lands with bushes, on rocky substrata and on earth.

In conclusion, the reserve's geographical position on the western spurs of the Apennines of Umbria and the Marches (the Sibillini Mountains) induces a montane characteristic into Torricchio's bryological flora, although the defining features are not particularly accentuated. In addition, the presence of environments of differing ecological conditions brings a notable contribution to the area's bryological biodiversity.

REFERENCES

1. Aleffi M., Cortini Pedrotti C., 2001 - Considerazioni biogeografiche sulla flora briologica italiana. Braun-Blanquetia, **31**: 7-13.
2. Cortini Pedrotti C., 1996 - Aperçu sur la bryogéographie de l'Italie. Bocconea, **5**: 301-318.
3. Cortini Pedrotti C., 2001 - Flora dei muschi d'Italia. Ed. Antonio Delfino, Roma.
4. Deiana G., Pieruccini U., 1976 - Geologia e geomorfologia della Montagna di Torricchio. La Riserva Naturale di Torricchio, **1**: 27-76.
5. Dierßen K, 2001 - Distribution, ecological amplitude and phytosociological characterization of European bryophytes. Bryophytorum Bibliotheca, **56**: 1-289.
6. Düll R., 1983 - Distribution of European and Macaronesian Liverworts (Hepaticophytina). Bryol. Beitr., **2**: 1-115.
7. Düll R., 1984 - Distribution of European and Macaronesian Mosses (Bryophytina). Part. I. Bryol. Beitr., **4**: 1-113.
8. Düll R., 1985 - Distribution of European and Macaronesian Mosses (Bryophytina). Part. II. Bryol. Beitr., **4**: 110-232.
9. Kwiatkowski W., Venanzoni R., 1994 - Carta dei suoli della Riserva naturale di Torricchio. La Riserva Naturale di Torricchio, **9**: 15-21.
10. Pedrotti F., 1976 - La Riserva naturale di Torricchio. La Riserva Naturale di Torricchio, **1**: 5-20.
11. Rivas-Martínez S., Sanchez-Mata D., Costa M., 1999 - North American boreal and western temperate forest vegetation. Itinera Geobotanica, **12**: 5-316.

12. Schofield W.B., 1992 - Bryophyte distribution patterns. In: Bates J.W., Farmer A.M. (eds.), Bryophytes and lichens in a changin environment, pp. 103-130. Clarendon Press, Oxford.
13. Sérgio C., Casas C., Brugués M., Cros R.M., 1994 - Lista Vermelha dos Briófitos da Península Ibérica. Museu, Laboratório e Jardim Botânico da Universidade de Lisboa (MLJB), Instituto da Conservação da Natureza (ICN), Lisboa.
14. Venanzoni R., 2003 - Prime valutazioni dei dati climatici della Riserva naturale di Torricchio. La Riserva Naturale di Torricchio, **11**: 437-444.
15. Venanzoni R., Kwiatkowski W., 1994 - Carta delle serie di vegetazione della Riserva Naturale di Torricchio (Appennino centrale). La Riserva Naturale di Torricchio, **9**: 23-30.

FIVE NEW SPECIES FOR THE ROMANIAN LICHEN FLORA

Florin CRIŞAN
"Babeş-Bolyai" University, Department of Taxonomy and Ecology, 42 Republicii Str.,
400015 Cluj-Napoca (ROMANIA); e-mail: florincrisan1964@yahoo.com

Abstract: Between 1993 and 1995 I identified in the Pădurea Craiului Mountains five species new for the Romanian lichen flora. Of these species, four are saxicolous and occur mainly on limestone: *Collema subflaccidum* Degel., *Leptogium teretiusculum* (Wallr.) Arnold, *Phaeophyscia chloantha* (Ach.) Moberg and *P. hirsuta* (Mereschk.) Moberg. The other species – *Parmelina pastilifera* – is corticolous and was found on the nutrient-rich bark of old trees (*Fraxinus* sp. and *Acer pseudoplatanus*). The specimens are lodged in the Herbarium of the "Alexandru Borza" Botanical Garden, Cluj-Napoca. For each species a description, synonymy, general distribution, ecology and chorology in the Pădurea Craiului Mountains is given.

Introduction

The Pădurea Craiului Mountains are situated in the western part of Romania, occupying an area between the Iadului Valley (east), the Beiuşului Depression (south) and the Vadului Depression (north).

The geological substrate is composed of impermeable (lime-free) rocks and limestone. The basic feature of the edaphic cover of the studied area is the predominance of clay-illuvial soils and the occurrence of rendzina, terra rossa, etc. as intrazonal soils. The Pădurea Craiului Mountains have a foothill and montane climate.

The greatest part of the Pădurea Craiului Mountains is covered with deciduous forests – pure stands of beech (*Fagus sylvatica*), mixed woods of hornbeam (*Carpinus betulus*) and *Fagus sylvatica*, and stands of sessile oak (*Quercus petraea*) – which account for 57% of the area [10].

Material and Methods

During field research undertaken in 1993–1995 in the Pădurea Craiului Mountains, foliose and fruticose lichen specimens were collected. For a more complete analysis of the lichen flora I have used chorological data [2, 3, 4, 5, 6, 7, 16], as well as specimens from the Herbarium of Liége University, the Herbarium of the Botany Institute of Uppsala University and the Herbarium of the "Alexandru Borza" Botanical Garden in Cluj-Napoca.

The specimens were verified by Dr. Roland MOBERG of Uppsala University (Sweden) and Dr. Emmanuel SÉRUSIAUX of Liège University (Belgium), and are lodged in the Herbarium of the "Alexandru Borza" Botanical Garden. For species identification and characterization I have followed several works [1, 8, 9, 11, 12, 13, 14, 15, 17, 18, 19, 20, 21].

For each species a description, synonyms, ecology, world distribution and chorology in the Pădurea Craiului Mountains (Fig. 1) is given.

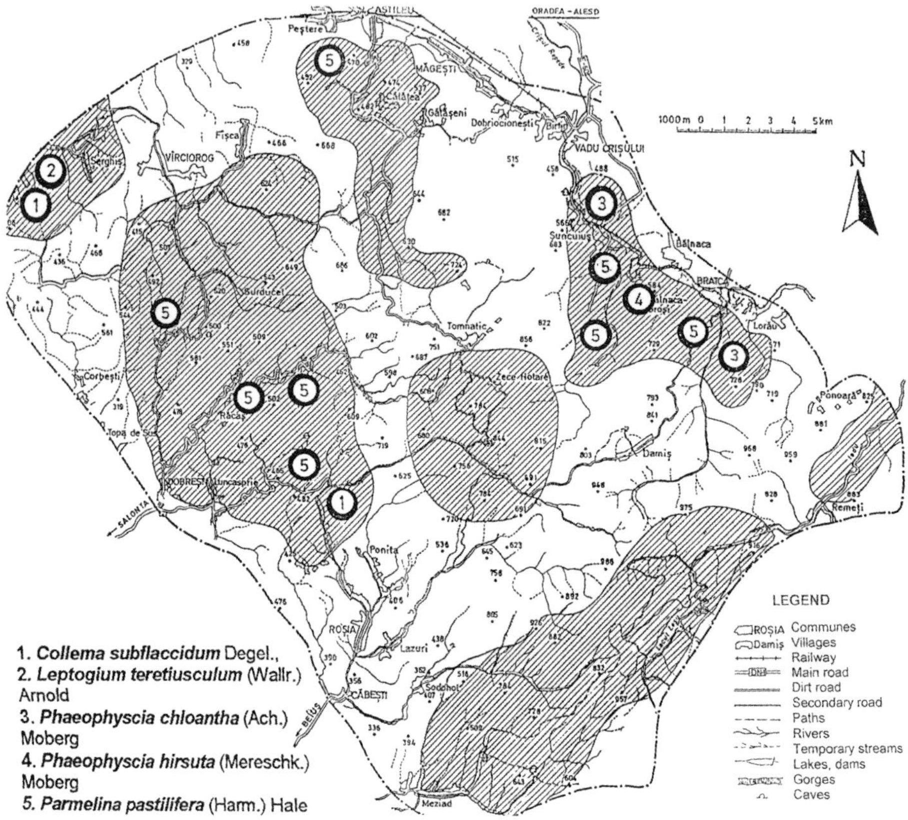

Fig. 1: Distribution of the five new species for the Romanian lichen flora in the Pădurea Craiului Mountains. The shaded patches represent the main areas of interest for lichenological studies.

Results

The five lichen species presented below have not been reported previously from Romania.

1. ***Collema subflaccidum*** Degel. - *C. subfurvum* Degel., Bot. Not., 1948, p. - *Synechoblastus flaccidus* v. *subfurvus* Muller Arg., Proceed. Roy. Soc. Edinburgh, vol. XI, 1882, p. 457 - *Collema rupestre* var. *subfurvum* Zahlbruckner, Cat. Lich. Univ., vol. III, 1925, p. 61.

Thallus foliose, 3–6 cm across, ± deeply and very irregularly lobate, characterized by the numerous small, densely distributed, laminal isidia, which are mostly globular or cylindrical, very rarely becoming flattened or squamule-like, giving a scruffy appearance to the upper surface. On tree bark, especially old ash (*Fraxinus*) in relatively moist, shady places, more rarely on rocks.

Ecology: moderate ombrophilous, hygrophilous, micro-thermal, sub-neutrophilous.

World distribution: rather frequent in Europe (from Norway to Portugal, also in the Mediterranean region; it has not been reported from the central and eastern part of the continent), North America, Africa, Asia, Hawaii.

Chorology in Pădurea Craiului Mountains: Cheile Albioarei (close to Ţarina) on rocks, N-NE-facing, near the road, at 400 m altitude; Valea Pietroasa (close to Şerghiş), in oak-hornbeam (*Quercus cerris*-*Carpinus betulus*) forest, on moist rocks, N-facing, at 350 m altitude.

2. *Leptogium teretiusculum* (Wallr.) Arnold, Zur Lich.-Flora Munchen in Bericht. Bayr. Botan. Gesellsch., vol. II, Anhang, 1892, p.26 - *Collema teretiusculum* Flk. apud. Wallr., Flora Cryptog. German., vol. III, 1831, p.551 - *Parmelia teretiuscula* Wallr., Flora Cryptog. German., vol. III, 1831, p. 551 - *L. microscopicum* Nyl. in Act. Soc. Linn. Bordeaux, vol. XXI, 1856, p.272

Thallus brown, at first with small, radiating, narrow, ± branched, flat, adpressed lobes that at their margins and apices develop long, cylindrical, often coralloid isidia-like extensions, which become crowded to form dense clusters or cushions; at maturity, the primary squamules are mostly obscured except sometimes at the edge of the cushions. Apothecia very rare. On rough, basic bark of old trees, especially sycamore (*Acer*), ash (*Fraxinus*) and elm (*Ulmus*), more rarely on basic rocks, old walls and coastal turf.

Ecology: hygrophilous, eurythermic, sub-neutrophilous, photo-ombrophilous.

World distribution: a frequent but often overlooked species in Europe and North America.

Chorology in Pădurea Craiului Mountains: Valea Pietroasa (near Şerghiş), N-facing, on moist rocks close to the bifurcation toward Bucuroaia, at 360 m altitude.

3. *Phaeophyscia chloantha* (Ach.) Moberg, Bot. Notiser Stockholm, nr. 131, 1978, p.259-262 - *Parmelia chloantha* Ach., Synops. meth. lich., 1814, p. 217 - *Physcia luganensis* Mereschk., Ann. Conserv. Jard. Bot. Genève, vol XXI, 1919, p.145-216 - *P. labrata* var. *intermedia* Mereschk. Ann. Conserv. Jard. Bot. Genève, vol XXI, 1919, p.187 - *P. pragensis* Nádv., Stud. Bot. Cechoslov., vol. VIII, p. 69-124.

Thallus orbicular to irregular, usually confluent with other thalli, easy distinguished from other *Phaeophyscia* species by its distinct, lip-shaped soralia, pale underside with only a few pale rhizinae and lower cortex prosoplectenchymatous. It has no close relatives in Europe and no other species has a prosoplectenchymatous lower cortex. A corticolous species which occasionally grows on rocks.

Ecology: xero-mesophilous, eurithermic, moderate acidophilous-subneutrophilous, photophilous.

World distribution: the species is known from the southern part of Europe and from Central North America.

Chorology in Pădurea Craiului Mountains: Valea Brătcuța (close to Bratca) on rocks, SE-facing slope, lat./long.: 46°53' N/22°35' E, 550 m altitude; on the left bank of Crişul Repede river (to the issue from Şuncuiuş toward Oradea), S-facing slope, above the railway, on semi-shaded rocks, lat./long.: 46°57' N/22°32' E, at 300 m altitude.

4. ***Phaeophyscia hirsuta*** (Mereschk.) Moberg. comb. nov., Bot. Notiser Stockholm, nr. 131, 1978, p.259-262 - *P. hirsuta* Mereschk. Ann. Conserv. Jard. Bot. Genève, vol XXI, 1919, p.181 - *P. labrata* Mereschk., Ann. Conserv. Jard. Bot. Genève, vol XXI, 1919, p.183 - *P. labrata var. olivacea* Mereschk. Ann. Conserv. Jard. Bot. Genève, vol XXI, 1919, p.184.

With a thallus orbicular, usually confluent with other thalli, ± loosely adpressed, grey-brown to dark-brown, *P. hirsuta* is separated from other European *Phaeophyscia* species by the hyaline hairs on the upper side and the short ascending lobes with lip-shaped soralia. Underside black almost to the very tips, with abundant black rhizinae. Apothecia and pycnidia rare. Predominantly corticolous, often found growing mixed with *P. chloantha.*

Ecology: xero-mesophilous, mesothermal-moderate thermophilous, subneutrophilous, photophilous.

World distribution: the species is known from southern Europe and central North America.

Chorology in Pădurea Craiului Mountains: Cheile Mişidului (close to Şuncuiuş) on SE-facing slope, on limestone, lat./long.: 46°56' N/22°32' E, at 300 m altitude.

5. ***Parmelina pastilifera*** (Harm.) Hale, Smithsonian Contrib. Bot., vol. XXXIII, 1976, p. 1-60 - *Parmelia pastillifera* (Harm.) R. Schub. & Klem. - *P. tiliacea v. pastillifera* (Harm.) Grumann - *P. scortea f. borealis* Lynge, Stud. Lich. Flor. Norway, 1921, p. 167 - *P. scortea v. pasttilifera* Harm., Lich. de France, fasc. IV, (1909) 1910, p. 558.

Thallus, 4–8(20) cm across, rather unevenly attached, often wavy towards the center with closely adpressed or ascending margins, the upper surface pale bluish-gray, ± shiny, often entirely closely adpressed. Lower surface black, brown towards lobe margins; rhizines to the lobe edges, simple or forked. Apothecia rare. Distinguished by the distinctive, scattered, blue-black, knob-like isidia and the C+ red medulla. On free-standing trees, on nutrient-rich bark, commonly ash (*Fraxinus* sp.) and sycamore (*Acer pseudoplatanus*), rarely on siliceous rocks, in well-lit situations.

Ecology: meso-hygrophilous, micro-mesothermal, moderate acidophilous, photo-ombrophilous.

World distribution: British Isles, western Europe to Norway, southern and central Europe.

Chorology in Pădurea Craiului Mountains: on bark of oak (*Quercus petraea*) - Valea Mişidului (close to Şuncuiuş) at 420 m altitude, Sclavu Pleş (close to Luncasprie) at 400 m altitude, Dâmbul Hodişanului (near Răcaş) at 380 m altitude, Valea Vida (near Răcaş) 400 m altitude, Valea Brătcuţa (close to Bratca) at 420 m altitude, Dealul Vasului (between Vârciorog and Corbeşti) 400 m altitude; on bark of Turkey oak (*Quercus cerris*) - Dealul Şerbota (close to Aştileu) at 360 m altitude; on bark of *Acer pseudoplatanus* - Valea Vida (near Cantonul Vida) 400 m altitude, Cheile Mişidului (close to Şuncuiuş) at 400 m altitude.

REFERENCES

1. Bartók K., Crişan F., Coroi A.M., 1999 - Genul *Collema* în România. Studii şi cercetări, Biologie, Muzeul Bistriţa-Năsăud, **5**: 77-98.
2. Ciurchea M., 2004 - Determinatorul lichenilor din România, Ed. BIT, Iaşi.
3. Codoreanu V., 1966 - Flora şi vegetaţia rezervaţiei naturale "Defileul Crişului Repede". Flora lichenologică. Contr. Bot., Cluj, **I**: 83-110; 164-172.
4. Cretzoiu P., 1941 - Flora lichenilor folioşi şi fruticuloşi epidendri şi epixili din România, An. I.C.E.F., vol. II, Bucureşti.
5. Crişan F., 1998 - Lichenoflora din Cheile Mişidului. An. Univ. Oradea, fasc. Biol., **IV**: 142-162.
6. Crişan F., 1998 - Vegetaţia lichenologică din Cheile Mişidului. An. Univ. Oradea, fasc. Biol., **V**: 213-232.
7. Crişan F., 2001 - Studii corologice, ecologice şi cenologice asupra lichenilor foliacei şi fruticuloşi din Munţii Pădurea Craiului (Jud. Bihor). Teză de doctrorat, Univ. "Babeş-Bolyai", Cluj-Napoca.
8. Dobson F.S., Hawskworth D.L., 1976 - *Parmelia pastillifera* (Harm) Schub. & Klem. and *P. tiliacea* (Hoffm.) Ach. in the British isles. Lichenologist, **8**: 47-59.
9. Esslinger T.L., 1978 - Studies on the lichen family Physciaceae, II. The genus *Phaeophyscia* in North America. Mycotaxon, **23**: 209-212.
10. Groza G., 1999 - Vegetaţia Munţilor Pădurea Craiului. Studiu fitocenologic, ecologic şi bioeconomic, Teză de doctorat, Univ. Babeş-Bolyai Cluj-Napoca.
11. Hale M.E., 1976 - A monograph of the lichen genus *Parmelina* (Parmeliaceae), Smithsonian Contr. Bot., **33**: 1-60.
12. Jørgensen P.M., 1994 - Further notes on European taxa of the lichen genus *Leptogium*, with emphasis on the small species Lichenologist, **26**: 1-29.
13. Jørgensen P.M., James P.W., 1983 - Studies on some *Leptogium* species of Western Europe. Lichenologist, **15**: 109- 125.
14. Moberg R., 1978 - Overlooked names and new combinations in *Phaeophyscia* (Lichenes). Bot. Not., **131**: 259-262.
15. Moberg R., Holmåsen I., 1992 - Flechtenflora von Nord- und Mitteleuropa, Ein Bestimmungsbuch, Gustav Fischer Verlag, Stuttgart.

16. Moruzi C., Petria E., Mantu E., 1967 - Catalogul lichenilor din România. Acta Bot. Horti Bucurestiensis, Bucureşti.
17. Purvis Q.W., Coppins B.J., Hawksworth D.L., James P.W., Moore D.M., 1992 - The Lichen Flora of Great Britain and Ireland. Natural History Museum Publications, London.
18. Santesson R., 1993 - The Lichen and Lichenicolous Fungi of Sweden and Norway. Lund.
19. Sérusiaux E., 1977 - Les lichens foliicoles: concept, classification écologique et position systématique. Naturaliste Belges, **58**: 111-118.
20. Van Halluwyn C., Lerond M., 1993 - Guide des lichens, Ed. Lechevalier, Paris.
21. Wirth V., 1995 - Die Flechten Baden-Wurttembergs, Teil 1, 2, Verlag Eugen Ulmer, Stuttgart.

THE ECOLOGICAL ROLE OF DEADWOOD IN NATURAL FORESTS

Stelian RADU
Str. Aurel Vlaicu, nr. 11, **330005 Deva (ROMANIA)**

Abstract: Veteran and dead trees at different stages of decay have an important ecological role to play in conserving forest biodiversity and are recognized as Pan-European indicators of a sustainable management.

Dying and dead trees, either standing or fallen and at different stages of decay, are valuable habitats (providing food, shelter and breeding conditions, etc.) for a large number of rare and threatened species: saproxylic insects, invertebrates, lichens, bryophytes, birds and mammals. Hollow trees, in particular, provide suitable microhabitats for mammals (e.g. rodent species, squirrels, martens, bat species and wild cats) and cavity-nesting birds (e.g. woodpecker, tit and owl species).

In natural forests in Romania, deadwood is associated with several relict, rare and protected animal species (e.g. *Rhysodes sulcatus, R. americanus, Cerambyx cerdo, Lucanus cervus, Rosalia alpina, Camponotus herculeanus*). Many lichens (*Parmelia, Peltigera* and *Lobaria* species) indicate the continuity of natural forests and the presence of more than 20 species of fungi contribute to the ever-present processes of decay. Large quantities of deadwood (50-130 m^3/ha) also play a significant role in forest nutrient cycles, carbon budgets, soil morphology and natural regeneration. Natural forests in Romania provide exceptional conditions for research into the ecological role of deadwood habitats.

Introduction

From a nature conservation perspective, trees increase in ecological value with age and are particularly important during maturity, senescence and periods of decay. Old trees, dead/dying trees and fallen branches from storm damage are a priority habitat in terms of maintaining biodiversity because they provide food and shelter for mammals (particularly bats), birds, and less visible forest-dwelling biota such as invertebrates (especially beetles), fungi and lichens. Deadwood is a constant and critical subsystem component in natural forests, playing a key role in sustaining forest productivity and environmental functions such as carbon storage.

For many years, in managed forests, foresters considered deadwood to be a threat – a source of infection for healthy trees. For this reason, damaged and dying trees were extracted on a periodic basis in a management practice known as 'tending the stands'.

However, over the past 80 years perceptions of deadwood have dramatically changed. North American ecologists were amongst the first to recognize and highlight the importance of the ecological role and biodiversity value of deadwood in the forest ecosystem. An impressive resource of literature has now accumulated on the ecological role of deadwoods in forests (e.g. symposia held in Chambéry, France in 2004, Mantova in Italy in 2003 and in Reno-Nevada, USA, in 1999; studies in the UK [4, 10], Germany [1, 2, 17],

Poland [6] and Scandinavia [9]). A useful reference resource is the WWF booklet 'Deadwood – living forests', published in 2004. Unexploited natural forests still contain a significant amount of deadwood, in varying stages of decomposition, providing a rich resource for scientists researching the ecological role and biodiversity value of this important habitat.

Romania's natural forests – nature's sanctuaries and treasures of biodiversity

Romania still has approximately 300 000 ha of primeval forests, remnants of the large former virgin forests. These remnants have escaped cutting and other human influences because of their location in 'inaccessible hydrographic basins' – steep terrain in which there is an absence of roads.

The publication of two monographs, 'Banater Urwälder' [18] and 'Les forêts vierges de Roumanie' [8] has drawn the attention of the wider European research community to Romania's natural forests. At national level, the Institute of Forest Research and Management (ICAS) has undertaken an inventory and ecological valuation for the purpose of conserving these valuable habitats as part of a PINMATRA Project. This project has confirmed the exceptional heritage and importance of natural forest in Romania.

Natural forests, characterized by outstanding diversity in the composition of their structure and importance in the landscape, shelter a very rich fauna, notably the large carnivores – bears, wolves and lynx – that have disappeared elsewhere in Europe. However, the tree species present have high values of wood productivity and quality, and are therefore suitable for logging. Natural forests will inevitably be lost unless they are conserved within strictly protected areas [7, 8, 15].

The full range of development stages can be found in these forests but in stand senescence the amounts of deadwood (standing or fallen) are at their highest. Even after their death, veteran trees function for many years in the ecosystem, passing through the phases of decay and providing ecological niches for many organisms (fungi, invertebrates and vertebrates).

Romanian foresters have now begun to pay attention to the ecological function of dead wood as part of the sustainable management of Romanian forests. In this respect, Giurgiu [7] has recommended that a minimum number of old, hollow and dead trees should be kept in every stand. In addition, Radu [13] has published a review of the biological importance of deadwood, which is in the process of being officially recognized as a Pan-European indicator of sustainable forest management [15].

The ecological importance of deadwood in Romanian forests: some preliminary results

Research on deadwood is in a preliminary phase, and in order to be successful in the future, must be continued in a systematic and multidisciplinary way using standardized research methodologies. The following points summarize the findings of the inventory, ecological evaluation and mapping of Carpathian virgin forests undertaken as part of the PINMATRA project:

- Deadwood volume varies widely according to the tree species, site fertility and stand development phase. In five beech stands, volumes of deadwood range between 49 and 128 m^3/ha. In the Izvoarele Nerei Forest Reserve (5028 ha), one of the largest beech forests in Europe (where trees can exceed 130-160 cm in diameter, 55-58 m in height and an age of 300-350) the living tree volume is 603-868 m^3/ha and the deadwood has been estimated to be between 78 and 121 m^3/ha [3].
- A team of German entomologists have identified 350 coleopteran species in the dead wood of Valea Cernei Forest Reserve, 60 of which are relict species specific to virgin forests. Some of these relict species, e.g. *Rhysodes sulcatus*, *Rhysodes americanus* and *Bothrideres contractus*, are now extinct in Western Europe. The Valley of Cerna is considered to have one of the richest saproxylic coleopteran faunas in Europe [5]. Many other rare and protected insect species, such as *Cerambyx cerdo*, *Lucanus cervus*, *Rosalia alpina* and *Camponotus herculeanus*, occur in old and decaying trees [12, 17].
- Twenty fungus species are associated with the processes of deadwood decay.
- There are numerous corticolous lichens (e.g. *Parmelia* spp., *Peltigera* spp. and *Lobaria* spp.) considered to be indicators and evidence of the continuity of the natural forest.
- A preliminary list [16] for cavity (primary and secondary) nesting birds in virgin forests includes 38 species, nine of which are woodpecker species, seven are tit species, eight are species of the family *Strigidae*, and four are species of flycatchers (*Muscicapidae*).
- Fourteen bat species frequently (and eight species occasionally) use the cavities of trees for shelter.
- A further 16 mammal species use the hollows, cavities, roots, fallen branches and deadwood of trees including bear, lynx, fox, martens, squirrels and many small rodents [16].
- Deadwood also provides a favourable environment for the natural regeneration of plant species in natural forests. On an 8 m spruce trunk felled in a storm in the Gemenele Reserve, we observed the frequent occurrence of green moss cushions (*Hylocomium splendens, Pleurozium schreberi* and *Eurynchium striatum*), layers of litter, ferns, many *Oxalis acetosella* plants, one fungus fruiting body and one shrub (*Lonicera nigra*). On the same trunk 39 spruce saplings and two of *Sorbus aucuparia* occurred. The ages of the spruce saplings indicated three temporal stages: 1-4, 4-7 and 10-12 years [11].

Conclusion and recommendations

The biological importance of deadwood dependent upon several factors and, in particular, on tree species composition and size, vertical and horizontal position, the decay stage and micro-environmental conditions. In the Romanian forests, the longest complete decay process was observed in oaks and Swiss stone pine (*Pinus cembra*) – all hardwood species distributed on the borders of forests. Willow, birch and poplar have a much shorter decay period [14]. The impressive biodiversity of deadwood confirms the ecological importance of these forests – a national, European and global heritage. Formerly viewed as a threat to forest health, deadwood is now recognized as a crucial habitat for many threatened species. This is acknowledged in the fact that the habitat has been recently recognized as a Pan-European indicator of sustainable forest management. The practice of leaving a certain proportion of deadwood in managed forests has marked an important move towards a more sustainable and ecological silviculture in Romania and elsewhere.

REFERENCES

1. Albrecht L., 1991- Die Bedeutung des toten Holzes im Wald. Forstw. Cbl. **110**: 106-113.
2. Ammer U., 1991- Konsequenzen aus den Ergebnissen der Totholzforschung für die forstliche Praxis. Forstw. Cbl. **110**: 149-157.
3. Biriş I.A., Radu S., Coandă C., 2002 - Pădurile virgine din România: sanctuare ale naturii şi comori ale biodiversităţii. ICAS, Bucureşti.
4. Elton C.S., 1966 - Dying and dead wood. In: The patterns of animal communities, pp. 279-305. John Willey and Sons, New York.
5. Fabricius K., 1994 - Pădurile României, valori inestimabile! Academica, **1** (49): 26-29.
6. Falinski J.B., 1997 - La forêt et son millieu. Braun-Blaquetia, **20**: 21-47.
7. Giurgiu V. (ed.), 1995 - Protejarea şi dezvoltarea durabilă a pădurilor României. Arta grafică, Bucureşti.
8. Giurgiu V., Doniţă N., Bândiu C., Radu S., Cenuşă R., Dissescu R., Stoiculescu C., Biriş I.A., 2001 - Les forêts vierges de Roumanie. ASBL Forêt Wallone, Louvain-la-Neuve.
9. Hansson L., 1992 - Ecological principles of nature conservation. Elsevier Science Publishers.
10. Kirby K.J., Drake C.M. (ed.), 1993 - Dead wood matters - the ecology and conservation of saproxylic invertebrates in Britain. English Natural Science.
11. Radu S., 1994 - Natural regeneration in forest ecosystems in the Retezat National Park. Conservation of forests in Central Europe. WWF Workshop, Zvolen, Arbora Publishers: 105-108.
12. Radu S., 1995 - Definitivarea "Listei roşii" a speciilor de plante şi animale rare, endemice şi periclitate din ecosistemele forestiere. Ref. şt. Mss. ICAS, Bucureşti.
13. Radu S., 1996 - Rolul şi importanţa lemnului mort în ecosistemele forestiere naturale. In: Ciochia V. (ed.), a 3-a Conferinţa naţională pentru protecţia mediului

prin metode şi mijloace biologice şi biotehnice, pp. 342-345. Universitatea "Transilvania", Braşov.
14. Radu S., Coandă C., Burza E., 2000 - Cercetări asupra biodiversităţii în ecosistemele de pădure cu structuri particulare (ecosisteme cvasivirgine şi naturale din Parcurile Naţionale). Ref. şt. Mss. ICAS, Bucureşti.
15. Radu S., Bândiu C., Coandă C., Doniţă N., Biriş I.A., Teodorescu M.E., 2004 - Conservarea pădurilor virgine (suport de curs). GEEA, Bucureşti.
16. Radu S., Hodor C., 2004 - Liste des oiseaux cavernicoles, des chauves-souris et d'autres mammifères qui utilisent le bois mort ou à cavités. Mss. Administraţia Parcului Naţional Retezat, Deva.
17. Speight M.C.D., 1989 - Les invertébrées saproxyliques et leur protection. Collection Sauvegarde de la Nature (Strasbourg, Conseil de l'Europe), **42**: 1-78.
18. Smejkal G.M., Bândiu C., Vişoiu-Smejkal D., 1995 - Banater Urwälder. Mirton Verlag, Temeswar.

A SYNTAXONOMIC REVIEW OF THERMOPHILOUS SHRUB COMMUNITIES (*SYRINGO–CARPINION ORIENTALIS*) IN SW ROMANIA

Gicu-Gabriel ARSENE, Ioan COSTE, Alina-Georgeta NEACŞU, Octavian-Ioan AVRĂMUŢ, Alina FĂRCĂŞESCU
Universitatea de Ştiinţe Agricole şi Medicină Veterinară a Banatului din Timişoara, Calea Aradului 119, **300645 Timişoara (ROMANIA)**; e-mail: arsene_g_2000@yahoo.fr

Abstract: We describe the biotope conditions, synchorology, floristic composition, coenotaxonomy and syndynamics of three thermophilous community types from SW Romania: *Asplenio–Syringetum* Jakucs 1959, *Syringo–Carpinetum orientalis* Jakucs 1959, *Syringo–Genistetum radiatae* Maloş 1972. The synthetic tables associated with the first two syntaxa contain original data as well as data of other authors. All phytocoenoses studied are characterized by high floristic richness. These three plant associations reach in Romania the northern limits of their distribution. We consider the origin and dynamics of communities for each plant association. Both their high species richness and important ecological function make these shrub communities a valuable natural resource worth preservation.

Introduction

Shrub associations formed by southern European and Balkan thermophilous species with their northern limit in Romania constitute important examples of vegetation for establishing the post-glacial evolution of European plant cover. They represent variants of some southern associations poor in thermophilous species and the more we advance towards the north, the more they interfact with central and eastern European vicariants. Such associations, formed by *Syringa vulgaris*, *Carpinus orientalis*, *Fraxinus ornus* and *Cotinus coggygria*, often co-dominant, can be found in Banat, Oltenia, and Dobrudja.

Most of these shrubs are widespread on the hills and low mountain areas on a limy substrate and less so on other rock types (Locvei Mountains, the Rudăriei Pass), where they cover different biotopes on steep slopes or small plateaux between 80 m in the Danube Narrows and 1350 m in the Piatra Cloşanilor Mountains.

The characteristic climate of SW Romania is moderate continental (Fig. 1); within this climate one can differentiate local micro-climates, with sub-Mediterranean influences, characterized by early springs and periods with temperatures above the regional averages (i.e. 10 ºC) and with a strong moisture deficit doubled by slope run-off and the mother rock which is mainly lime-rich and close to the surface. The soils on which these shrubs grow are of rendzina and terra rossa types.

Our research work aims to draw up a synthesis of different associations from *Syringo–Carpinion orientalis*, thus eliminating redundant names.

Material and Methods

In the synthesis we have used both original and published data in the form of community samples/relevés according to phytosociological methodology [5, 15]. The sample areas range from tens of square metres to several hundreds of square metres. The field records include the species list and, for each species, a score of abundance-dominance (degree of cover), estimated visually according to the Braun-Blanquet scale. The data have been combined in synthetic tables, in which the species were sorted according to their growth form (woody versus herbaceous) and constancy (general frequency). Valid species names are taken from the online *Flora Europaea Database* [57]. The species recorded by only one author are listed at the foot of the table. In the synthetic table, for each species and each author we give the minimum and maximum scores of abundance-dominance, followed by the number of relevés in which the species occurs (e.g. "+-3/4' means that the species has an abundance-dominance between "+" and "3" and is present in 4 relevés).

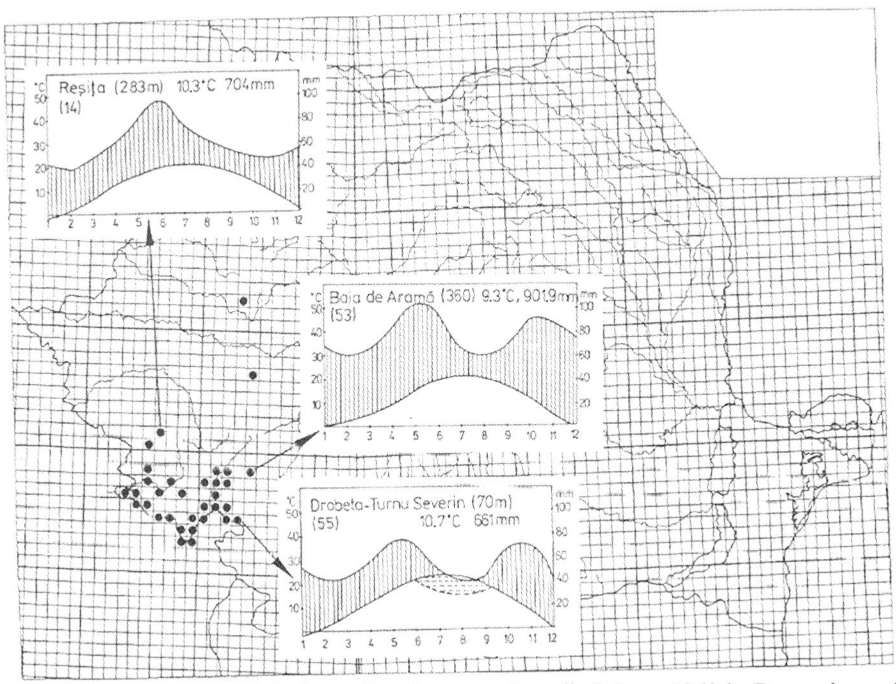

Fig. 1: Synchorology of *Syringo–Carpinetum orientalis* Jakucs 1960 in Romania, and the Walter-Lieth climatic diagrams for three representative biotopes [14] (UTM grid, 10 x 10 km).

Principal Components Analysis (PCA) and Cluster Analysis (CLAN) were performed (using STATGRAPHICS PLUS) starting from a matrix of species in rows and authors in columns. The data sources for constructing the synthetic table employed in the numerical analysis are Table 1 (in appendix) for *Asplenio–Syringetum*, Table 2 (in appendix) for *Syringo–Carpinetum orientalis*, and the association table presented by Maloş [38] for *Syringo–Genistetum radiatae*. We have employed species (general) frequencies in numerical analyses, as the use of cover percentages converted from abundance-dominance scores leads to a larger approximation. The species recorded by only one author have been removed, thus resulting in a reduced matrix of 30 columns by 245 rows to be analysed. In both PCA and CLAN the data have been standardized, as the number of relevés differs between authors. The single linkage method in CLAN was applied on a matrix of squared Euclidean distances in order to build a dendrogram.

The description of the three plant associations follows a common pattern that includes references to their biotope, chorology, floristic composition and coenotaxonomy, and the dynamics of the herbaceous layer.

The synchorological map has been presented in UTM format according to Lehrer [36], using various bibliographic references.

Results and Discussions

Based on our observations and those of other authors, as well as over 250 relevés, we have distinguished 10 thermophilous shrub associations in SW Romania. Their comparative study allows us, on the one hand, to reveal some nomenclatural synonymy and characteristic species combinations for the study region; and, on the other hand, to classify these communities within the context of SE European vegetation:

Quercetea pubescenti–petraeae (Oberd. 1948) Jakucs 1960
Orno–Cotinetalia Sóo 1960
Syringo–Carpinion orientalis Jakucs 1960
Asplenio–Syringetum Jakucs 1959
Syringo–Carpinetum Jakucs 1959
Syringo–Genistetum radiatae Maloş 1972

In accepting this classification, we simplify the scheme proposed in previous publications [50, 51].

This classification is based not only on differences in both biotope features and dominant species cover, but it is also partly validated by the outcomes of PCA (Fig. 2). The relevé groups described by different authors are ordered quite distinctively within the space determined by the two first PCA components (Fig. 2).

However, it is difficult to ascribe any accurate significance to the two axes as they have composed significances. Undoubtedly they express floristic differences, first of all because we have employed frequencies and not abundance-dominance data.

The dendrogram (Fig. 3) does not conspicuously reveal such a discrimination of the relevé groups as does PCA.

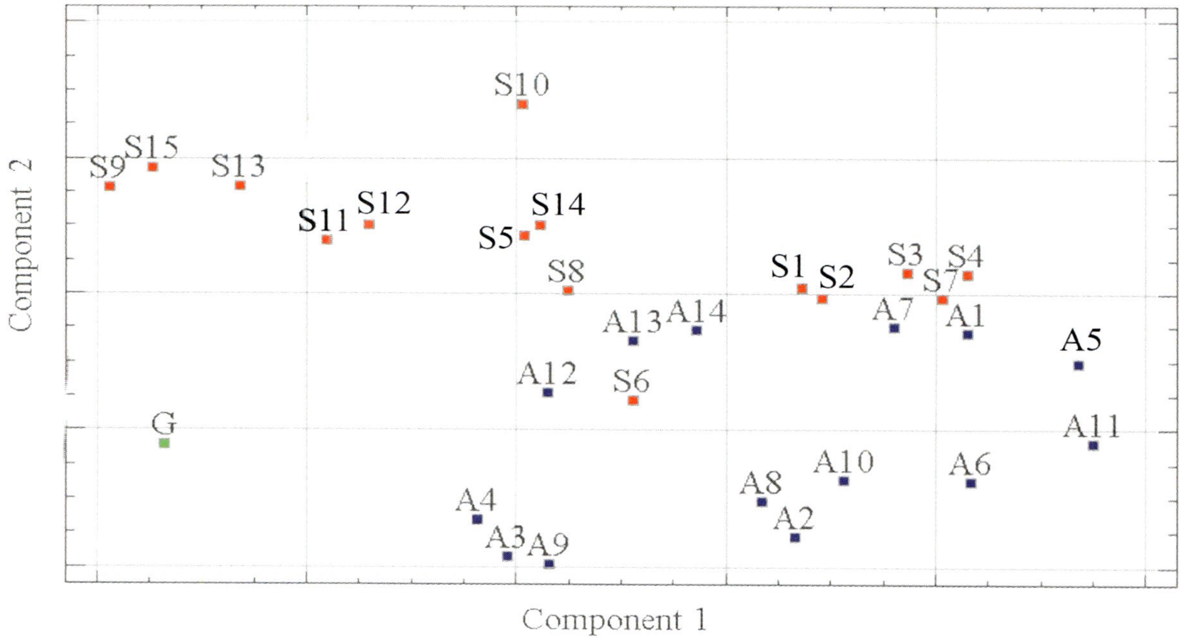

Fig. 2: Ordination of the plant communities studied, in the space determined by first two PCA components. ***A*** – phytocoenoses of *Asplenio–Syringetum*; ***S*** - phytocoenoses of *Syringo–Carpinetum orientalis*; ***G*** – phytocoenoses of *Syringo–Genistetum radiatae*; *A1 ...A14* correspond to the columns from Table 1 (see the appendix); *S1...S15* correspond to the columns from Table 2 (see the appendix); *G* corresponds to the table from Maloş [38]. The components 1 and 2 explains each 29.01 % and 8.45 %, respectively, of the total variance.

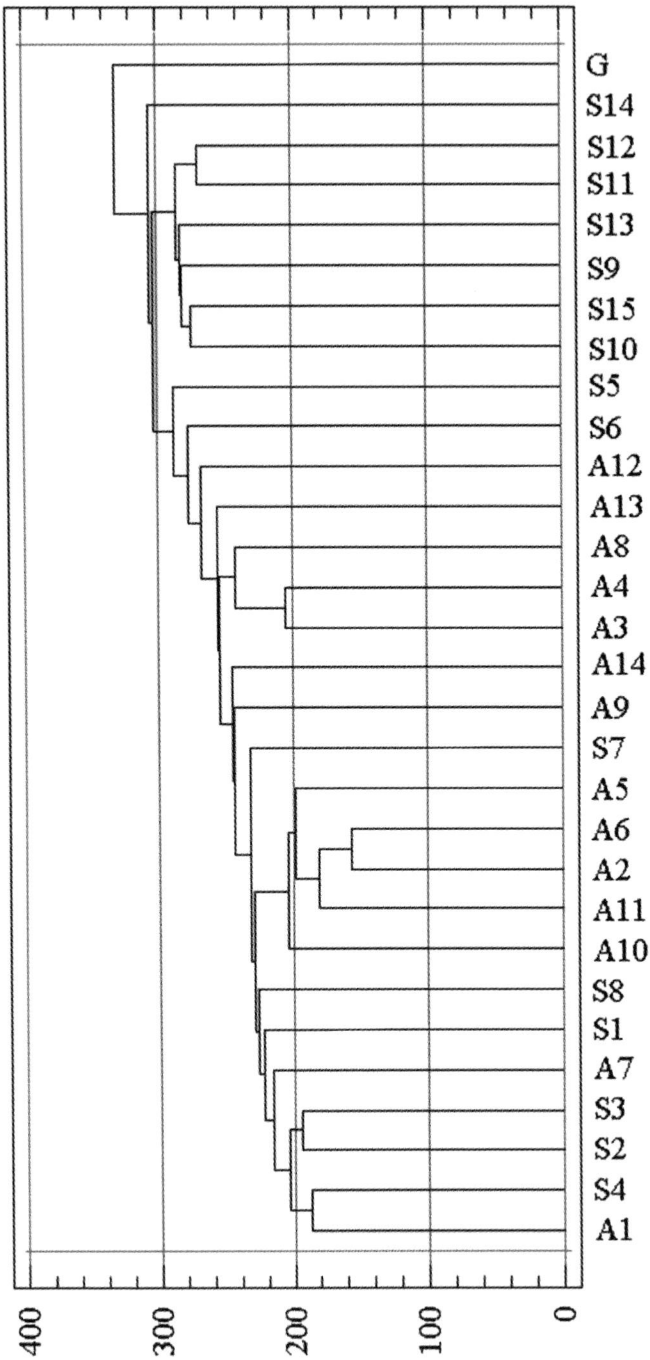

Fig. 3: The dendrogram of the plant communities studied, based on the frequencies of species. Abbreviations as in Fig. 2.

***1. Asplenio–Syringetum vulgaris* Jakucs 1959**
(syn. p.p. *Syringo–Fraxinetum orni* (Borza 1958) Bujorean *et al.* 1969, *Syringo–Cotinetum coggygriae* Resmeriță 1971)

Biotope and chorology. This association includes the phytocoenoses of *Fraxinus ornus* and *Syringa vulgaris*, extending along relatively narrow valleys on steep limy slopes where rocks are close to the soil surface, on rendzina and terra rossa soils. These phytocoenoses are to be found in the Danube Narrows and tributary river valleys, the limy gorges of the Cerna, Țesna, Caraș, Miniș and Bulz rivers, on hills and low-mountain areas at altitudes of 80–1350 m.

The climate of these regions is characterized by annual average temperatures of 10–11°C, with average minimal temperatures that do not drop below 1°C and with annual rainfall of 660–700 mm, distributed unevenly, with periods of drought in July–September accentuated by the lime-rich substrate and slope run-off.

Floristic composition. The association exhibits high floristic richness (Table 1 in the appendix), which can be explained by the variety of micro-biotopes and the interface at this level with other vegetation units. The shrub layer, 4–6 m tall, is composed of *Syringa vulgaris* and *Fraxinus ornus* with relative participation of the species *Cotinus coggygria*, *Prunus mahaleb*, *Cornus mas*, *Crataegus monogyna*, *Acer campestre* and, more rarely, *Carpinus orientalis*, *Acer monspessulanum*, *Corylus colurna*, *Cotoneaster nebrodensis* and *Rhamnus catharticus*.

The herbaceous layer is extremely heterogeneous, dominated by an ensemble of Balkan and Mediterranean species (*sensu lato*) circumscribing the thermophilous vegetation in SE Europe: *Scabiosa columbaria*, *Stachys recta* subsp. *leucoglossa*, *Dianthus giganteus* subsp. *banaticus*, *Echinops bannaticus*, *Cephalaria laevigata*, *Hypericum rochellii*, *Alyssum petraea* and *Acinos alpinus* subsp. *majoranifolius* (*Calamintha alpina* subsp. *hungarica)*; these species constitute, along with *Syringa vulgaris* and *Fraxinus ornus*, the core characteristics of this association. In the phytocoenoses around Hunedoara [52] and in the upper course of the Crișul Alb River [1], the number of thermophilous species is significantly reduced, these phytocoenoses marking the north-eastern limit of the association.

This herbaceous layer has quite frequently been composed of rupicolous species such as: *Asplenium ruta-muraria*, *A. trichomanes*, *A. ceterach*, *Draba lasiocarpa*, *Sesleria rigida*, *Dianthus petraeus*, *Seseli rigidum* and *Melica ciliata*, which indicate the heterogeneity of the substrate, and by transgressive plants from the neighbouring forests, such as *Helleborus odorus* and *Geranium robertianum*.

The phytogeographical spectrum of the association is characterised by the following elements: Euro-Asiatic – 20.4%, European – 12.5%, Central-European – 7.9%, Mediterranean and sub-Mediterranean – 17.1%, Balkan and Ponto-Mediterranean – 28%, Alpine-Carpathian-Balkan – 4.9%, continental – 1.6%, circumpolar – 3.9%; this spectrum underlines the almost Balkan character of the association.

Coenotaxonomy. The association was first described in the Cheile Nerei-Beuşniţa nature reserve under the name *Syringo–Fraxinetum orni* Borza 1958, without floristic tables (*nomen nudum*) but mentioning some characteristic species and biotope elements. Later it was mentioned as such with the floristic tables necessary for a proper identification in Valea Mare-Moldova Nouă [8], Ponoare-Mehedinţi [41], Ţesna Valley [47], Nera Valley [53], Depresiunea Hunedoarei [52], Miniş Valley – Bozovici [45] and Bulz Valley [1].

Phytocoenoses similar to those included in the above association were described in the Parâng and Domogled Mountains, under the name *Asplenio–Syringetum* Jakucs 1959, an association of less thermophilous character than the previous and with a marked interface between shrub and rupicolous vegetation. This association, validly described (with complete floristic tables and coenotaxonomic ranging), is considered to be synonymous with the previous association, cited from the Cerna Valley [7] and the Danube Narrows [6], without mentioning its relationship with *Syringo–Fraxinetum orni* Borza 1958 n.n. Analysing the vegetation of the Cheile Nerei–Beuşniţa reserve, Schrött [53] differentiates the two associations, mentioning that *Syringo–Fraxinetum orni* Borza 1958 covers the foot of the slopes and has a xero-mesophilous character with numerous transgressive species of *Phylitidi–Fagetum* Moor 1952 and *Colurno–Fagetum* Borhidi, while *Asplenio–Syringetum* Jakucs 1959 covers the middle and top of the slopes, including numerous transgressive species from the rupicolous associations in *Asplenietea* Meir & Br.-Bl. 1934.

Scrub from the Ţesna Valley that is floristically and ecologically similar, but with abundant *Cotinus coggygria*, have been described under the name *Syringo–Cotinetum coggygriae* Resmeriţă 1971 [47].

The comparative study of the phytocoenoses included within the above associations does not allow for their ecological and floristic differentiation. This is due to the high heterogeneity of the biotopes, to the interfaces and probably to the sampling procedures. As a consequence, we consider that their assimilation within the broader association *Asplenio–Syringetum vulgaris* Jakucs 1959 is justified, at least for the present.

Scrub with *Syringo vulgaris* and *Fraxinus ornus* is the result of the primary colonizing of the rocky slopes, where the scrub has a relatively constant composition, while the neighbouring vegetation has undergone profound changes, induced by the evolution of the climate and by processes of self-genesis characteristic of the post-glacial period [7].

The association *Corno–Fraxinetum orni* Pop & Hodişan 1964 [46] is described as a vicariant of *Asplenio–Syringetum* Jakucs 1959 from the Cheile Godineşti – Zam, and included in *Quercion pubescenti–petraeae* Br.-Bl. 1931 em. Tx. 1931.

2. *Syringo–Carpinetum orientalis* Jakucs 1959

(syn. *Carpinetum orientalis* Borza 1931 *n.n.*; *Cotino–Carpinetum orientalis*

Csürös *et al.* 1968; *Aristello bromoides–Carpinetum orientalis* Ştefureac & Popescu 1970)

Biotope and chorology. This association includes scrub of *Carpinus orientalis* present on moderate gradients at the foot of the slopes and more rarely on the plateaux. It grows on shallow rendzina and clayish-alluvial soils that have evolved over lime-rich and crystalline rocks. These phytocoenoses are mentioned from the hilly areas of the South-Banat and Oltenia (Danube Narrows, Cerna Valley, Nera Valley and their tributaries).

Floristic composition (Table 2 in the appendix). It displays the same constant heterogeneity as the previous association. The tree layer, 4–6 m tall and formed mainly by *Carpinus orientalis*, also contains some thermophilous species such as *Cotinus coggygria* and *Syringa vulgaris*, which become co-dominant in some phytocoenoses. They are frequently joined by *Fraxinus ornus*, *Cornus mas*, *Crataegus monogyna*, *Prunus mahaleb* and *Euonymus verrucosus*, but with a lower abundance-dominance. The tree layer is often dominated, in deeper soil biotopes, by isolated trees of species such as *Quercus cerris*, *Q. frainetto*, *Q. pubescens*, *Q. petraea* and *Tilia tomentosa*; which suggests, on the one hand, the interference of the association with zonal forests in the area (*Quercion farnetto* Horv. 1954) and, on the other hand, a possible regrowth of scrub after deforestation.

The herbaceous layer contains a higher number of thermophilous species, which with a few exceptions are not highly abundant, but give the association its specific Balkan aspect: *Echinops bannaticus*, *Ruscus aculeatus*, *Campanula lingulata*, *Stachys recta* subsp. *leucoglossa*, *Bupleurum praealtum*, *Arabis turrita*, etc. Among these species, *Ruscus aculeatus* has a special status and on some plateaux or moderate slopes it constitutes compact facies [8, 53]. The heterogeneity of the herbaceous layer is strongly accentuated by transgressive forest plants (*Lithospermum purpuro-caeruleum*, *Sedum telephium* subsp. *maximum*, *Brachypodium silvaticum*, *Vincetoxicum hirundinaria*, *Poa nemoralis*, etc.) and by rupicolous species (*Centaurea atropurpurea, Allium flavum, Asplenium trichomanes, Teucrium montanum* and *Asplenium ceterach*). These are joined by xerophilous meadow species that facilitate evolution towards this type of vegetation in case of deforestation: *Chrysopogon gryllus*, *Phleum montanum*, *Elymus hispidus*, *Andropogon ischaemum* and *Euphorbia cyparissias*.

The phyto-geographical spectrum is composed of the following elements: Euro-Asiatic – 26.6%, European – 9.7%, Central-European – 7.9%, Mediterranean and sub-Mediterranean – 23.6%, Balkan and Ponto-Mediterranean – 27.2%, circumpolar – 4.3%, cosmopolitan – 0.8%. Similar to the previous association, the spectrum displays relatively numerous Mediterranean (*sensu lato*) and Balkan species, probably connected with the higher degree of gradation into continental elements.

Coenotaxonomy. *Carpinus orientalis* scrub is mentioned in SW Romania under different names, varying between authors: *Syringo–Carpinetum orientalis* Jakucs 1959 [7, 8, 49], *Carpinetum orientalis* Borza 1931 [8], *Carpinetum orientalis* Borza 1946 [53], *Cotino–Carpinetum orientalis* Csürös *et al.* 1968 [13, 16], *Aristello bromoides–Carpinetum orientalis* Ştefureac & Popescu 1970 [56], included within *Syringo–Carpinion orientalis* Jakucs 1960 or *Querco–Carpinion orientalis* Csürös *et al.* 1968. Despite the heterogeneity of the phytocoenotic data already existing, we consider possible to assimilate these associations within *Syringo–Carpinetum orientalis* Jakucs 1959, which has a broad content and includes all scrub dominated by *Carpinus orientalis* on rocky slopes and more rarely on plateaux with shallow soil.

In the Mehedinţi Plateau phytocoenoses of *Carpinus orientalis* [49], with a forest-like physiognomy, are described on relatively deep soils under the name *Oryzopsis holciformis–Carpinetum orientalis* Jakucs & Zólyomi 1960 and circumscribed by *Orno–Cotinion* Sóo 1964. This association reflects an ecological and syndynamic situation different from the previous; it has a considerable number of species characteristic of oak forests in the area (*Quercion farnetto* Horvát. 1954) and it may secondarily result from the proliferation of the species associated with *Carpinus orientalis* after deforestation, often forming sub-storeys [13]. Under progressive erosion of the soil, the association changes into *Syringo–Carpinetum orientalis* Jakucs 1959, with which it interfaces floristically and topographically in some areas.

3. *Syringo–Genistetum radiatae* Maloş 1972

Biotope. The association, represented by *Syringa vulgaris* scrub, was described in the Piatra Cloşanilor Mountains (the southern Carpathians) at an altitude of 1100–1350 m, on moderate limy slopes and on a rendzina soil. The local microclimate exhibits a dry period (July–September) under the influence of wind and the lime-rich substrate, which favour rapid percolation and evaporation of water.

Floristic composition. The association is composed of *Syringa vulgaris*, which forms a discontinuous shrub layer 1 m tall, accompanied by a few species of similar size, among which we can mention *Cotinus coggygria* and *Rosa canina*. A second layer, 30–50 cm tall, with a share of 90% is formed by *Genista radiata* (*Cytisanthus radiatus*). The small size of the shrub layer and the high abundance of the last species give the association a special physiognomy remarked particularly in early July, when the shrubs are in bloom.

The herbaceous cover [37] consists of numerous Mediterranean (*sensu lato*) and Balkan species, such as *Teucrium montanum*, *T. chamaedrys*, *Thymus montanum*, *Dianthus kitaibeli*, *Piptatherum virescens*, *Arabis turrita*, *Cerastium bannaticum*, *Alyssum petraeum*, etc., which indicate syngenetic links with SE European vegetation, but endemic and Dacian elements are also present: *Dianthus henteri*, *Draba lasiocarpa*, *Sempervivum schlehanii* and *Sesleria rigida*.

The spectrum of phyto-geographical elements consists of the following components: Euro-Asiatic – 17.8%, European – 4.4%, Central-European –7.8 %, Mediterranean and sub-Mediterranean – 41.0%, continental – 4.4%, circumpolar – 2.2%.

Coenotaxonomy. The association has floristic affinities with *Asplenio–Syringetum* Jakucs 1959, without the possibility of being assimilated within it; its ecology, physiognomy, and partly its floristic composition are arguments in favour of its being considered a distinct coenotaxon. This association is probably a vicariant of the *Eryngio–Syringetum* Diklič 1965 association, described from similar habitats in Serbia, but also with different characteristic species.

Phytocoenoses in which *Genista radiata* is co-dominant along with *Syringa vulgaris* hare also been described in SE Romania and at lower altitudes [7, 45], but we consider that they are circumscribed by *Syringo–Fraxinetum orni* Jakucs 1959, in which the floristic composition is very similar.

The association *Syringo–Genistetum radiatae* Maloş 1972 represents an example of a coenotic assemblage of relict character, from the inter-glacial period in SE Europe.

Conclusions

1. Our analysis offers good evidence to include the phytocoenoses from *Syringo–Carpinion orientalis* Jakucs 1960, described from SW Romania, in three associations: *Asplenio–Syringetum* Jakucs 1959, *Syringo–Carpinetum orientalis* Jakucs 1959 and *Syringo–Genistetum radiatae* Maloş 1972. This simplified classification appears to be validated by PCA performed on a relative large relevé set.
2. The phytocoenoses analyzed are characterized by great floristic richness: 270 taxa (the great majority being species) in 122 relevés of the *Asplenio–Syringetum* association; 314 taxa (the great majority being species) in 125 relevés, for *Syringo–Carpinetum orientalis*; and 88 taxa (mostly species) in 7 relevés, for *Syringo–Genistetum radiatae.*
3. The phytocoenoses studied contain plant species included in the Red Lists compiled for Romania: *Atamantha hungarica* (*Athamantha turbith* subsp. *hungarica*), *Corylus colurna*, *Dianthus giganteus* subsp. *banaticus*, *Dianthus henteri*, *Fritillaria orientalis, Linum uninerve*, *Micromerion pulegium*, *Myrrhoides nodosa*, *Orchis simia*, *Paeonia officinalis* subsp. *banatica*, *Ruscus aculeatus* and *Scutellaria velenovskyi*.
4. Besides the formation of a community environment for certain rare species in Romania, these phytocoenoses represent a defining element of the SW Romanian landscape.
5. Similar numerical analyses also need to be performed with large relevé sets for *Quercetea pubescenti–petraea*, in order to eliminate redundant associations and clarify syntaxonomic relationships. The inclusion of some environmental variables and quantitative community parameters in numerical comparisons would greatly fine-tune the outcome of this vegetation classification.

Acknowledgements: We are grateful to Prof. Rodica Zehan and Prof. Dan Gafta for comments and suggestions to improve the manuscript.

REFERENCES

1. Ardelean A., 1980 - Flora şi vegetaţia din Valea Crişului Alb, între izvoare şi oraşul Ineu. Teză de doctorat, Universitatea „Babeş-Bolyai", Cluj-Napoca.
2. Borza A., 1931 - Botanischer Fürher durch die Ungebung von Băile Herculane (Herculesbad) bis an die Donau. Guide de la sixième excursion phytogéographique internationale en Roumanie, Cluj: 1-55.
3. Borza A., 1958 - Contribuţii la flora şi vegetaţia din răsăritul României. Contr. Bot., Cluj: 127-158.
4. Borza A., 1958 - Vegetaţia rezervaţiei Beuşniţa. Ocrotirea Naturii, **3**, Bucureşti: 117-128.
5. Borza A., Boşcaiu N., 1965 - Introducere în studiul covorului vegetal, Ed. Academiei RPR, Bucureşti.
6. Boşcaiu N., Lupşa V., Resmeriţă I., Coldea G., Schneider E., 1971 - Vegetaţia lemnoasă mezoxerotermă (*Orno - Cotinetalia*) în defileul Dunării. Ocrotirea Naturii, Bucureşti, **15** (1): 49-55.
7. Boşcaiu N., 1971 - Flora şi vegetaţia Munţilor Ţarcu, Godeanu şi Cernei, Ed. Academiei, Bucureşti.
8. Bujorean G., Grigore S., Coste I., 1969 - Assoziationen in den Locvei Bergen, Ber. der Int. Symp. „Vegetation und Substrat", Hinteln: 433-441.
9. Bunescu A., Doniţă N., 1960 - Răspândirea cărpiniţei (*Carpinus orientalis*) în R.P. Română. Studii şi Cercetări de Biologie, seria Biol. Veg., Bucureşti, **12** (3): 331-342
10. Călinescu R., 1957 - Contribuţii la studiul şibliacului în R.P. Română. Revista Pădurilor, Bucureşti, **70** (2): 76-84.
11. Călinescu R., 1933 - Repartiţia liliacului (gen. *Syringa* L.) în România. Buletinul Societăţii Regale Române de Geografie, **LI** (1932): 100-196.
12. Ciocârlan V., 1968 - Flora şi vegetaţia bazinului subcarpatic al Slănicului de Buzău. Teză de doctorat, Universitatea din Bucureşti.
13. Coste I., 1975 - Flora şi vegetaţia Munţilor Locva. Teză de doctorat, Universitatea „Babeş-Bolyai", Cluj-Napoca.
14. Coste I., Avrămuţ I., Arsene G.-G., 1996 - Stabilirea şi caracterizarea unităţilor de vegetaţie şi a tipurilor de biotop din sud-vestul României (Banat). Raport cercetare CNCSIS, nr. 5022, tema 149, U.S.A.M.V.B. Timişoara.
15. Cristea V., Gafta D., Pedrotti F., 2004 - Fitosociologie. Ed. Presa Universitară Clujeană, Cluj-Napoca.
16. Csürös Ş., Pop I., Hodişan I, Csürös-Kaptalan M., 1968 - Cercetări floristice şi de vegetaţie între Orşova şi Eşelniţa. Contr. Bot., Cluj: 277-311.
17. Csürös Ş., Popescu M., 1974 - Elementul balcano-dacic în flora României. Contrib. Bot., Cluj-Napoca: 80-84.
18. Csürös Ş., Popescu. M., 1975 - Elementul balcano-dacic în flora României (Nota 2). Contrib. Bot., Cluj-Napoca: 71-76.
19. Dihoru G., Doniţă N., 1970 - Flora şi vegetaţia Podişului Babadag, Ed. Academiei R.S.R., Bucureşti.
20. Domin K., 1932 - Domugled Kazanski sontesky, Ada-Kaleh à Vârciorova. Publications de la Faculté de Sciences de l'Université Charles, Praha: 122-144.
21. Doniţă N., 1958 - *Carpinus orientalis* Mill. la nord de Iaşi. Contr. Bot., Cluj: 193-196.

22. Georgescu C.C., 1965 - Consideraţii asupra unor elemente termofile, în special sudice din flora României, Studii şi Cerc. Biol., ser. Bot., **17** (4-5): 397-402.
23. Georgescu C.C., 1934 - Studii fitogeografice în bazinul inferior al Văii Cernei (Băile Herculane). Analele Institutului de Cercetări şi Exploatări Forestiere, Bucureşti **1**: 71-135.
24. Georgescu G., Ciucă M., 1952 - Contribuţii la studiul răspândirii scumpiei (*Cotinus coggygria* Scop.) în R.P. Română. Buletinul Ştiinţific al Academiei Române – seria Biologie, Ştiinţe Agricole, Geobotanică, Geografie, Bucureşti, **4** (2): 409-414.
25. Gergely I., 1958 - Contribuţii la cunoaşterea vegetaţiei din jurul oraşului Petroşani. Contr. Bot., Cluj: 165-168.
26. Grebenschikoff S.C., 1963 - On distribution of common lilac and its low forest in South–Eastern Europe. Biul. Mosk. Olese Isp. Prir. Otd. Biol., Moskva, **LXIII** (1): 63-72.
27. Grigore S., Coste I., 1974 - Aspecte botanice din rezervaţia naturală Valea Mare, Moldova Nouă. Ocrotirea Naturii, Bucureşti, **18** (2): 159-166.
28. Grigore S., Coste I., 1978 - Cercetări asupra vegetaţiei dintre Moldova Veche şi Pescari (jud. Caraş-Severin), Banatica – Caiete de Studii Naturale, Muzeul Judeţean Reşiţa, **7**: 173-189.
29. Grigore S., Coste I., 1979 - Contribuţii la studiul vegetaţiei xerofile de stâncării din Cheile Caraşului şi Cheile Gârliştei, Tibiscus – Muzeul Banatului, Ştiinţe Naturale, Timişoara: 35-49.
30. Haralamb M.A., 1938 - Câteva staţiuni noi de mojdrean. Analele Institutului de Cercetări şi Experimentaţie Forestieră, **IX**: 248-251.
31. Horvat I., Glavač V., Ellenberg H., 1974 - Vegetation Süd-Osteuropas, Gustav Fisher Verlag, Jena.
32. Iana S., 1957 - Contribuţii la studiul răspândirii scumpiei în regiunea Bucureşti, Analele Universităţii Bucureşti, seria Ştiinţe Naturale, **16**: 189-199.
33. Jakucs P., 1959 - Uber die oskalkanischen Flieder Buschwalder. Acta Botanica Hungarica, Budapest, **5**: 357-390.
34. Jakucs P., 1961 - Die phytozonologischen Verhalnisse der Flaumeichen Buchwalder Sud-Ostmitteleruopas, Budapest.
35. Jakucs P., 1967 - Gedanken zur Höheren Systematik der Europäischen Laubwälder. Contrib. Bot., Cluj-Napoca: 159-166.
36. Lehrer A., 1977 - Codul biocartografic al principalelor localităţi din R.P. România. Ed. Dacia, Cluj-Napoca.
37. Maloş C., 1968 - Contribuţii la studiul florei şi vegetaţiei din bazinul superior al Văii Motrului. Buletinul Ştiinţific, Universitatea Craiova, **X**: 72-83.
38. Maloş C., 1972 - Cercetări asupra fitocenozelor cu liliac (*Syringa vulgaris* L.) din Oltenia. Studii şi Cercetări de Biologie, s. Botanică, Bucureşti, **24** (3): 189-198.
39. Maloş C., 1974 - Şibliacul în bazinul superior al Motrului şi problema evoluţiei sub influenţa factorului antropogen. Studii ale Academiei Române, seria Biologie - Ştiinţe Agricole, Bucureşti, **8** (1): 209-219.
40. Morariu I., Ciucă M., 1956 - *Fraxinus ornus* L. – Sistematica, răspândirea în R.P.R. şi întrebuinţările lui. Buletinul Ştiinţe al Academiei – Biologie şi Ştiinţe Agricole, Bucureşti, **8** (1): 209-219.
41. Măgălie E., 1970 - Pădurea de liliac de la Ponoare. Ocrotirea Naturii, **14** (2): 181-186.
42. Nedelcu A.G., Sanda V., 1982 - Vegetaţia lemnoasă din zona lacului de acumulare

Porţile de Fier (Baziaş, Drobeta-Turnu-Severin). Acta Bot. Horti Buc.: 147-158.
43. Păun M., Popescu G., 1978 - Aspecte din vegetaţia Văii Sohodolului (jud. Gorj). Studii şi Cercetări, Târgu-Jiu: 93-97.
44. Păun M., Popescu G., Cârţu M., Cârţu D., 1970 - Aspecte din vegetaţia dintre Berzasca şi Pescari (jud. Caraş-Severin). Analele Universităţii Craiova, seria Biologie-Ştiinţe Agricole, Craiova: 61-70.
45. Peia P., 1981 - Asociaţii vegetale noi în Munţii Banatului. Contr. Bot., Cluj-Napoca: 103-109.
46. Pop I., Hodişan I., 1964 - Contribuţii la cunoaşterea vegetaţiei calcarelor de la Godineşti-Zam (reg. Hunedoara). Contr.. Bot., Cluj: 229-239.
47. Resmeriţă I., 1972 - Vegetaţia lemnoasă din Valea Ţesnei (jud. Mehedinţi). Studii şi Cerc. Biol., ser. Bot., Bucureşti, **24** (4): 277-294.
48. Rodwell J.S., Schaminée J.H.J., Mucina L., Pignatti S., Dring J., Moss D., 2002 - The diversity of European Vegetation – An overview of phytosociological alliances and their relationship to EUNIS habitats, Wageningen.
49. Roman N., 1974 - Flora şi vegetaţia din sudul Podişului Mehedinţi. Ed. Academiei RSR, Bucureşti.
50. Sanda V., Popescu A., Arcuş M., 1999 - Revizia critică a comunităţilor de plante din România. Ed. „Tilia Press International", Constanţa.
51. Sanda V., Popescu A, Barabaş N., 1998 - Cenotaxonomia şi caracterizarea grupărilor vegetale din România. Studii şi comunicări, ser. Biologie vegetală, Complexul Muzeal de Ştiinţele Naturii, Bacău, **14**.
52. Sanda V., Popescu A., Pelcea I., 1972 - Contribuţii la cunoaşterea vegetaţiei din judeţul Hunedoara, Studii şi Cerc. Biol., ser. Biologie vegetală, Bucureşti, **24** (4): 295-317.
53. Schrött L., 1972 - Flora şi vegetaţia rezervaţiei naturale Beuşniţa-Cheile Nerei, Munţii Aninei. Teză de doctorat, Universitatea din Bucureşti.
54. Şerbănescu I., 1934 - Noi localităţi pentru *Syringa vulgaris* în jud. Buzău. Buletinul Societăţii de Ştiinţe Naturale din România, Bucureşti, **5**.
55. Şerbănescu I., 1959 - Cercetări asupra vegetaţiei din estul Câmpiei Române. Dări de Seamă ale Societăţii Române de Geobotanică, Bucureşti, **XLII** (1954-1955): 469-508.
56. Ştefureac T., Popescu A., 1970 - Recherches sur les phytocénoses à *Stipa aristella* L. du sud-ouest de la Roumanie. Revue Roumaine de Biologie, série de Botanique, Bucarest, **16** (5): 323-335.
57. * * *, http://rbg-web2.rbge.org.uk/. Flora Europaea Database.

Appendix

Table 1: ***Asplenio - Syringetum*** **Jakucs 1959**

Reference/study area*	**1**	**2**	**3**	**4**	**5**	**6**	**7**	**8**	**9**	**10**	**11**	**12**	**13**	**14**
Number of relevés	**12**	**5**	**10**	**6**	**20**	**10**	**15**	**11**	**5**	**6**	**5**	**5**	**7**	**5**
Syringa vulgaris	+-3/12	1-2/5	+-3/8	+-2/6	+-1/20	2-3/10	+-1/13	+-3/11	1-3/5	1-4/10	+-4/5	3-4/5	+-3/7	1-3/5
Fraxinus ornus	+-3/12	1-2/5	1-4/8	1-2/5	1-4/20	+-1/8	+-3/15	+-1/9	+-5/3	2-3/10	2-3/5	2-3/5	2-3/7	2-5/5
Asplenium ruta-muraria		+/5	+/3	+/4	+/2	+/3	+-1/4	+/7	+/4	+/7	+/1	+/4	+/2	
Asplenium trichomanes	+/4	+/5		+/1	+/6	+/6	+/2		+/3	1-4/10	+-4/3	+-2/5	+/1	+-1/3
Crataegus monogyna	+/10				+/19	+/6	+/9	+/1		+-1/8	+/1	+/4	+/7	+-2/4
Cornus mas	+/4			+/1	+-2/6	+/8	+/12			+-1/8	+/2	+/5	+/3	+-2/5
Cotinus cogyggria	+/7	+-2/4	+/1	2-3/6	+/8	+/7	+-3/11	+-1/3	+/3		+/4			
Prunus mahaleb	+/6	+/4		+-1/5		+/7	+/9			+/4	+/3	+/2		
Carpinus orientalis	+-1/10	+/2	+/2			+/4	+-4/14				+/2	+/1		
Viburnum lantana	+/5				+/9		+/2		+/1	+/5		+/1	+/4	+/2
Corylus avellana			+/3	+-1/3			+/3	+-1/6			+/1	+/4	+/2	+/1
Clematis vitalba	+/2	2/1	+/1	+/1	+/1		+/6	+/1	+/1	+/4	+/1		+/2	+/1
Acer campestre			+/2		+/7		+/5					+/3	+-1/3	
Evonymus verrucosus					+/9					+-1/7	+/1		+/2	+/1
Tamus communis	+/2	+/1			+/2		+/3		+/2	+/4	+/1	+/5		
Berberis vulgaris					+/4	+/1				+/9	+/2		+/3	
Cotoneaster nebrodensis						+/4	+/9	+/1	+/2					
Hedera hedelix	+/6				+/5			+/4						
Rhamnus catharticus	+/2				+/4					+/3			+/5	+/1
Ruscus aculeatus	+-4/5						+/5			+-1/2	+/1			
Sorbus torminalis					+/2		+/1			+/4	+/1	+/2	+/1	+/2
Pyrus pyraster	+-1/5				+/2					+/4			+/1	
Chamaecytisus hirsutus	+/3		+/1			+/5								+/1
Rhamnus saxatilis subsp. *tinctoria*									+/3	+-2/7				
Staphylea pinnata	+/3						+/7							
Ligustrum vulgare	+/2				+/2								+/2	+/1
Quercus cerris					+/2								+/1	+/4
Rosa canina	+/4											+/2	+/1	

Reference/study area*	1	2	3	4	5	6	7	8	9	10	11	12	13	14
Sorbus graeca						+/2	+/3							
Tilia tomentosa				+/1	+/3						+/1			
Fagus sylvatica				+/1			+/3							
Chamaecytissus ciliatus			+/1										+/1	
Cotoneaster vulgaris				+/1									+/1	
Asplenium ceterach		+-1/5	+/8	+/2	+/13	+/10	+/6	+/2	+/4	+/9	+/2	+-1/5		
Aurinia petraea	+/3	+/4	+/6	+/1	+-1/9	+/6		+/6	+/3	+/4	+-1/3			
Draba lasiocarpa		+-1/5	+/2		+/13	+/8		+/2	+/4	+/9	+/2			
Sedum telephium subsp. *maximum*	+/7				+/8	+/1	+/7	+/2	+/3	+/6	+/1		+/2	+/3
Gallium album subsp. *album*			+/1	+/4	+/14			+/5	+/4	+/6		+/1		+/4
Seseli libanotis			+/4	+/3		+/3	+/9	+/6	+/3	1-2/10				
Coronilla varia	+/3	+/1	+/6	+/2	+/9	+/1		+/3					+/1	+/2
Sesleria rigida	+/2	+/3	1-2/2	+/1		+-2/10		+/2	1-2/4	+-2/3	+/1			
Centaurea atropurpurea	+/5	+/5			+/5	+/6					+/1		+/6	
Dactylis glomerata subsp. *aschersoniana*	+/4		+/1		+/12		+/10							
Arabis turrita	+/4				+/7	+/2	+/6	+/3	+/3		+/1			
Scabiosa columbaria	+/4	+-1/3	+/3		+/1	+/1	+/5		+/4	+/3	+/2			
Dianthus petraeus		+/1	+/4			+/5		+/9	+/3	+/3				
Teucrium chamaedrys			+/4		+-1/6		+/7	+/3				+/1	+-1/2	+-2/2
Piptatherum virescens	+/2				+-2/5	+/7	+/3		+/1	+-1/6				
Seseli rigidum		+/1	+/1	+/2		+/6		+/4	+/2				+/8	
Digitalis grandiflora	+/3		+/1		+/7		+/7					+/1		+/3
Stachys recta subsp. *leucoglossa*	+-2/2		+/4	+/5	+/6			+/1		+/1		+/3		
Allium flavum	+/4	+/1			+/1			+/5	+/4	+/3			+/3	
Echinops bannaticus	+/5				+/1	+/4	+/9			+/2				
Euphorbia epythimoides	+/7						+/2	+/8		+/4				
Silene italica			+/3	+/3			+/6			+/4	+/1	+/4		
Thalictrum aquilegiifolium	+/3		+/2	+/1		+/2			+/1	+/6	+/1			+/5

Reference/study area*	**1**	**2**	**3**	**4**	**5**	**6**	**7**	**8**	**9**	**10**	**11**	**12**	**13**	**14**
Vincetoxicum hirundinaria	+/2						+/11					+/5		+/3
Dianthus giganteus subsp. *banaticus*	+/2	+/2	+/2	+/1	+/4	+/3	+/6							
Veronica chamaedrys					+/5		+/3			+/6	+/1	+/3		+/2
Brachypodium silvaticum	+/2				+-2/5		1-2/6				+/1	+-3/4		+/1
Heleborus odorus					+/6	+/1	+/11				+/1			
Teucrium montanum	+/5	+/1	+/6	+/5					+/2					
Acnatherum calamagrostis			+-2/10	+-2/5				+/3						
Moehringia muscosa		+/3	+/3		+/1			+/4		+/2	+/1	+/2		+-1/2
Silene vulgaris subsp. *alpina*	+/4			+/7		+/5					+/2			
Veronica austriaca subsp. *teucrium*	+/2	+/1		+/3	+/4	+/2	+/1	+/1	+/3				+/1	
Polygonatum odoratum	+/3	+/1			+/4		+/3			+/3	+/1	+/1	+/1	
Cardaminopsis arenosa			+/3	+/3			+/4					+/4		+/2
Lythospermum purpurocaeruleum	+/3				+-2/5		+/4				+/1			+/3
Melica ciliata subsp. *ciliata*		+/4	+/1			+/3		+/3					+-1/3	+/2
Melica uniflora	+/2	+/1			+/5					+/4	+/1	+/3		
Silene saxifraga			+/5	+/3	+/1			+-1/7						
Campanula sibirica					+/7						+/2		+/4	+/2
Eryssimum odoratum			+/3		+/2	+/1			+/3	+/3	+/1		+/1	+/1
Origanum vulgare	+/3					+/1	+/6	+/5						
Orlaya grandiflora	+/3	+/1			+/1	+/2		+/1	+/2	+/1			+/4	
Fragaria vesca					+/6		+/2				+/1	+/5		
Glechoma hirsuta	+/5	+/1			+/5							+/3		
Sedum hispanicum	+/6		+/1		+/2					+/2			+/1	+/2
Sempervivum marmoreum								+/6	+/4	+/4				
Euphorbia cyparissias		+/1	+/1	+/3	+/2								+/4	+/2
Campanula divergens			+/2	+/4		+/4	+/2							
Jurinea mollis	+/4	+/2				+/6								
Lychnis coronaria		+/1			+/5		+/2					+/4		

Reference/study area*	**1**	**2**	**3**	**4**	**5**	**6**	**7**	**8**	**9**	**10**	**11**	**12**	**13**	**14**
Athamanta hungarica			+/4	+/2					+/2	+/3				
Campanula persicifolia	+-2/5						+/1					+/3		+/2
Clinopodium vulgare					+/5					+/3			+/3	
Fragaria collina	+/3		+/1										+/3	+-1/4
Melityis melissophylum					+/2		+/5			+/3		+/1		
Sedum album			+/1			+/1		+/5				+/2	+/2	
Viola hirta					+/5	+/2				+/3			+/1	
Acinos alpinus subsp. *majoranifolius*			+/2						+/3		+/1	+/1	+/3	
Aremonia agrimonoides							+/6					+/4		
Campanula grossekii					+/7		+/1				+/2			
Carex humilis		+/1	+-1/3	+-1/5		+/1								
Ferulago sylvatica	+/2			+-2/4						+/3	+/1			
Geranium robertianum	+/2	+/1	+/1		+/3				+/2				+/1	
Seseli gracile	+/2	+/4	+/1											+/3
Cephalaria laevigata			+/1			+/2			+/3	+-1/3				
Cystopteris fragilis			+/1							+/3		+/5		
Hypericum perforatum	+/3							+/1					+/2	+/3
Scutellaria altissima	+/3	+/1							+/3		+/1	+/1		
Valeriana officinalis subsp. *collina*				+/1	+/4								+/2	+/2
Veronica spicata			+/6	+/1			+/2							
Acinos arvensis	+/5		+/1									+/1	+/1	
Arabis hirsuta	+/1		+/3	+/1								+/3		
Gallium flavescens			+/4	+-1/4										
Hypericum rochelii		+/1	+/1	+/2	+/1					+/3				
Inula conyza					+/2			+/1		+/1			+/4	
Laser trilobum			+/1	+/1			+/6							
Leontodon crispus			+/3	+/1		+/2							+/2	
Helianthemum canum													+/5	+/2
Lathyrus venetus							+/1				+/1	+/5		
Micromerion pulegium			+/2	+/1				+/4						
Primula columnae var. *serratifolia*			+/6	+/1										

Reference/study area*	1	2	3	4	5	6	7	8	9	10	11	12	13	14
Sesleria rigida var. *haynaldiana*			1-2/2	+/1					1-2/4					
Asperula tenella	+/4					+/2								
Genista radiata								2-3/1	2-4/5					
Inula ensifolia			+/3	+/3										
Lunaria annua	+/2	+/1								+/3				
Orchis simia							+/2					+/4		
Phleum montanum					+/1				+/3				+/2	
Polypodium vulgare	+/3				+/2								+/1	
Potentilla recta	+/2				+/1					+/3				
Stachys recta			+/4						+/2					
Aster amellus			+/1	+/2							+/1		+/1	
Astragalus glycyphyllos	+/2											+/1		+/1
Dicanthium ischaemum					+/2								2-3/3	
Erythronium dens-canis					+/1					+/3		+/1		
Ferula heuffeli		+/2				+/3								
Lembotropis nigricans							+/2						+/3	
Melampyrum bihariense			+/1		+/2						+/1			+/1
Pimpinella saxifraga	+/2									+/2			+/1	
Trifolium medium	+/3			+/2										
Veronica spicata subsp. *barrelieri*			+/1	+/1					+/3					
Allium paniculatum subsp. *fuscum*										+/3	+/1			
Asperula tinctoria			+/1				+/3							
Biscutella laevigata			+/1	+/3										
Brachypodium pinnatum			+/1	+/1										+/2
Bupleurum praealtum	+/2					+/1						+/1		
Campanula crassipes		+/2						+/2						
Cerastium banaticum		+/1							+/3					
Fritillaria orientalis	+/2								+/1	+/1				
Mycelis muralis			+/1				+/3							
Parietaria officinalis			+/1							+/3				
Piptatherum holciforme						+/2	+/2							
Saxifraga paniculata			+/2		+/2									

Reference/study area*	1	2	3	4	5	6	7	8	9	10	11	12	13	14
Stipa pennata subsp. *eriocaulis*					+/3			+/1						
Veronica spicata subsp. *orchidea*			+/1	+/3										
Aristolochia pallida				+/1		+/2								
Campanula rapunculoides	+/1										+/1	+/1		
Campanula trachelium			+/1									+/1	+/1	
Euphorbia amygdaloides			+/1	+/1			+/1							
Festuca panciciana									+/2				+/1	
Hepatica nobilis			+/1									+/1		+/1
Kernera saxatilis			+/2	+/1										
Myrhoides nodosa		+/2							+/1					
Ornithogalum umbellatum					+/2								+/1	
Peucedanum longifolium		+/1	+/1								+/1			
Ranunculus bulbosus			+/1							+/2				
Campanula kladkiana			+/1	+/1										
Dorycnium pentaphyllum subsp. *germanicum*	+/1													+/1
Epipactis helleborine								+/1						+/1
Galium schultesii		+/1		+/1										
Lilium martagon							+/1					+/1		
Silene otites	+/1												+/1	

Species in one sample (between paranthesis, the number of the author / column):
Acanthus balcanicus (1); *Acer monspessulanum* (2); *Acer platanoides* (13); *Acer tataricum* (13); *Adoxa moschatellina* (12); *Agrimonia eupatoria* (2); *Ajuga genevensis* (7); *Allium albidum* subsp. *albidum* (13); *Anemone nemorosa* (12); *Anthemis tinctoria* (13); *Anthericum ramosum* (13); *Artemisia campestris* (13); *Arum maculatum* (7); *Arum orientale* (12); *Asperula cynanchyca* (2); *Asperula taurina* (7); *Bromus riparius* (5); *Calamagrostis epigejos* (13); *Campanula lingulata* (6); *Campanula rotundifolia* (3); *Cardamine bulbifera* (12); *Carex brevicollis* (10); *Carex pilosa* (5); *Centaurea apiculata* subsp. *spinulosa* (5); *Centaurea biebersteinii* subsp. *biebersteinii* (13); *Cleistogenes serotina* (13); *Cornus sanguinea* (13); *Coronilla emerus* (2); *Corydalis solida* (12); *Corylus colurna* (4); *Crepis biennis* (13); *Cruciata laevipes* (12); *Cytissus multiflorus* (13); *Daphne mezereum* (7); *Dianthus armeria* (7); *Erigeron annuus* (13); *Eryngium campestre* (13); *Euonymus latifolius* (2); *Euphorbia lingulata* (12);*Festuca rupicola* (13); *Festuca valesiaca* (1); *Festuca xanthina* (3); *Galium humifusum* (13); *Galium kitaibelianum* (3); *Galium verum* (13); *Gallium mollugo* (13); *Genista tinctoria* (1); *Geranium dissectum* (13); *Geranium lucidum* (12); *Helleborus purpurascens* (12); *Hordelymus europaeus* (7); *Hypochoeris maculata* (12); *Isopyrum*

thalictroides (12); *Juglans regia* (4); *Koeleria macrantha* (13); *Lactuca perennis* (11); *Lamiastrum galeobdolon* (5); *Leontodon autumnalis* (3); *Leontodon hispidus* (1); *Linaria genistifolia* (6); *Linum uninerve* (3); *Listera ovata* (12); *Melilotus officinalis* (13); *Melissa officinalis* (3); *Milium effusum* (1); *Muscari comosum* (13); *Onosma arenaria* (13); *Ophioglossum vulgatum* (12); *Orobanche caryophyllacea* (5); *Peltaria alliacea* (8); *Petrorhagia prolifera* (14); *Pinus nigra* subsp. *palasiana* (3); *Poa angustifolia* (5); *Poa nemoralis* (1); *Potentilla cinerea* (13); *Potentilla micrantha* (12); *Primula veris* (3); *Quercus pubescens* (13); *Ranunculus oxyspermus* (13); *Rosa corymbifera* (13); *Salvia pratensis* (1); *Saxifraga rotundifolia* (12); *Scorzonera hispanica* (13); *Scrophularia heterophylla* subsp. *laciniata* (8); *Senecio jacobaea* (13); *Senecio rupestris* (3); *Silene dichotoma* (13); *Silene flavescens* (8); *Silene nutans* (3); *Spiraea altissima* (1); *Spiraea chamaedryfolia* (11); *Stipa capillata* (13); *Tanacetum corymbosum* (4); *Thymus pannonicus* (1); *Thymus praecox* subsp. *skorpilii* (10); *Torilis arvensis* (13); *Trifolium alpestre* (1); *Valerianella locusta* (10); *Verbascum chaixii* (13); *Veronica austriaca* subsp. *austriaca* (1); *Veronica officinalis* (13); *Viola odorata* (12); *Viola tricolor* subsp. *subalpina* (10).

Authors and sampling areas: **1** - Bujorean *et al.*, 1969 – Valea Mare, Moldova Nouă; **2** – Boşcaiu *et al.*, 1971 – Defileul Dunării; **3** - Resmeriță, 1972 – Valea Țesna, Băile Herculane (*Syringo – Fraxinetum orni*); **4** – Resmeriță, 1972 – Valea Țesnei, Băile Herculane (*Syringo – Cotinetum coggygriae*); **5** – Grigore & Coste, 1979 – Moldova Veche; **6** – Schrött, 1972 – Cheile Nerei (*Asplenio - Syringetum*); **7** - Schrött, 1972 – Cheile Nerei (*Syringo – Fraxinetum orni*); **8** – Boşcaiu, 1971 – Valea Cernei; **9** – Peia, 1981 – Cheile Minişului; **10** – Peia, 1978 – Cheile Minişului; **11** – Coste *et al.*, 1990 (unpublished) – Cheile Caraşului; **12** – Măgălie, 1970 – Ponoare, Mehedinți; **13** - Sanda *et al.*, 1972, Hunedoara; **14** – Ardelean, 1980 – Cheile Bulz – Gurahonț.

Table 2: *Syringo – Carpinetum orientalis* Jakucs 1959

Reference/ /study area*	1	2	3	4	5	6	7	8	9	10	11	12	13	14	15
Number of relevés	**8**	**7**	**10**	**12**	**3**	**6**	**20**	**9**	**3**	**6**	**1**	**6**	**14**	**10**	**10**
Carpinus orientalis	+-3/8	+-4/7	2-5/10	2-5/12	4-5/3	+/6	3-5/20	1-2/9	+-2/3	1-5/6	3	3/6	2-5/14	3/10	+-3/10
Fraxinus ornus	+-2/8	+-2/7	1-2/10	+-3/10	+-2/2		+-3/20	+-1/7	+-1/2	+/6	+	+-1/6	+-1/13	+-1/2	+-2/6
Syringa vulgaris	1-3/8	+-1/6	+-1/10	+-3/12	+-3/3	1-2/6	+-4/12	2-5/9						+-2/4	
Crataegus monogyna	+/4		+/7	+/9	+-2/2	+/2	+-3/10	+/6		+/6	+	+/5	+/2	+-3/9	+/6
Cotinus coggygria	+/2	+/2	--1/7	+/6		1-3/6	+-4/12	+-1/4	3/3	3-5/4	+	+-1/6	+-1/8	+/4	
Cornus mas	+/6	+-2/5	+-1/7	+/4	+/1	+/2	+-3/15	+/1		+/1	+		+-1/6	+/5	+/2
Prunus mahaleb	+-2/3		+/7	+/5		+/2	+-3/14	+/4					+-1/10		
Tamus communis	+-3/5	+/4	+-1/9	+/2				+/5		+/1			+/10		
Ruscus aculeatus	+/1	-2/4	+-1/8	+-4/10			+-4/2	+/1		+/2		+/6			

Reference/ /study area*	1	2	3	4	5	6	7	8	9	10	11	12	13	14	15
Euonymus verrucosus	+/7	+/3	+/5	+/3			+/11	+/1					+/1		
Quercus pubescens	+/2		+/1		+/1			+-1/4		+-1/4			+/8		+-2/9
Clematis vitalba	+/5	+-1/4	+-2/3	+/3			+-2/5				+		+/1		+/6
Hedera helix			+-1/5	+/6		+/5	+-2/1	+/3			+				
Acer campestre		+-1/6	+-3/6			+/3				+/1		+/2			+/2
Tilia tomentosa		+/3	+-1/6	+/4			+/3				+	+/1			
Corylus avellana		+-1/4	+/1			+/4	+-2/5				+			+/3	
Quercus petraea	+/2		+/1				+-2/6	+/1	+/1	+/1			+-1/4		
Rosa canina	+/5			+-1/4	+/1			+/1		+/1				+/1	+/1
Sorbus aria	+/2						+/11								
Crataegus pentagyna									+/2		+	+/9			
Quercus cerris	+/1	+/2					+/1	+/1	+/2	+/1			+/2	+/1	
Staphylea pinnata	+/1	+-1/4	+-2/3	+/3											
Cotoneaster nebrodensis	+/3							+/3	+-2/2						
Ailanthus altissima			+/3	+/4											
Berberis vulgaris	+/1		+/1			+/2								+/3	
Pyrus pyraster			+/1	+-1/5											+/1
Quercus frainetto			+/1						+-2/2			+/3			+/1
Vitis silvestris			+/4							+/1			+/1		+/1
Acer monspessulanum	+/1		+/5												
Carpinus betulus				+/3				+/3							
Cornus sanguinea				+/2					+/2		+	+/2			
Juglans regia	+/1							+/5							
Viburnum lantana				+/5			+/1								
Ligustrum vulgare			+/1	+/2					+/2						
Rhamnus catharticus	+/3		+/2												
Sorbus domestica										+/1			+/4		
Chamaecytissus ciliatus			+/2										+/1		
Euonymus europaeus	+/1		+/1				+/1								
Sorbus torminalis	+/1		+/1							+/1					

Reference/ /study area*	**1**	**2**	**3**	**4**	**5**	**6**	**7**	**8**	**9**	**10**	**11**	**12**	**13**	**14**	**15**
Brachypodium silvaticum	+-1/4		+/4	+/2			+-2/11		+-1/3	+-1/7			+/8		+-2/8
Sedum telephium subsp. *maximum*	+/5	+/4	+/4	+/7	+/1		+/11	+/5		+/1	+		+/1		+/2
Lithospermum purpuro-caeruleum	+/1	+-1/3	+/4	+/3	+/1		+-4/15	+/1		+/3			+/5		
Piptatherum virescens	+-1/4	+/3	+-1/8	+/2			+-3/12			+/1		+/3			
Asplenium ceterach	+/4		+/4	+/4		+/4	+/15	+/1							
Echinops bannaticus	+/2	+-1/2	+/1	+/2			+-1/12	+/6		+/2			+/1	+/1	+/3
Teucrium chamaedrys	+-1/7						+/14	+/4	+/2				+/1		+/3
Scabiosa columbaria	+-1/8	+/1	+/7	+/3				+/6					+/1		+/4
Dactylis glomerata subsp. *ascheroniana*	+/5		+/1	+/4						+-1/4			+-1/11		+/2
Aurinia petraea	+/4		+/1	+/3			+/1	+/6			+	+/4			+-2/4
Orlaya grandiflora	+/1			+/3	+/1			+-3/4	+/2	+/1		+/2		+/6	+/4
Euphorbia epithymoides	+/2		+/5	+/7				+/7		+/2					
Poa nemoralis	+/2		+/1	+/3	+/1			+/2	+-1/3	+-1/6		+/3	+/2		
Polygonatum odoratum	+/6		+/4	+/3			+/7	+/3							
Campanula divergens	+/4						+/1	+/7					+/10		
Origanum vulgare	+/2			+/3	+/1	+/2	+/10	+/3						+/1	
Stachys recta subsp. *leucoglossa*	+	+-1/6	−/1	+-2/2			+/2			+/2		+/2			+/6
Arabis turitta	+/3	+/3	+/1	+/4			+/6	+/1		+/1		+/2			
Euphorbia cyparissias	+/1		+/1		+/1	+/2	+/1			+/3			+/1	+/1	+/10
Coronilla varia	+-2/8	−/2	+/2	+/3			+/1	+/3							+/1
Geranium robertianum	+/5		+/3				+/11							+/1	
Arum maculatum (+ A. orientale)	+/2		+/4					+/1					+/11		+/1
Chrysopogon gryllus					+-1/1		+/1	+/3					+/5	+/1	+-1/8
Veronica chamaedrys	+/2						+/1	+/5					+/6	+/3	+/2
Acinos arvensis					+/1		+/16								+/1

Reference/ /study area*	1	2	3	4	5	6	7	8	9	10	11	12	13	14	15
Asplenium trichomanes	+/2		+/2	+/4			+/5					+/4			+/1
Centaurea atropurpurea	+/5		+/4	+/4			+/1	+/4							
Glechoma hirsuta	+/1		+/6	+/2						+/1			+/5	+/3	
Delphinium fissum	+/5		+/4					+-1/7							
Lunaria annua	+-1/1		+-3/4	+/2				+/3					+-1/6		
Teucrium montanum	+/2			+/5	+/1	+/4						+/4		+/1	
Eryssimum odoratum	+/5		+/1	+/5			+/1	+/2					+/1		
Geum urbanum	+/2		+/4							+/1			+/5	+/3	
Piptatherum holciforme								+/1					+-3/14		
Scutellaria altissima	+/1		+/4	+/3				+/6		+/1					
Achillea coarctata								+/1					+/5		+/8
Achillea crithmifolia			+/1					+/7	+/1			+/3	+/1		+/1
Cardamine graeca			+-1/2					+-2/3					+-2/9		
Digitalis grandiflora	+/4		+/2	+/3			+/1	+/3						+/1	
Ferula heuffelii	+/2			+/3				+/8					+/1		
Hypericum perforatum	+/3			+/3			+/1	+/6					+/1		
Phleum montanum	+/1							+-1/4		+/3			+/2		+/4
Thalictrum aquilegiifolium	+/4		+/3	+/3			+/1	+/2						+/1	
Allium flavum	+/3		+/1	+/4			+/1	+/3			+				
Bromus riparius	+-3/8							+/5							
Campanula sparsa subsp. *sphaerotrix*	+/1												+/1	+/2	+/9
Dictamnus albus	+/1		+/1									+/2	+/9		
Asparagus tenuifolius								+/1					+/11		
Galium aparine			+/4	+/2								+/2			+/4
Inula conyza							+/10						+/1	+/1	
Melica ciliata	+/6					+/4	+-2/1	+/5						+/2	
Veronica austriaca subsp. *teucrium*		+/1		+/2			+/1		+/2	+/2		+/4			
Campanula grossekii			+/1				+/1						+/9		

Reference/ /study area*	**1**	**2**	**3**	**4**	**5**	**6**	**7**	**8**	**9**	**10**	**11**	**12**	**13**	**14**	**15**
Carduus candicans	+/1				+/1			+/5	+/2	+/1	+				
Dianthus giganteus subsp. *banaticus*	+/1			+/2			+/1	+/6						+/1	
Jurinea glyacantha	+/3		+/1	+/4								+/3			
Melica uniflora	+/2		+/6	+/2			+/1								
Sedum hispanicum				+/7				+/1		+/2				+/1	
Silene vulgaris subsp. *alpina*	+/1		+/4	+/4									+/2		
Acanthus balcanicus				+/2				+/8							
Campanula lingulata			+/1		+/1					+/4			+/4		
Clinopodium vulgare	+/3						+/1		+/2				+/4		
Lychnis coronaria					+/1		+/1			+/1		+/2	+/5		
Potentilla argentea				+/2						+/2				+/6	
Potentilla micrantha		+/2	+/1										+/5		
Tanacetum corymbosum	+/2			+/2				+/3		+/3					
Vincetoxicum hirundinaria	+/3		+/1	+/4			+/1						+/1		
Asperula tenella	+/2			+/4				+/1						+/1	+/1
Campanula persicifolia				+-2/5			+/2						+/2		
Centaurea biebersteinii subsp. *biebersteinii*	+/1													+/4	+/4
Elymus hispidus	+/1								+/3	+/2		+/3			
Festuca valesiaca				+/6	+-1/2										+/1
Helianthemum nummularium	+/1			+/2		+/2						+/3		+/1	
Medicago sativa subsp. *falcata*						+/2		+/7							
Seseli rigidum		+/1	+/4	+				+/3							
Coronilla emerus	+/1		+-2/6												+/1
Euphorbia amygdaloides	+/1		+/6				+/1								

Reference/ /study area*	**1**	**2**	**3**	**4**	**5**	**6**	**7**	**8**	**9**	**10**	**11**	**12**	**13**	**14**	**15**
Galium mollugo				+/2			+/1	+/3		+/1					+/1
Silene viridiflora										+/1			+/7		
Asplenium septentrionale														+/3	+/4
Bupleurum praealtum	+/1			+/2						+/1			+/3		
Dicanthium ischaemum					+-2/2					+-1/3					+-1/2
Thymus pannonicus				+/2			+1								+/4
Asplenium ruta-muraria			+/2			+/3	+/1								
Campanula rapunculus				+/3				+/3							
Convolvulus cantabrica								+/1					+/3		+/2
Fragaria collina	+/1			+/1	+/1			+-1/3							
Heleborus odorus								+/2						+/4	
Leontodon hispidus			+/1	+/2		+/3									
Verbascum phoeniceum								+/1							+/5
Vinca herbacea					+/1		+/1								+/4
Alliaria petiolata				+/4				+/1							
Astragalus glycyphyllos			+/2	+/2						+/1					
Gallium flavescens	+/2			+/2				+/1							
Lathyrus venetus		+/2	+/1										+/2		
Orchis simia							+/1						+/4		
Verbascum lychnitis				+/1				+/3						+/1	
Veronica austriaca	+/1			+/3									+/1		
Acinos arvensis subsp. *majoranifolius*	+/3		+/1												
Campanula glomerata	+/2						+/1							+/1	
Dianthus petraeus	+/2		+/2												

Reference/ /study area*	**1**	**2**	**3**	**4**	**5**	**6**	**7**	**8**	**9**	**10**	**11**	**12**	**13**	**14**	**15**
Dorycnium pentaphyllum subsp. *germanicum*														+/1	+/3
Linaria genistifolia						+/3		+/1							
Potentilla recta	+/1			+/2						+/1					
Sesleria rigida			+/2	+/2											
Athamanta hungarica	+/1			+/2											
Glechoma hederacea	+/1								+/1					+/1	
Polypodium vulgare	+/1				+/1			+/1							
Asperula tinctoria			–/1				+/1								
Bromus sterilis												+/1			+/1
Calystegia sylvatica	+/1		+/1												
Crocus moesiacus								+/1					+/1		
Helianthemum canum													+/1	+/1	
Koeleria macrantha										+/1					+/1
Lepidium campestre						+/1									+/1
Medicago lupulina				+/1											+/1
Myrrhoides nodosa			+/1										+/1		
Ornithogalum collinum	+/1													+/1	
Peltaria alliacea	+/1												+/1		
Seseli libanotis	+/1						+/1								
Viola alba							+/1					+/1			
Viola tricolor					+/1							+/1			

Species in one sample (between paranthesis, the number of the author / column):
Acer tataricum (9); *Achillea millefolium* (14); *Achillea setacea* (15); *Agrostis capillaris* (7); *Ajuga genevensis* (13); *Ajuga laxmannii* (8); *Althaea cannabina* (14); *Anthoxanthum odoratum* (14); *Anthriscus cerefolium* (13); *Arabidopsis thaliana* (13); *Arabis hirsuta* (3);*Arenaria serpyllifolia* (5); *Aristolochia pallida* (14); *Asparagus officinalis* (8); *Ballota nigra* (15); *Bellis perennis* (9); *Brachypodium pinnatum* (1); *Briza media* (14); *Bromus commutatus* (14); *Bromus erectus* (14); *Bromus japonicus* (15); *Bromus squarrosus* (8); *Cachrys ferulacea* (8); *Calamintha nepeta* subsp. *glandulosa* (3); *Camelina microcarpa* (8); *Campanula rapunculoides* (1); *Campanula sibirica* (14); *Campanula trachelium* (3); *Cardaminopsis arenosa* (5); *Carduus hamulosus* (15); *Carex digitata* (2); *Carex muricata* (1); *Carthamus lanatus* (15); *Celtis australis* (9);*Centaurea cuneifolia* subsp. *pallida* (8); *Centaurea rocheliana* (4); *Cerastium banaticum* (6); *Cerastium pumilum* (15); *Chamaecytisus hirsutus* (1); *Chamaecytisus supinus* (1); *Cleistogenes serotina* (10);

Corylus colurna (3); *Crepis sancta* (13); *Crucianella angustifolia* (15); *Cruciata glabra* (12); *Cuscuta monogyna* (8); *Cynoglossum officinale* (14); *Cynosurus cristatus* (7); *Cynosurus echinatus* (15); *Dasypirum villosum* (15); *Dianthus armeria* (7); *Dianthus giganteus* (7); *Diplotaxis muralis* (6); *Draba lasiocarpa* (1); *Echium vulgare* (14); *Epipactis helleborine* (2); *Eryngium campestre* (15); *Euonymus latifolius* (2); *Fallopia dumetorum* (6); *Ferulago silvatica* (1); *Filago pyramidata* (14); *Fritillaria orientalis* (8); *Fumaria officinalis* (6); *Galeopsis speciosa* (4); *Galium verum* (15); *Genista radiata* (1); *Genista tinctoria* (4); *Geranium dissectum* (15); *Geranium pusillum* (15); *Geranium rotundifolium* (9); *Geranium sanguineum* (11); *Hieracium praealtum* subsp. *bauhinii* (15); *Hypericum rochelii* (1); *Hypochoeris radicata* (9); *Inula ensifolia* (1); *Inula hirta* (15); *Iris reichenbachii* (6); *Lactuca viminea* (1); *Lamium garganicum* subsp. *laevigatum* (3); *Lamium purpureum* (15); *Lapsana communis* (1); *Lathyrus sphaericus* (1); *Lavatera thuringiaca* (10); *Leontodon crispus* subsp. *crispus* (15); *Leucanthemum vulgare* (4); *Linum uninerve* (1); *Lotus corniculatus* (1); *Lychnis viscaria* (8); *Marrubium vulgare* (15); *Medicago arabica* (15); *Melampyrum nemorosum* (1); *Melica nutans* (1); *Micromeria pullegium* (1); *Milium vernale* (14); *Minuartia setacea* (1);*Muscari comosum* (15); *Muscari neglectum* (6); *Myosotis ramosissima* (13); *Myosotis stricta* (9); *Onobrychis alba* (6); *Onopordum acanthium* (15); *Paeonia officinalis* subsp. *banatica* (10); *Petrorhagia prolifera* (1); *Peucedanum officinale* (4); *Peucedanum oreoselinum* (1); *Plantago lanceolata* (15); *Poa angustifolia* (5); *Ranunculus bulbosus* (15); *Ranunculus illyricus* (8); *Rosa tomentosa* (4); *Rubus caesius* (15); *Rubus discolor* (9); *Salvia pratensis* (4); *Sambucus ebulus* (14); *Saxifraga rotundifolia* (1); *Scutellaria velenovskyi* (8); *Sedum album* (1); *Sempervivum schlehanii* (1); *Senecio vulgaris* (5); *Sherardia arvensis* (5); *Sideritis montana* (6); *Silene italica* (7); *Silene otites* (6); *Silene vulgaris* (2); *Smyrnium perfoliatum* (14); *Sorbus graeca* (7); *Spiraea media* (1); *Stachys recta* (8); *Stipa capillata* (14); *Sysimbrium orientale* (15); *Taraxacum officinalis* (6); *Thlaspi perfoliatum* (14); *Thymus glabrescens* subsp. *glabrescens* (14); *Thymus pulegioides* (13); *Tordylium maximum* (15); *Tragopogon balcanicus* (6); *Trifolium alpestre* (4); *Trifolium diffusum* (8); *Trifolium montanum* (5); *Ventenata dubia* (1); *Verbascum banaticum* (15); *Verbascum speciosum* (15); *Veronica austriaca* subsp. *austriaca* (4); *Veronica serpyllifolia* (4); *Veronica spicata* (1); *Vicia cracca* (5); *Vicia grandiflora* (15); *Vicia sativa* subsp. *nigra* (15); *Vicia sepium* (1); *Viola arvensis* (15); *Viola hirta* (1); *Xeranthemum annuum* (15).

* Authors and sampling areas: **1**- Jackucs, 1959 – Valea Cernei, Domogled; **2** - Boşcaiu, 1971 – Valea Cernei, Domogled; **3** - Boşcaiu *et al*., 1971 – Defileul Dunării; **4** - Bujorean *et al*., 1969 – Valea Mare – Moldova Nouă; **5** - Grigore & Coste, 1974 – Moldova Nouă; **6** - Păun *et al*., Defileul Dunării (Pescari - Berzeasca); **7** – Schrött, 1972 – Cheile Nerei; **8** – Roman, 1974 – Podişul Mehedinţi; **9** – Csürös *et al*., 1968 – Defileul Dunării (Orşova - Eşelniţa); **10** – Coste, 1975 – Baziaş, Munţii Locvei; **11** – Coste, 1994 – Baziaş (unpublished); **12**- Păun *et al*., 1970 – Defileul Dunării (Pescari - Berzeasca); **13** – Roman, 1974 – Podişul Mehedinţi; **14** - Schrött, 1972 – Cheile Nerei; **15** – Ştefureac & Popescu, 1970 – Defileul Dunării.

FOREST COMMUNITIES FLORISTICALLY SPECIFIC TO EASTERN ROMANIA

Toader CHIFU[*], Nicolae ŞTEFAN, Oana ZAMFIRESCU, Ciprian MÂNZU, Ştefan ZAMFIRESCU
"Al. I. Cuza" University, Faculty of Biology, no. 20A, Carol I Bvd.,
700505 Iaşi (ROMANIA); e-mail: chifutoader@yahoo.com

Abstract: A series of forest communities from the eastern part of Romania (Moldova) are presented synthetically. Phytosociologically, these communities appertain to the following associations: *Aro orientalis - Carpinetum* (Dobrescu et Kovacs 1973) Täuber 1992 (*typicum* **subass. nova** hoc loco; *quercetosum pedunculiflorae* Chifu *et al.* 2002); *Dentario quinquefoliae - Carpinetum* (Dobrescu et Kovacs 1973) Täuber 1992 (*typicum* Chifu et Zamfirescu 2001; *fraxinetosum excelsioris* Chifu et Zamfirescu 2001); *Lathyro veneti - Fagetum* (Dobrescu et Kovacs 1973) Chifu 1995 (*typicum* (Chifu 1995) Chifu et Zamfirescu 2001; *quercetosum dalechampii* (Chifu 1995) Chifu *et al.* 1999). Floristic analysis reveals the dominance of Pontic and Balkan elements compared to Central European species. The reason for this pattern is the location of the studied communities in the zone where Central European and Pontico-South-Siberian regions overlap. Based on the similarity between them, the relevés studied fall into three clusters that correspond to the three above-mentioned associations of the same coenotaxon – suballiance *Aro orientalis - Carpinion*. The grouping of *Lathyro veneti - Fagetum* with *Dentario quinquefoliae - Carpinetum* relevés results from the overlapping altitudinal distribution of the two forest types. *Quercus dalechampii* dominance-abundance scores distinguish the communities of *Dentario quinquefoliae - Carpinetum*, whereas *Fraxinus excelsior* discriminates between the two subassociations. The communities belonging to *Aro orientalis - Carpinetum* distinguish themselves through the high cover scores of *Quercus robur* and *Q. pedunculiflora*. Furthermore, *Quercus pedunculiflora* has higher cover scores in southern relevés (Vaslui County), whereas *Q. robur* dominates in the northern relevés (Iaşi County). The group of relevés of *Lathyro veneti - Fagetum* is clearly discriminated by the high cover of *Fagus sylvatica*.

Introduction

From 1950 several botanists, especially from Iaşi Academic Center, have carried out investigations that reveal the specificity of the vegetation cover in the east of Romania [5, 6, 7, 8, 9, 10, 11, 14, 19, 26, 29]. Thus, on the Plateau of Bârlad, the forest associations are different compared with those from other Romanian forests.

This specificity results from the occurrence of certain Pontic elements (including Ponto-Balkan and Pontico-Mediterranean elements), Balkan elements, Mediterranean elements and Central European-Mediterranean elements, such as: *Allium ursinum* subsp. *ucrainicum*, *Arum orientale*, *Asparagus tenuifolius*, *Carex brevicolis*, *Corydalis cava* subsp. *marschalliana*, *Dentaria quinquefolia*, *Fagus orientalis*, *Fagus sylvatica* subsp. *moesiaca*, *Fagus taurica*, *Fraxinus coriariaefolia*, *Lathyrus aureus*, *Melica picta*, *Quercus dalechampii*, *Quercus pedunculiflora*, *Tilia tomentosa*, etc.

Our research emphasised the high frequency of *Fagus taurica*, *Quercus dalechampii*, *Quercus pedunculiflora*, species considered as rare for this area.

Study area

Moldavia is a historical province situated in E-NE Romania. The altitude in this area decreases from the Eastern Carpathians towards the Danube.

The Moldavian Plateau extends in the north and centre of the Moldavian Platform, up to the Bârladului Basin. The relief has resulted from the shaping of Sarmatic and Pliocene surface deposits.

The Moldavian climate is temperate-continental with extreme tendencies on the Moldavian Plateau. Annual rainfall is 500–700 mm, which brings about periods of drought.

The beech and hornbeam forests studied occur in the Central Moldavian Plateau, as far as the Tutovei Hills. Here the hornbeam forests expand further south, as far as the Fălciului and Covurluiului Hills. The altitude ranges between 100 m and 400 m for the hornbeam forests, and between 250 m and 400 m for the beech forests.

The soils in this area have a great variety and zonal distribution. The plant communities studied occur on cambic chernozem and brown forest soils. In addition, hornbeam forests can appear on clays.

These forests thrive in a warm, dry climate with an annual average temperature of 8–9°C, and annual average rainfall of 450–600 mm.

From the phytogeographical point of view, the forest area in which these associations were identified corresponds to the Central European Region – the Moldavian Plateau Province [18].

Material and Methods

The phytosociological analysis followed the principles of the Central-European geobotanical school (Zürich–Montpellier). The floristic composition of the coenotaxa is presented by coenotic category, which includes species characteristic of associations, suballiances, alliances, orders, and classes of vegetation. In addition, the table comprises the presence of each species in the plant communities, expressed by the constancy score (K).

Row abundance-dominance (+ – 5) code data were converted in proportions (0.1; 5; 17.5; 37.5; 62.5; 87.5) and, afterwards, transformed, using the arcsin analytical function. The similarity between relevés was described through an agglomerative hierarchical classification (Bray-Curtis index, average distance linkage) and through a principal components analysis. The latter can also give information on the species that differentiate best between the clusters of relevés previously distinguished.

Results

The studies of our predecessors and our own studies allowed the identification of the following coenotaxa on the Plateau of Bârlad (Table 1):

Cl. *Querco – Fagetea* Br.-Bl. et Vlieger 1937
Ord. *Fagetalia sylvaticae* Pawlowski in Pawlowski *et al.* 1928
Al. *Lathyro hallersteinii – Carpinion* Boşcaiu 1974
Subal. *Aro orientalis – Carpinenion* (Dobrescu et Kovacs 1973) Täuber 1991-1992

1. *Aro orientalis – Carpinetum* (Dobrescu et Kovacs 1973) Täuber 1992
   - *typicum* subass. nova hoc loco
   - *quercetosum pedunculiflorae* Chifu *et al.* 2002
2. *Dentario quinquefoliae – Carpinetum* (Dobrescu et Kovacs 1973) Täuber 1992
   - *typicum* Chifu et Zamfirescu 2001
   - *fraxinetosum excelsiori* Chifu et Zamfirescu 2001
3. *Lathyro veneti – Fagetum* (Dobrescu et Kovacs 1973) Chifu 1995
   - *typicum* (Chifu 1995) Chifu et Zamfirescu 2001
   - *quercetosum dalechampii* (Chifu 1995) Chifu *et al.* 1999

Table 1: Forest associations of the suballiance *Aro orientalis – Carpinenion* on the Bârlad Plateau.

Forest association	**1a**	**1b**	**2a**	**2b**	**3a**	**3b**
Altitude (m)	100-390	100-260	120-400	150-320	220-370	200-440
Aspect (dominant)	N, NE	S	SW, SE	-	N, NW	E, SE
Slope (degrees)	5-15	0-5	5-15	-	10-25	10-20
Tree layer cover (%)	70-95	75-100	70-90	70-90	70-85	70-90
Shrubs / saplings layer cover (%)	5-30	15-40	10-20	10-50	0-10	10-15
Herbaceous layer cover (%)	5-55	10-90	10-45	5-70	5-10	5-35
Number of species	210	122	213	109	170	228
Number of relevés	100	16	121	13	35	149
Ass. characteristic sp.						
Arum orientale	IV	IV	II	II	II	II
Dentaria quinquefolia	-	-	II	-	I	I
Lathyrus venetus	II	III	III	I	IV	III
Subass. differential sp.						
Quercus pedunculiflora	-	V	I	I	-	I
Quercus dalechampii	II	I	V	I	-	V
Fraxinus excelsior	III	III	II	V	III	III
Aro orientalis – Carpinenion						
Asparagus tenuifolius	II	II	II	I	I	I
Carex brevicolls	I	-	II	I	I	I
Carpesium cernuum	I	-	II	-	I	II
Carpinus betulus	V	V	V	V	V	V
Corydalis cava subsp. cava	-	-	I	I	I	I
Corydalis cava subsp. marschalliana	II	I	III	II	III	III
Fagus taurica	I	-	I	I	V	V
Fraxinus coriariifolia	-	-	I	-	-	I
Fritillaria orientalis	I	I	I	-	-	-

Forest association	**1a**	**1b**	**2a**	**2b**	**3a**	**3b**
Laser trilobum	I	-	I	-	-	-
Lathyrus aureus	-	-	II	-	II	II
Melica picta	I	-	I	-	-	I
Physospermum cornubiense	-	-	I	-	-	-
Scopolia carniolica	-	-	I	-	-	I
Scutellaria altissima	II	III	IV	III	I	II
Symphytum tauricum	-	-	I	-	-	-
Tilia tomentosa	V	V	V	V	V	V
Lathyro hallersteinii – Carpinenion						
Aposeris foetida	-	-	I	-	-	-
Dentaria glandulosa	I	-	I	-	III	III
Festuca drymea	-	-	-	-	-	I
Melampyrum bihariense	I	-	I	-	-	II
Galio schultesii – Carpinenion						
Campanula trachelium	II	II	III	I	I	III
Carex pilosa	III	II	V	II	III	V
Cerasus avium	III	IV	III	IV	III	IV
Dactylis polygama	IV	V	IV	IV	I	III
Festuca heterophylla	-	-	I	-	-	-
Galium schultesii	III	I	IV	II	I	III
Glechoma hirsuta	III	IV	V	IV	II	IV
Lathyrus vernus	III	III	III	III	II	IV
Melampyrum nemorosum	-	-	I	-	-	-
Omphalodes scorpioides	I	-	-	-	-	-
Primula acaulis	I	I	-	-	-	-
Ranunculus cassubicus	I	I	I	I	I	I
Stellaria holostea	IV	III	III	III	I	I
Tilia cordata	IV	IV	III	I	II	IV
Asperulo taurinae – Carpinenion						
Asperula taurina subsp. leucanthera	-	I	-	-	-	I
Coronilla elegans	-	-	I	-	-	-
Fagus sylvatica subsp. moesiaca	-	-	-	-	-	II
Potentilla micrantha	-	-	-	-	-	I
Tamus communis	-	-	I	-	-	-
Symphyto – Fagion						
Acer pseudoplatanus	I	-	II	II	II	III
Cephalanthera damasonium	-	-	II	I	I	II
Cephalanthera longifolia	I	-	I	-	I	I
Cephalanthera rubra	-	-	I	I	-	I
Epipactis atrorubens	-	I	-	-	-	-
Epipactis helleborine	I	I	II	I	II	III
Epipactis purpurata	-	-	-	-	-	II
Hordelymus europaeus	II	-	II	I	I	II
Listera ovata	-	-	I	-	I	III
Lunaria annua subsp. pachyrhiza	I	-	I	-	I	I
Polystichum aculeatum	-	-	-	-	I	I
Ulmus glabra	I	I	I	I	I	I
Veronica officinalis	I	I	II	I	I	II
Veronica urticifolia	-	-	-	-	I	I
Fagetalia sylvaticae						
Actaea spicata	I	-	I	-	I	III
Anemone ranunculoides	II	I	II	I	II	III
Asarum europaeum	III	II	IV	IV	III	IV
Campanula rapunculoides	II	II	II	I	I	II
Carex sylvatica	II	I	II	I	III	IV
Corydalis solida	II	I	II	I	I	III

Forest association	**1a**	**1b**	**2a**	**2b**	**3a**	**3b**
Daphne mezereum	-	-	-	-	-	I
Dryopteris carthusiana	-	-	I	-	I	II
Euphorbia amygdaloides	III	II	V	III	IV	IV
Fagus sylvatica subsp. sylvatica	I	I	II	II	V	V
Galanthus nivalis	-	-	I	-	-	II
Galium odoratum	III	III	IV	IV	IV	V
Geranium robertianum	II	III	II	III	III	IV
Isopyrum thalictroides	I	I	I	-	II	III
Lamium galeobdolon	II	I	II	-	I	III
Lathraea squamaria	I	-	I	-	II	II
Lilium martagon	I	-	I	I	I	I
Maianthemum bifolium	I	-	I	-	I	II
Mercurialis perennis	II	II	II	III	II	III
Milium effusum	II	I	II	I	I	I
Paris quadrifolia	I	-	I	I	I	III
Platanthera bifolia	II	-	I	I	II	II
Polygonatum multiflorum	II	-	II	-	-	II
Pulmonaria obscura	I	II	I	-	I	I
Pulmonaria officinalis	IV	II	III	IV	III	IV
Rubus hirtus	I	-	I	-	II	III
Salvia glutinosa	II	II	I	-	I	IV
Sanicula europaea	II	I	II	II	III	III
Scrophularia nodosa	II	I	III	II	I	III
Vicia sylvatica	I	-	I	-	I	I
Viola mirabilis	II	I	II	I	I	II
Tilia platyphyllos	I	I	I	I	II	I
Alno – Fraxinetalia et Alnion incanae						
Aegopodium podagraria	III	III	II	II	II	III
Carex pendula	I	I	I	-	I	I
Carex remota	I	I	I	II	I	I
Circaea lutetiana	II	I	I	II	II	III
Evonymus nanus	I	I	-	-	-	-
Festuca gigantean	I	I	I	I	I	I
Frangula alnus	I	-	I	-	-	I
Fraxinus angustifolia	I	I	-	-	I	-
Geranium phaeum	I	II	I	I	I	I
Glechoma hederacea	I	III	I	I	I	II
Humulus lupulus	-	-	I	-	-	I
Impatiens noli tangere	I	-	-	-	I	-
Lamium maculatum	II	II	II	I	I	II
Malus sylvestris	I	I	I	-	-	I
Pyrus pyraster	I	I	II	I	I	I
Rumex sanguineus	I	-	I		I	I
Ranunculus ficaria subsp. calthifolius	I		I	III	II	II
Sambucus nigra	II	II	I	II	II	II
Solanum dulcamara	I	-	I	-	-	-
Stachys sylvatica	II	III	III	III	III	IV
Stellaria nemorum	I	I	I	I	I	I
Ulmus minor	III	III	II	I	I	II
Querco – Fagetea						
Acer campestre	IV	V	IV	V	III	III
Acer platanoides	III	III	III	IV	IV	IV
Ajuga reptans	II	-	II	I	II	I
Anemone nemorosa	I	-	I	I	-	II
Athyrium filix-femina	I	I	I	I	I	I

Forest association	1a	1b	2a	2b	3a	3b
Brachypodium sylvaticum	IV	IV	IV	V	II	III
Bromus benekeni	-	I	II	II	-	I
Carex digitata	I	-	I	-	I	II
Convallaria majalis	III	III	III	II	I	II
Cypripedium calceolus	I	-	-	-	-	I
Dentaria bulbifera	III	II	III	I	IV	IV
Dryopteris filix-mas	I	-	II	I	I	III
Epilobium montanum	I	I	I	-	I	II
Geum urbanum	III	V	II	IV	III	II
Hedera helix	III	I	IV	II	II	IV
Hepatica nobilis	I	-	I	-	I	I
Hieracium murorum	-	-	I	-	-	I
Lonicera xylosteum	I	-	-	-	-	-
Melica nutans	I	-	II	-	I	I
Melica uniflora	III	IV	V	V	I	III
Moehringia trinervia	I	I	I	-	I	II
Mycelis muralis	III	I	III	II	III	IV
Neottia nidus-avis	I	I	I	-	II	III
Orthilia secunda	-	-	-	-	I	I
Poa nemoralis	III	I	III	I	II	III
Polygonatum latifolium	II	IV	I	III	II	I
Primula veris	I	-	I	-	I	I
Quercus petraea	I	-	II	I	II	II
Quercus robur	V	V	III	I	I	III
Ranunculus auricomus	III	II	III	I	I	III
Scilla bifolia	I	-	II	I	I	III
Staphylea pinnata	II	-	II	I	I	II
Ulmus procera	I	I	I	III	I	I
Vicia sepium	II	-	I	I	I	I
Viola reichenbachiana	IV	V	IV	V	III	IV
Quercetea pubescentis						
Acer tataricum	II	III	III	II	-	-
Asparagus officinalis	I	-	I	I	-	-
Campanula bononiensis	I	-	-	-	-	I
Campanula persicifolia	I	-	II	-	-	I
Campanula rapunculus	I	-	I	-	-	I
Clinopodium vulgare	I	I	I	I	I	I
Cornus mas	II	IV	IV	V	I	IV
Cotinus coggygria	-	-	I	-	-	-
Lactuca quercina	I	-	II	I	-	-
Lathyrus niger	II	II	III	-	I	III
Lithospermum purpureo-caeruleum	I	III	II	I	-	I
Polygonatum odoratum	II	-	I	I	II	I
Quercus polycarpa	I	I	II	I	-	-
Quercus pubescens	-	-	I	-	-	-
Sorbus domestica	I	-	I	-	-	I
Sorbus torminalis	II	I	III	I	I	II
Viola hirta	I	-	I	I	I	I
Viola suavis	-	-	I	-	I	I
Prunetalia et Sambucetalia racemosae						
Betula pendula	I	-	-	-	I	I
Clematis vitalba	I	I	I	I	I	I
Cornus sanguinea	III	II	III	II	I	II
Crataegus monogyna	IV	IV	IV	IV	II	II
Evonymus europaeus	II	IV	II	IV	I	II
Evonymus verrucosus	IV	III	IV	III	-	II

Forest association	1a	1b	2a	2b	3a	3b
Fragaria vesca	II	I	III	I	I	II
Ligustrum vulgare	II	I	III	I	I	I
Physalis alkekengi	I	-	I	I	-	I
Populus tremula	I	-	II	-	II	III
Prunus spinosa	I	I	I	I	-	I
Rhamnus cathartica	I	-	I	-	-	-
Rosa pimpinellifolia	-	-	I	-	-	-
Rubus candicans	-	-	-	-	-	I
Salix caprea	I	-	I	-	I	I
Viburnum lantana	IV	III	II	III	I	II
Epilobietea angustifolii						
Arctium nemorosum	I	-	I	I	-	I
Atropa belladonna	I	-	-	-	I	II
Digitalis grandiflora	-	-	-	-	-	I
Epilobium collinum	I	-	I	I	-	-
Eupatorium cannabinum	I	-	I	-	I	I
Galeopsis speciosa	I	I	I	-	I	II
Galeopsis tetrahit	I	-	I	-	I	I
Hypericum hirsutum	I	-	I	-	-	II
Torilis japonica	I	I	I	I	I	I
Trifolio – Geranietea						
Agrimonia eupatoria	I	-	I	-	I	I
Astragalus cicer	I	-	-	-	-	-
Astragalus glycyphyllos	I	I	I	I	-	I
Astrantia major	-	-	-	-	I	-
Carex divulsa	I	-	III	II	I	II
Clematis recta	I	-	-	-	-	-
Coronilla varia	-	-	-	-	-	I
Inula conyza	I	-	II	-	-	II
Lithospermum officinale	-	-	I	-	-	-
Melampyrum cristatum	I	-	-	-	-	I
Origanum vulgare	I	-	-	-	I	I
Trifolium medium	-	I	-	-	I	I
Valeriana wallrothii	-	-	I	-	-	-
Veronica chamaedrys	I	I	II	I	I	I
Vicia dumetorum	I	-	I	-	I	II
Galio – Urticetea						
Aethusa cynapium	I	-	-	I	I	-
Alliaria petiolata	II	III	I	II	I	I
Anthriscus cerepholium subsp. trichosperma	I	I	I	-	-	-
Aristolochia clematitis	I	-	-	-	-	-
Cardamine impatiens	I	I	I	-	I	II
Chaerophyllum aromaticum	I	-	I	II	I	I
Chaerophyllum aureum	-	-	-	-	-	I
Chaerophyllum bulbosum	-	-	-	-	I	I
Chaerophyllum hirsutum	I	-	I	-	-	-
Chaerophyllum temulum	II	-	-	-	I	I
Galium aparine	II	I	II	I	III	II
Heracleum sphondylium	I	-		-	-	I
Lapsana communis	II	III	III	III	II	II
Myosotis sparsiflora	I	-	-	-	-	-
Polygonum dumetorum	II	II	I	I	I	I
Rubus caesius	I	I	I	I	-	I
Sambucus ebulus	I	-	I	-	-	-
Telekia speciosa	-	-	-	-	-	I

Forest association	**1a**	**1b**	**2a**	**2b**	**3a**	**3b**
Urtica dioica	I	III	I	II	I	I
Veronica hederifolia	I	-	I	-	I	I
Viola alba	-	III	I	III	-	I
Viola odorata	I	I	I	I	I	I
Molinio – Arrhenatheretea						
Hypericum maculatum	I	-	I	-	-	-
Lolium perenne	I	I	-	-	-	-
Lysimachia nummularia	I	-	I	I	I	I
Ophioglossum vulgare	-	-	-	-	-	I
Plantago lanceolata	I	-	-	-	-	I
Poa pratensis	-	I	-	-	-	-
Prunella vulgaris	I	-	I	I	I	I
Ranunculus repens	I	-	-	-	-	-
Rorippa sylvestris	I	-	-	-	-	-
Stachys officinalis	-	-	I	-	I	-
Taraxacum officinale	I	-	I	-	-	I
Trifolium repens	I	-	-	-	-	-
Valeriana officinalis	I	-	-	-	-	-
Festuco – Brometea						
Achillea nobilis subsp. neilreichii	I	-	-	I	-	I
Ajuga genevensis	-	-	I	-	I	I
Anthericum ramosum	I	-	-	-	-	-
Asperula cynanchica	-	I	-	-	-	-
Bupleurum affine	I	-	I	-	-	-
Campanula glomerata	-	-	I	-	-	-
Festuca valesiaca	-	I	-	-	-	-
Galium verum	I	-	-	-	I	-
Plantago media	-	-	I	-	I	-
Poa angustifolia	I	-	I	-	-	-
Teucrium chamaedrys	I	-	I	-	-	-
Koelerio – Korynephoretea						
Hieracium pilosella	I	-	-	-	-	-
Hypericum perforatum	I	-	I	I	I	I
Sedum acre	-	-	I	-	-	-
Sedum maximum	I	I	II	-	I	I
Asplenietea trichomanis						
Asplenium ruta-muraria	-	-	-	-	-	I
Asplenium scolopendrium	-	-	-	-	I	I
Asplenium trichomanes	-	-	-	-	-	I
Cystopteris fragilis	-	-	-	-	-	I
Polypodium vulgare	-	-	-	-	I	I
Artemisietea vulgaris						
Anthemis tinctoria	I	I	I	-	-	-
Arctium tomentosum	I	-	-	-	-	-
Ballota nigra	I	-	I	I	-	I
Chelidonium majus	I	-	I	I	I	I
Cirsium vulgare	I	-	I	-	I	I
Linaria vulgaris	-	-	I	-	-	-
Melilotus officinalis	I	I	-	-	-	-
Tussilago farfara	-	-	-	-	I	I
Stellarietea mediae						
Cirsium arvense	-	-	-	-	I	I
Erigeron annuus	I	-	-	-	-	-
Euphorbia platyphyllos	-	-	-	-	-	I
Lamium purpureum	-	-	I	-	I	I
Stellaria media	I	-	I	I	II	I

Forest association	1a	1b	2a	2b	3a	3b
Torilis arvensis	I	I	II	I	-	I
Salicetea purpureae						
Populus alba	-	-	-	-	-	I
Salix alba	-	-	-	-	I	I

Number of relevés, sampling sites and corresponding references:

1. *Aro orientalis – Carpinetum*

typicum subass. nova hoc loco: 5 Bârnova – Repedea [29]; 9 Buciumeni Forest [2]; 5 Adjud [24]; 5 Perchiu Hill [3]; 30 Moldavian Central Plateau [19]; 2 Tazlăul Basin [1]; 5 Roman [25]; 2 Ceornohal Forest [22]; 1 Corbasca Forest [28]; 10 Moldavian Central Plateau [10]; 5 Moldavian Plain [11]; 5 Moldavian Central Plateau [6]; 5 Lungani Forest [14]; 5 Lungani Forest [15]; 5 Jijia Basin [23];

b. *quercetosum pedunculiflorae*: 1 Bârnova – Repedea [29]; 10 Popeşti and Cenuşa Forests [13]; 5 Tutovei Hills [7].

2. *Dentario quinquefoliae – Carpinetum*

a. *typicum* Chifu et Zamfirescu 2001: 27 Bârnova – Repedea [29]; 2 Bârnova – Repedea [20, 21]; 48 Moldavian Central Plateau [19]; 9 Moldavian Central Plateau [16]; 10 Moldavian Central Plateau [11]; 10 Moldavian Central Plateau [8]; 5 Popeşti, Cenuşa, Bârnova Forests [15]; 10 Vaslui County [13];

b. *fraxinetosum excelsioris*: 3 Bârnova – Repedea [29]; 10 from different localities [13].

3. *Lathyro veneti – Fagetum*

a. *typicum*: 4 Bârnova – Repedea [20, 21]; 4 Buciumeni Forest [2]; 16 Domniţa – Voineşti Forest [26]; 6 Moldavian Central Plateau [17]; 5 Moldavian Central Plateau [8];

b. *quercetosum dalechampii*: 3 Bârnova – Repedea [29]; 10 Bârnova – Repedea [20, 21]; 7 Buciumeni Forest [2]; 2 Adjud [24]; 85 Moldavian Central Plateau [19]; 5 Bacău [27]; 9 Domniţa – Voineşti Forest [26]; 1 Corbasca Forest [28]; 4 Moldavian Central Plateau [17]; 5 Moldavian Central Plateau [8]; 8 Fălciului and Tutovei Hills [13]; 10 Tutovei Hills [7].

The dendrogram (Fig. 1) outlines that the relevés studied fall into three groups that correspond to the three forest associations.

The inclusion of *Lathyro veneti – Fagetum* relevés within the same cluster as *Dentario quinquefoliae – Carpinetum* relevés can be explained by the altitudinal partial overlap of the two forest types.

The ordination diagram (Fig. 2) shows groups of similar relevés and the species that differentiate them.

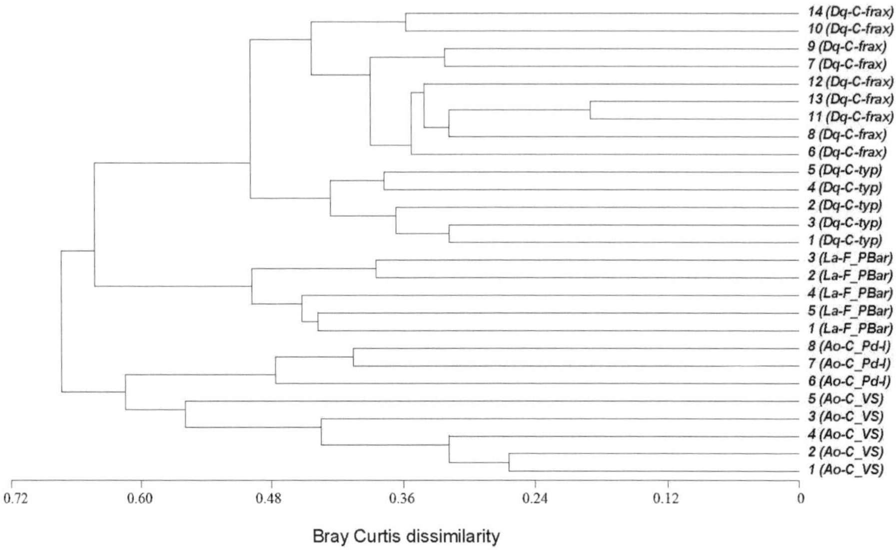

Fig. 1: The dendrogram of relevés assigned to the sub-alliance *Aro orientalis – Carpinenion*: (*Ao-C* = *Aro orientalis – Carpinetum*; *La-F* = *Lathyro veneti – Fagetum*; *Dq-C-typ* = *Dentario quinquefoliae – Carpinetum typicum*; *Dq-C-frax* = *Dentario quinquefoliae – Carpinetum fraxinetosum excelsioris*; VS = Vaslui; Pd-I = Podu-Iloaiei; PBar = Bârlad Plateau).

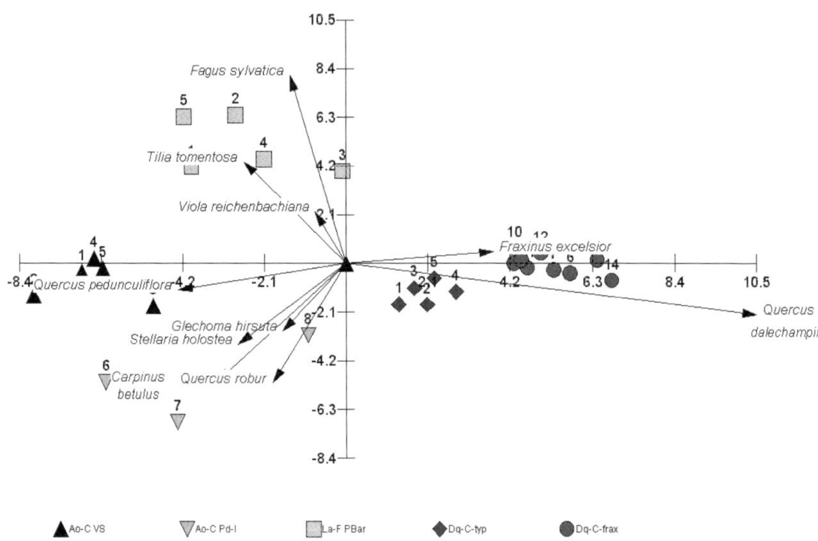

Fig. 2: Ordination diagram of the relevés assigned to the suballiance *Aro orientalis – Carpinenion*. Abbreviations as in Fig. 1.

Conclusions

Thus, *Quercus dalechampii* differentiates very well the communities of *Dentario quinquefolia-Carpinetum*, whereas *Fraxinus excelsior* discriminates between the two sub-associations – *fraxinetosum* and *typicum*. The communities belonging to *Aro orientalis - Carpinetum* distinguish themselves through the high constancy and cover of *Quercus robur* and *Q. pedunculiflora*. Furthermore, *Quercus pedunculiflora* has higher cover scores in southern relevés (Vaslui County), whereas *Q. robur* dominates in the northern relevés (Iaşi County). The group of relevés of *Lathyro veneti-Fagetum* is clearly discriminated by the high abundance-dominance scores of *Fagus sylvatica*.

REFERENCES

1. Barabaş N., 1974 - Contribuţii la studiul vegetaţiei din bazinul Tazlăului. Stud. şi Comunic., Muz. Şt. Nat. Bacău: 93–178.
2. Bârcă C., 1969 - Consideraţii geobotanice asupra pădurii Buciumeni-Tecuci. Stud. şi Comunic., Muz. Şt. Nat. Bacău: 107–123.
3. Burduja C., Mititelu D., Sârbu I., Barabaş N., 1971 - Flora şi vegetaţia Dealului Perchiu. Stud. şi Comunic., Muz. Şt. Nat. Bacău: 755–784.
4. Chifu T., 1995 - Contributions à la syntaxonomie de la vegétation de la classe *Carpino-Fagetea* (Br.-Bl. et Vlieg. 1937) Jakucs 1960 sur le territoire de la Moldavie (Roumanie). An. şt. Univ. "Al. I. Cuza" Iaşi, s. II, Biol. veget., **XLI**: 61-66.
5. Chifu T., 1997 - Biodiversitatea floristică şi fitocenologică în Podişul Central Moldovenesc. Bul. Grăd. Bot. Iaşi, **7**: 57-62.
6. Chifu T., Davidescu G., Cazacu V., 1999 - Caractérisation des complexes phytocénologiques de la végétation de Moldavie (Roumanie). Anuar. Muz. Nation. Bucovina, **XV**: 5-34.
7. Chifu T., Sârbu I., 2003 - O nouă contribuţie la studiul fitosociologic al pădurilor din Moldova (România). Bul. Grăd. Bot. Iaşi, **11**: 107–122.
8. Chifu T., Sârbu I., Ştefan N., Huţanu M., Şurubaru B., 1999 - Phytocénoses de l'alliance *Lathyro hallersteinii-Carpinion* Boşcaiu 1979 sur le territoire de la Moldavie. An. şt. Univ. „Al. I. Cuza" Iaşi, s. II, Biol. veget., **XLV**: 141-152.
9. Chifu T., Sârbu I., Ştefan N., Şurubaru B., 1999 - Contribuţii la fitocenologia făgetelor colinare şi a cărpinetelor din Podişul Central Moldovenesc. Bul. Grăd. Bot. Iaşi, **8**: 49-69.
10. Chifu T., Ştefan N., Coroi M., 1995 - Étude phytocoénologique et de la biomasse de l'association *Aro orientalis-Carpinetum* Taüber 1991/1992 du Plateau Central Moldave. Rev. Roum. de Biol., Sér. Biol. végét., **40** (1): 21-31.
11. Chifu T., Ştefan N., Sârbu I., 1996 - Contribuţii la studiul vegetaţiei din clasa *Carpino-Fagetea* (Br.-Bl. et Vlieg. 1937) Jakucs 1960 de pe teritoriul Moldovei (România). Stud. şi cerc., Muz. Şt. Nat. Piatra Neamţ, **VIII**: 295-326.
12. Chifu T., Ştefan N., Sârbu I., 2001 - L'association *Irido pseudocyperae-Quercetum pedunculiflorae*, un nouveau coenotaxon dans la végétation de la Roumanie. An. şt. Univ. „Al. I. Cuza" Iaşi, s. II, Biol. veget., **XLVII**: 137-143.

13. Chifu T., Ştefan N., Sârbu I., Mânzu C., Şurubaru B., 2002 - Nouvelles contributions à l'étude phytosociolique des forets de Moldova, An. şt. Univ. "Al. I. Cuza" Iaşi, s. II, Biol. veget., **XLVIII**: 103-118.
14. Chifu T., Şurubaru B., 1999 - Biomasa şi productivitatea aeriană a stratului arborescent în cărpinetele din pădurea Lungani-Iaşi, Bul .Grăd. Bot. Iaşi, **8**: 71-82.
15. Chifu T., Zamfirescu O., 2000 - O nouă contribuţie la sintaxonomia pădurilor din clasa *Querco-Fagetea* Br.-Bl. et Vlieger in Vlieger 1937 de pe teritoriul Moldovei (România). Bul. Grăd. Bot. Iaşi, **10**: 85–98.
16. Chifu T., Ştefan N., Coroi M., 1993 - Etude phytocoenologique et de la biomasse de l'association *Aro orientalis - Carpinetum* Täuber 1992 du Plateau Central Moldave. Rev. Roum. Biol. Série Biol. végét., **40** (1): 21 – 31.
17. Chifu T., Ştefan N., Huţanu M., Coroi M., Coroi A.M., 1997 - Biomasse et productivité annuelle aérienne de la strate arborescente des hêtraies collinaires du Plateau Central Moldave. St. Cerc. Biol., Seria Biol. veget., **49** (1-2): 43-56.
18. Ciocârlan V., 2000 - *Flora ilustrată a României.* Ed. Ceres, Bucureşti.
19. Dobrescu C., Kovacs A., 1973 - Contribuţii la fitocenologia pădurilor de *Fagion* din Podişul Central Moldovenesc. Rev. Păd., **88** (11): 592-599.
20. Dobrescu C., Bârca C., Lazăr M., 1964 - Contribuţii floristice şi geobotanice asupra masivului forestier Bârnova - Repedea Iaşi (I). An. şt. Univ. "Al. I. Cuza" Iaşi, s. II, Biol., **10**: 147-158.
21. Dobrescu C., Bârca C., Lazăr M., 1964 - Contribuţii floristice şi geobotanice referitoare la masivul forestier Bârnova - Repedea Iaşi (II). An. şt. Univ. "Al. I. Cuza" Iaşi, s. II, Biol., **10**: 323-357.
22. Horeanu C., Horeanu G., 1981 - Flora şi vegetaţia rezervaţiei forestiere Ceornohal (jud. Botoşani). Ocrot. Nat. şi a Med. Înconj., **25** (2): 211-220.
23. Huţanu M., 2004 - Diversitatea florei vasculare, a vegetaţiei şi a macromicetelor din bazinul Jijiei (jud. Botoşani). Edit. Gh. Asachi, Iaşi.
24. Mititelu D., Barabaş N., 1970 - Flora şi vegetaţia împrejurimilor oraşului Adjud. Stud. şi Comunic., Muz. Şt. Nat. Bacău: 75-112.
25. Mititelu D., Barabaş N., Nechita F., 1978 - Flora şi vegetaţia împrejurimilor municipiului Roman (jud. Neamţ). Stud. şi Comunic., Muz. Şt. Nat. Bacău: 281–316.
26. Mititelu D., Chifu T., (eds.), 1977 - Cercetări ecologice în pădurea (*Carpino-Fagetum*) Domniţa-Voineşti. Stud. şi Comunic., Muz. Şt. Nat. Bacău: 361-434.
27. Mititelu D., Barabaş N., 1978 - Flora şi vegetaţia judeţului Bacău. St. Com., Muz. St. Nat. Bacău, **9 - 10**: 193-272.
28. Mititelu D., Stratulat-Matei, J., 1994 - Flora şi vegetaţia comunei Corbasca. St. Com., Complex. Muz. Şt. Nat. "Ioan Borcea" Bacău, **13**: 61- 66.
29. Raclaru P., Bârcă C., 1959 - Studii asupra vegetaţiei regiunii păduroase de la sud-est de Iaşi. Stud. şi Cerc. Şt. Biol. şi Şt. Agr., **10** (1): 85-95.

DRY GRASSLANDS OF THE CORHAN HILL – SĂBED VILLAGE (MUREŞ COUNTY)

Silvia OROIAN[1*], Mihaela SĂMĂRGHIŢAN[2]
[1] University of Medicine and Pharmacy, Department of Pharmaceutical Botany 35, Gh. Marinescu Str., **540140 Tg-Mureş (ROMANIA)**; e-mail: osilvia@umftgm.ro
[2] The Mureş County Museum, 24 Horea Str., **540050 Tg-Mureş (ROMANIA)**

Abstract: This paper presents a description of vegetation from the Corhan Hill-Săbed village, Mureş County. The study area is situated in a part of the Transylvanian Depression of special relief. Climate and aspect have determined the development of dry vegetation of the order *Festucetalia valesiacae*. Two plant associations were identified on Corhan Hill: *Stipetum pulcherrimae* Soó 1942 and *Medicagini minimae-Festucetum valesiacae* Wagner 1941. An analysis of these plant associations was carried out, in relation to ecological factors (humidity, temperature and soil). Some rare, endemic and endangered plants of the Romanian flora were identified in this area.

Introduction

Several floristic studies on Corhan Hill and Săbed Village area have been carried out by Oroian [9, 10]. However, until now no syntaxonomical conclusions have been published.

Deforestation in Transylvania of the woods of durmast oak with hornbeam has led to a dominance of xerophilous phytocoenoses with strong steppic characters [1]. A great number of species can be noticed in their composition, in particular those belonging to *Festucion* and *Festucetalia valesiacae*: *Festuca valesiaca, Festuca rupicola, Campanula sibirica, Polygala major, Teucrium chamaedrys, Veronica spicata* ssp. *orchidea, Onobrychis viciifolia* etc.

The overall objective of this research is to contribute to a better knowledge of the flora and vegetation of this area. The specific objectives have been to investigate plant communities and assess the presence of certain endangered species in these grasslands.

Study area

Săbed village belongs to Ceuaşu de Câmpie commune, situated in southeastern part of the Transylvanian Depression, 20 km N-NW from Tg-Mureş.

The relief is particular to the Transylvanian Depression. Corhan hill is part of the Mădăraşului hills and is the highest point (500 m) in the Săbed neighbourhood. Concerning the climate, the annual average temperature oscillates around 8-9°C and mean annual rainfall is 550 mm/m^2 (Tg-Mureş Meteorological station).

Material and Methods

Two plant associations were analysed belonging to the order *Festucetalia valesiacae* [6]: *Stipetum pulcherrimae* Soó 1942 and *Medicagini minimae-Festucetum valesiacae* Wagner 1941. In the analysis we took into account the main characteristics of relief, soil and climate. The studies have been based on field observation, with data from the study area analyzed using the Braun-Blanquet approach. We also recorded the percentage cover of each vascular plant species in each 1-m^2 plot within five cover classes. We composed the phytocoenotical tables for each association and designated it to the appropriate class [7, 11]. We followed Mucina [7] for nomenclature and delimitation of class. The present study includes all phytosociological data, and aims to produce a phytosociological classification of these grasslands.

Results and Discussion

Two plant associations were identified on the Corhan Hill. These were syntaxonomically included in:

FESTUCO-BROMETEA Br.-Bl. et R.Tx. ex Klika et Hadač 1944

Festucetalia valesiacae Br.-Bl.et R.Tx.ex Br.-Bl.1949

Festucion valesiacae Klika 1931

Stipetum pulcherrimae Soó 1942

Medicagini minimae-Festucetum valesiacae Wagner 1941

Stipetum pulcherrimae Soó 1942

Stipetum pulcherrimae communities can be found on the Transylvanian Plateau [2], from the northern side of the 'plain' to the southern border of the Târnave Plateau. These phytocoenoses generally occur on the upper third of steep slopes of south, south-west and south-east aspect. The soils vary from slope chernozem, partially eroded, to slope light brown soil, highly eroded. The analysis of Transylvanian phytocoenoses has shown the succession of the vegetation from thermophylic woods to xerothermic shrubs of *Prunus tenella* and *Prunus fruticosa*, a grassy skirt consisting of *Peucedanum cervaria, P. oreoselinum, Aster villosus, A. linosyris, Dictamnus albus, Rosa gallica* and islands of steppic xerothermic vegetation (*Stipa lessingiana* and *S. pulcherrima*). Among the species assemblages characteristic of the association, besides *Stipa pulcherrima* we recorded the presence of *Vinca herbacea, Salvia austriaca, Festuca valesiaca, Centaurea biebersteinii* ssp. *biebersteinii, Jurinea mollis, Inula ensifolia* and *Verbascum phoeniceum* (Table 1). Widespread species characteristic of the dry grasslands of the class Festuco-Brometea, such as *Euphorbia cyparissias, Teucrium chamaedrys, Eryngium campestre, Medicago falcata*, are of constant occurrence in this area.

It is a secondary association with a heterogeneous floristic composition, and the coenoses present a full coverage (100%). The phytocoenoses of this association are less widespread in Mureş County, occupying sunny slopes of south and southwest exposure.

The moderate conditions in which this association develops facilitate a great number of xero-mesophilous species (70.65%) and besides these, some xerophilous species (20.65%). The thermic conditions favour the micro-mesothermous (40.22%) and moderate thermophilous (38.04%) species. Most species of this association are low acid-neutrophilous (61.96%) in respect of their preference for the soil reaction (Fig. 1). In the life forms spectrum (Fig. 2), hemicryptophytes are dominant (73.91%); therophytes are present in considerable numbers (10.87%). In the floristic spectrum (Fig. 4) Eurasian species are at the greatest frequency (26.08%); beside these, numerous European (16.30%) and Eurosiberian (14.13%) species are present. Because of the xerophilous character of these grasslands, in their structure a significant percentage of Ponto-Pannonian (10.87%) and Mediterranean (8.69%) species can be found. Caryological study indicates a high number of diploid species (55.43%). Polyploid species represent 36.96% and diplo-polyploid 4.35% (Fig. 3).

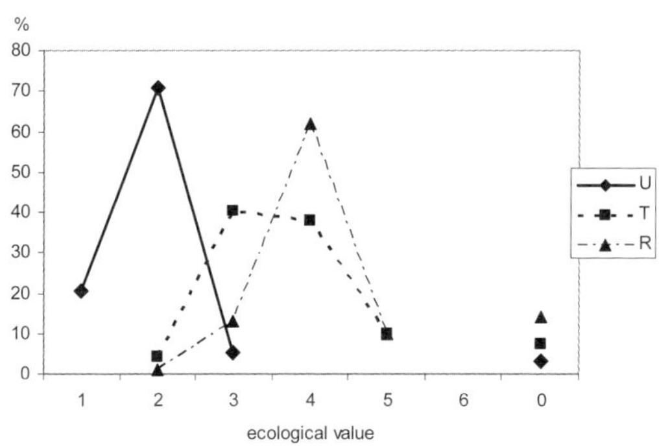

Fig. 1: Ecological indicator values of *Stipetum pulcherrimae* association.

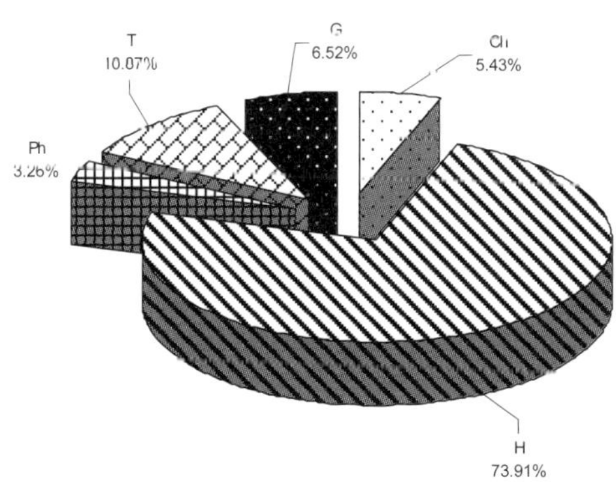

Fig. 2: Life form spectrum of *Stipetum pulcherrimae* association.

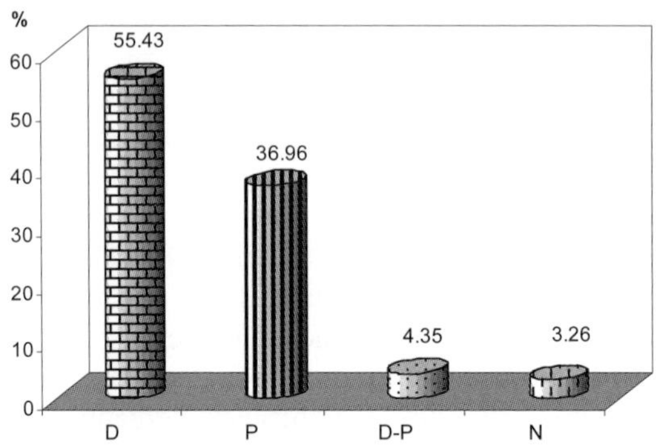

Fig. 3: Caryologic spectrum of *Stipetum pulcherrimae* association.

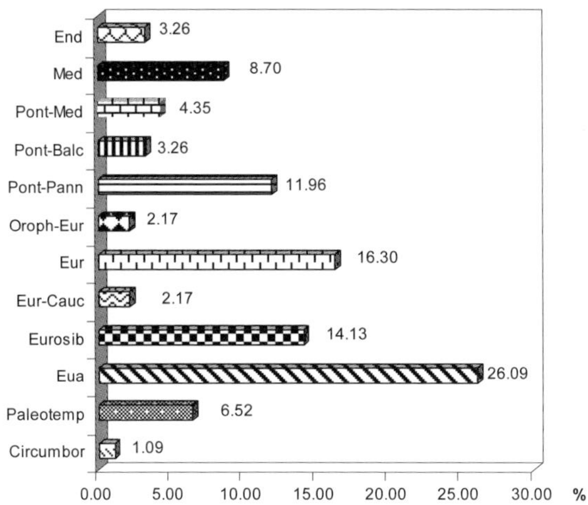

Fig. 4: Floristic elements of *Stipetum pulcherrimae* association.

Abbreviations:

Circumbor.	-	Circumboreal	U	-	Humidity
Paleotemp.	-	Paleotemperate	T	-	Temperature
Eua.	-	Eurasian	R	-	Soil reaction
Eurosib.	-	Eurosiberian	H	-	Hemycryptophyte
Eur.-Cauc.	-	European-Caucasian	Th	-	Therophyte
Eur.	-	European	G	-	Geophyte
Oroph.-Eur.	-	Orophyte European	Ph	-	Phanerophyte
Pont.-Pann.	-	Pontic-Pannonian	Ch	-	Chamephyte
Pont.-Balc.	-	Pontic-Balcanian	D	-	Diploide
Pont.-Med.	-	Pontic- Mediterranean	P	-	Polyploide
Med.	-	Mediterranean	D-P	-	Diplo-Polyploide
End.	-	Endemic	N	-	Unknown

Table 1: ***Stipetum pulcherrimae*** Soó 1942.

Area (m^2)	**100**	**100**	**100**	**100**	**100**	
Aspect	**SE**	**S**	**S**	**S**	**S**	
Inclination (degrees)	**20**	**40**	**40**	**40**	**45**	
Altitude (m)	**476**	**476**	**476**	**476**	**476**	
Releves	**1**	**2**	**3**	**4**	**5**	**K**
Stipion lessingianae						
Stipa pulcherrima	3	4	4	4	4	V
Festucion rupicolae						
Astragalus exscapus var. *transsilvanicus*	+	+	-	1	+	IV
Crambe tatarica	-	-	-	-	+	I
Falcaria vulgaris	+	-	-	+	+	III
Festuca rupicola	-	+	-	+	-	II
Inula ensifolia	+	+	-	-	-	II
Knautia arvensis	+	+	+	+	+	V
Salvia austriaca	+	+	-	-	+	III
Festuco-Brometea						
Ajuga genevensis	-	+	-	+	-	II
Anthericum ramosum	-	+	-	+	+	III
Anthyllis vulneraria	+	-	-	+	-	II
Artemisia austriaca	-	-	+	-	-	I
Artemisia campestris	+	-	-	-	+	II
Asparagus officinalis	+	-	-	+		II
Astragalus monspessulanus	+	+	+	+	+	V
Carex caryophyllea	-	-	+	-	-	I
Carlina vulgaris	-	-	-	-	+	I
Cephalaria radiata	+	-	+	-	-	II
Coronilla varia	+	+	+	+	+	V
Dianthus carthusianorum	+	-	+	+	+	IV
Euphorbia cyparissias	+	+	-	+	+	IV
Filipendula vulgaris	+	+	-	+	-	III
Galium verum	+	+	+	+	+	V
Hypericum perforatum	-	-	-	+		I
Lavathera thuringiaca	-	-	-	-	1	I
Linaria vulgaris	-	-	-	-	+	I
Linum austriacum	+	+	-	+	-	III
Linum tenuifolium	+	-	-	+	-	II
Medicago falcata	+	+	+	-	-	III
Pimpinella saxifraga	-	-	+	+	+	III
Plantago lanceolata	+	+	+	-	+	IV
Plantago media	+	-	+	+	-	III
Potentilla arenaria	+	+	+	-	+	IV
Prunella laciniata	+	-	-	+	-	II
Salvia verticillata	+	+	+	+	+	V
Scabiosa ochroleuca	+	-	+	+	+	IV
Seseli annuum	-	-	-	+	-	I

Teucrium chamaedrys	+	+	+	+	+	V
Thlaspi perfoliatum	+	-	-	-	-	I
Trifolium arvense	+	-	+	-	-	II
Trifolium campestre	+	+	+	-	-	III
Trifolium ochroleucon	+	+	-	+	+	IV
Veronica teucrium	+	-	-	+	-	II
Festucetalia valesiacae						
Achillea setacea	-	+	-	+	+	III
Adonis vernalis	1	+	+	+	+	V
Ajuga laxmanni	+	-	-	+	+	III
Anchusa barrelieri	+	-	-	-	+	II
Artemisia pontica	+	+	+	+	+	V
Asperula cynanchica	+	+	-	+	+	IV
Astragalus onobrychis	+	+	-	+	-	III
Bupleurum falcatum	+	+	-	+	+	IV
Campanula sibirica	+	-	+	-	+	III
Carduus hamulosus	-	+	-	-	+	II
Centaurea biebersteinii ssp. *biebersteinii*	-	+	-	-	+	II
Centaurea apiculata ssp. *spinulosa*	+	-	-	-	-	I
Dorycnium pentaphyllum *ssp.herbacea*	+	+	+	+	+	V
Eryngium campestre	+	-	-	+	-	II
Festuca valesiaca	1	-	-	+	+	III
Fragaria viridis	+	+	+	-	-	III
Galium glaucum	+	+	+	+	+	V
Iris aphylla	-	-	+	+	+	III
Iris pumila	+	+	+	+	+	V
Jurinea mollis	+	+	-	+	-	III
Muscari tenuiflorum	+	+	+	+	+	V
Nonea pulla	+	+	+	+	+	V
Onobrychis viciifolia	+	-	+	-	+	III
Oxytropis pilosa	+	-	-	+	+	III
Peucedanum ruthenicum	+	+	+	-	+	IV
Phlomis tuberosa	+	+	+	+	+	V
Pulsatilla pratensis ssp *.nigricans*	+	-	-	-	-	I
Pulsatilla vulgaris ssp. *grandis*	+	-	-	-	-	I
Salvia nemorosa	-	-	+	-	-	I
Salvia transsylvanica	-	-	+	-	-	I
Scorzonera hispanica	-	+	-	-	+	II
Senecio jacobea	-	-	-	+	-	I
Serratula radiata	1	+	+	+	-	IV
Stachys recta	+	+	+	+	+	V
Thalictrum minus	+	+	-	+	-	III
Thymus pannonicus	+	+	-	+	+	IV
Veronica spicata ssp. *orchidea*	+	+	+	+	+	V

Vinca herbacea	+	+	+	+	+	V
Vincetoxicum hirundinaria	-	+	-	-	-	I
Molinio-Arrhenatheretea						
Agrostis capillaris	+	-	+	-	-	II
Eryngium planum	+	+	-	-	-	II
Primula veris	-	-	-	-	+	I
Rhinanthus angustifolius	-	-	+	-	+	II
Tragopogon pratensis ssp. *orientalis*	+	+	-	+	-	III
Rhamno-Prunetea						
Crataegus monogyna	+	+	+	+	+	V
Prunus tenella	-	+	+	+	-	III
Rosa gallica	-	-	-	+	-	I
Variae syntaxa						
Agrimonia pilosa	+	+	+	-	+	IV
Cruciata laevipes	+	+	-	+	+	IV

Date and place of releves: Săbed, Corhan Hill (N 46°39' ; E 24°25'), 25.04.2004

Medicagini minimae-Festucetum valesiacae Wagner 1941

Grasslands of Volga fescue with bur medick occupy the sunny side of the hill with 15-45° slope. They develop on levigate chernozem and semi-carbonated soils of low acid to neutral reaction. The 123 vascular plants of this association mostly belong to coenotaxa of the class *Festuco-Brometea* (Table 2). Southern thermophilous elements are well represented in the grasslands analyzed.

The continental climate favours xero-mesophilous (62.60%), micro-mesothermic (42.28%) and moderate thermophilous (32.52%) species, the edaphic conditions favouring a great number of low acid neutrophilous species (62.60%) (Fig. 5). The hemycryptophytes (73.98%) are the dominant lifeforms and the great number of therophytes (13%) is a result of grazing and haymaking in these grasslands (Fig. 6). The phytogeographical structure of this association shows the composition in different percentages, for some categories of elements of diverse origin. A basic Eurasian (23.58%) background can be distinguished, on which Eurosiberian (15.45%) and European (15.45%) species overlapped in different phyto-historical periods of time (Fig. 8). The significant number of Ponto-Pannonian (13%) and Mediterranean (9.76%) species confirms the xero-mesophilous character of these grasslands. The predominance of diploid populations (55.28%) in this area indicates the age of this flora, guaranteeing favourable potential for future phyto-evolution (Fig. 7).

Botanists recognize this area of the plain for its high research potential. It is under a great peril because of human activities such as deforestation, agricultural expansion (arable), and building development.

A few decades ago, Borza [3] mentioned: "*There are two types of overloaded vegetation climax in the Transylvanian Plain: the forest climax and the discontinuous herby dry steppe climax ...*"

Table 2: ***Medicagini minimae-Festucetum valesiacae*** **Wagner 1941**

Area (m²)	**100**	**100**	**100**	**100**	**100**	
Aspect	**SE**	**S**	**S**	**S**	**SE**	
Inclination (degrees)	**20**	**45**	**15**	**20**	**20**	
Altitude (m)	**476**	**476**	**476**	**476**	**476**	
Releves	**1**	**2**	**3**	**4**	**5**	**K**
Festucion rupicolae						
Acinos arvensis	+	-	-	-	-	I
Ajuga chamaepitys	+	-	+	-	-	II
Alium ericetorum	-	-	+	-	-	I
Astragalus austriacum	+	-	+	-	+	III
Astragalus exscapus var. *transsilvanicus*	+	-	+	+	+	IV
Crambe tatarica	+	+	+	-	+	IV
Falcaria vulgaris	+	+	+	-	+	IV
Festuca rupicola	-	+	-	+	-	II
Inula ensifolia	+	+	-	-	-	II
Knautia arvensis	+	+	+	+	+	V
Salvia austriaca	+	+	-	-	+	III
Stipion lessingianae						
Stipa pulcherrima	+	+	-	-	-	II
Cirsio-Brachypodion						
Centaurea scabiosa	+	-	+	-	+	III
Linum austriacum	+	+	-	+	-	III
Festucetalia valesiacae						
Achillea setacea	+	+	+	+	+	V
Adonis vernalis	1	+	-	+	+	IV
Ajuga laxmanni	+	+	+	+	+	V
Anthemis tinctoria	-	-	+	-	-	I
Asperula cynanchica	+	+	-	+	+	IV
Astragalus onobrychis	+	+	+	+	-	IV
Bupleurum falcatum	+	+	-	+	+	IV
Campanula sibirica	+	+	+	+	+	V
Centaurea biebersteinii ssp. *biebersteinii*	-	-	-	+	-	I
Centaurea apiculata ssp. *spinulosa*	+	+	+	+	+	V
Centaurea rhenana	+	-	-	+	-	II
Chamaecytisus albus	+	-	+	-	-	II
Dorycnium pentaphyllum ssp. *herbacea*	-	-	+	+	-	II
Eryngium campestre	+	+	-	+	-	III
Fragaria viridis	+	+	+	+	+	V
Galium glaucum	+	+	-	+	+	IV
Iris hungarica	-	-	+	-	-	I
Iris pumila	+	+	-	+	+	IV
Jurinea mollis	+	+	-	+	-	III
Linum flavum	+	-	+	-	+	III
Linum hirsutum	-	-	+	-	+	II
Medicago minima	+	+	1	+	+	V

Muscari tenuiflorum	+	-	+	+	+	IV
Nonea pulla	+	+	+	-	+	IV
Onobrychis viciifolia	+	-	+	-	+	III
Oxytropis pilosa	+	+	-	-	-	II
Peucedanum cervaria	+	+	+	+	+	V
Phlomis tuberosa	+	+	-	-	+	III
Polygala major	+	+	-	+	+	IV
Pulsatilla vulgaris ssp. *grandis*	+	-	-	+	-	II
Pulsatilla pratensis ssp. *nigricans*	+	-	+	-	-	II
Salvia nemorosa	-	-	+	+	+	III
Salvia nutans	+	1	+	+	1	V
Salvia transsylvanica	-	-	-	+	-	I
Scorzonera hispanica	-	+	-	-	+	II
Senecio jacobea	-	-	-	+	-	I
Serratula radiata	1	-	+	+	-	III
Stachys recta	+	-	+	-	+	III
Thalictrum minus	+	+	-	-	+	III
Thymus pannonicus	+	-	+	+	+	IV
Trifolium montanum	-	-	-	+	+	II
Veronica spicata ssp. *orchidea*	+	+	+	+	+	V
Vinca herbacea	+	+	+	-	+	IV
Vincetoxicum hirundinaria	-	+	+	+	+	IV
Festuco-Brometea						
Ajuga genevensis	-	+	-	+	+	III
Anthyllis vulneraria	+	-	-	+	-	II
Artemisia austriaca	-	+	-	+	-	II
Artemisia pontica	+	+	+	-	-	III
Asparagus officinalis	+	-	+	+	+	IV
Astragalus monspessulanus	+	+	+	+	-	IV
Campanula glomerata	-	-	-	+	+	II
Carex caryophyllea	-	+	+	-	-	II
Carlina vulgaris	-	+	-	+	-	II
Cephalaria radiata	-	+	+	-	-	II
Cerinthe minor	+	-	+	-	-	II
Cirsium pannonicum	+	-	-	-	-	I
Coronilla varia	+	+	+	+	+	V
Dianthus carthusianorum	+	-	+	+	+	IV
Echium vulgare	-	+	-	-	+	II
Euphorbia cyparissias	+	-	+	+	+	IV
Festuca valesiaca	4	4	5	5	4	V
Galium verum	+	+	+	+	+	V
Hieracium pilosella	-	-	+	-		I
Hypericum perforatum	-	+	+	+	+	IV
Lavathera thuringiaca	-	+	-	+	-	II
Linaria vulgaris	-	+	-	-	-	I
Linum perenne	-	+	-	-	-	I

Linum tenuifolium	+	-	-	+	-	II
Lotus corniculatus	+	-	+	-	+	III
Medicago falcata	+	+	+	+	-	IV
Nepeta nuda	+	-	+	-	-	II
Orchis morio	+	-	-	-	+	II
Pimpinella saxifraga	-	-	+	+	+	III
Plantago lanceolata	+	+	+	-	+	IV
Plantago media	+	+	+	-	+	IV
Potentilla arenaria	+	-	+	+	-	III
Prunella grandiflora	+	-	-	+	-	II
Salvia pratensis	+	-	+	-	-	II
Salvia verticillata	+	+	+	+	+	V
Scabiosa ochroleuca	+	-	+	+	+	IV
Silene otites	+	-	-	+	-	II
Stachys germanica	-	+	+	-	+	III
Teucrium chamaedrys	+	+	+	+	+	V
Thlaspi perfoliatum	+	-	-	+	-	II
Trifolium arvense	+	-	+	-	-	II
Trifolium campestre	+	+	+	-	-	III
Trifolium ochroleucon	+	+	-	+	+	IV
Veronica teucrium	+	-	-	+	-	II
Viola hirta	-	-	+	-	-	I
Molinio-Arrhenatheretea						
Agrostis capillaris	+	-	+	-	-	II
Anthoxanthum odoratum	+	-	-	-	+	II
Briza media	-	+	+	-	+	III
Dactylis glomerata	-	+	-	-	+	II
Eryngium planum	+		+	-	+	III
Festuca pratensis	-	+	-	-	+	II
Leucanthemum vulgare	+	+	-	+	+	IV
Linum catharticum	+	-	+	-	+	III
Medicago lupulina	-	+	-	+	-	II
Primula veris		+	+	-	+	III
Rhinanthus angustifolius	-	-	+	-	+	II
Tragopogon pratensis ssp. *orientalis*	+	+	+	-	+	IV
Trifolium pratense	-	+	-	+	-	II
Rhamno-Prunetea						
Crataegus monogyna	+	-	-	-	-	I
Dictamnus albus	+	+	+	+	+	V
Prunus tenella	-	+	+	-	-	II
Rosa gallica	-	-	+	-	-	I
Variae syntaxa						
Agrimonia pillosa	+	+	-	-	+	III
Cruciata laevipes	+	+	+	+	-	IV
Stachys officinalis	-	-	+	-	+	II

Date and place of releves: Săbed, Corhan Hill (N 46°39' ; E 24°25'), 25.04.2004

The conclusions drawn from the present study confirm those of Borza [3], as long as both types of vegetation climax can be found in Săbed. Here the forest is divided by dry vegetation islands that lie on the upper third of the hills and are dominated by grasses and various xerophilous dicotyledons.

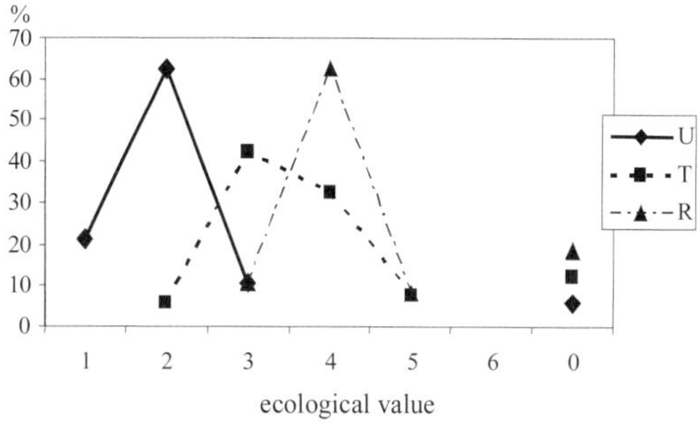

Fig. 5: Ecological indicator values of *Medicagini minimae-Festucetum valesiacae* association.

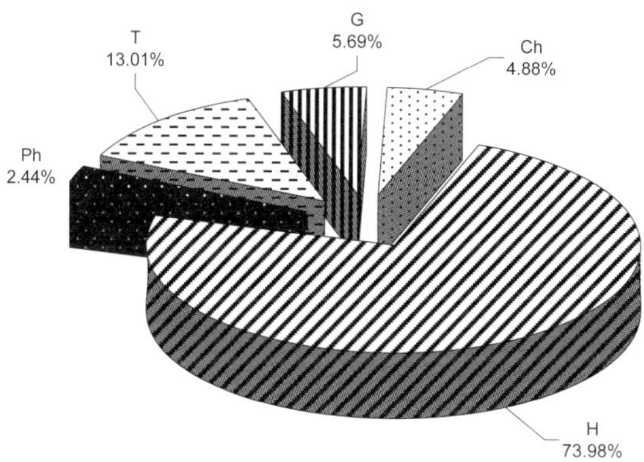

Fig. 6: Bioforms spectrum of *Medicagini minimae-Festucetum valesiacae* association.

Rare, vulnerable, endemic and endangered plants from studied area

Seven European threatened species occur in the study area: *Agrimonia pilosa, Crambe tatarica* (Fig. 10), *Pulsatilla pratensis* ssp. *nigricans*, *Pulsatilla vulgaris* ssp. *grandis*, *Astragalus exscapus* var. *transsilvanicus* (Fig. 9), *Cephalaria radiata,* and *Salvia transsylvanica.*

The associations include also taxa from the Romanian Red Lists [4, 5, 8]: *Adonis vernalis* R; *Serratula radiata* R; *Iris aphylla* R; *Stipa pulcherrima* R; *Salvia nutans* V (Fig. 11); *Dictamnus albus* VU (Fig. 12); *Prunus tenella* VU (Fig. 13) and *Orchis morio* R.

A part of Corhan Hill has been transformed into an orchard and the grasslands described in this study are intensely grazed.

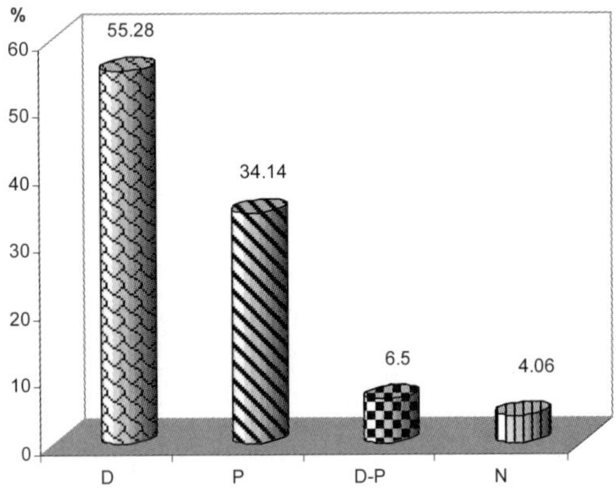

Fig. 7: Caryological spectrum of M*edicagini minimae-Festucetum valesiacae* association.

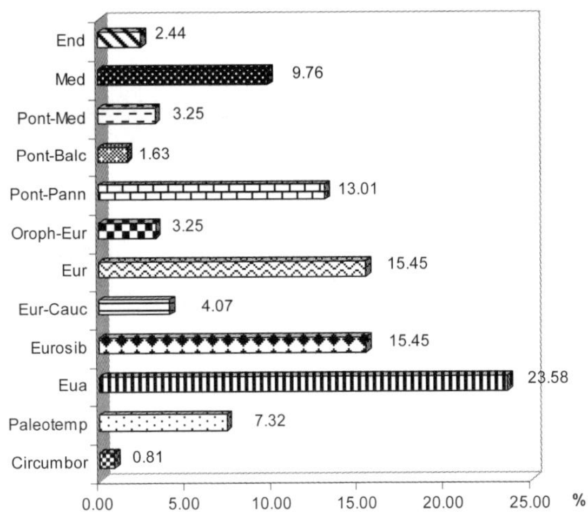

Fig. 8: Floristic elements of M*edicagini minimae-Festucetum valesiacae* association.

Fig. 9: ***Astragalus exscapus* var. *transsilvanicus***

Fig. 10: *Crambe tatarica*

Fig. 11: *Dictamnus albus*

Fig. 12: *Salvia nutans*

Fig. 13: *Prunus tenella*

REFERENCES

1. Borza A., 1929 - Vegetaţia şi flora Ardealului. Schiţă geobotanică. Atelierele grafice „Cultura Naţională", Bucureşti.
2. Borza A., 1931 - Botanic excursion through „The Câmpia". Guide de la sixième excursion phytogeographique internationale en Roumaine. Institutul de Literatură şi Tipografie „Minerva" S.A., Cluj.
3. Borza A., 1936 - Câmpia Ardealului. Studiu geobotanic. Tipografia cărţilor bisericeşti, Bucureşti.
4. Boşcaiu N., Coldea G., Horeanu C., 1994 - Lista roşie a plantelor vasculare dispărute, periclitate, vulnerabile şi rare din flora României. Ocrot. Nat. Med. Înconj., **38** (1): 45-56.
5. Dihoru G., Dihoru A., 1994 - Plante rare, periclitate şi endemice în Flora României – Lista roşie. Acta Bot. Horti Bucurest., (1993-1994): 173-197.
6. Doniţă N., Ivan D., Coldea G., Sanda V., Popescu A., Chifu T., Paucă-Comănescu M., Mititelu D., Boşcaiu N., 1992 - Vegetaţia României, Ed. Tehnică Agricolă, Bucureşti.
7. Mucina L., Grabherr G., Ellmauer T., 1993 - Die Pflanzengesellschaften Österreich, Teil I, VEB Gustav Fischer Verlag, Jena.
8. Oltean M., Negrean G., Popescu A., Roman N., Dihoru G., Sanda V., Mihăilescu S., 1994 - Lista roşie a plantelor superioare din România. Studii, sinteze, documentaţii de ecologie, **1**. Acad.Română, Inst. de Biologie, Bucureşti.
9. Oroian S., 1983 - Cercetări fitotaxonomice pe dealul Corhan-Săbed, judeţul Mureş şi posibilităţi de valorificare a florei. Marisia, Târgu-Mureş, **XI-XII** (1): 47-74.
10. Oroian S., 1991 - Aspecte din flora şi vegetaţia pădurii Săbed (jud. Mureş). Ocrot. Nat. Med. Înconj., Bucureşti, **35** (1-2): 35-41.
11. Sanda V., Popescu A., Stancu D.I., 2001 - Structura cenotică şi caracterizarea ecologică a fitocenozelor din România, Ed. Conphis, Piteşti.

THREATENED PLANT COMMUNITIES AS AN INDICATOR OF FISHPONDS VALUE: AN EXAMPLE FROM SILESIA (SW POLAND)

Krzysztof SPAŁEK
Department of Biosystematics, Division of Phytosociology and Plant Geography, University of Opole, Oleska str. 22, **45-052 Opole (POLAND)**;
e-mail: kspalek@uni.opole.pl

Abstract: This paper presents the first results of studies of fishpond plant communities of Silesia (SW Poland). On fishponds, many rare and threatened plant associations were found, especially within the aquatic, mud and rush communities. Among the most interesting are: *Lemno minoris-Salvinietum natantis*, *Trapetum natantis*, *Eleocharetum ovatae* and *Scirpetum radicantis*. The majority of the above-mentioned associations occur only in artificial fishpond reservoirs within the region. Thus, to reach an effective level of conservation, co-operation between biologists and owners, and integration of conservation and agricultural activities, have to be achieved. This would be the only way to stop the impoverishment of the biological diversity of these hydrophilous ecosystems. Also it seems to be essential to establish the legally based protection system for the most valuable fishponds. Phytosociological methods of evaluation were effectively used to find the priority areas.

Introduction

Increasing anthropogenic pressure significantly contributes to the destruction of shallow water and marsh ecosystems. Fishponds are a typical element of the landscape of Silesia, SW Poland. They play an important role in biodiversity conservation in this agricultural and industrial region. The natural value of ponds increases with the cessation of fish farming.

Material and Methods

Fieldwork was conducted during the growing seasons 2000-2004. Plant communities were studied following the Zurich-Montpellier School of Phytosociology [2]. Phytosociological nomenclature and syntaxonomical attachment are based on Matuszkiewicz's guide [7]. Species names are given according to Mirek *et al.* [8].

Results

In the Silesian fishponds many plant associations occur that are rare and threatened, at the level of Poland as well as the whole of Europe. The most interesting are connected with the following syntaxonomical classes: *Lemnetea minoris* R. Tx. 1955, *Potametea* R. Tx. et Prsg. 1942, *Phragmitetea* R. Tx. et Prsg. 1942 and *Isoëto-Nanojuncetea* Br.-Bl. et R. Tx. 1943. Some of them are under special conservation, on the base of Habitat Directive 92/43/EEC from 21st May 1992 on the conservation of natural habitats and of wild fauna and flora.

Associations from the *Lemnetea minoris* R. Tx. 1955 class belong to simple assemblages built by plants, which passively float on the surface of running or stagnant water. This plant community prefers mainly eu- and mesotrophic water bodies. Rarely it occurs also in dystrophic waters. Quite frequently this plant association occurs within a mosaic with more complex water phytocoenoses from the *Potametea* R. Tx. et Prsg. 1942 class, of freshwater macrophytes [6, 7, 9, 15, 17, 18, 21]. Communities from these classes are commonly distributed all over Poland. But some of them are considered to be not enough investigated and relatively rare [7, 20]. The most interesting is the assemblage built by *Salvinia natans - Lemno minoris-Salvinietum natantis* (Slavnić 1956) Korneck 1959 from the class *Lemnetea minoris*, which still occurs quite commonly in the Silesian fishponds. In Poland it is considered as a rare and threatened plant association. More and more in retreat in fishpond reservoirs is *Trapetum natantis* Müll. et Görs 1969. It is one of the most threatened plant communities in Poland, with the majority of occurrences in the south part of the country [13].

Among rush communities from the *Phragmitetea* class, the most interesting seems to be *Scirpetum radicantis* Hejný in Hejný et Husák 1978 *em.* Zahlh. 1979. This plant community was first recognized in the south of the Czech Republic in the littoral zone of fishponds [4]. Subsequently, it was found in Austria [1], Germany [17, 22], Moravia and the Danube valley in Slovakia [10] and in Poland [19]. This association occupies clayey and sandy soils in mesotrophic fishponds, and is usually abundant in peat sediments in the littoral [4]. In Germany it was recorded on fertile and muddy river banks of the Odra and old Danube riverbeds [17, 22]. In Slovakia *Scirpetum radicantis* is reported from standing waters or periodically inundated areas with clayey basement [11]. In Poland *Scirpetum radicantis* was known from one locality. It was discovered in fishponds in south-western Poland in Silesia [19]. This rush community is recognized as endangered at both regional and Central European level [1, 11, 16, 17, 19].

Mud associations from the *Isoëto-Nanojuncetea* Br.-Bl. et R. Tx. 1943 class belongs to the group of small, central European terophytes, blooming in late summer or autumn, and develop on periodically wet mineral soils, with a short period suitable for plant growth [7, 9, 12, 14, 15, 17]. Most of these associations, because of the environmental changes, have a regressive character and are registered as threatened [7]. One of the rarest examples observed on the Silesian fishponds is *Eleocharetum ovatae* Hayek 1923 n. n. Its phytocoenoses develop on the bottoms of the dried ponds or within the shore zone of the other reservoirs. But the changing economy of Poland is resulting in drastic intensification of fisheries, resulting in alterations in fish breeding cycle and water management of ponds. The most significant is the shortening of the period without water in ponds, which makes the development of vegetation from *Isoëto-Nanojuncetea* class impossible. Only under special conditions, e.g. rebuilding or deepening the pond, water shortage allows a normal germination of the mud plant phytocoenoses. *Eleocharetum ovatae* finds the best environmental conditions for growing on the fishpond bottoms in the south part of Western Europe [e.g. 3, 5]. In Poland it is considered a rare and threatened plant association [14, 23].

Conclusions

The time of fishponds existence depends on the profits given by fish farming and therefore if they are not profitable enough they are usually dried off. This time however proves to be long enough for fish-ponds to preserve their biodiversity in a particular area and to become a kind of refugium for many rare and threatened biota. Thus, to reach the effective level of conservation, the co-operation between biologists and owners, and integration of the conservation and agricultural activities, have to be achieved. This would be the only way to stop the impoverishment of the biological diversity in the hydrophilous ecosystems. Also it seems to be indispensable to establish the legally based protection system for the most valuable fishponds. Phytosociological methods of evaluation were effectively used to find the priority areas. Considering the great importance of fishponds as valuable habitats for their wildlife and educational qualities, it seems to be useful to set aside the ecologically interesting lands and to establish a network of natural reserves.

REFERENCES

1. Balátová-Tuláčková E., Mucina L., Ellmauer T., Wallnöfer S., 1993 - *Phragmiti-Magnocaricetea*. In: Grabherr G., Mucina L. (eds.), Die Pflanzengesellschaften Österreichs. Teil II, Natürliche waldfreie Vegetation., pp: 79-130. G. Fischer Verlag, Jena.
2. Braun-Blanquet J., 1964 - Pflanzensoziologie, Gründzüge der Vegetationskunde. Dritte Auflage. Springer Verlag, Wien-New York.
3. Ellenberg H., 1986 - Vegetation Mitteleuropas mit den Alpen in ökologischen Sicht. E. Ulmer Verlag, Stuttgart.
4. Hejný S., Husák Š., 1978 - Higher plant communities. In: Dykyjová D., Kvét J. (eds.), Pond littoral ecosystems, pp. 23-64, 93-95. Springer Verlag, Berlin.
5. Klika J., 1935 - Die Pflanzengesellschaften des entblössten Teichbodens in Mitteleuropa. Beih. Bot. Zentralbl., **53 B**: 286-310.
6. Landolt E., 1982 - Distribution pattern and ecophysiological characteristics of the European species of the *Lemnaceae*. Ber. Geobot. Inst. ETH Stiftung Rübel Zürich, **49**: 127-145.
7. Matuszkiewicz W., 2001 - Przewodnik do oznaczania zbiorowiska roślinnych Polski (Guide to the determination of plant communities in Poland). Wydawnictwo Naukowe PWN, Warszawa (in Polish).
8. Mirck Z., Piękoś-Mırkowa H., Zajac A., Zając M., 2002 - Flowering plants and pteridophytes of Poland. A checklist. Biodiversity of Poland, **1**: 9-442.
9. Oberdorfer E. (ed.), 1977 - Süddeutsche Pflanzengesellschaften. 2 Aufl. 1. G. Fischer, Stuttgart.

10. Otahelová H., 1995 - K výskutu asociácie *Scirpetum radicantis* Hejný in Hejný et Husák 1978 na Slovensku (Distribution of *Scirpetum radicantis* Hejný in Hejný et Husák 1978 in Slovakia). Bull. Slov. Bot. Spol., Bratislava, **17**: 123-125 (in Slovakian).
11. Otahelová H., 2001 - *Oenanthetalia*. In: Valachovič M. (ed.), Rastlinné společenstvá Slovenska. 3. Vegetácia mokradlí, pp: 148-160 (Plant communities in Slovakia. 3. Vegetation of wetlands). Academic Electronic Press, Bratislava (in Slovakian).
12. Pietsch W., 1973 - Beitrag zur Gliederung der europäischen Zwergbinsengesellschaften (*Isoëto-Nanojuncetea* Br.-Bl. et R. Tx. 1943). Vegetatio, **28** (5-6): 401-438.
13. Piórecki J., 1980 - Kotewka orzech wodny (*Trapa* L.) w Polsce. (*Trapa natans* L. in Poland). Biblioteka Przemyska, **13**: 5-159 (in Polish).
14. Popiela A., 1997 - Occurrence of the *Isoëto-Nanojuncetea*-class communities in Poland. Monogr. Bot., **80**: 3–59 (in Polish).
15. Pott R., 1995 - Die Pflanzengesellschaften Deutschlands. 2 Aufl. E. Ulmer, Stuttgart.
16. Rennwald E. (ed.), 2000 - Rote Liste der Pflanzengesellschaften Deutschlands mit Anmerkungen zur Gefährdung. Schr.-R.f. Vegetationskunde, **35**: 393-592.
17. Schubert R., Hilbig W., Klotz S., 1995 - Bestimmungsbuch der Pflanzengesellschaften Mittel- und Nordostdeutschlands. G. Fischer, Jena.
18. Schwabe-Braun A., Tüxen R., 1981 - *Lemnetea minoris*. In: R. Tüxen (ed.), Prodromus der europäischen Pflanzengesellschaften, 4, 141 pp. Lief. J. Cramer, Vaduz.
19. Spałek K., Nowak A., 2003 - *Scirpetum radicantis* Hejný in Hejný et Husák 1978 em. Zahlh. 1979, a plant association new to Poland. Acta Soc. Bot. Pol., **72** (4): 347-350.
20. Tomaszewicz H., 1979 - Roślinność wodna i szuwarowa Polski (Water and rush vegetation of Poland). Rozpr. Uniw. Warsz., Warszawa (in Polish).
21. Wołek J., 1991 - Synusial assemblages of pleustonic plants of genera: *Lemna*, *Spirodela*, *Wolffia*, *Salvinia*, *Hydrocharis*, *Riccia* and *Ricciocarpus*. Ber. Geobot. Inst. ETH Stiftung Rübel Zürich, **57**: 193-202.
22. Zahlheimer W., 1979 - Vegetationsstudien in den Donauauen zwischen Regensburg und Straubing als Grundlage für den Naturschutz. Hoppea, Denkschr. Regensb. Bot. Ges., **38**: 3-398.
23. Zając M., Zając A., 1988 - Communities of the class *Isoëto-Nanojuncetea* on the bottoms of drying-up ponds in the S part of the Oświęcim Basin (southern Poland). Zesz. Nauk. Uniw. Jagiell., Pr. Bot., **17**:155-160 (in Polish).

THE SAXON VILLAGES OF SOUTHERN TRANSYLVANIA: CONSERVING BIODIVERSITY IN A HISTORIC LANDSCAPE

John R. AKEROYD[1], Nat PAGE[2]
[1] Plant Talk, Lawn Cottage, West Tisbury, **Salisbury, Wiltshire SP3 6SG, UK 6SG**;
e-mail: akeroyd@dial.pipex.com
[2] ADEPT, Upper Leigh Farm, East Knoyle, **Salisbury, Wiltshire SP3 6AP, UK**;
e-mail: npage@copac.org.uk

Abstract: The landscape and biodiversity of the Saxon villages of southern Transylvania comprise a remarkable fragment of an older Europe, where species-rich plant and animal communities thrive alongside traditional agriculture. Not only are the wildflower meadows, probably the best that survive in lowland Europe, a living link with medieval times, but they are also a genetic resource of forage crops, especially legumes. Within Romania, many of the plants, animals and habitats of the region are not particularly rare or threatened, but from a European perspective their survival in such substantial numbers and extent considerably alleviates their loss elsewhere. Rarity may not always be the best criterion for assessing conservation needs, and a holistic approach is required to conserve this ecosystem in its geographical, cultural and biological entirety. To this end, a consortium of conservation NGOs, in conjunction with the Government of Romania, is working towards the establishment of a Nature Park or similar protected area to conserve traditional villages, countryside and biodiversity. Organic and non-intensive mixed farming, eco-tourism and promotion of village crafts and skills are vital components of the complex jigsaw of sustainable development.

Introduction

The Saxon Villages region or Târnave Plateau of southern Transylvania is one of Europe's last extant medieval landscapes [1, 8]. The region, lying within the southern bend of the Carpathians, exhibits a remarkable diversity of habitat types, from natural and semi-natural woodland to dry and semi-dry grassland, damp grassland and wetlands [2]. An eroded plateau of often steep valleys and gently rolling hills to 600–700 m or more, the well-wooded countryside has well-dispersed settlements and few roads. Agriculture is largely un-mechanized with little use of agrochemicals. The region retains and nurtures both an ancient human culture and an abundance of once plentiful wildflowers and wildlife, including significant numbers of large mammals such as brown bears and wolves that have disappeared from much of modern Europe. However, the Târnave plateau has not attracted the international attention of other regions and habitats of Romania, for example the montane forests of the Carpathians or the reed-swamps of the Danube Delta.

The Saxon lands of southern Transylvania lie mostly within the area bounded by the medieval fortress-towns of Sibiu, Sighişoara and Brasov. They straddle parts of three counties: southern Mureş, around Sighişoara; north-west Brasov, north of Fagaraş; and eastern Sibiu. The region provides an astonishing glimpse of biodiversity-rich rural landscapes lost over most of Europe, with its traditional villages, fortified medieval churches (several now UNESCO-designated), extensive woodlands and flowery meadows, and wealth of vertebrate and invertebrate wildlife. In particular, in summer the orchards, arable strips, hay meadows and pastures are a display of wildflowers on a scale unseen in northern Europe for a generation. This is how Europe may have looked in the 18th or even the 14th century, a fragile ecosystem that needs to be conserved as a geographical, cultural and biological entity. It is a landscape that most of Europe has now lost, where rich animal and plant diversity thrives alongside traditional and non-intensive agriculture.

The region is underlain by sedimentary strata dating from the Middle Tertiary onward, mostly calcareous clays or marls, with some deposits of fine sand, tufa (volcanic ash) and narrow limestone beds. Soils range from moderately rich brown earth to podsols and more acid or lime-poor soils over the sands. The clays give rise to soils that tend to dry out in summer. Thin-bedded limestones give rise to base- shallow soils resembling rendzinas, especially on steeper slopes. There are also pockets of yellow loess soil. The best soils, for example at Angofa and Sapartoc just south of Sighişoara, are chernozems or steppe soils which probably derive from an ancient dry climatic period and are extremely fertile.

This landscape is a product of a distinctive history. From the mid-12th century, immigrants ('Saxons') from northwestern Germany settled here at the invitation of the Hungarian kings who then ruled the region, to defend their eastern marches against Tatars and other invaders from the steppes. Farmers, craftsmen and merchants, the Saxon colonists were part of a mass movement of German people and improved agricultural techniques such as 3-crop rotation and the scythe into eastern Europe in the 12th and 13th centuries. They built the famous churches and spacious planned villages (*Hüfen*), and nurtured the land over generations of careful husbandry. Their frontier existence as farmer-warriors encouraged independence, isolation and self-sufficiency. A distinctive entity in the ethnic and cultural mix of southern Central Europe, for 800 years these Saxon communities formed an inward-looking and intrinsically conservative but well-ordered and prosperous society.

Traditional and un-mechanized agriculture, without intensive land-use, has enabled ancient patterns of European agrarian and village life to survive, modified but substantially intact, to the present day. Viscri, for example, has 400 inhabitants and 85 working horses. Mowing is largely by scythe and weeding by hoe, and the lack of herbicides has, for example, enabled rarer arable weeds to survive. In the village itself wayside weeds are a living historical link – many of them old medicinal herbs, of a sort once used all over Europe. Alas, most of the Saxon population emigrated to Germany in the early 1990s, but their cultural legacy endures, notably their elegant 18th century farmhouses and magnificent churches [4]. However, the landscape they created, along with its rich biodiversity, faces a precarious future at a time of considerable economic and social change, especially as Romania prepares to accede to the European Union. It is important that steps be taken to conserve the rich biodiversity within the framework of sensitive agriculture and sustainable development that will also enable local people to enjoy a more secure future.

Methods

A preliminary survey [1, 8] identified a number of significant woodland, wetland and grassland communities. Detailed examination and assessment of grassland communities has since been carried out by Jones [7] in relation to soil chemistry at Mălâncrav (Sibiu county), Angofa (Mureş county) and Viscri (Braşov county), and by Oroian *et al.* [9] from point of view of Phytosociology at several sites in Mureş county, together with a general habitat survey and compilation of records from the floristic literature and Red Data books [3]. Discussions about the ecological and cultural value and the possible future of the region have been held with scientific colleagues in Romania and abroad, with local, regional and national government staff, and with agronomists and conservation scientists.

Discussion: a wealth of plant diversity

The Saxon Villages Region still supports extensive stands of semi-natural vegetation, which is species-rich and, in the case of the woodlands, closely resembling the natural habitats that occupied the Transylvanian foothills of the Carpathians prior to human impact. At the same time the region supports habitats that have evolved in intimate association with human agriculture and other activities. Several of the habitats present, and individual species, are localized in distribution and highly characteristic of this part of Central Europe. This makes the southern Saxon Villages region of considerable international value.

Diverse and often pristine habitats support more than 1000 plant taxa in over 100 families, more than 30% of the Romanian flora. A total of 40 species of trees, 36 species of shrubs, and over 900 species and subspecies of herbaceous plants have been identified so far, including 18 orchids. This richness is a result of geographical position, diversity of relief, varied climatic conditions and soils, and traditional land-use.

From a floristic point of view, 87 taxa are listed for protection and conservation at national and international level, including the globally threatened Polish Larch (*Larix decidua* subsp. *polonica*) and nine taxa threatened in Europe and included in the Annexes of the EU Habitats Directive and the Bern Convention. A further 77 taxa are threatened at national level and included in the Romanian Red List. Just over half occur in meadow-steppe communities, and all are included in IPA lists for Romania. Several are rare and decreasing in Europe. More than 50 of the native plants are related to cultivated or crop plants and constitute a potential resource for plant breeding, notably forage legumes such as Sainfoin (*Onobrychis viciifolia*) and Red Clover (*Trifolium pratense*). Some village fruit trees may represent old varieties or cultivars, especially plums and pears, and the wild pears too are a natural gene-pool.

Not only do the flora and vegetation of the Târnave plateau exhibit a high level of diversity, but also the whole landscape is of high conservation value. This buffers the best 'hot-spots', including species and communities rare at regional, national or European level. Some vegetation types have been nominated for protection by international legislation (Table 1), and several grassland and woodland habitats would greatly complement the EU's 'NATURA 2000' network and Council of Europe's 'EMERALD' network. As a first step, available evidence suggests at least two districts – Criţ and the hills from Saschiz to Viscri and just to the east – probably qualify as Important Plant Areas (IPAs) of Europe [10].

The woodland communities can be summarized as follows [8]:

- **Dacian *Lathyrus hallersteinii* oak-hornbeam forest**, *Lathyro hallersteinii-Carpinetum* association (*Querco petraeae-Carpinetum transsilvanicum*) on more acidic leached brown earth soils of sheltered slopes. In the most typical examples of this forest, *Lathyrus hallersteinii*, endemic to Romania and the central part of the Balkan peninsula, is frequent.
- **Dacian *Melampyrum bihariense* oak-hornbeam forest**, *Melampyro bihariensis-Carpinetum* association, which tends to replace *Lathyrus hallersteinii* oak-hornbeam woodland on the more base-rich soils that cover much of the region, on the deeper soils of depressions and gentle slopes.
- **Pannonic downy oak woodland**, *Quercion pubescentis* alliance, protected in two small forest reserves near Sighişoara: 11.9 ha Wolzen Hill, and 6.8 ha Gotca Mare (near Daneş). The valley of the small River Mălincrav, which flows north from Mălincrav, also has stands of Downy or White Oak (*Quercus pubescens*) on ridges and steeper and drier slopes.
- **Dacian hairy sedge beech-hornbeam forest** of the *Lathyro-Carpinion* alliance, with forests of Beech (*Fagus sylvatica*), usually with Hornbeam (*Carpinus betulus*), appear to be the main type of beech forest in the region, relatively common along the ridges and steeper slopes, for example above Saschiz and near Sighişoara. *Carpino-Fagetum* woods, often with Small-leaved Lime (*Tilia cordata*) and Wild Cherry (*Prunus avium*), occur in similar conditions, but mostly on better-drained and shallower soils.
- **Peri-Pannonic hawthorn-blackthorn scrub**, comprising noda of the order *Prunetalia* (notably within the *Prunion spinosae* alliance), including the associations *Pruno spinosae-Crataegetum*, *Ligustro-Prunetum spinosae* and *Euonymo-Sambucetum nigrae* – any one of several shrubs may be dominant. These communities, such as on the Huluba hills just north-east of Viscri, are floristically rich and include several rare grassland species; e.g. Burning Bush (*Dictamnus albus*), Red Viper's Bugloss (*Echium russicum*), and the endemic wallflower, *Erysimum witmannii* subsp. *transsilvanicum*.

- **Eastern European poplar-willow forest**, of the alliance *Salicion albae*, in gallery woodland along watercourses, dominated by White Willow (*Salix alba)*, Crack Willow (*S. fragilis*) and Black Poplar (*Populus nigra*).
- **Almond willow-osier scrub**, of the association *Salicetum triandro-viminalis*, a shorter, denser scrub typically composed of Almond Willow (*Salix triandra*) and Purple Willow (*S. purpurea* subsp. *lambertiana*).

A 70 ha reserve on the Breite Plateau, 1.5 km south-east of Sighişoara protects a remarkable stand of *Quercus pedunculiflora*. The oaks, many of great age and size, grow as widely spaced individuals or clumps in a broad grassy clearing surrounded by oak-hornbeam-beech woodland [2]. The ground flora is a mixture of plants of heathy and damp grassland. A similar habitat, but with most of the trees apparently removed, occurs at Apold [2], and there is a similar wood-pasture at Saschiz (A. Gota, pers. comm.). Some woodland habitats are protected at national and international level, and five woodland communities, including the Breite, have been proposed as Important Plant Areas (IPAs) of Europe.

The most obvious manifestation of Transylvania's astounding richness of plant and animal diversity is the wildflowers of the traditionally managed grasslands. These are probably the best lowland hay-meadows and pastures left in Europe; so extensive that you can walk through them for several hours. The colourful and varied flora of these grasslands comprises a mixture of western and central European plants, but with a significant element of steppic species. This species-rich 'meadow-steppe' has retreated throughout Europe [5], even in Poland, Czech Republic and Slovakia. Wiry grasses dominate the sward, and the species-rich communities often include 20–30 or more species of legumes, notably Sainfoin *(Onobrychis viciifolia*), several dwarf brooms (*Chamaecytisus* and *Genista* spp.) and numerous clovers (*Trifolium* spp.), a characteristic floristic element of dry grasslands in Transylvania [11].

Drier grasslands in Transylvania frequently and characteristically have 20–30 wild legume species [11], and apparently almost all the grassland legumes of Romania have been recorded in the Buneşti district (T. Maruşca, pers. comm. 2003). On hot, dry south-facing slopes, the flora is distinctly steppic, with Pontic-Sarmatian elements such as *Adonis vernalis*, *Crambe tatarica*, *Linum flavum* and *Salvia nutans*, and Mediterranean elements such as *Muscari comosa* and *Vinca herbacea*.

Jones [7] has identified five broad grassland types:

1. Degraded, over-grazed pasture or former arable land, with plentiful White Clover (*Trifolium repens*) and Wild Carrot (*Daucus carota* subsp. *carota*), and a high proportion of ruderals such as Field Bindweed (*Convolvulus arvensis*).
2. Short grazed pasture dominated by *Festuca pseudovina* and Yarrow (*Achillea millefolium*). These pasture communities are locally species-rich.

3. Tall hay-meadow and pasture dominated by Tor-grass (*Brachypodium pinnatum*) and Sainfoin (*Onobrychis viciifolia*). These species-rich communities correspond to the *Cirsio-Brachypodion pinnati* alliance recognized by Romanian botanists [6]. They include both semi-natural grassland on marginal land, with especially species-rich noda such as *Polygalo majalis-Brachypodietum pinnati* [9] and moister hay-meadows associated with villages, with species such as *Sanguisorba officinalis* and *Colchicum autumnale*, now often ploughed or improved.
4. Semi-steppic grassland, with the sages *Salvia nutans* and *S. transsilvanica* and other forbs (non-grasses), characterized by a high level of often rare or local Pontic-Sarmatian floristic elements but relatively low species diversity. These communities correspond to the *Festucion valesiacae* alliance recognized by Romanian botanists [6].
5. A heathy, somewhat acid grassland that is relatively species-poor, with plants such as a wild strawberry (*Fragaria viridis*) and Dyer's Greenweed (*Genista tinctoria*). This 'heathy grassland' community probably deserves formal phytosociological designation.

One of the most interesting and significant findings of Jones's study was the low nutrient status of the soils, especially on the Huluba hills. Generations of villagers have transferred nutrients to the valleys as hay or animal dung and there has been almost no input of nutrients to the upper pastures. This correlates with the great species diversity, the richest grassland communities (more than 40 species per 1/2 m^2 relevé) being on medieval 'ridge and furrow' fields along high slopes. In other parts of Europe, nutrient enrichment has done untold damage to similar ancient grasslands.

To summarize the ecological and conservation importance of the habitats of the southern Saxon Villages region (see also Table 1):

- The woodlands clearly derive from the original forests of the region, and their ground flora shelters species and noda of restricted world distribution. Distinctive oak wood-pastures are a local feature.
- The grasslands and their biodiversity are of considerable importance at a European level, and are particularly rich in Dacio-Pannonic, Pontic-Sarmatian and Mediterranean floristic elements. They represent a major resource of a habitat that has contracted or disappeared over much of Europe through agricultural intensification.
- Many of the wetlands, both floodplains and flushes, remain hydrologically intact, with an almost natural zonation of habitats.
- The floristically rich habitats contain substantial populations of vertebrate and invertebrate animals that are increasingly rare over much of Europe.
- The architecturally outstanding villages are an integral part of this landscape in intimate association with the rich biodiversity.
- Not only are the habitats important for conservation, but also they provide biologists with a model of historical ecological patterns and processes and how these maintain high levels of biodiversity.

Table 1: EU Habitats Directive Annex I habitat types present in the south-east Transylvania Saxon villages region.

Name	Comments
* Sub-continental Peripannonic scrub	Generally well-maintained but locally threatened by neglect or over-grazing.
* Semi-natural dry grasslands & scrubland facies on calcareous substrates (*Festuco-Brometalia*) with important orchid sites	Generally well-maintained but locally threatened by neglect or over-grazing.
* Sub-Pannonic steppic grasslands	Generally well-maintained but locally threatened by neglect or over-grazing.
Lowland hay meadows (*Alopecurus pratensis, Sanguisorba officionalis*)	In danger since they are easily accessible from villages and many have been ploughed or otherwise altered.
* Calcareous fens with *Cladium mariscus* and species of *Caricion davallianae*	Rich in wild fauna, for example corncrake (*Crex crex*)
* Pannonic woods with *Quercus petraea* and *Carpinus betulus*	Generally well maintained
* Pannonian woods with *Quercus pubescens*	Well maintained
Dacian Beech forests (*Symphyto-Fagion*)	Generally well maintained

* *indicate priority habitats according to Annex I of Habitats Directive.*

The plant life of the region is a priceless scientific and a cultural component of a special landscape. More than 1000 plant taxa have been recorded, over 30% of the entire Romanian flora. Many of these plants and the communities in which they occur are now rare or threatened in Europe, and they will be of particular significance for conservation strategies in an expanded European Union. Not only are the wildflower meadows, probably the best that survive in lowland Europe, a living link with medieval times, but they are also a genetic resource for future selection of forage crops, especially sainfoin, clovers and other legumes. Drier grasslands contain more than 20 species on the Romanian Red Data List and at least nine (see Table 2) listed on Appendix 2 of the EU Habitats Directive. The assemblage is more important than any individual component plant species, as such species-rich, dry meadow-steppe grassland has disappeared over most of Europe.

Potential threats to the flora and vegetation

The grasslands are flourishing at present, but the survival of these plant-rich but nutrient-poor plant communities faces potential threats from nutrient enrichment, both over-grazing and cessation of grazing and shrub encroachment. *Just a single application of fertilizer would have catastrophic effects on the survival of these remarkable grasslands and their special assemblages of plants and animals.* Woodlands are also flourishing, but there is increasing evidence of local clear-felling which needs to be monitored.

Table 2: EU Habitats Directive Annex II Species present in the Saxon south-east Transylvania villages region.

Group	Species	Group	Species
Mammals	*Canis lupus* *	Insects	*Astacus astacus*
	Ursus arctos *		*Lucanus cervus*
	Lutra lutra		*Callimorpha quadripunctaria*
	Myotis myotis		*Eriogaster catax*
	Barbastella barbastellus		*Lycaena dispar*
			Maculinea teleius
Amphibians	*Triturus cristatus*		
	Rana dalmatina	Plants	*Echium russicum*
	Bombina variegata		*Crambe tataria*
	Rana temporaria		*Cypripedium calceolus*
			Pulsatilla pratensis ssp. *hungarica* *
Reptiles	*Lacerta agilis*		*Arnica montana*
	Natrix natrix		*Gentiana lutea*
	Emys orbicularis		*Angelica palustris*
			Lycopodium clavatum
Fish	*Barbus meridionalis petenyi*		*Pulsatilla vulgaris* ssp. *grandis*
	Gobio albipinnatus vladikovy		
	Rhodeus sericeus amarus		
	Cobitis taenia taenia		
	Sabanejewia aurata balcanica		
	Gobio uranoscopus frici		
	Gobio kessleri kessleri		

* *indicate priority species according to Annex II of Habitats Directive.*

The principal threats to the wild plants and vegetation of the region are:

- Uncontrolled agricultural expansion into grasslands, with nutrient over-enrichment and over-grazing, especially by sheep, and invasion by a ruderal flora of unpalatable species such as thistles and other invasive weeds;
- Abandonment or reduction of traditional grassland management such as grazing and scrub clearance;

- Unsustainable forestry practices such as planting with exotic trees, or clear-felling with loss of tree cover;
- Further spread of weeds, especially aggressive alien species such as Japanese Knotweed (*Fallopia japonica* var. *japonica*);
- Unsuitable and unsustainable infrastructure development for recreation and tourism, with new roads and buildings;
- Unsustainable exploitation of wild populations of plants, especially over-collection of medicinal plants such as gentians and *Adonis vernalis*;
- Loss of ruderal weed communities, rich in archaeophytes and old medicinal and other useful plants such as Elecampane (*Inula helenium*) and Marsh Mallow (*Althaea officinalis*) around the villages;
- Climate change, for example an increase in frequency and duration of prolonged spring and summer drought;
- Lack of public knowledge and information about the region's ecological value.

There is also a scientific paradox that may threaten the future of these habitats. Within Romania, many of the plants, animals and habitats of the Saxon villages region are not regarded as particularly rare or threatened. Several woodland communities are recognized as important at a national and international level, but much of the grassland does not fall within existing categories of protection. However, from a western European perspective the survival of grassland communities in such substantial numbers and extent considerably makes up for their loss elsewhere. The EU Habitats Directive is going to have to take account of a different range of habitats as countries from Eastern Europe accede to the Union.

Rarity on its own may not always be the best criterion for assessing conservation needs and a more holistic approach is required to protect such a sensitive and fragile ecosystem. The grasslands cannot be separated from the cultural landscape, of which they are a historical and integral element. Sites with the rarest and most interesting plants, for example a steep grazed slope kept clear of scrub through burning and with *Salvia nutans* and *Linum flavum* [7], were poor in species (c. 10 per relevé) but of inestimable ecological and conservation interest at a European level. Plant species diversity, although important in ecological terms, should not be always be considered in isolation as a measure of conservation value. Numbers of Red Data Book species or other threatened plants (and animals) may not also be an accurate measure of the value of a community or habitat.

Throughout most of Europe, traditional grasslands have suffered drastic shifts in management and are in a state of flux. This part of southeast Transylvania represents a still functioning historic landscape, with the fauna, flora and complement of soil microorganisms of an intact ancient ecosystem, in which extensive wildflower meadows still retain their role in agriculture. Such areas are rare in lowland Europe, and are therefore extremely valuable for conservation research and interpretation.

Conclusions: a way forward

This remarkable ecological and cultural landscape is more or less intact but extremely fragile. The brilliant carpet of wildflowers and associated dependent wildlife are but a single fertilizer application away from degradation and ultimate destruction. We in Britain, and other countries in western Europe, have allowed ourselves to lose so many similar habitats, where rich animal and plant diversity thrive alongside traditional agriculture; some 95% of species-rich lowland grassland has disappeared in Britain since 1945. It cannot be long before this ecosystem and others like it in Eastern Europe go the same way – recent years have seen the steady decline of species that were formerly common across this region. This emphasizes the importance of the habitats in Saxon Transylvania, which some biologists apparently take for granted.

To this end, a consortium of conservation NGOs from Romania, Germany, Norway and the UK, in conjunction with the Government of Romania, is working towards the establishment of a Nature Park or similar protected area to conserve traditional villages, countryside and biodiversity. Conservationists are working with local people, especially in the villages of Viscri and Mălincrav. From 1993 the Mihai Eminescu Trust has supported traditional building techniques to conserve the fabric and style of the ancient churches and village farmhouses, encouraged visitors and promoted local crafts, in collaboration with a core of dedicated Saxons who maintain the traditions of their diminished community [12].

Nature conservation and farming is the responsibility of ADEPT (Agricultural Development and Environmental Protection in Transylvania), an Anglo-Romanian project promoting organic farming and biodiversity conservation in the villages, and promoting the idea of establishing a Nature Park (IUCN 'Category V') or similar protected area to conserve the biodiversity of the area. ADEPT's aims are:

1. to continue to compile a floral and faunal inventory of an area of about 90,000 ha to the south and south-east of Sighişoara, with the assistance of colleagues from the University Babeş Bolyai - Cluj-Napoca, University of Medicine and Pharmacy - Târgu-Mureş, and Romanian Association of Botanic Gardens;
2. to design and publish environmental education booklets for secondary schools, and wider public distribution;
3. to carry out measures to encourage and support farming and other small enterprises that are consistent with conservation goals;
4. to establish a pilot project in a suitable commune (Saschiz, area 9,300 ha., population 2,100), including a centre for environmental information, education and training for local people, to promote economic development as well as environmental awareness.

The 90,000 ha area being investigated contains about 30 small villages and has a population of about 19,000. Measures 2 and 3 above are expected to increase local support for the statutory protection of all or part of this area. In an inhabited, anthropogenic landscape such as this, local support is a prerequisite for successful conservation, and to obtain local support the villagers must obtain tangible benefits as a direct result of the conservation programme. The Romanian Ministry of Environment and Water Management supports some form of protection for the area.

In the wider countryside, the key to conserving a living landscape and ensuring sustainable development is sensitive farming, employing modern techniques but continuing the careful and prosperous husbandry that have nurtured the biodiversity-rich woodlands, grasslands and wetlands for hundreds of years. This agricultural tradition yielded the prosperity that created the famous fortified churches and village architecture. Organic and non-intensive mixed farming, eco-tourism and promotion of village crafts, skills and country products are vital components of the complex jigsaw of sustainable development. The interaction of conservation and a possible secure economic future for local people makes the Saxon Villages a useful model for other regions of Europe with similar problems. There is no reason why the best aspects of an older way of life should not remain relevant and viable at this time of change and uncertainty. To save a substantial fragment of this ecological and cultural landscape for posterity, the Saxon lands need long-term investment and support to encourage sustainable activities: organic farming, traditional but living crafts, and village tourism – harnessing the numerous skills of local people. The villages will regenerate and prosper only if we build upon their greatest assets by encouraging small-scale economic and tourist development in sympathy with the landscape, its intangible sense of place and atmosphere of history.

Acknowledgements: Initial studies of the flora and vegetation of the southern Saxon Villages were funded by the Mihai Eminescu Trust (2000–2001). The authors thank the World Bank/International Finance Corporation and the Global Environment Facility, Dinu Patriciu and the Rompetrol Group, Virginia Gheorghiu, Michael Radomir and Ben Goldsmith for funding 2002–2004.

They also record their thanks to Dr. Andrew Jones (Grassland Trust, UK), Dr. Owen Mountford (NERC Centre for Ecology and Hydrology, Abbots Ripton, UK), Prof. Anca Sârbu (Association of Romanian Botanic Gardens, Bucharest), Prof. Silvia Oroian and Ing. Mariana Hirițiu (University of Târgu Mureș), and Dr. Mihaela Sămărghițan (County Museum, Târgu-Mureș), for use of unpublished data and for valuable discussion; and to Cristi Gherghiceanu and family in Viscri who have helped and supported us during our field work.

REFERENCES

1. Akeroyd J.R., 2002 - Protecting Romania's Lost World. Plant Talk, **30**: 19–24.
2. Akeroyd J.R., 2003 - A Transylvanian wood-pasture. Plant Talk, **34**: 34–37.
3. Akeroyd J.R., Sârbu A., Oroian S., Hirițiu M., 2003 - Botanical assessment of the Saxon Villages region (Târnave plateau) of southern Transylvania, Romania. Unpublished report to International Finance Corporation, Washington DC, USA.

4. Blacker W., 1997 - The plight of the Saxons of Transylvania and their fortified churches. Mihai Eminescu Trust, London, UK.
5. Cerovsky J., 1995 - Endangered plants. London: Sunburst Books. IUCN (1994) Parks for Life: Action for Protected Areas in Europe. IUCN, Cambridge, UK.
6. Coldea G., Sârbu I., Sârbu A., Negrean G., 2001 - Ghid pentru identificarea şi inventarierea pajiştilor seminaturale din România. Editura Alo, Bucureşti!, Bucureşti.
7. Jones A., 2003 - Grassland Management Strategy – Transylvanian grasslands. Unpublished report to ADEPT project, Salisbury, UK.
8. Mountford J.O., Akeroyd J.R., in press - A biodiversity assessment of the Saxon villages region of Transylvania. Transylvanian Review of Systematical and Ecological Research.
9. Oroian, S., Sămărghiţan, M., Hiriţiu, M., 2003 - Information concerning the Târnave Plateau vegetation. Unpublished report to ADEPT project, Salisbury, UK.
10. Plantlife International, 2004 - Identifying and protecting the world's most important plant areas. Plantlife International, Salisbury, UK.
11. Puşcaru-Soroceanu E. *et al.* (eds.), 1963 - Păşunile şi fîneţele din Republica Populară Română. Editura Academiei Republicii Populare Române, Bucureşti.
12. Wilkie K., 2001 - The Saxon villages of Transylvania, Romania. Kim Wilkie Associates, Richmond-upon-Thames, UK.

BIOTOPE TYPES OF THE TREELINE OF THE CENTRAL GREATER CAUCASUS

George NAKHUTSRISHVILI*, **Otar ABDALADZE**, **Maia AKHALKATSI**
Institute of Botany, Georgian Academy of Sciences, Kodjori Road 1,
0105 Tbilisi (GEORGIA); e-mail: nakhutsrishvili@yahoo.com

Abstract: Some characteristics (habitat, distribution, characteristic species, ecological importance, practical use and danger factors) are given of 17 typical treeline biotopes in the Kazbegi region, situated on the north-facing macro slope of the central part of the Main Watershed Range of the Greater Caucasus. The diversity of species composition is mainly due to peculiar structural properties of the vegetation of these biotopes, rather than ecological (altitude, moist or dry conditions,) and anthropogenic (grazing, haymaking) factors. The majority of the studied biotopes are referred to the II stage of hemeroby, but untouched shrub communities (dominated by *Rhododendron caucasicum*), elfin crooked-stemmed birch forests (dominated by *Betula litwinowii*) and fragments of tall herbaceous vegetation should be referred to the I stage of hemeroby (natural and close to natural). Current global climate change will cause the most drastic changes in moist (tall herbaceous vegetation), snowline (humid broad-leaved meadows dominated by *Trollius ranunculinus*), broad-leaved mesophilous meadows (dominated by *Anemone fasciculata*), elfin crooked-stemmed birch forest and scrub biotopes. The numerous biotopes are characterized by high sensitivity, which is caused by the following factors: landscape mainly devoid of forests, high degree of relief and, consequently, bare soil cover, low and unstable snow cover, frequent solifluction, etc. The following biotopes should be regarded as priority habitat types: tall herbaceous vegetation, scrub and elfin crooked-stemmed birch forest.

Introduction

The treeline is a natural ecotone between woodlands and scrub. In humid mountains it is the most noticeable and important altitudinal boundary. There are various reasons for the formation of the treeline: solar radiation, scarcity of heat and water, overbalance of water, influence of winds, properties of substrate/soil, snow cover in winter, avalanches, mud-streams, relief and aspect [4].

The ecological importance of treeline vegetation is immense. This ecotone is not very wide; however it is distinguished by high diversity of species and biotopes. Further, the treeline vegetation regulates ecological balance; in particular, it gives protection against avalanches, landslides, mud-flows and debris-flows and, especially important, it creates a natural reservoir of fresh water [13, 12, 19, 20, 26]. At the same time, strong human impact can have dramatic consequences for life conditions in this very sensitive zone, especially under the influence of global warming [21].

Altitudinal zonation, typical for the temperate nemoral zone, is well marked in the Kazbegi region (east part of the Central Greater Caucasus), where the middle-mountain forest (1000-1500 m a.s.l.), the upper-mountain forest (1500-1750 m a.s.l.), the subalpine (from 1800-1900 up to 2400-2500 m a.s.l.), the alpine (from 2400-2500 up to 2950-3000 m a.s.l.), the subnival (2950-3000 up to 3600-3750 m a.s.l.) and the nival (above 3750 m a.s.l.) belts are distinguished. However, the boundaries between some altitudinal belts are not clear, due to strong human impact and specific topographic features in this alpine region. Thus, for example, in some areas of the Kazbegi region it is rather difficult to draw a demarcation line between the subalpine and alpine belts. This can be explained by the destruction of forests, the considerable descent of meadows (to about 1750 m a.s.l.), and also by the invasion of alpine vegetation into the subalpine belt.

On the Central Greater Caucasus the natural treeline ecotone consists of open and elfin crooked-stemmed (Krummholz) birch forests, tall herbaceous vegetation, scrub and meadows. The treeline vegetation in this part of the Caucasus is strongly degraded and lowered (at the average by 200-400 m) because of long-term overgrazing, tree cutting, etc. [6, 7, 8, 9, 25]. The treeline in the Central Caucasus is characterized by high level of plant diversity and endemism [17, 32, 11]. In the Kazbegi region, the ecological state of the treeline ecotone can be assessed as normal only on certain massifs. These forests have been protected because of their religious significance and they are called "Holy Forests". According to the degree of naturalness of these fragments, they should be attributed to the first level of hemeroby: natural and close to natural [27].

History of the Forest Vegetation of the Kazbegi Region. On the east-facing slopes of the Kazbegi volcanic massif (Fig. 1 in Appendix) is situated the Tkvarsheti lateral volcano (3360 m a.s.l.). While the volcano was erupting, the lava stream dammed up the river Tergi and a temporary lake appeared [16]. The forests, which had covered the banks of the river Tergi, were flooded. Later the river destroyed the barrier, the dammed water flooded out and 4 m of thick alluvial deposits covered the forests. The absolute age of the fossil trees is 5950±60 and 6290±90 years, which confirms the post-glaciation age of the Tkvarsheti volcano. In the same area – in Tkvarsheti (1980 m a.s.l.) – 2 m of thick fossil peat was found. It has been ascertained that the peat formation took place in the Middle Holocene; however, due to erosive processes enhanced by the intensive drainage of surface waters, the process stopped. Palynological analysis of this peat has shown that it contains pollen of some woody plants which occur neither in Tkvarsheti nor in the surroundings of the Kazbegi region today. These species are: *Abies nordmanniana, Picea orientalis* and *Castanea sativa*. This peat also contains pollen of certain woody plants (for instance, *Pinus kochiana*), which in the Kazbegi region have survived only on inaccessible rocks, in the Dariali Canyon (Fig. 2 in Appendix). Pollen of *Fagus orientalis* is also found in this peat. This plant now occurs only on humid slopes of some lateral gorges of the Dariali Canyon. The presence of chestnut, lime, hornbeam, etc. pollen in the palynological spectra may be explained by the wider distribution of these plants in Georgia and adjacent areas in the Middle Holocene and, consequently, by easiness of the pollen penetration into the Kazbegi region. According to the data reported by the same authors, in the period of the climatic optimum (the Middle Holocene) the treeline passed along higher altitudes and, pine and birch forests were more widespread in the Kazbegi region than they are now. Alder forests and thickets of *Hippophaë rhamnoides* covered river banks. The palaeobotanical material obtained from travertine at the sources of river Tergi also proves the wide distribution of subalpine forests in the Kazbegi region in the Middle Holocene. According to these data, *Fagus orientalis, Quercus macranthera* and some other species were distributed in the region besides *Betula litwinowii*. In the late Holocene forests have disappeared in the Kazbegi region probably due to the worsening of climatic conditions and strong human impact [16, 23].

Phytoecological and ecophysiological studies concerning the stability of treeline vegetation in the Kazbegi region have been undertaken since 2002 [2, 3, 27, 28]. This work is considered in the context below.

The goal of our study was to reveal and characterize the main treeline biotopes of the Kazbegi region (the eastern part of the Central Greater Caucasus).

Material and Methods

Study Site

The studies have been carried out in the Kazbegi region which is situated on the north-facing major slope of the Main Watershed Range of the Greater Caucasus (N 42°39'; E 44°37'), on the valley of the river Tergi. This region is located on the highest and geomorphologically the most complex central part of the Greater Caucasus. The topography is formed by Jurassic rocks, Palaeozoic and even older granites, young lava and moraines. The mountain massives of the Kazbegi volcanic area are overlayed by Quaternary and Contemporary glacial or river deposits and stone falls as well as strong accumulation of calcareous tuffs and travertine. Glacial and other continental Würm deposits occur in many places. The elevation range in the region varies from 1210 m up to 5033 m a.s.l. (the highest peak is Mkinvartsveri); the average elevation is 2850 m a.s.l.. Brown and light brown skeletal soils of medium and shallow depth and degraded forest and secondary meadow soils are predominant in the treeline ecotone.

In the Kazbegi region the natural treeline ecotone is located between 2450 and 2550 m a.s.l. The climate between 1900 and 2600 m a.s.l. is moderately humid. Winter is relatively dry and cold, and summer is short. The mean temperature of the warmest months (July, August) ranges from 10°C to 14°C (the absolute maximum is 33°C). The mean temperature of the coldest month (January) is -11°C (the absolute minimum is -32.5°C). The number of days with freezing temperature is 124 per year. Stable snow cover persists for 5–7 months and reaches its maximum depth (115-120 cm) in March. The average annual precipitation is 1000-1200 mm, and the summed precipitation during May-August is about 100 mm. The mean air humidity in summer is 75%. Mist is frequent in the area (135 misty days per year), especially in summer. Winds of mountain-gorge type prevail. Duration of growing season is six months [25, 26].

Methods

Phytosociological surveys have been carried out according to the method of Braun-Blanquet [5] during 1961-2004. For biotope classification we were guided by the works of Holzner [14], Pott [30] and Pedrotti [29]. We followed the work of Pott [30] to individualise the hemeroby levels. Priority habitat types were distinguished according to 92/43 EEC from September 1993 [30]. Mean and standard deviation were determined for every data set.

Results

The total area of the subalpine zone in the Kazbegi region amounts to 245.75 km^2. About 1100 species of vascular plants are recorded in the whole region. The species number in the subalpine zone (from 1800-1900 up to 2400-2500 m a.s.l.) is 595, of which 33.2% are endemics. The total area covered by forest is 8707 ha.

Below are given some characteristic features of the main treeline biotopes of the Kazbegi region (the Central Greater Caucasus).

Biotope complex: biotopes of woody plants

<u>Birch forest (*Betula litwinowii*).</u> The range of the birch, which belongs to the periglacial plants, was remarkably diminished in the xerothermic period of the postglacial epoch and the species has remained only in the subalpine zone [6, 9]. At present the birch is dispersed at the treeline of the whole Caucasus, although the areas occupied by the species are not large. Individual trees grow at an elevation of 2550 m in some places and single specimens broken down by avalanches may be found even in the middle-mountain forest belt. ***Habitat:*** Humid north-facing slopes (10-25° of inclination) with stable snow cover. Deep and slightly skeletal brown soils covering volcanic rocks, light, loamy; humus content in the upper 10-20 cm layer is about 7%. ***Distribution:*** fragmentary (between 1850 and 2200 m): the Liphu forest (above the village Gergeti), the Sno gorge, the surroundings of the village Sioni and village Djuta; the forest is best developed in the Devdoraki and Khde gorges; a fragment is found on northwest-facing slopes in the Gudauri region (2000-2300 m). ***Characteristic species:*** *Betula raddeana, Salix caprea, Heracleum roseum, Aconitum nasutum, A. orientale, Swertia iberica, Geranium silvaticum, Campanula latifolia, Dolichorrhiza caucasica, Senecio propinquus, Aquilegia caucasica, Vicia balansae, Lathyrus roseus, Cephalanthera longifolia, Platanthera chlorantha.* ***Ecological importance:*** Water-regulatory, protection against avalanches and erosion. ***Threats:*** Tree felling, intensive grazing, construction of roads, climate global change (warming).

<u>Elfin crooked-stemmed birch forest (*Betula litwinowii*).</u> These forests occur at the treeline, particularly, in areas with abundant snow in winter (Fig. 3 in Appendix). The elfin form is due to the weight of the snow covering trees in winter. Such a life form helps woody plants to adjust to severe winter conditions. In the period following the Ice-age the range of the birch forests diminished drastically; they have remained only in the subalpine belt [8, 9]. Its upper distribution limit does not exceed the level of the 11°C isotherms of August, while in the areas where it is unprotected by snow coat, the 9.5°C isotherms [25]. ***Habitat:*** North-facing slopes with deep snow cover as well as various inclinations (15-70°). Peat brown soils of medium depth, which are situated mainly on volcanic rocks. ***Distribution:*** The upper line of the subalpine belt between 2350 and 2500 m; in the Sno gorge, on the Qvena Mt., in the Devdoraki gorge. ***Characteristic species:*** *Sorbus caucasigena, Salix kazbekensis, Rhododendron caucasicum, Vaccinium myrtillus, Anemone fasciculata, Swertia iberica, Aconitum nasutum, Calamagrostis arundinacea, Dolichorrhiza renifolia, D. caucasica, Cicerbita racemosa, Cephalanthera longifolia.* ***Ecological importance:*** Water-regulatory function, protection of slopes from avalanches, mud-streams, debris-flows, landslides and erosion. ***Threats:*** Tree felling, intensive grazing, construction of roads, climate global change (warming).

Scrub with dominant *Rhododendron caucasicum*. *Rhododendron caucasicum* - a semi-prostrate evergreen shrub, vegetatively spreading by subterranean creeping stems - is a dominant species in an ecosystem dependent on snow cover. It occurs almost over the whole Caucasus forming dense thickets, in which the cover of this species is usually not less than 90-100%. *R. caucasicum* forms an endotrophic mycorrhiza, which enables this shrub to successfully colonize poor, acidic soils. ***Habitat:*** Slightly inclined (10-15°) as well as steep (40-70°) north- facing hillsides with deep snow cover in winter. Peat soils of shallow depth (30-40 cm); peat layer located at the depth of 10-15 cm; humus content in the upper soil layer is 23%. ***Distribution:*** Almost in all the gorges of the region (2300-2900 m a.s.l.); individual specimens of *Rhododendron caucasicum* are brought down to 1800-1900 m a.s.l. by avalanches (*e.g.* on the right bank of the river Bidara). ***Characteristic species:*** *Vaccinium myrtillus, V. vitis-idaea, Empetrum caucasicum, Daphne glomerata, Pyrola minor, P. rotundifolia, Anemone fasciculata, Calamagrostis arundinacea.* Besides the characteristic species noted, the following plants can also be found in the *R. caucasicum* scrub: *Juniperus sabina, Salix kazbekensis.* ***Ecological importance:*** Protection of slopes from erosion, avalanches, landslides and mud- and debris-streams; water-regulatory function. ***Threats:*** Intensive grazing, climate global change (warming), uncontrolled collection of leaves and shoots for medicinal purposes.

Low scrub community with dominant *Dryas caucasica*. The areas occupied by these communities are not large. ***Habitat:*** Skeletal and stony, calcareous soils; slaty humid steep (20-50°) slopes with North and North-West aspects. ***Distribution:*** Relatively wide in the Truso gorge and on the Mt. Kuro slopes; between 2000-2600 m a.s.l. ***Characteristic species:*** *Deschampsia flexuosa, Daphne glomerata, Vaccinium vitis-idaea, Selaginella Helvetica, Primula amoena, Polygonum viviparum, Leontodon danubialis, Parnassia palustris.* ***Ecological importance:*** Protection of humid slopes from erosion. ***Threats:*** Overgrazing, climate warming.

Biotope complex: biotopes of herbaceous plants

Subalpine tall herbaceous vegetation (Megaphorbia). The subalpine tall herbaceous vegetation constituted by tall herbs (2.0-2.5 m, rarely 3.0 m tall) is mainly distributed on the western and central parts of the Greater Caucasus, although fragments of this vegetation are scattered all over the Caucasus. Species richness of an individual community is not high. One of the principal characteristics of these plants is their rather rapid development during the growing season. In May and June the vegetation comprises only one layer of low herbs, whilst in the middle of the growing season (the end of July and the beginning of August) it already consists of tall herbs. Tall herbaceous vegetation is typically three-layered; these three layers are constituted not only by one and the same species, but also by one and the same individual plants. Thus, the first layer consists almost completely of generative stems; leaves of the same plants form the second layer, which actually "controls" species content of the community and suppresses the development of other species and individual plants. Strong shading prevents the growth of other autotrophic plants in the third layer except some early spring plants; only stems and elongated leaf stalks are developed in this layer. The main constituents of the tall herbaceous vegetation are forbs; the ecosystem is stenotopic. It is principally distributed in the subalpine belt; however, avalanches often take plants of the community down to lower belts [6, 10]. The main factors in the development of the tall herbaceous vegetation are high air humidity and soil moisture as well as gradual (and not rapid) thawing, slight fluctuations of daily temperatures, frequent fog. ***Habitat:*** Humid north- and west-facing slopes of low inclination (until 10-15°); near subalpine open birch woodlands or in strongly rarefied birch forests; the communities are especially dense on river banks. ***Distribution:*** Between 1900 and 2300 m a.s.l. Best developed in the Devdoraki gorge; fragments occur in the Liphu forest, the Sno gorge, near the village Sioni, in the Khde gorge. ***Characteristic species:*** *Heracleum sosnowskyi, Aconitum nasutum, A. orientale, Cephalaria gigantea, Angelica tatiannae, Cicerbita macrophylla, Senecio rhombifolius, Agasyllis latifolia, Doronicum macrophyllum.* ***Ecological importance:*** Water-balance regulatory function. ***Threats:*** Grazing, tree felling and carry out of firewood, uncontrolled collection of medicinal herbs (especially *Senecio rhombifolius*).

Meadows with dominant *Bromopsis variegata*. These grasslands are widespread. They belong to rare-turf meadows. The dominant species - *Bromopsis variegata* - flowers in the first half of June. The degree of coverage amounts to 95-100%. ***Habitat:*** Plateaus and south-facing hillsides of low and medium degree of sloping (5-15°) as well as alluvial cones. Moderately dry skeletal soil. ***Distribution:*** Between 1850 and 2700 m a.s.l. in the whole Caucasus. ***Characteristic species:*** *Agrostis tenuis, Anthoxanthum odoratum, Festuca ovina, Koeleria luerssenii, Trifolium ambiguum, T. trichocephalum, Ranunculus oreophilus, Alchemilla sericata, Leontodon hispidus, Lotus caucasicus, Platanthera chlorantha.* ***Ecological importance:*** Protection from erosion. ***Practical use:*** Hay-meadows. ***Threats:*** Overgrazing, construction of roads.

Dry meadows with dominant *Agrostis tenuis*. These meadows are widespread in the subalpine belt, but occur in the alpine belt too. Very often *Agrostis tenuis* is a co-dominant of *Bromopsis variegata*. ***Habitat:*** Plateaux or south-facing hillsides of medium slope (10-12°). Humid northwest-facing slopes in the alpine belt. Rare-turf meadows. ***Distribution:*** Between 1800 and 2600 m a.s.l. in all the gorges and on all the ranges. ***Characteristic species in the subalpine zone:*** *Bromopsis variegata, Festuca ovina, Phleum phleoides, Koeleria luerssenii, Helictotrichon asiaticus, Pedicularis chroorrhincha, Ranunculus oreophilus, R. caucasicus, Trifolium ambiguum, Alchemilla sericata, Gymnadenia conopsea, Coeloglossum viride.* ***Characteristic species in the alpine zone:*** *Poa alpina, Phleum alpinum, Carum caucasicum, Leontodon caucasicus, Taraxacum confusum, Sibbaldia semiglabra.* ***Ecological importance:*** Protection from erosion. ***Practical use:*** Hay-meadows (in the subalpine zone), pastures (in the alpine zone). ***Threats:*** Overgrazing, construction of roads.

Dense-turf xerophilous meadows with dominant *Festuca varia*. These meadows are widespread in the high mountains of the Caucasus. The meadows develop mainly on very steep slopes of any aspect (more often on south-facing slopes). *Festuca varia* is a densely tufted, summer-wintergreen plant. ***Habitat:*** Very steep (30-50°) stony south-facing slopes. If grazed, they also occur on north-facing slopes. Unlike in the subalpine belt, *Festuca varia* develop in cold and dry environment in the alpine belt, where fluctuations of daily temperatures are sharp. Mountain meadow soil, soddy turf, skeletal. ***Distribution:*** In all the gorges and on all mountain ranges of the region, both in the subalpine and alpine belts. Fragments occur in the subnival belt too. ***Characteristic species in subalpine zone:*** *Helictotrichon asiaticus, H. pubescens, Calamagrostis arundinacea, Oxytropis cyanea, Betonica macrantha, Inula orientalis, Polygonum carneum, Pyrethrum roseum;* ***Characteristic species in alpine zone:*** *Kobresia schoenoides, Carex tristis, Alopecurus dasyanthus, Anthoxanthum odorarum subsp. alpinum, Alchemilla caucasica, Festuca ruprchechtii, Bromopsis riparia, Polygonum carneum, Primula amoena, Chaerophyllum roseum, Helictotrichon asiaticus, H. pubescens, Podospermum alpigenum, Cerastium purpurascens, Betonica macrantha.* ***Ecological importance:*** Protection of slopes from erosion. ***Practical use:*** Grazed well by cattle and satisfactorily by sheep until flowering (while leaves are tender). Hard for grazing after leaves have toughened [31]. ***Threats***: None.

Humid broad-leaved meadows with dominant *Trollius ranunculinus*. These meadows are distributed in humid places; they are found in forest openings as well as on slightly sloping hillsides (until 10^0) and small depressions of both subalpine and alpine belts. The areas occupied by these meadows are not large. They occur as patches scattered over the other vegetation belts. *Trollius ranunculinus* blooms in May and June. ***Habitat:*** Slopes and small depressions of high as well as medium humidity. ***Distribution:*** Patchy distribution in almost all the gorges of the region; between 1800 and 2800 m a.s.l.; these meadows cover relatively large areas on the Kolteshi range, in the Gudauri region. ***Characteristic species:*** *Veratrum lobelianum, Dactylorhiza euxina, D. urvilleana, Poa alpina,*

Swertia iberica, Deschampsia flexuosa, Pedicularis crassirostris. ***Threats:*** Extensive grazing, climate global change (warming).

Broad-leaved mesophilous meadows with dominant *Anemone fasciculata.* These meadows are distributed almost throughout the Caucasus. ***Habitat:*** Moderately humid North and North-West slopes. ***Distribution:*** They occupy quite small areas in the region and develop mainly in the subalpine belt (on the Kolteshi range, in the Devdoraki gorge) between 2000 and 2300 m a.s.l. *Anemone fasciculata* occurs at forest edges, in *Rhododendron caucasicum* scrub, meadows dominated by *Calamagrostis arundinacea* and *Festuca varia* growing on humid slopes. ***Characteristic species***: *Trollius ranunculinus, Geranium ibericum, Scabiosa caucasica, Betonica macrantha, Veratrum lobelianum, Polygonum carneum.* ***Threats:*** Extensive grazing, tree felling, climate global change (warming).

Dense-turf meadows with dominant *Nardus stricta.* These meadows are widely spread in the subalpine and alpine belts of the Caucasus. The subalpine meadows made up of *Nardus stricta* are generally secondary, developed as a result of overgrazing. ***Habitat:*** In level places as well as on hillsides of medium (10-15°) and high degree (20-30°) of slope, on slopes of any aspect (except very steep, south-facing). ***Distribution:*** In the subalpine and alpine belts between 2000 and 2800 m a.s.l.; in the subalpine belt the *Nardus stricta* meadows replace the forb-grass meadows, on all the mountain ranges of the region [31]. ***Characteristic species:*** *Agrostis tenuis, A. planifolia, Luzula pseudosudetica, Anthoxanthum odoratum, Trifolium trichocephalum, T. ambiguum, Leontodon danubialis, Phleum alpinum, Poa alpina, Sibbaldia semiglabra, Hieracium pilosella, Carum caucasicum, Dactylorhiza euxina, D. urvilleana.* ***Ecological importance:*** Protection of slopes from erosion. ***Threats:*** Over-grazing.

Meadows with dominant *Carex tristis.* These meadows are widespread in the alpine belt of the Caucasus. Some fragments of these meadows are also present in the subnival belt. ***Habitat:*** The meadows occupy moderately humid slopes with low (3-5°) and medium (10-15°) degree of inclination (of different aspects) and plane areas. ***Distribution:*** In all the gorges and on all mountain ranges of the region; between 2300 and 2900 m a.s.l. ***Characteristic species:*** *Kobresia capilliformis, Thalictrum alpinum, Poa alpina, Gnaphalium supinum, Nardus stricta, Luzula spicata, L. multiflora, Festuca supina, Anthennaria caucasica, Polygonum viviparum, Alchemilla caucasica.* ***Ecological importance:*** Protection of slopes from erosion. ***Threats:*** Over-grazing.

Meadows with the *Kobresia capilliformis.* These meadows are especially widely distributed in the Central and East Caucasus. The meadows occupy quite small areas in the West Caucasus on limestone. Their invasion of the Caucasus coincides with the xerothermic period after the glacial epoch. ***Habitat*:** Glacial relief, ridges, windy places with thin snow cover; mainly on carbonate soil. ***Distribution*:** Widely distributed between Qwena Mt. and Betlemi Pass, in the Truso gorge and Kolteshi mountain range; between 1900 and 2800 m a.s.l. ***Characteristic species:*** *Kobresia persica, Alchemilla elisabethae, Thalictrum* *alpinum, Polygonum viviparum, Carum caucasicum, Campanula biebersteiniana.* ***Ecological importance:*** Protection of slopes from erosion. ***Threats:*** Extensive grazing.

Mesophilous rare-turf meadows with dominant *Calamagrostis arundinacea.* ***Habitat:*** Level places, slightly inclined (2-5°) as well as steep (20-25°) slopes; mostly within the range of the birch. ***Distribution:*** limited – the Liphu forest, the Sno birch forest, the Devdoraki gorge. ***Characteristic species:*** *Agrostis planifolia, Helictotrichon asiaticus, Deschampsia flexuosa, Geranium ibericum, Anemone fasciculata.* ***Threats:*** Tree felling, global climate change (warming).

Biotope complex: biotopes of rocks and scree

Distinct fluvio-glacial topography, intense present-day denudation process, granitoids bared in consequence of Tergi erosion, black shale and quartzite of Lower Lias age remained uncovered with soil on steep slopes, volcanic lava of the Quaternary with glacial, alluvial and colluvial layers – all these create the environmental conditions necessary for the development of the rock and scree biotopes.

The biotopes of rocks and scree of the Kazbegi region are distinguished by quite high species richness. According to the data reported by Ivanishvili [15], 25% of the Kazbegi flora is represented by the species of the biotopes mentioned and half of them are stenotopic endemics. It is worth mentioning that there are several mono- and oligotype endemic genera in the flora of the rocks and scree biotopes of the Kazbegi region.

Biotope of moist rocks. ***Habitat:*** Shaded rocks moistened with horizontal precipitation or water flowing down from hillsides. ***Distribution:*** The Devdoraki, Dariali, Kazbegi, Sno, Truso gorges, etc. ***Characteristic species:*** *Parietaria micrantha, P. judaica, Campanula sosnowskyi, C. hypopolia, Diphasium alpinum, Cryptogamma crispa, Polypodium vulgare, Woodsia fragillis, Dryopteris pumila, Primula darialica.* ***Threats:*** Climate warming.

Biotope of marly and slaty scree. ***Habitat:*** Dry slaty scree, north-facing slopes. ***Distribution:*** In the subalpine belt, in the whole region. ***Characteristic species:*** *Silene lacera, Erysimum ibericum, E. substrigosum, Linaria vulgaris, L. meyeri, Thalictrum foetidum, Salvia verticillata, Scutellaria leptostegia, Thymus collinus, Bromopsis riparia, B. biebersteinii, Trigonocaryum involucratum.* ***Threats:*** Avalanches and landslides.

Biotope of stones. Rather different plants settle on stones of various origin (glacial, volcanic, etc.); the abundance of the plants growing on stones is often quite high. Many of the species settling on stones are the constituents of the meadow vegetation and communities of rocks and scree. ***Habitat:*** South-facing stones. ***Characteristic species:*** *Sempervivum pumilum, Campanula bellidifolia, Silene ruprechtii, Thymus collinus, Pulsatilla violacea, Festuca ovina, Koeleria caucasica, K. luerssenii, Carex buschiorum, Sedum oppositifolium.* ***Threats:*** Out of danger.

The lowest index of species number is found in the biotope of tall herbaceous vegetation, which is caused by the specific feature of its canopy structure and microclimate (Tab. 1). Other biotopes are characterized by rather high value of species richness.

Table 1: Species number in the various treeline biotopes of the Central Greater Caucasus (n = 14).

Biotopes	Minimum	Maximum	Mean±S.D.
	per 25 m^2		
Elfin crooked-stemmed forest dominated by *Betula litwinowii*	24	29	27,0±1,6
Scrub dominated by *Rhododendron caucasicum*	16	38	25,7±9,3
Low scrub community dominated by *Dryas caucasica*	23	35	30,3±6,4
Tall herb vegetation dominated by *Heracleum sosnowskyi*	6	9	7,5±1,3
Meadow dominated by *Calamagrostis arundinacea*	27	38	31.6±3,9
Meadow dominated by *Festuca varia*	17	43	31,2±7,8
Meadow dominated by *Nardus stricta*	20	47	32,3±7,9
Meadow dominated by *Carex tristis*	25	28	26,5±2,1
Meadow dominated by *Kobresia capilliformis*	22	34	27,7±4,1
Meadow dominated by *Agrostis tenuis*	24	47	33,0±8,5
Meadow dominated by *Bromopsis variegata*	26	38	32,0±8,4

Discussion

The treeline biotopes of the Kazbegi region are distinguished by quite high diversity [32, 11], which is expressed in the contrasting species composition and caused by various ecological conditions [8, 25, 26]. Here, in a relatively small area, one can find biotopes which are very rich in plant species. This is mainly due to peculiar structural properties of the vegetation of these biotopes, rather than ecological (altitude, moisture, aspect, stoniness) and anthropogenic (grazing, mowing) factors. This is illustrated by the biotopes with a relatively simple structure (a dry biotope, *Bromopsietum variegatae* and moist ones: *Nardetum strictae* and *Caricetum tristis*, etc.) and the moist tall herbaceous vegetation. In the first case the number of species amounts to 35-40 on 1 m^2, whereas in the tall herbaceous vegetation this number does not exceed 5–7. Species number is also small in biotopes with pronounced dominance of *Rhododendron caucasicum* [26].

The majority of the biotopes studied are referred to the II stage of hemeroby. Scrub formed at the treeline (dominated by *Rhododendron caucasicum*) and biotopes of the elfin birch forests (dominated by *Betula litwinowii*) should be referred to the I stage of hemeroby, because they have remained almost untouched, as "Holy Forests" [27]. A great part of the biotopes in the Kazbegi region is characterized by high sensitivity. This is caused by the following factors: the landscape mainly devoid of forests, high inclination of the relief (30-75%) and, consequently, not well-formed soil cover, insignificant snow cover, frequent solifluction, etc.

The droughts that have occurred during recent years (1996-2001) have particularly affected the chionophiles (*Daphne glomerata, Rhododendron caucasicum, Vaccinium myrtillus, V. vitis-idaea*) and tall herb vegetation. Therefore, noticeable changes have taken place in some biotopes; these changes have manifested themselves first of all by elimination of some species from these biotopes. For instance, *Botrychium lunaria* has disappeared from the biotopes of the lower part of the treeline (meadows dominated by *Bromopsis variegata, Festuca rubra*). Withering of leaves of *Rhododendron caucasicum*, noticeable decline of *Daphne glomerata* and, what is the most interesting, expansion of 6- to 8-year saplings of *Betula litwinowii* are observed at the treeline [28].

Some other factors provoke the significant suppression of *Rhododendron caucasicum* vitality. Intensive solar radiation in spring makes snow thaw rapidly at the edge of a *Rhododendron caucasicum* shrubbery, warmed air makes leaves of this shrub start transpiration and photosynthesis, whilst the soil is yet frozen and cannot provide plants with water. Therefore, plants at the margins of the community die ("winter drought" phenomenon). These changes may be connected with the absence of snow cover during dry winter. The second reason for the damage of *Rhododendron caucasicum* leaves is a process of photo-inhibition. The shrub often collides with a strong radiation shock in spring, which results in a damage of photosynthetic pigments and thylakoid structures of chloroplasts [22]. *Vaccinium myrtillus* and *V. vitis-idaea* grow predominantly in the *Rhododendron* communities, where due to the dense shrub cover these plants easily bear humidity and temperature fluctuations. A different pattern is observed for plants that do not grow in the dense vegetation, *i.e.* in the absence of the facilitation effect [1, 18]. For example, in the wet Devdoraki gorge (2000-2200 m), at the birch forest margin, the above-mentioned dwarf shrubs were observed withered in the dry summer of 2000.

Subalpine tall herbaceous vegetation is very sensitive to the fluctuations of temperature and humidity. Recent droughts have had negative influence on the development of such typically tall herbaceous species as *Senecio rhombifolius, Aconitum nasutum, Swertia iberica*, in particular, plant do not grow properly (nanism), undeveloped inflorescence and flowers open prematurely and early senescence and even complete wilting of leaves occurred. Plants growing near streams and springs were in better shape.

Conclusions

The biotopes of the treeline of the Kazbegi region are, first of all, distinguished by their contrasting nature: from saxicolous dry meadows and scree communities to moist tall herbaceous vegetation and elfin crooked-stemmed birch forests.

Relatively high species richness is characteristic of subalpine meadows and high specific endemism is a feature of the tall herbaceous vegetation, elfin crooked-stemmed birch forests and rock (scree) communities.

Current global climate change will induce the most drastic changes in moist tall herbaceous vegetation, snowline broad-leaved meadows dominated by *Trollius ranunculinus*, elfin crooked-stemmed birch forest and scrub.

The following biotopes should be regarded as priority habitat types: 1) tall herbaceous vegetation, 2) elfin crooked-stemmed birch forest and 3) scrub.

Acknowledgements: The work has been supported by the Civilian Research Development Foundation (CRDF) (award # 3322) (Proposal # 12220) / Project title: "Alpine Tree-line Stability in a Changing Global Environment: Mechanism of Tree Seedling Establishment".

REFERENCES

1. Abdaladze O., Kikvidze Z., 1991 - Vli'yanie Travosto'ya na Intensivnost' Pogloshcheni'ya CO_2 Kleverom Skhodnim (*Trifolium ambiguum* M.B.) v Subalpiiskom po'yase Centralnogo Kavkaza [Canopy Impact on CO_2 assimilation in *Trifolium ambiguum* M.B. in the subalpine belt of the central Caucasus]. Ecology, **1**: 33-37.
2. Akhalkatsi M., Abdaladze O., Smith W.K., Nakhutsrishvili G., 2004a - Mechanisms of Treeline Stability in the Caucasus Mountains of Georgia: Low-Temperature Photoinhibition and Interspecific Facilitation. Abstracts of scientific meeting "Botany 2004", Alpine Diversity: Adapted to the Peaks, Snowbird, Utah: 55.
3. Akhalkatsi M., Abdaladze O., Nakhutsrishvili G., Smith W.K., 2004b - Global Warming Effect on Birch Seedling Establishment at Tree Line in the Central Caucasus. Botanikertagung 05 bis 10 September, 2004. Braunschweig: 386.
4. Burga C.A., Klötzli F., Miehe G., 2004 - Waldgrenze: Phänomen und globaler Vergleich. In: Burga C.A., Klötzli F. & Grabherr G. (eds.), Gebirge der Erde: Landschaft, Klima, pp: 37-44. Eugen Ulmer, Stuttgart.
5. Braun-Blanquet J., 1964 - Pflanzensoziologie. Grundzüge der Vegetationskunde. 3. Aufl. Springer, Wien.
6. Dolukhanov A., 1966a - Zakonomernosti Geograficheskogo Raznoobrazia Rastitel'nosti I Verkhnaya Granitsa Lesa V Gorakh Zakavkaz'ya [Regularities of Geographical Diversity of Vegetation and Timberline in the Trans-Caucasian Mountains]. Probl. Bot., **8**: 196-207.
7. Dolukhanov A., 1966b - Rastitel'nost' [Vegetation]. In: Gerassimov I.P. (ed.), Kavkaz [The Caucasus], pp. 223-251. Nauka, Moscow.
8. Dolukhanov A.G., 1978 - Timberline and subalpine belt in Caucasus Mountains, USSR. Arct. Alp. Res., **10** (2): 409-422.
9. Dolukhanov A., 1989 - Rastitelnost' Gruzii t. I: Lesna'ya Rastitelnost' Gruzii [Vegetation of Georgia vol. I: Forest Vegetation of Georgia]. Metsniereba, Tbilisi.
10. Gagnidze G., 1974 - Botanicheskii i Geograficheskii Analiz Florotsenoticheskogo Kompleksa Vysokotrav'ya Kavkaza [Botanical and Geographical Analysis of the

Florocoenotic Complexes of Tall Herb Vegetation of the Caucasus]. Metsniereba, Tbilisi.

11. Gagnidze R., 2000 - Diversity of Georgia's Flora. In: Berutchashvili N., Kushlin A., Zazanashvili N. (eds.), Biological and Landscape Diversity of Georgia, pp. 21-33. WWF Georgia Country Office, Tbilisi.
12. Grabherr G., 1997 - The High Mountain Ecosystems of the Alps. In: Wielgolaski F.E. (ed.), Ecosystems of the World, vol. 3: Polar and Alpine Tundra, pp. 97-121. Elsevier, Amsterdam.
13. Grace J., 1989 - Tree lines. Philosophical Transactions of the Royal Society of London, B 324: 233-245.
14. Holzner W., 1989 - Biotoptypen in Österreich. Umweltbundesamt, Wien.
15. Ivanishvili M., 1998 - The Thorn-Cushion Vegetation in the Caucasus. In: Nakhutsrishvili G., Abdaladze O. (eds.), Plant Life in High-Mountains, pp. 43-50. Diogen, Tbilisi.
16. Janelidze C., Margalitadze N., 1977 - K Voprosu Istorii Lesnoi Rastitelnosti Kazbegskogo Raiona v Golotsene [History of Forest Vegetation in Holocene]. In: Nakhutsrishvili G. (ed.), Alpine Ecosystem of Kazbegi, pp. 17-21. Moscow-Tbilisi.
17. Kharadze A., 1948 - O Periglacial'noyi Rastitelnosti Kavkaza [On periglacial vegetation of the Caucasus]. Bull. Acad. Sci. Georgia, **9-10**: 615-622.
18. Kikvidze Z., Nakhutsrishvili G., 1998 - Facilitation in the subnival vegetation patches. J. Veg. Sci., **9**: 222-226.
19. Körner C., 1998 - A re-assessment of high elevation treeline positions and their explanation. Oecologia, **115**: 445-459.
20. Körner C., 1999 - Alpine Plant Life. Springer, Berlin.
21. Körner C., 2002 - Alpine Ecosystems. In: Encyclopedia of the Life Sciences, pp. 392-396. Macmillan Publishers Ltd, Nature Publishing Group, London.
22. Larcher W., 2003 - Physiological Plant Ecology: Ecophysiology and Stress Physiology of Functional Groups. Springer, Berlin.
23. Margalitadze N., 1998 - Results of the Palaenological Study of the High-Mountain Holocene Deposits in Georgia. In: Nakhutsrishvili G., Abdaladze O. (eds.), Plant Life in High-Mountains, pp. 138-140. Diogen, Tbilisi.
24. Nakhutsrishvili G., 1998 - Hochgebirgsvegetation Georgiens (Kaukasus). In: Nakhutsrishvili G., Abdaladze O. (eds.), Plant Life in High-Mountains, pp. 93-100. Diogen, Tbilisi.
25. Nakhutsrishvili G., 1999 - The Vegetation of Georgia (Caucasus). Braun-Blanquetia **15**: 5-74.
26. Nakhutsrishvili G., 2003 - High Mountain Vegetation of the Caucasus Region. In: L. Nagy, G. Grabherr, C. Körner, D.B.A. Thompson (eds.), Alpine Biodiversity in Europe, Ecological Studies, vol. 167, pp. 93-103. Springer, Berlin.
27. Nakhutsrishvili G., Abdaladze O., Akhalkatsi M., 2004a - Concerning the Tree Line Vegetation of the Kazbegi Region (the Central Caucasus). Bull. Georg. Acad. Sci., **169** (1): 122-125.
28. Nakhutsrishvili G., Abdaladze O., Akhalkatsi M., 2004b - Global Warming and Treeline. Proc. Acad. Sci. Biol. Ser. B., **2** (3-4): 87-90.

29. Pedrotti F., 1998 - La cartographie géobotanique des biotopes du Trentin (Italie). Ecologie, **29** (1-2): 105-110.
30. Pott R., 1996 - Biotoptypen: Schützenswerte Lebensräume Deutschlands und angrenzender Regionen. Ulmer, Stuttgart.
31. Sakhokia M., 1983 - Agrobotanicheskii Obzor Pastbishch i Senokosov Kazbegskogo Raiona [Agrobotanical Review of Pastures and Hay Meadows of the Kazbegi Region]. Metsniereba, Tbilisi.
32. Sakhokia M., Khutsishvili E., 1975 - Conspectus Florae Plantarum Vascularium Chewii. Metsniereba, Tbilisi.

Appendix:

Fig. 1: Mt. Kazbegi - 5033 m a.s.l. (photo by P. Ozenda).

Fig. 2: Rock pine forest - *Pinus kochiana* (photo by O. Abdaladze).

Fig. 3: Timberline (2450 m a.s.l.) with *Betula litwinowii* and *Rhododendron caucasicum* (photo by G. Nakhutsrishvili).

THE WATER RESERVOIRS OF THE CONFLUENCE OF THE BISTRIȚA WITH THE SIRET RIVER (EASTERN ROMANIA) - AN IMPORTANT BIRD AREA

Cătălin Petre RANG[1], **Florin FENERU**[2]
[1] University of Bacău, Faculty of Science - Biology,
Calea Mărășești no. 157, **600115 Bacău (ROMANIA)**; e-mail: rang@ub.ro
[2] Museum of Nature Sciences "Ion Borcea" **Bacău (ROMANIA)**

Abstract: The confluence zone between the Bistrița and Siret rivers lies in the foothills of the Carpathian mountains. Five reservoirs have been created there for generating electrical energy. The water bodies have a total length of 46.5 km and an area of 5.391 ha (Lilieci: 262 ha - 6 km; Bacău II: 202 ha - 5 km; Galbeni: 1.123 ha - 7 km; Răcăciuni: 2.004 ha - 13 km; Berești: 1.800 ha – 15.5 km). The number of bird species identified in the area, about 210 in all, represents only 36% of the total. The species that also have in their phenology migratory aspects are in a proportion of 90%. This shows that these lakes are important to the full development of autumn and spring migration in the eastern Carpathians escarpment, for populations in northern Europe. For some species, the zone has begun to represent an important station for wintering. Agglomerations are exceeding over 100,000 individuals for each of the lakes. Over 60% of bird species belong to the European and Transpalearctic fauna, and 20% are Siberian and Arctic. With regard to the importance that these wetlands represent for the bird population of northern and eastern Europe, they have been declared an Important Bird Area.

Introduction

The Siret river drains the eastern part of Romania, its catchment derived from the eastern Carpathian Mountains and Moldavian Plateau. The area is known to be an important passage for European bird migration – the Carpathian passage. The creation of artificial accumulation lakes for generating electrical energy has caused an important change in residence structure. The new wetlands offer proper conditions for food and shelter for many bird species that are on passage or in their wintering quarters. The confluence zone between Bistrița and Siret rivers includes five lakes (Lilieci, Bacău II, Gârleni, Răcăciuni, Berești) and supports a very large bird fauna. This study marks out especially a particular diversity regarding bird fauna in the area.

Study Area and Methods

The first two lakes in the Bistrița-Siret confluence zone, Lilieci (6 km long and with a water surface of 262 ha) and Bacău II (5 km long and a water surface of 202 ha) were filled in 1966. Three other lakes followed. Galbeni (7 km long, 1123 ha), covering the confluence of the Bistrița and Siret, was filled in 1983; and Răcăciuni (13 km long, 2004 ha) and Berești (15.5 km long,1800 ha) were filled in 1986. The appearance of vast areas of open water caused important changes to the general aspect of the area (Fig. 5). These water bodies had flooded a large area of forest and agriculture, and replaced more than 45 km of running water.

The main study method has been to make successive observations at different seasons over many years.

Results

The results of the observations are summarized in Figure 1.

Regarding species diversity in the area of the accumulation lakes, the total number of species identified was (Fig. 1): Lilieci – 87; Bacău II –78; Gârleni – 98; Răcăciuni – 82; and Bereşti – 91. In other zones beside the lakes (forest, agricultural land, running water, etc.), 108 species were identified. In the main habitats, the distribution of species was (Fig. 2): open water: 49 species (15%), shore areas: 79 species (24%), reed beds: 32 species (10%), running water: 28 species (9%), forest habitats: 83 species (26%), agricultural fields: 53 species (16%). The percentages were calculated bearing in mind that a species may be encountered in many other habitats.

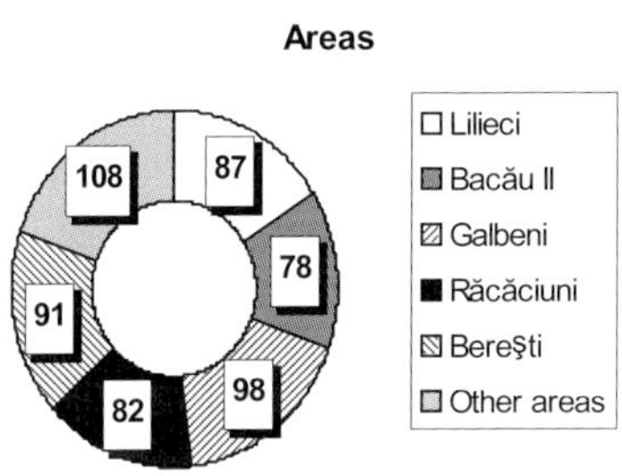

Fig. 1: Number of bird species observed in different areas.

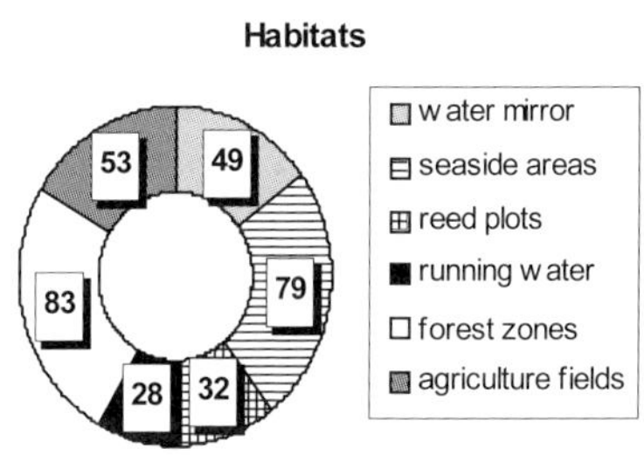

Fig. 2: Number of bird species observed in different habitats.

Regarding bird mobility in the studied area, the species identified have been classified (Fig. 3) as: sedentary: 30 (10%); migratory: 77 (26%); passage species: 151 (51%), winter visitors: 39 (13%). There are species that could be included in many categories of mobility.

Regarding the type of fauna to which the identified bird species belong, this can be established (Fig. 4) as: European – 92 species (44%); Transpalearctic – 42 species (20%); Siberian – 25 species (12%); Mongolian – 20 species (10%); Mediterranean –11 species (5%); Arctic – 17 species (8%); Chinese – 2 species (1%); Tibetan – 1 species.

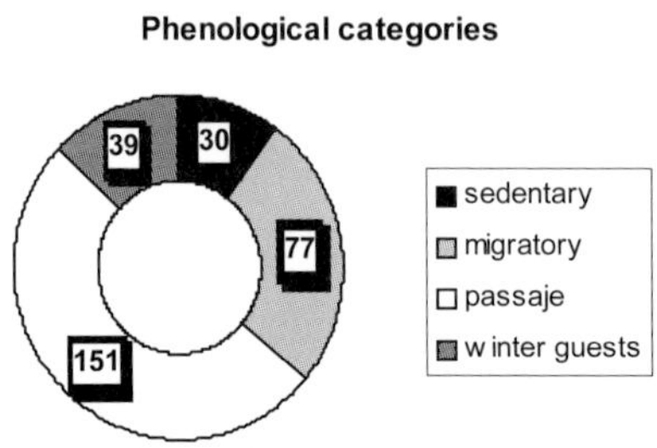

Fig. 3: Number of bird species by phenological categories.

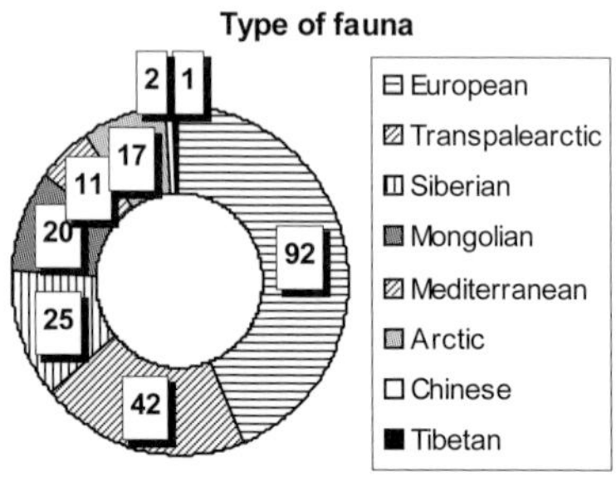

Fig. 4: Number of bird species belonging to the different geographical origins.

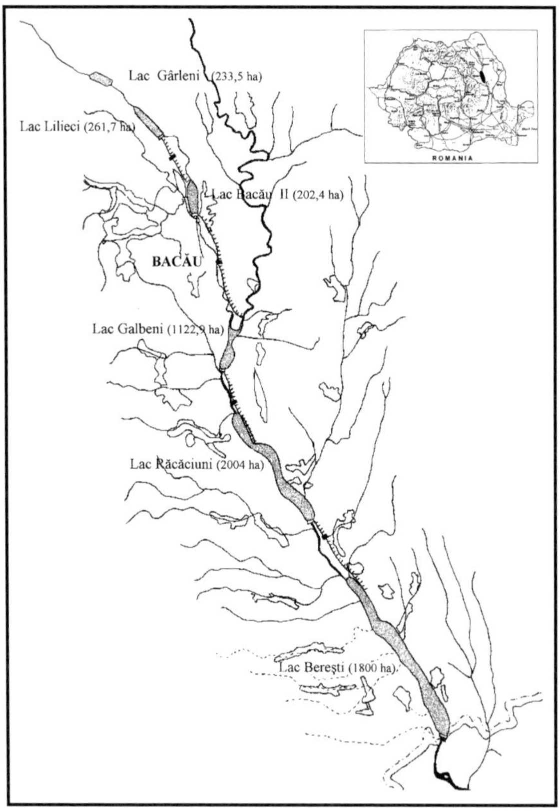

Fig. 5: Protected areas of the confluence zone of the rivers Bistrita and Siret.

Table 1: Bird species from the confluence zone of Bistrița and Siret rivers.

Species	Type of fauna	Phenology				Habitats						Observation point*					
		Sedentary	Summer visitor	Passage	Winter visitor	Water area	Seaside	Reed plots	Running water	Forest	Crops	Lilieci	Bacău II	Galbeni	Răcăciuni	Berești	Other areas
0	**1**	**2**	**3**	**4**	**5**	**6**	**7**	**8**	**9**	**10**	**11**	**12**	**13**	**14**	**15**	**16**	**17**
Accipiter gentilis	Tp	+						+		+	+	+	+	+	+	+	+
Accipiter nisus	Tp	+						+		+	+	+	+	+	+	+	+
Actitis hypoleucos	Tp			+			+		+			+	+	+		+	+
Acrocephalus arundininaceus	E		+	+				+				+	+	+	+		+
Acrocephalus palustris	E		+						+	+	+						+
Acrocephalus schoenobaenus	E		+	+				+				+	+	+	+	+	
Acrocephalus scirpaceus	E		+	+				+				+				+	
Aegithalos caudatus	Tp	+								+							+
Alauda arvensis	Mo		+	+							+						+
Alcedo atthis	E	+				+	+					+	+	+	+	+	+
Anas acuta	S			+		+						+	+	+	+	+	
Anas clypeata	Tp			+		+						+	+	+	+	+	
Anas crecca	Tp			+	+	+	+		+			+	+	+	+	+	+
Anas penelope	S			+		+						+	+	+	+	+	+
Anas platyrhynchos	Tp		+	+	+	+	+		+		+	+	+	+	+	+	+
Anas querquedula	Tp		+	+	+	+	+					+	+	+	+	+	+
Anas strepera	Tp			+		+						+		+	+		
Anser albifrons	A			+		+					+			+	+	+	
Anser anser	Mo			+		+					+			+	+	+	
Anser erythropus	A			+		+					+			+			
Anthus campestris	Mo			+			+				+			+			+
Anthus pratensis	E			+			+					+	+	+			
Anthus trivialis	E		+	+						+							+
Apus apus	E		+								+						+
Aquila pomarina	E			+						+							+
Ardea cinerea	Tp			+			+	+	+			+	+	+	+	+	
Ardea purpurea	M			+			+	+	+			+	+	+	+	+	
Ardeola ralloides	M			+			+					+		+			
Arenaria interpres	A			+					+							+	+
Asio flammeus	Tp			+	+			+				+	+	+	+		
Asio otus	Tp		+	+	+					+							+
Athene noctua	Mo	+								+	+						+
Aythya ferina	E		+	+	+	+						+		+	+	+	
Aythya fuligula	S		+	+	+	+						+					
Aythya marila	A			+		+						+		+			
Aythya nyroca	E			+	+							+		+	+	+	
Bambycilla garrulus	S			+	+					+							+
Botaurus stellaris	Mo							+		+		+		+	+	+	
Branta ruficollis	A			+		+					+			+			
Bubo bubo	Tp	+								+							+
Bucephala clangula	S				+	+						+	+	+	+	+	
Burchinus oedicnemus	M			+							+						+
Buteo buteo	Tp		+	+						+	+						+

0	1	2	3	4	5	6	7	8	9	10	11	12	13	14	15	16	17
Buteo lagopus	A				+		+				+						+
Calidris alpina	A			+		+								+		+	
Calidris minuta	A			+			+							+		+	
Calidris temminkii	A			+			+							+			
Caprimulgus europaeus	E		+	+						+							+
Carduelis cannabina	E		+	+			+				+		+	+			+
Carduelis carduelis	E		+	+	+					+	+						+
Carduelis chloris	E		+	+						+	+						+
Carduelis flammea	S				+						+		+				+
Carduelis spinus	E			+	+					+	+						+
Certhia familiaris	E	+								+							+
Charadrius dubius	Mo		+	+		+			+			+	+	+	+	+	
Charadrius hiaticula	A			+		+								+		+	
Chlidonias hybrida	M			+		+	+					+	+	+	+	+	
Chlidonias leucopterus	E			+		+	+					+	+	+	+	+	
Chlidonias niger	E			+		+	+					+	+	+	+	+	
Ciconia ciconia	E		+	+			+		+		+	+	+	+	+	+	+
Ciconia nigra	E		+	+						+		+	+			+	
Cinclus cinclus	E				+				+								+
Circaetus gallicus	E			+			+					+					
Circus aeruginosus	Mo			+			+	+	+		+	+	+	+	+	+	+
Circus cyaneus	E			+	+		+	+	+		+	+	+	+	+	+	+
Clangula hyemalis	A				+	+						+	+	+			
Coccothraustes coccothraustes	E		+							+							+
Columba oenas	E			+						+	+						+
Columba palumbus	E			+						+	+						+
Coracias garrulus	E		+	+						+							+
Corvus corax	Tp	+					+	+		+	+	+	+	+	+	+	+
Corvus cornix	E	+					+	+			+	+	+	+	+	+	+
Corvus frugilegus	E		+	+	+					+	+						+
Corvus monedula	E	+								+	+						+
Coturnix coturnix	E		+	+			+				+						+
Crex crex	E		+	+							+						+
Cuculus canorus	Tp		+	+				+		+							+
Cygnus cygnus	S			+	+	+	+					+	+	+	+	+	
Cygnus olor	E		+	+	+	+	+					+	+	+	+	+	
Delichon urbica	Tp		+								+						+
Dendrocopos major	Tp	+								+							+
Dendrocopos medius	E	+								+							+
Dendrocopos minor	Tp	+								+							+
Egretta alba	Ch			+			+	+	+			+	+	+	+	+	
Egretta garzetta	M			+			+	+	+			+	+	+	+	+	
Emberiza citrinella	E		+	+	+		+				+	+	+				+
Emberiza hortulana	E		+							+							+
Emberiza schoeniclus	Tp		+	+	+			+				+	+	+	+	+	
Eremophila alpestris	Ti				+		+							+			
Erithacus rubecula	E		+	+						+							+
Falco peregrinus	Tp			+			+				+						+
Falco subbuteo	Tp			+			+	+		+	+						+
Falco tinnunculus	Tp		+	+							+						+
Falco vespertinus	Mo		+	+						+							+

0	1	2	3	4	5	6	7	8	9	10	11	12	13	14	15	16	17
Ficedula albicollis	E		+	+						+							+
Ficedula hypoleucos	E			+						+							+
Fringilla coelebs	E		+	+	+					+							+
Fringilla montifringilla	S				+					+							+
Fulica atra	Tp		+	+		+	+	+				+	+	+	+	+	
Galerida cristata	Mo	+									+						+
Gallinago gallinago	E			+			+					+	+	+	+	+	
Gallinago media	E			+			+							+	+		
Gallinula chloropus	E		+	+			+	+				+	+	+	+	+	
Garrulus glandarius	E	+								+							+
Gavia arctica	S			+	+	+			+			+	+	+	+	+	
Gavia stellata	A			+		+								+			
Grus grus	E			+							+						+
Haematopus ostralegus	Tp			+			+		+						+	+	+
Haliaetus albicilla	Tp			+		+	+					+	+				
Himantopus himantopus	Mo			+			+							+		+	
Hirundo rustica	Tp		+	+			+					+	+	+	+	+	
Ixobrychus minutus	E		+	+			+	+				+	+	+	+	+	
Jynx torquilla	Tp		+	+						+							+
Lanius collurio	E		+	+						+							+
Lanius excubitor	Tp				+						+						+
Lanius minor	E		+							+							+
Larus argentatus	Tp		+	+	+	+	+					+	+	+	+	+	
Larus canus	S			+		+						+		+	+		
Larus fuscus	A			+		+						+	+	+	+	+	
Larus minutus	S			+		+								+		+	
Larus ridibundus	Tp			+	+	+	+				+	+	+	+	+	+	+
Limicola falcinellus	S			+			+						+	+			
Limosa limosa	Mo			+			+							+	+		
Locustela luscinoides	E		+					+				+	+				
Luscinia megarhynchos	E		+	+						+							+
Lymnocriptes minimus	S			+			+					+	+	+	+	+	
Mergus albellus	S			+	+	+						+	+	+	+	+	
Mergus merganser	Tp			+	+	+						+	+	+	+	+	
Mergus serrator	S					+						+					
Merops apiaster	M		+	+						+	+						+
Miliaria calandra	E		+								+						+
Milvus migrans	E			+		+	+					+					
Motacilla alba	E		+	+			+		+			+	+	+	+	+	
Motacilla cinerea	E			+			+		+			+	+	+	+	+	
Motacilla flava	Tp		+	+			+		+			+	+	+	+	+	
Muscicapa striata	E		+	+						+							+
Netta rufina	M		+	+		+	+	+					+	+			
Numenius arquata	E			+			+			+		+	+	+	+	+	
Nycticorax nycticorax	M			+		+	+			+		+	+	+	+	+	+
Oenanthe oenanthe	Tp		+	+			+		+		+						+
Oriolus oriolus	E		+	+						+							+
Pandion haliaetus	Tp			+			+					+					
Panurus biarmicus	Mo		+					+				+	+	+	+	+	
Parus ater	E	+								+							+
Parus caeruleus	E	+								+							+

0	1	2	3	4	5	6	7	8	9	10	11	12	13	14	15	16	17
Parus major	E	+								+							+
Parus palustris	S	+								+							+
Passer domesticus	Tp	+								+	+						+
Passer montanus	Tp	+								+	+						+
Pelecanus onocrotalus	Mo			+			+								+		
Perdix perdix	E	+							+		+						+
Phalacrocorax carbo	Tp			+	+	+	+								+	+	
Phalacrocorax pygmaeus	M			+		+	+								+	+	
Phalaropus lobatus	A			+			+					+	+	+			
Phasianus colchicus	Ch	+					+	+		+	+	+	+	+		+	+
Philomachus pugnax	S			+			+		+			+	+	+	+	+	
Phoenicurus ochruros	Mo		+							+							+
Phoenicurus phoenicirus	E		+							+							+
Phylloscopus collybita	Tp			+						+							+
Phylloscopus trochilus	E			+						+							+
Pica pica	E	+					+	+	+	+	+	+	+	+	+	+	+
Picus canus	E	+								+							+
Picus viridis	E	+								+							+
Platalea leucorodia	E			+			+							+	+	+	
Plegadis falcinellus	M			+			+									+	
Pluvialis apricaria	A			+			+		+					+		+	
Pluvialis squatarola	A			+			+		+					+	+	+	
Podiceps cristatus	Tp		+	+	+	+						+	+	+	+	+	
Podiceps griseigena	E			+		+						+	+	+	+	+	
Podiceps nigricollis	E			+		+						+	+	+	+	+	
Podiceps ruficollis	E			+	+	+			+			+	+	+	+	+	
Porzana parva	E		+	+			+	+				+	+	+	+	+	
Porzana porzana	E		+	+			+	+				+	+	+	+	+	
Prunella modularis	E			+						+							+
Pyrrhula pyrrhula	S			+	+					+							+
Rallus aquaticus	E		+	+			+	+				+	+	+	+	+	
Regulus ignicapillus	E				+					+							+
Regulus regulus	E				+					+							+
Remiz pendulinus	Mo		+							+							+
Riparia riparia	Tp		+	+				+				+	+	+	+	+	
Saxicola rubetra	E		+							+	+						+
Saxicola torquata	Mo		+					+		+	+	+					+
Scolopax rusticola	E			+						+							+
Sitta europaea	Tp	+								+							+
Stercorarius parasiticus	A			+		+									+		
Sterna albifrons	E			+		+	+										
Sterna hirundo	E		+	+								+	+	+	+	+	
Streptopelia decaocto	M	+									+						+
Streptopelia turtur	E		+	+						+	+						+
Strix aluco	E	+								+							+
Strix uralensis	S		+	+						+							+
Sturnus vulgaris	E		+	+				+		+		+	+	+	+	+	+
Sylvia atricapilla	E		+							+							+
Sylvia communis	E		+							+							+
Sylvia curruca	E		+							+							+
Tadorna ferruginea	Mo			+			+							+			

0	1	2	3	4	5	6	7	8	9	10	11	12	13	14	15	16	17
Tringa erythropus	S			+			+							+	+	+	
Tringa glareola	S			+			+		+			+	+	+	+	+	
Tringa nebularia	S			+			+						+		+		
Tringa ochropus	S			+			+		+			+	+	+	+	+	
Tringa stagnatilis	Mo			+			+						+				
Tringa totanus	Mo			+			+		+			+	+	+	+	+	
Troglodytes troglodytes	E	+								+						+	
Turdus merula	E		+	+						+						+	
Turdus philomelos	E		+	+						+						+	
Turdus pilaris	S		+	+	+					+						+	
Tutrdus viscivorus	E			+	+					+						+	
Upupa epops	E		+	+						+	+					+	
Vanellus vanellus	Mo		+	+			+					+	+	+	+	+	
Xenus cinereus	S			+			+									+	

Abbreviations: S – Siberian; Tp – Transpalearctic; E – European; M – Mediterranean; Mo – Mongolian; A – Arctic; Ch – Chinese; Ti – Tibetan.

* See Fig. 5 for the location of the observation points.

Discussion

The large number of species (108) recorded in areas other than the accumulation lakes themselves is determined by the diversity of existing habitats (forest, crops, running waters). Most of these habitats existed before the accumulation lakes, but over larger areas.

The greatest diversity was identified on Galbeni lake. This is probably because it is situated at the confluence of the river Bistrița with the Siret. At the same time, one can add the fact that this lake is one of the most visited. The depth of water and the silt bottom are favourable finding the food for many sea-birds.

The number of species that prefer maritime habitats (79) is among the highest in the central zone. The species that prefer open water (49) and those that prefer reed beds (32) represents half of the total number and their presence is strictly linked to the existence of special habitats created by the new accumulation lakes.

The large number of passage species (151) and winter visitors (39) is associated with the existence of these lakes. The visitation is determined by the existence of a broad water area, fairly deep, with a silty bottom, favouring the existence of a strongly varied and abundant invertebrate fauna.

The alternating process of water controls is inducing important variations of the depth of lakes and the result is a temporary receding of water over extensive areas. A large amount of food is now accessible for many other bird species apart from swimming species. The most favoured in this case are those that prefer maritime habitats.

The same daily variations of the depth of lakes are also important during winter. Then the ice on large areas is broken and carried by water streams leaving some important areas ice-free. The winter visitor species find food in these places (39). Sometimes agglomerations on the ice-free areas of lakes exceed a hundred thousand. Even in the most frozen winters with temperatures under -30° C, the water is not entirely frozen.

The presence of 42 bird species belonging to the Siberian and Arctic fauna shows how important this zone with lakes is for northern species on passage or over-wintering.

Conclusions

The creation of accumulation lakes in the confluence zone of the rivers Bistriţa and Siret has determined a large diversity of habitats. A total of 210 bird species were identified in the area.

The use of lake water for generating electrical energy has determined fluctuations of the depth of water, especially in the sea zones where important areas of parallel zones are becoming accessible for the search of food by many bird species.

Because of this exploitation of lakes, during winter some important areas remain unfrozen.

The important food sources which are daily accessible on the accumulation lakes are drawing a large diversity of species on passage and during winter, favouring the survival of northern species.

After documentation and a proposal provided with the approval of Commission of Nature Monument of the Romanian Academy, the Council of Bacău County has designated the five lakes, with a total area of 5391 ha and total length of 46.5 km, as an Important Bird Area.

REFERENCES

1. Feneru F., 2002 - Studiul avifaunei acvatice din bazinul mijlociu al Siretului. Teză de doctorat. Univ. „Al. I. Cuza", Iaşi, 190 pp.
2. Munteanu D., 1969 - Bird migration in Romania. Bulletin of the British Ornithologists' Club, **8** (2): 33-37.
3. Rang C., 1968 - Contribuţii la cunoaşterea avifaunei văii mijlocii a Siretului în perioadele de pasaj. St. şi Com., Muz. Şt. Nat. Bacău, **1**: 79-90.
4. Rang C., 1971 - Beitrage zur Kenntnis der Vogelfauna der Stauseen Gârleni, Bacău I, Bacău II, auf den Bistritza Fluss. St. şi Com., Muz. Şt. Nat. Bacău: 285-300.
5. Rang C., 2002 - Studiul dinamicii unor comunităţi de păsări din bazinul mijlociu al râului Siret, incluzând zonele lacurilor de acumulare. Teză de doctorat. Publ. Societăţii Ornitologice Române, Cluj-Napoca, **13**: 249 pp.

CONSERVATION OF STEPPE BIRDLIFE IN ITALY

Francesco PETRETTI
Vicolo del Borghetto 15a, **00187 Roma (ITALY)**; e-mail: okapia.studium@virgilio.it

Abstract: The paper deals with the status of bird species living in dry lowland grasslands in Italy, many of which are endangered or vulnerable due to land use changes. The main lowland dry grasslands are distributed in four Italian regions: Latium, Apulia, Sardinia and Sicily. Although true Mediterranean dry steppes have been drastically reduced due to agricultural changes and urban development, they still cover 200,000 ha, which are of outstanding importance for maintaining populations of Little Bustard – *Tetrax tetrax* (2,000-2,500 individuals), Stone Curlew – *Burhinus oedicnemus*, Bee-eater – *Merops apiaster*, Calandra Lark – *Melanocorypha calandra*, Black-eared Wheatear – *Oenanthe hispanica*, raptors and a significant number of plants and invertebrates.

Introduction

In last century Italy has lost nine species of breeding birds: White-headed Duck *(Oxyura leucocephala)* – last breeding record in Sardinia in 1977, White-tailed Sea Eagle *(Haliaetus albicilla)* – last breeding record in Sardinia in 1956, Black Vulture (*Aegypius monachus)* – last breeding record in Sardinia in 1961, Hen Harrier *(Circus cyaneus)* – last breeding record in the Po Valley in 1935, although there was a breeding attempt in 1998, Osprey (*Pandion haliaetus*) – last breeding record in Sardinia in 1965 and in Sicily in 1968, *Turnix sylvatica* – last breeding record in Sicily in the 1920s, Crane (*Grus grus)* – last breeding record in Veneto in the 1920s, Black Wheatear (*Oenanthe leucura*) – last breeding records in the 20th century, and Aquatic warbler (*Acroceophalus paludicola*) – last breeding record in the 1950s.

Most of these species still occur in Italy during migration and the wintering season and some (White-headed Duck, Osprey) are involved in conservation projects which rely on captive breeding to reintroduce the species in the wild in the forthcoming years.

Bird species have been classified in the Red List of Italian Vertebrates (WWF 1998) in the following categories, according to the IUCN classification:

"EX" = Extinct
"EW" = Extinct in the Wild
"CR" = Critically Endangered
"EN" = Endangered
"VU" = Vulnerable
"LR" = Lower Risk
"DD" = Data Deficient
"NE" = Not Evaluated

The Red Data Book lists 164 out of the 250 species which nest in Italy, plus 18 subspecies. Among the most endangered species it is worth mentioning the birds living in grassland ecosystems, most of which occur in extensively farmed areas and dry lowland pastures.

This paper aims to update the information given in previous reports [7, 8] and to present the results of the first conservation measures undertaken to preserve a sample of the lowland dry grasslands so far identified.

Field work was carried out from 1982 to 2005 in four regions (Latium, Apulia, Sicily and Sardinia) within the framework of studies sponsored by WWF Italy and LIPU. The bird species concerned are those listed by Goriup and Batten [3], with the addition of a small group of typical grassland birds of the Mediterranean.

Study area

Grasslands in the Mediterranean basin represent the result of the impact of man and livestock on more complex native vegetation types. According to official statistics summarized by Angle [1], in Italy the herbaceous ecosystems used to produce fodder consist of 3,800,000 ha of permanent pasture and 1,150,000 ha of permanent grassland.

For the purpose of the present study, only true grasslands never ploughed are considered. This category comprises 205,300 ha of true lowland "steppes". The steppic habitats so far identified include the highest quality habitats with undisturbed ground and rich flora grazed by free-ranging livestock (mainly sheep).

All over the country, lowland steppic habitats, which rise up to 800 m in the Sardinian plateau, share some common features. They are characterized by a Mediterranean climate, with autumn and winter rainfall not exceeding 800 mm (rainfall ranges from 300 mm in the driest area of Salento, Apulia to 800 mm in the wettest area of Tolfa, Latium) and by rather shallow soils overlying calcareous rocks. The maximum average daily temperature, recorded in July, is 25–28°C.

The dominant vegetation is composed of annual grasses *(Poa, Stipa, Alopecurus, Triticum, Hordeum, Avena),* Compositae and perennial plants with succulent roots (asphodels, lilies, irises, orchids, *Ferula, Urginea).* A few areas show scattered bushes *(Rubus, Pyrus, Prunus, Crataegus* and evergreen Mediterranean scrub). A tiny percentage is represented by azonal vegetation that occurs on alkaline soils (*e.g.*, *Limonium* and *Suaeda* communities of dry mud, close to wetlands). Although covering only small surfaces, these habitats are critical for the conservation of invertebrates, plants and bird-life.

Results

Apulia accounts for 65,500 ha of steppe (31.9%) and Sardinia for 102,100 ha (49.7%). Smaller areas occur in Latium and Sicily, where major agricultural changes took place soon after World War II. Bird-life represents the most striking aspect of the wildlife inhabiting the steppe areas, which provide shelter and food to a rich community of invertebrates, reptiles, amphibians and plants.

These ecosystems are important breeding areas for Stone Curlew (*Burhinus oedicnemus)*, Calandra Lark *(Melanocorypha calandra)*, whose population probably exceeds 100,000 pairs, Spectacled Warbler (*Sylvia conspicillata*), Bee-eater (*Merops apiaster*), with a population exceeding 10,000 pairs, and Lanner Falcon (*Falco biarmicus*) with about 50 pairs in Apulia and Sicily, the largest populations in Europe.

Little Bustard – a flagship species

In Sardinia 15 subpopulations of Little Bustard (*Tetrax tetrax*) are known to occur in the inner and coastal dry lowland ecosystems of the islands, each supporting from a minimum of 20 to a maximum of 200 individuals [9] to build up the total figure of 400–700 displaying males (1,500–2,000 individuals in post breeding population). The sizes of the Sardinian population in historical times are unknown, but the analysis of the few available bibliographical records confirms the population in that island probably did not experience a significant decrease in the last 50 years.

The situation of the Little Bustard in Apulia is dramatically different. That population probably numbered between 10,000 and 100,000 specimens at the beginning of the 20th century, but began to decrease in the 1940s when farmers moved from the extensive and transhumant sheep economy to intensive farming. In the 1960s the population probably still numbered more than 1,000 individuals, and the Little Bustard was widespread from the river Fortore, north of Gargano, to the river Ofanto, south of Gargano, in the area classified as the Tavoliere plateau, which is the largest plain in Southern Italy covering 7,500 km^2. Soon afterwards, it dropped dramatically to 500–1,000 individuals in the 1970s and finally to 15–20 individuals in 2005. The current population in Apulia could be less than 1% of that occurring in the 1960s and could be considered the smallest isolated nucleus in the whole range.

The habitat of the Little Bustard in Italy consists mainly of extensive farmland exploited for sheep and cattle breeding, combined with extensive cereal crops. This habitat falls in the broad category of "dry lowland grasslands", being characterized by low rainfall and a long and dry summer. I monitored 24 leks of males in Sardinia and 15 in Apulia. Male density ranged from 2.0 males/100 ha to 2.8 males/100 ha in Sardinia (mean = 2.4 males/100 ha ± 0.32) and from 1.4 males/100 ha to 2.0 males/100 ha in Apulia (mean = 1.5 males/100 ha ± 0.43). The combined mean is 1.9 males/100 ha (±0.58). The habitat composition of the monitored lek area was: 81.3% dry stony pastures, 13.4% oat/barley fields and 5.3% wheat fields. Although Little Bustard males made some use of the available herbaceous ecosystems (including intensive cereal crops), it is clear that they showed a strong preference for permanent pastures grazed by cattle and sheep.

These are the only grasslands that remain unploughed and include also cultivated fields, which remain unploughed for at least three years and then develop into permanent pastures. Cereal crops and related fallow lands are harvested once a year in late May/early June and are suitable to bustards only between late autumn and early summers. In the range of the Little Bustard there are small marshes, gravel pits, maquis and garigues that are not directly exploited by birds, but contribute to build up the high biodiversity value of the area and the trophic resources for the bustards.

Bird community is dominated by passerine species as: larks, Tawny Pipit (*Anthus campestris*), Stonechat (*Saxicola torquata*), Spectacled Warbler (*Sylvia conspicillata*), Fan-tailed Warbler (*Cisticola juncidis*), Grey Strike (*Lanius minor*), Woodchat Strike (*Lanius senator*), Magpie (*Pica pica*), Hooded Crow (*Corvus cornix*), Linnet (*Carduelis cannabina*) and Corn Bunting (*Miliaria calandra*). Non-passerine birds include Montagu's Harrier (*Circus pygargus*), Lanner Falcon (*Falco biarmicus*), Lesser Kestrel (*Falco naumannii*), Quail (*Coturnix coturnix*), and Stone Curlew (*Burhinus oedicnemus*).

Discussion

Most endangered species are legally protected according to the Italian legislation on game hunting, but hunting pressure is still very high in the current range of many species in Central and Southern Italy between September and March. Many individuals are thus shot by poachers or are accidentally killed by hunters looking for other game species.

Permanent pastures on dry soils and herbaceous ecosystems not actively managed by man still represent a viable proportion of farmland in Italy, but their area is being reduced by agricultural improvement and expanding towns. On the other hand, the areas still suitable for birds have been progressively fragmented and the common pattern is now a mosaic of small 'islands' of steppe scattered among intensive cereal crops. This proved to have deleterious effects on the largest and most mobile and shy species of birds as shown by the case of the Little Bustard in Apulia. Male birds were chased from very good habitats due to the construction of roads and houses with the attendant human disturbance.

The application of set aside rules does not seem to be enforced in Italy at the moment. Most farmers apply to the Ministry of Agriculture to get set aside compensations just for the usual fallow period of the cereal crop cycle, which is something very different from permanent or semi-permanent uncultivated areas. Many grants are still available to transform seasonal grazing (*transumanza*) into stable and more intensive forms of stock farming.

Large irrigation schemes will provide water for agriculture in dry areas around the Coghinas Lake in Sardinia and the Tolfa and Santa Severa grasslands in Latium. The result will be the development of vineyards, kiwi-fruit plantations and vegetable cultivation in greenhouses.

The greatest difficulty faced by conservation organizations seems to be represented by the lack of the status of 'natural site' for grassland ecosystems: they are still treated as second-class farmland. In the recent past it was the same for wetlands, which are now considered to be areas of great economic value and ecological importance. A campaign to implement a convention for the preservation of dry lowland ecosystems could play a major role for the conservation of these habitats in the UE countries. A preliminary step was the inclusion of some grassland ecosystems in the IBA report and the inclusion of Campeda plateau in the list of the five Italian areas that require priority funds from the UE.

REFERENCES

1. Angle G., 1990 - Habitat: guida alla gestione degli ambienti naturali. Ministry of Forestry and Agriculture and WWF Italy, Rome.
2. Birdlife, 2001 - European Union Species Action Plan: Little Bustard. Birdlife, Cambridge.
3. Goriup P.D., Batten L.A., 1990 - The conservation of steppic birds - a European perspective. Oryx, **24**: 215-223.
4. Goriup P.D., 1994 - Little Bustard. *Tetrax tetrax.* In: Tucker G.M. & Heath M.F. (eds.), Birds in Europe: their conservation status, pp. 236-237. BirdLife Conservation Series 3. BirdLife International, Cambridge.
5. Petretti F., 1985 - Preliminary data on the status of the little Bustard in Italy. Bustard Studies, **2**: 165-170.
6. Petretti F., 1986 - An inventory of steppe habitats in Southern Italy. Steppe Symposium, ICBP International Conference, Kingston (Ontario).
7. Petretti F., 1988 - An inventory of steppe habitats in southern Italy. In: Ecology and conservation of grassland birds, pp. 125-143. ICBP Technical Publication 7. Cambridge.
8. Petretti F., 1991 - Status of lowland dry grasslands and birds in Italy. In: Goriup P.D., Batten L.A., Norton J.A. (eds.), The conservation of lowland dry grassland birds in Europe, pp. 69-76. The Joint Nature Conservation Committee, Peterborough.
9. Petretti F., 2005 - Piano di azione per la gallina prataiola (*Tetrax tetrax*) in Italia. Lipu, Parma.

IV

CONTRIBUTIONS TO CONSERVATION BIOLOGY

CONSERVATION OF DECIDUOUS TREE SPECIES IN EUROPE: PROJECTING POTENTIAL RANGES AND CHANGES

Elgene O. BOX, Michael MANTHEY
University of Georgia, Geography Department, **30602-2502 Athens, Georgia (USA)**
e-mail: boxeo@uga.edu ; manthey@uga.edu

Abstract: Conservation in settled fragmented landscapes, especially with climates changing, requires planning as well as physiological information not yet available. In the absence of the latter, climatic envelopes provide an initial means of projecting possible range changes, as well as potential range expansions on other continents. In this paper, climatic envelopes for major European tree species and hypothetical counterparts in eastern North America are evaluated, projected and cross-projected in order to demonstrate a hopefully useful methodology but also to suggest some initial questions of importance to conservation efforts in Europe, especially the Balkan region. Cross-projections of some counterpart species to the other continent are shown and may help improve understanding of limiting mechanisms and values, thus improving the robustness of climatic envelopes under change scenarios. Potential range displacements in Europe are also shown for selected tree species, based on a warming-only scenario (no net drying) that appears plausible, based on new warming estimates.

Introduction

Conservation in the 21st Century requires not only a focus on protecting current species habitats but also planning for fundamental changes in environmental conditions, from climate change to the more rapid changes that occur in fragmented and increasingly exploited landscapes. Global change, in other words, includes not only global warming, and drying in many areas, but also the often drastic, perhaps unexpected, usually ecologically detrimental changes in land use that result from an unrestrained free market, opportunistic land acquisition, economic and cultural predation, overpopulation, and economic pressures due to globalization. Planning for climate change is in some ways easier than planning for these other threats. Networks of protected areas can be designed to provide suitable habitat under foreseeable climatic changes and putative species responses. Once in the landscape, these may also provide obstacles, at least to some extent, to unwanted land 'development' and conversion to biologically hostile uses.

One requirement for successful species conservation involves understanding species ecology, including reproduction and dispersal, symbiotic relationships, life-cycle history, potential lifespan, potential and actual geographic ranges, and environmental factors that limit range and functional processes. Limiting factors, their quantitative values and relative importance can be studied in various ways. One effective way, at least for climatic factors, is to express hypothetical limiting factors in a climatic envelope model and project the climate space thus defined to test the hypotheses involved. Projections to other continents, hereafter called cross-projections, may be especially useful for supporting hypotheses of limiting mechanisms -- or perhaps more commonly for demolishing them and suggesting other limiting mechanisms or at least other limiting values. In the absence of the vast amount of physiological knowledge needed in order to anticipate species responses to changed climatic conditions, climatic envelopes also provide the only real way to estimate possible consequences of climatic changes geographically.

In this paper we look at some possible consequences of climatic and other changes in the Balkans, especially Romania and its vicinity, and their implications for conservation planning. The south-eastern United States provides a climatic counterpart to the Balkans and will be used for cross-projections to explore the potential range limits of selected major Balkan tree species. The purposes of this paper then, specifically, are: 1) to develop and explore the accuracy of climatic envelopes for some major European deciduous tree species found in the Balkans, especially oaks and beech; 2) to cross-project these climate spaces (envelopes) to eastern North America, in order to evaluate them further and suggest errors in limiting factors; 3) to explore how a simple climatic warming scenario might affect potential ranges of these species in Europe; and 4) to identify implications for conservation in the Balkan landscape.

Methodology and Data

The methodology for this study involves development, calibration, and testing of climatic envelopes for selected major tree species of Europe and eastern North America, followed by application of these envelopes to regional climatic data to predict potential species ranges, on their home continents and on the other continent.

A climatic envelope is defined by the values of important climatic variables that correspond to the geographical boundaries within which a taxon is considered to grow and reproduce under natural conditions [5, 7]. Each envelope is thus defined by upper and lower limits for the variables selected. These variables must involve climatic data that are readily available throughout the study region or indices that can be calculated from such data. On the other hand, variables should also be closely related with and be thought to constrain basic plant functions, such as metabolism, reproduction and dispersal [2]. Climatic limits are estimated empirically by identifying the local values of climatic variables at the species range boundaries. For this study we used envelopes, for major tree species in both regions, that had already been developed and had worked relatively well [6]. Sample envelopes are shown in Table 1, along with definition of the climatic variables involved. These variables represent seasonal and annual values for temperature and precipitation, as well as extreme low temperatures and a climatic moisture balance.

Climatic data are required both for envelope development and for mapping predicted results. These data were obtained from local and surrounding meteorological stations, accumulated over the years from various sources but especially the on-line worldwide data-base of the National Climatic Data Center (Asheville, North Carolina). The data represent various periods of measurement, from the late 1800s to the present. They have been checked repeatedly for errors and for artifacts that may result from such factors as short periods of measurement or changes in the locations of the meteorological observatories. The basic data include site location, elevation and name, long-term monthly values for mean temperature and average precipitation, and the number of years of observation. Also included, where available, are the mean night-time minimum temperature of the coldest month and the absolute minimum temperature (lowest temperature ever recorded), plus number of years of observation for the latter. Where these last two values were not available, they were estimated by triangulation from their relationships to Tmin at several nearby weather stations (program TXTRAP; Box, unpublished). This method has been applied worldwide to produce both global site coverage and a 0.5 x 0.5-degree pixel field for absolute minimum temperature, used in various global models [1]. Potential evapo-transpiration (PET), as needed for the annual moisture index, is estimated from air temperature using a geographical methodology that has been tested globally and used in various global and regional models [1].

Table 1: Estimated climatic envelopes for *Fagus sylvatica s.l.* and *Quercus robur*. Asterisks indicate upper limits assumed not to be relevant. The significance of these and other climatic factors in limiting species ranges was discussed by Box [2, 3] and Box *et al.* [5].

Species		Tmax	Tmin	Tmmin	Tabmin	MIy	Pmin	Pmtmax
Fagus sylvatica s.l.								
Tolerance	Upper	28.0	10.0	5.0	0.0	*****	****	*****
	Lower	15.5	-4.5	-15.0	-40.0	0.85	25.0	30.0
Quercus robur								
Tolerance	Upper	28.0	10.0	5.0	0.0	*****	****	*****
	Lower	14.0	-20.0	-30.0	-45.0	0.70	15.0	20.0

Tmax = mean temperature of the warmest month (°C)
Tmin = mean temperature of the coldest month (°C)
Tmmin = mean night-time minimum temperature of the coldest month (°C)
Tabmin = absolute minimum temperature (°C: coldest ever measured, generally over at least 30 years)
MIy = annual moisture index: precipitation / potential evapotranspiration
Pmin = average precipitation of the driest month (mm)
Pmtmax = average precipitation of the warmest month (mm)

Data are also needed for species ranges, preferably in the form of range maps. For the North American species, maps are readily available in the volumes compiled by Little [13, 14, 15]. For the European species, range maps were obtained from Meusel *et al.* [16, 17], Meusel and Jäger [18], and Hultén and Fries [10]. In some cases additional distributional information was available (other books or field experience and relevés); in such cases envelopes were calibrated to these wider concepts of species range. Some limiting values in the earlier envelopes were slightly improved, based on these improved estimates of actual species ranges.

Recent evaluations of the potential for global warming have revised previous estimates upward, to global average values of at least 1.5°C and perhaps as much as 5–6°C [9, 20]. We have chosen a warming scenario for Europe that is intermediate, based on annual average warming of 2°C. This scenario also includes the provision that winter warming is greater, about double the annual warming, and that summer warming is less. The scenario chosen is also a warming-only scenario, since wetter or drier conditions are much more difficult to foresee accurately. This criterion of no net drying is insured by raising annual precipitation (proportionately for the months) just enough to keep the annual moisture balance the same, i.e. precipitation versus potential evapo-transpiration. An example of the effect of this warming scenario, for Cluj-Napoca (Romania), is shown in Table 2.

Projection of predicted species ranges is done by applying the climatic envelopes to as dense a network as possible of site data (from available meteorological stations), in this case a data-base of 1813 sites for Europe and western Asia. If any site datum falls outside the climatic envelope of a species, then that species is assumed not to occur at that site. The results were mapped as potential range maps using ArcView 3.3 software [8]. Results for the warming scenario were processed similarly, after first applying the warming scenario to each site in the European-West Asian data-base.

Ranges and Limiting Factors of Balkan Trees and Forest Types

Two of the trees studied are *Fagus sylvatica* (European Beech, *sensu lato*, i.e. including *F. orientalis*) and *Quercus robur* (Pedunculate Oak, Truffle Oak), both of which occur widely throughout Europe, including the Balkan area (see climatic envelopes in Table 1). Both species extend from Atlantic western Europe into continental eastern Europe. They both thus have relatively low requirements for summer warmth (Tmax) as well as needs and tolerance of winter cold (Tmin, Tmmin and Tabmin), especially the oak (see Table 2 for abbreviations). Beech also requires a generally moist climate, higher than for most oaks, with no extreme dry month (Pmin at least 40 mm).

As a test of envelope accuracy, each species studied was mapped alongside what appeared to be the best map of the species' actual range (see Methods section). The predicted and actual ranges of *Fagus sylvatica* (s.l.) and *Quercus robur* are shown juxtaposed in Figure 1.

Table 2: Climate scenario based on 2°C annual warming, with more in winter, no net drying.

Temperature

Warming increases with latitude and continentality (globally)

T → T + 2° C (mean annual temperature)

- Winter increases more: about 3°C in North Africa to as much as 5°C in interior northern France
- Summer increases less: generally not more than about 1°C anywhere
- Extreme minima: increase as does the temperature of the coldest month

Precipitation

P increases to maintain the same annual climatic moisture balance (P/PET)

- If PET increases 10% with T + 2°, then P increases 10%
- P increase (%) is applied equally to all months

For example, the current mean monthly temperatures in Cluj-Napoca and those under the warming scenario (denoted T + 2w) are as follows:

(degrees C)	Jan	Feb	Mar	Apr	May	Jun	Jul	Aug	Sep	Oct	Nov	Dec	Annual	Tmmin	Tabmin
Current	**-4.4**	**-2.1**	**3.4**	**8.9**	**14.4**	**17.4**	**19.2**	**18.3**	**14.2**	**8.9**	**3.1**	**-1.6**	**8.3**	**-7.9**	**-30.6**
T + 2w	**0.3**	**2.3**	**7.0**	**11.6**	**16.3**	**18.9**	**20.4**	**19.6**	**16.1**	**11.6**	**6.7**	**2.7**	**11.1**	**-3.2**	**-25.9**

The T+2w warming scenario used herein is based on recent estimates by the Intergovernmental Panel on Climate Change [9], which sees 2°C global average warming as a likely value. T+2w involves greater warming in winter than in summer but no net drying or humidity increase, since patterns of precipitation change are much less predictable. In reality, however, climate change in the Mediterranean region will most likely also be accompanied by net drying, since the subtropical high-pressure belt will probably shift northward.

T = mean annual temperature; Tmmin = mean minimum temperature in winter; Tabmin = absolute minimum temperature

P = average annual precipitation; PET = average annual potential evapotranspiration

The match is relatively good in each case, though *Q. robur* is predicted further north in Fennoscandia and both are predicted further south in eastern Spain than is indicated by their range maps.

One effective way to test the robustness of estimated species limits in an envelope model is to project the climate space thus defined to other continents. Such cross-projections may be especially useful for supporting hypotheses of limiting mechanisms -- or perhaps more commonly for demolishing them and suggesting other limiting mechanisms or at least limiting values. The value of this procedure is illustrated in Figure 2, in which an earlier version of our envelope for *Fagus sylvatica*, based entirely on European data, is projected to eastern North America. When we looked at this map, one question was obvious: why is *F. sylvatica* not predicted in the northeastern US and eastern Canada? Of course the answer is that this region represents conditions not found in Europe and so was outside our experience. But are these conditions outside the potential tolerance limits of *Fagus sylvatica*? A previous study using surface pollen abundances has revealed very similar climatic tolerances for both species [11], and based on this we assume that the limiting value that precluded *Fagus* from northeastern North America, i.e. Tmin = -4.5°C, does not seem to be a value that should be limiting. Even so, the envelope for *Fagus sylvatica* was modified (causing essentially no effect of its predicted range in Europe). This kind of error in estimation of limits causes no problem if envelopes are used only in their home continent and only under current climatic conditions – but may represent a significant error for projections under an altered climate.

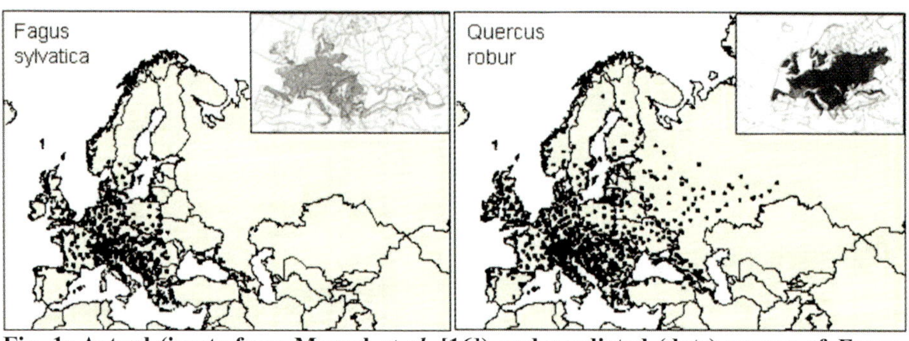

Fig. 1: Actual (insets from Meusel *et al.* [16]) and predicted (dots) ranges of *Fagus sylvatica* (*s.l.*) and *Quercus robur*.

Cross-Projections and European Counterparts in Eastern North America

We have just seen one useful result of envelope cross-projection to other continents. Such cross-projections may also be valuable for generating useful questions or hypotheses, and as an indication of the similarity of possible ecological vicariants. In order to explore these questions and uses, the potential ranges of several deciduous tree species pairs from Europe and eastern North America were cross-projected in both directions.

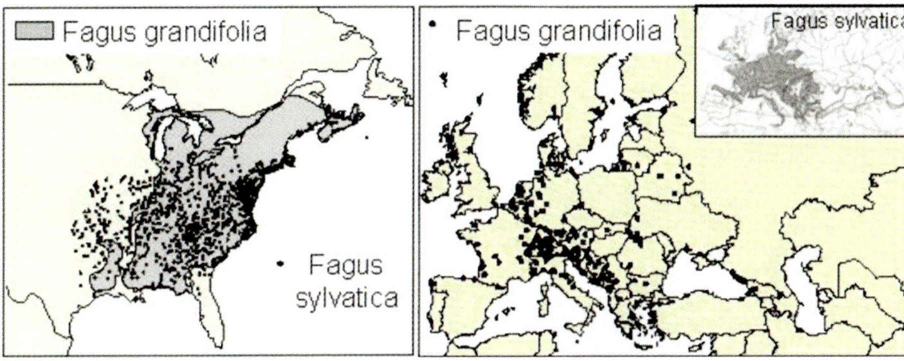

Fig. 2: Cross-projections for European and North American beech species: *Fagus sylvatica s.l.* and *Fagus grandifolia*; their native ranges are indicated in the inset (from Meusel *et al.* [16]) and by shading (from Little [13]), respectively.

Cross-projections of beech species were shown already in Figure 2. Each region has only one beech species, *Fagus sylvatica* in Europe (recently incorporating the former *F. orientalis*, Shen 1992) and *F. grandifolia* in eastern North America. Unlike European beech, the North American species is not so commonly a canopy dominant by itself but rather occurs with other deciduous canopy species such as *Quercus rubra* and *Q. alba*, *Fraxinus* spp. (ashes) and *Tilia* spp. (limes) – and in the South with evergreen *Magnolia grandiflora*. Nevertheless, both beech species seem to have the same general ecological requirements (as also do East Asian species), namely a simultaneous need for constant moisture but also drained, aerated soil. In North America, *F. sylvatica* is projected to go farther west into the drier (southern) Great Plains, e. g. Oklahoma as far west as Oklahoma City, definitely in the grassland region. This prediction may be an artifact, however, of the measure used for the climatic moisture balance, which probably does not adequately capture the differences in potential evapo-transpiration between more maritime Europe with its more consistently high humidity but low precipitation amounts, on the one hand, and the more continental east side of (also lower-latitude) North America, with its greater amounts of precipitation coming from short, mainly convective or frontal thunderstorms. In the other direction, *F. grandifolia* is not predicted over large parts of eastern Europe, perhaps for the same reason. More difficult to explain immediately is why *F. grandifolia* is not predicted in more of Britain, Denmark or southern Sweden (Tmax too low for beech threshold of 17°C).

Cross-projections of deciduous oak species are shown in Figure 3. In particular:

– *Quercus robur*, a pan-European species, is cross-projected and compared with *Q. macrocarpa*, an American species of rather more northern and continental distribution; and

– *Quercus frainetto*, a species especially characteristic of the Balkan peninsula, is cross-projected and compared with *Q. stellata*, a 'southern' species that extends and becomes more important well into the southern Great Plains.

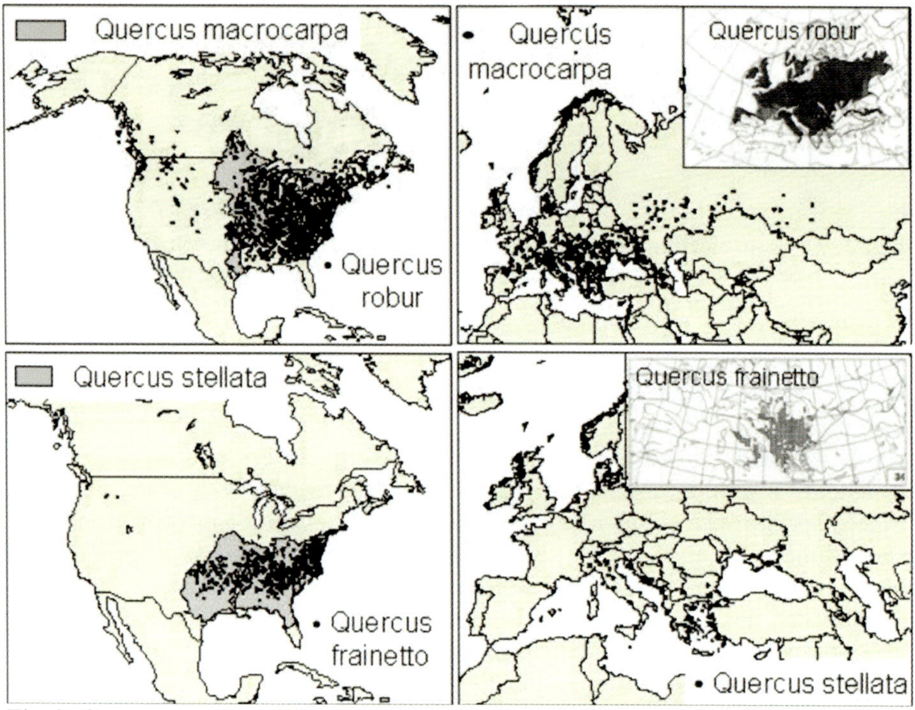

Fig. 3: Cross-projections for oak counterpart species from Europe and eastern North America: *Q. robur* and *Q. macrocarpa*, *Q. frainetto* and *Q. stellata*. Actual ranges (shaded) for both American species are from Little [13]; the inset range for *Q. robur* is from Meusel *et al.* [16] and that for *Q. frainetto* is from Jalas & Suominen [12].

These species were identified as possible vicariants, or at least as geographic counterparts, in an earlier study [6]. The projection of *Q. robur* to North America is especially interesting, since its predicted potential range completely spans North America, something no American broad-leaved tree species, oak or other, actually does. *Q. robur* is also projected to occur across the interior south-eastern USA, where *Q. macrocarpa* is absent. In the other direction, the projected potential range of *Q. macrocarpa* in Eurasia matches the range of *Q. robur* more closely, though it is not predicted to go into more maritime areas such as Britain, where *Q. robur* does occur and is important.

Cross-projections of two maple species are shown in Figure 4, *Acer platanoides* from Europe and another "hard maple", *Acer saccharum* (sugar maple), a species so important in northeastern North America that it appears on the national flag of Canada. These ranges also match relatively closely, but more so in Europe. In North America the European species is again predicted much further west than the range of the American species, and again this may be due to the inadequacy of our estimator for potential evapo-transpiration (see above). In the southeastern USA, the apparent 'overprediction' may not in fact be such, since *A. saccharum* has been removed from scattered locations outside the range shown on the map and a very closely related species, *A. floridanum*, with slightly smaller but otherwise identically shaped leaves, extends well south into Florida.

These and similar cross-projections may suggest useful questions about species ecology in their own regions. Such insights may have useful conservation applications but probably cannot be generalized and will be somewhat idiosyncratic for each species involved. Another conservation problem in many parts of the world involves the familiar phenomenon of range expansion by foreign species, sometimes 'explosively', well beyond what appear to be their environmental limits on their home continents. Cross-projection of sufficiently robust envelopes may also be useful for estimating such range expansions.

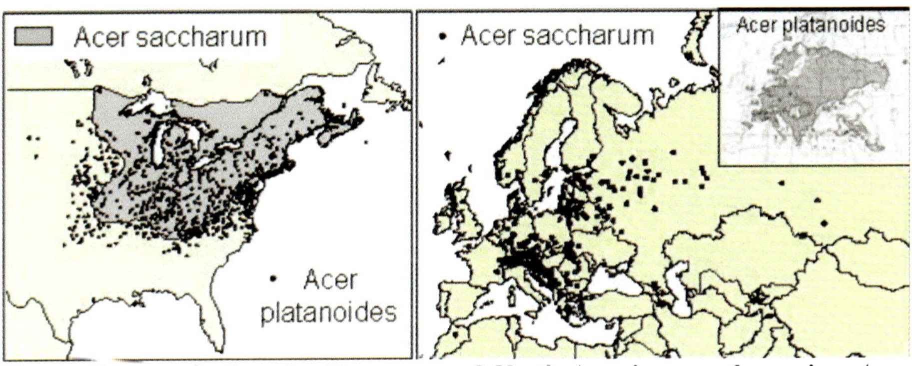

Fig. 4: Cross-projections for European and North American maple species: *Acer platanoides* and *Acer saccharum*; the actual ranges (shaded or inset) are from Little [13] for *A. saccharum* and from Meusel *et al.* [17] for *A. platanoides*.

Potential Range Shifts and Community Vulnerability under Simple Warming

Another, and perhaps more daunting conservation problem, involves potential range displacements due to environmental changes such as global warming. Based on the warming-only scenario described earlier (Methods section), potential range displacements in Europe are shown in Figure 5 for four species: *Fagus sylvatica* (s.l.), the Balkan-Submediterranean oak *Quercus frainetto*, pan-European *Carpinus betulus* (Hornbeam), and a primarily Balkan species *Tilia tomentosa*. Remember that the warming scenario (see Table 1) raises winter temperatures by as much as 4°C in central Europe and 3°C in the Mediterranean area.

Under this scenario, with no net drying, *Fagus sylvatica* is seen to expand (open circles) eastward in eastern Europe and northward in Scandinavia and around the Baltic Sea. Its losses (crosses) are mainly from parts of Britain and along the west coast of France and the north and northwest coasts of Spain. Beech is in a sense 'overpredicted' in Britain by its climatic envelope (Figure 1), but ecologists have long noted that beech may be absent from parts of Britain for historical reasons only. The large core area of the beech's range in Europe remains intact (blackened circles) from southern Sweden through central Europe to interior Italy, Transylvania, northernmost Greece and even in the Caucasus and Hyrcanian region of northern Iran.

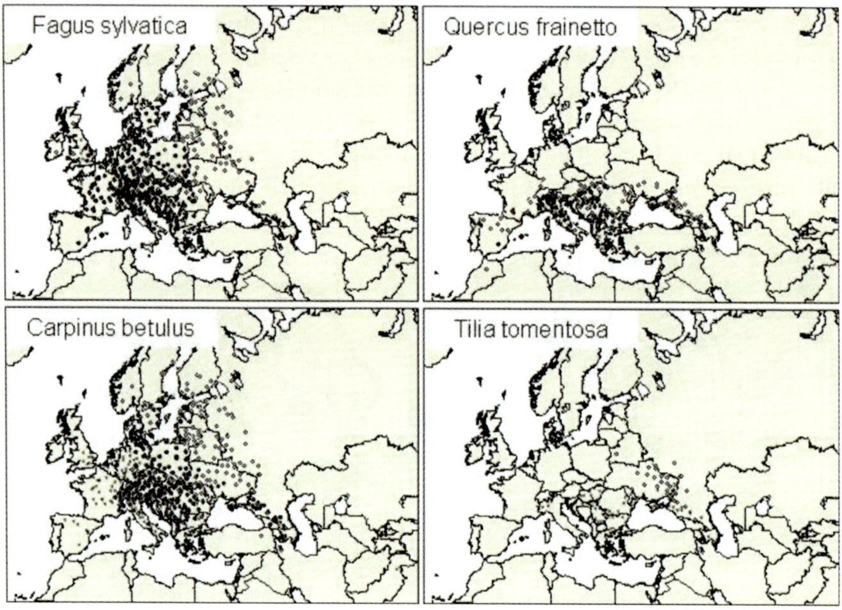

Fig. 5: Potential range displacements of four European tree species under 2°C global-average warming, based on their climatic envelopes. Details of the warming scenario are illustrated in Table 1. Sites at which the species are predicted to occur both now and under the warming scenario are shown by solid black dots. Predicted losses under the warming scenario are shown by crosses (x) and predicted gains by open circles.

The climatic envelope of *Q. frainetto* is predicted to disappear from much of the western Mediterranean coastal region of Europe, as well as the Adriatic coast, the west coast of Greece, and the Black Sea coast of Turkey and Colchis (areas where it does actually occur now, except the western Mediterranean). *Q. frainetto* appears to be potentially stable across northern Italy and the interior Balkan peninsula, also at some scattered sites on the Crimea and in the Caucasus.

Its climate space is predicted to expand northward, from the western Alps to Slovakia, various parts of Romania, across southern Ukraine, and southernmost Russia (inland from Black Sea coastal sites where it disappears). The reason for the retreat from Mediterranean and other coastal areas, in the model at least, is the sharp rise in winter temperatures that removes the winter cold, or vernalization, than may be necessary for temperate deciduous species. Two other oak species with very similar ranges and climatic envelopes are *Q. cerris* and *Q. pubescens* (the latter going further north). The ranges of these species may be displaced similarly under warming (without drying).

Carpinus betulus is an especially important species because it has a wide range and predictably replaces other species, such as beech, especially in more continental areas of eastern Europe, including the Balkan peninsula. *Carpinus* is seen to expand quite significantly into more of eastern Europe and southern Fennoscandia but to disappear from an equally large area of western Europe, from Spain to Britain to Denmark. In the model this is due to the higher winter temperatures, but it raises a perhaps useful research question: does *Carpinus* (and do also perhaps some other deciduous trees) really require such low winter temperatures? Some degree of vernalization is thought to be required by almost all temperate deciduous (summergreen) trees, but how much, and how much does it vary from species to species? Such information is really important for anticipating not only how species ranges may or may not be displaced by warming but also for anticipating the degree to which familiar natural plant communities may be dismantled and familiar landscapes and habitats disrupted by different responses to warming by different, currently co-occurring species.

Finally, consider the case of *Tilia tomentosa*, a species with a smaller natural range confined mainly to the Balkan peninsula but also the Crimea and Caucasus. Unlike the other species in Figure 5, *Tilia* is predicted to have no stable core area that remains intact under warming. Its entire range is displaced eastward and slightly northward from the Balkan area primarily to eastern Ukraine and southernmost Russia. This seems to be another case in which we may not know how much winter cold the species actually needs, for its own physiology as well as for protection against disease and other stresses that may increase with milder winters. These too are conservation questions.

Conclusion

Planning for consequences of environmental changes, including climatic but also land-use and other direct changes to the landscape, is a part of conservation activities. Suitable habitats for species must be preserved or provided, not only for species whose ranges may be displaced by climatic warming or drying, but also for whole communities or for successor communities that result from divergence of the formerly overlapping ranges of the main structural elements of communities.

Construction and testing of climatic envelopes may help to suggest tolerance limits and other limitation mechanisms that could determine the extent and direction of range displacement under climate change. Study of vicariant or other counterpart species on other continents, as well as projection of climatic envelopes to other continents, may help to identify erroneous estimates of climatic limits that would not be seen through local calibration and testing alone.

Finally, the potential range displacements shown in Figure 5 suggest that change may be significant, with unexpected threats to community integrity [4], even in well-buffered ecosystems such as deciduous forests.

REFERENCES

1. Box E.O., 1994 - Global Potential Natural Vegetation: Dynamic Benchmark in the Era of Disruption. In: Murai (ed.), Toward Global Planning of Sustainable Use of the Earth, pp. 77-95. Elsevier, Amsterdam.
2. Box E.O., 1995 - Factors determining distributions of tree species and plant functional types. Vegetatio, **121**: 101-116.
3. Box E.O., 1996 - Plant functional types and climate at the global scale. J. Vegetation Science, **7**: 309-320.
4. Box E.O., Choi J.-N., 2000 - Estimating species-based community integrity under global warming, with special reference to the western Mediterranean region. Phytocoenologia, **30**: 335-352.
5. Box E.O., Crumpacker D.W., Hardin E.D., 1993 - A climatic model for plant species locations in Florida. J. Biogeography, **20**: 629-644.
6. Box E.O., Manthey M., (in press). Oak and other deciduous forest types of eastern North America and Europe. Botanika Chronika.
7. Crumpacker D.W., Box E.O., Hardin E.D., 2002 - Use of plant climatic envelopes to design a monitoring system for biotic effects of climatic warming. Florida Scientist, **65**: 159-184.
8. ESRI, 2002 - ArcView 3.3. Environmental Systems Research Institute. Redlands (California).
9. Houghton J.T., Ding Y., Griggs D.J., Noguer M., van der Linden P.J., Dai X., Maskell K., Johnson C.A. (eds.), 2001 - Climate Change 2001: The Scientific Basis. Contribution of Working Group I to the Third Assessment Report of the Intergovernmental Panel on Climate Change. Cambridge University Press.
10. Hultén E., Fries M., 1986 - Atlas of North European vascular plants north of the Tropic of Cancer. Vol. 1-3. Koeltz Scientific Books, Königstein.
11. Huntley B., Bartlein P.J., Prentice I.C, 1989 - Climatic Control of the Distribution and Abundance of Beech (*Fagus* L.) in Europe and North America. Journal of Biogeography, **16**: 551-560.
12. Jalas J., Suominen J., 1972-1991 - *Atlas Florae Europaeae*: distribution of vascular plants in Europe. Vols 1-9. Akateeminen Kirjakauppa, Helsinki.
13. Little E.L., 1971 - Atlas of United States Trees: Conifers and Important Hardwoods. (Vol. 1). US Dept. of Agriculture, Misc. Publ. no. 1146. Washington.
14. Little E.L., 1977 - Atlas of United States Trees: Minor Eastern Hardwoods. (Vol. 4). US Dept. of Agriculture, Misc. Publ. no. 1342. Washington.
15. Little E.L., 1978 - Atlas of United States Trees: Florida. (Vol. 5). US Dept. of Agriculture, Misc. Publ. no. 1361. Washington.

16. Meusel H., Jäger E.J., Weinert E., 1965 - Vergleichende Chorologie der zentraleuropaeischen Flora. Volume I (Text und Karten). Fischer Verlag, Jena.
17. Meusel H., Jäger E.J., Rauschert S., Weinert E., 1978 - Vergleichende Chorologie der zentraleuropaeischen Flora. Volume II (Text und Karten). Fischer Verlag, Jena.
18. Meusel H., Jäger E.J., 1992 - Vergleichende Chorologie der zentraleuropaeischen Flora. Volume III (Text und Kartenteil). Fischer Verlag, Jena.
19. Shen C.F., 1992 - A monograph of the genus *Fagus* Tourn. ex L. (*Fagaceae*). Ph.D. Dissertation, City University of New York.
20. Wigley T.M.L., Raper S.C.B., 2001 - Interpretation of high projections for global-mean warming. Science, **293**: 451-454.

DETERMINISM, CHAOS AND STOCHASTICITY IN PLANT COMMUNITY SUCCESSIONS: CONSEQUENCES FOR PHYTOSOCIOLOGY AND CONSERVATION ECOLOGY

Guillaume DECOCQ
Université de Picardie Jules Verne - Department of Botany,
1 rue des Louvels, **F-80037 Amiens Cedex, (FRANCE)**;
e-mail: guillaume.decocq@u-picardie.fr

Abstract: Vegetation dynamics has long been studied from a deterministic perspective, leading to important concepts like climax, equilibrium and reversibility. In recent decades, the emergence of the theory of chaos has changed our vision of natural laws. New concepts such as non-equilibrium, heterogeneity, disturbance and irreversibility have increasingly become popular. Some ecological successions have even been considered as stochastic and thus their outcome a matter of chance. Here, I briefly review these three paradigms, within the special framework of vegetation typology and conservation ecology. I conclude that all successions are, at least partly, deterministic. The proportion of stochasticity depends mainly on the size and composition of the regional species pool, and is likely to occur in early-successional stages. When the lack of acute disturbance allows a succession to develop, the late-successional stages usually converge toward a 'strange attractor', i.e. the climax. Short-term and/or small-scale monitoring are common biases leading to conclude that determinism does not exist in plant community successions.

Introduction

Plant communities change over time. The study of succession, i.e. the process of directional change in communities, has been an important focus of community ecologists, since the pioneer work of Warming [39]. In the field, one can easily recognize a succession by a progressive change in vegetation physiognomy, corresponding to changes in species composition, diversity and, even if less evident to the field observer, function. For example, by monitoring vegetation in a former cultivated field in Europe, one is likely to observe the early colonisation by annuals and biennals, soon replaced by perennial grasses and forbs and, usually a few years later, by the first shrubs and juvenile trees. From an ecological point of view, succession is a dynamic process resulting from a balance between the colonizing ability of some species and the competitive ability of others. This ecological phenomenon may be considered within a larger physical context, in which a dynamic system is one whose state changes over time according to a process termed dynamics.

Ecosystems, as well as their constituent plant communities, are undoubtedly dynamic systems. Surprisingly, during almost the two last centuries, plant community dynamics has been only poorly taken into account in the description of plant communities, the latter being often described as 'stable species combinations' [18]. Of course it has been well recognized that succession proceeded through a series of seral stages from the pioneer stage to the so called 'climax' stage, the latter being the most stable of all stable communities. But, the 'seral states' have usually been considered as deterministic, discrete entities, that follow 'natural laws'. Hence, successional stages are highly predictable. However, in the second half of the 20th century, the emergence of the theory of chaos has changed our vision of the natural laws. These were traditionally associated with determinism and reversibility in time. But in the (special?) case of the newly recognized unstable systems, the natural laws express what is probable rather than what is certain [30]. According to this new paradigm, chance and hazard would become the rules in ecological systems. To certain authors it appeared that ecological successions were unpredictable since completely random [28]. For others, successions would be rather globally deterministic trajectories with probabilistic states and a more or less important proportion of chance (see [24] for a review). Imagine a dynamic system. From a given starting point, the succession will lead to a particular endpoint. Now, reinitialize the system from the same starting point a number of times. If the endpoint is different each time, the system is *stochastic*. If there are two or few possible endpoints, the system is *chaotic*. Conversely, if the succession always leads to the same endpoint, the system is *deterministic*. In the latter case, one should look at the intermediate points of the succession. If they are exactly the same for all reiterations, the system is *completely deterministic*, but if the roads to the endpoint change, the system is rather *chaotic*.

I believe that there is no general theory and any generalization is probably hazardous. It is even unlikely that any single theory may be able to account for all plant community successions and all types of ecosystem. In the following contribution, I propose to review briefly the three main theoretical frameworks of system dynamics – determinism, chaos and stochasticity – within the special case of plant community successions. I shall illustrate my argument with several examples from my own research. I shall conclude by deriving some implications for phytosociology (i.e. vegetation typology) and conservation ecology (i.e. the management of plant communities for conservation or restoration purposes).

Should we bury the determinism paradigm?

Determinism used to be the most influential paradigm in vegetation theory, at least for the last two centuries. For instance, in the field of vegetation description, either methods derived from the *continuum* theory [7, 20, 40] or those referring to discrete plant communities [1, 18, 19] focus on 'determinants' of vegetation composition and structure. Those determinants are mainly ecological, e.g. climate, soil properties, substrate, topography, biotic factors, and are expected to control vegetation patterns. Within the framework of vegetation dynamics, this paradigm postulates that species replace one another because at each stage they modify the environment to make it less suitable for them and more suitable for others. Thus, species replacement is orderly and predictable, and provides directionality for succession. This has been formalized by Connell and Slatyer [6] through their '*facilitation model*', in which the early species in succession facilitate the establishment of the later species. The final stage of succession, which is a stable community, is commonly called 'climax', that is a self-perpetuating community in equilibrium with the physical and biotic environment. The eminently deterministic concept of climax has been one of the most influential and debated during the 20th century, as shown by the numerous climax-derived terms one can find in the ecological literature. The *Monoclimax theory* of Clements [5] is the most deterministic of all. It postulates that every region has only one climax community, a function of its climate ('climatic climax'), toward which all communities are developing, whatever the initial habitat conditions. This highly controversial theory has been rapidly replaced by the *Polyclimax theory* [37]. The latter considers that many different stable communities may co-exist in a given area, controlled by soil properties, relief, microclimate and biotic factors. Thus, several 'subclimaxes' may exist into a region supporting a single climatic climax. But difference between these two theories is probably a question of scale for measuring stability.

Firstly, this is a matter of time-scale. The Monoclimax theory may be valid on a geological time-scale (but equilibrium can never be reached because the vegetation is subject to climatic variations!), while the more realistic Polyclimax theory rather considers an ecological time-scale [27]. During a primary or secondary succession, the rate of change in plant communities is rapid in early-stages, so that one can objectively describe these changes (e.g. in terms of plant species composition) from one decade to another or even from one year to another. Conversely, the rate of change is much slower in late-successional stages, particularly when the climax is a forest, so that one may consider that the steady state has been reached and the plant community is stable. At this climax stage and in the absence of major disturbance, plant communities only undergo fluctuations in species composition or relative abundance [16], but the whole structure and function are preserved. Hence, the ecosystem is nearly stable and, in a certain way, equilibrium is reached.

For example, studying the post-cultural succession in a semi-deciduous tropical forest of Côte d'Ivoire [26], we have shown that a secondary forest was able to establish as early as 25 years after land abandonment. However, species composition and structure were significantly different from those of the primary forest, indicating that a much longer time lapse is needed for the forest to recover completely. We have also demonstrated that despite different initial conditions (e.g. type and length of cultivation, distance from the primary forest, remnant density) both early- and late-successional stages were highly constant from an old field to another, indicating a strong deterministic process. The local species pool has been estimated at 670 species, and the biotope conditions were quite uniform over the studied area (i.e. peneplain supporting either ferralitic or hydromorphic soils). A number of discrete communities have been identified along the time sequence. Surprisingly, we found no significant difference in vegetation composition between dry and wet soils at any stage of the chronosequence, though the primary forest differed slightly between the two soil types, in accordance with the Polyclimax theory.

Secondly, this is a matter of spatial scale. Within a climax community, like a climatic forest for example, one can observe several dynamic stages forming a patch mosaic. Silvigenesis has been intensively studied by Oldeman [29] who called the forest as a whole, the 'silvatic mosaic'. This mosaic consists of treefall-induced gap, pioneer, post-pioneer, mature and senescent communities, randomly assembled. Each of these communities contributes to the structure and function (i.e. stability, resilience, homeostasis) of the climax ecosystem. This spatial heterogeneity results from temporal cyclic changes. In the literature, it is often referred to as 'patch dynamics' [27]. Within this framework, we considered that the definition of climax given by Gillet *et al.* [19] is a good one: "*the mature state of biocoenosis successions, consisting of a spatio-temporal complex of pioneer, transient and final stages, which maximizes its autonomy, homeostasis and resilience, and corresponds to a dynamical equilibrium with the periodical fluctuations of the habitat.*" Climax is thus a kind of stable 'metacommunity', that is a mosaic of patches repeatedly undergoing cyclic changes that are part of the internal dynamics of the community. The period of the cycle driving the internal dynamics primarily depends on the life span of the dominant species, but may be shortened by catastrophic external events.

For example, within a semi-deciduous tropical forest of Central African Republic (Denguéadhé & Decocq, unpublished data), we have shown that the climatic primeval forest consisted of a mosaic of patches with contrasted species composition and structure despite uniform abiotic conditions, in accordance with Oldeman's model. As a consequence, the climatic forest provided a patchiness structure, in which each patch corresponded to a successional stage. If the species composition of the forest as a whole (i.e. forest 'metacommunity') can be easily predicted once the regional species pool is known, one of the individual patches is much more variable, depending upon the age and size of the gap, and of the seed/sapling bank composition prior to the gap creation [3]. Thus, the smaller the spatial scale of assessment, the less predictable (i.e. less deterministic) the succession.

I conclude that vegetation dynamics is deterministic when species of the regional pool are quite few without significant niche overlap, particularly under stressful conditions and/or when several higher competitors are already present. The more selective the filters, the more deterministic the entire succession (i.e. from initial stage to climax).

Is Chaos the rule for vegetation dynamics?

While determinism in community ecology is connected to equilibrium, homogeneity and stability concepts, chaos is often linked to non-equilibrium, heterogeneity and disturbance concepts [31, 32]. Chaos is an ordered disorder that means that it is structured and determinist [21]. As I stated in the introduction, there are two ways for a dynamic system to be chaotic: either the succession may have two or few possible endpoints (*true chaotic system*), or the succession may follow a number of ways to reach the same endpoint (*apparent chaotic system*).

In a true chaotic system, mature plant communities could exist in multiple stable states, as suggested by Sutherland [36]. That means that there may be two or more climaxes for a single set of ecological conditions. Hence, changes in community composition that appear to be non-equilibrium may instead be the result of two or more alternative states for the same community. Considerable controversies exist about the frequency of such multiple stable states in natural ecosystems. Many cases can be rejected because the physical environment differs between the two stable states. Thus, one of the alternative states may persist only because artificial inputs are maintained, which means that the environmental filter is not uniformly the same over the whole area investigated.

In a series of studies aimed at identifying plant indicators of archaeological sites, my colleagues and I have studied the difference in plant species composition between former human settlements (from Roman to Medieval periods) and control sites within ancient forest ecosystems [10]. Each time we found significant differences despite apparently similar ecological conditions (i.e. soil profile, climate, topography, biotic factors). We spoke about 'phytosociological anomalies' since the plant communities observed on the archaeological sites were in contravention of the predictive phytosociological laws [17]. Such results indicated that two alternative steady states could coexist within a same area governed by the same ecological conditions (but with different histories!). However, soil analyses have shown that the soil properties differed slightly between control and archaeological sites. The latter supported soils of higher pH (due to buried building materials, mainly composed of calcareous rock), higher nutrient content and soil compaction (due to prolonged human settlement). Hence, as the initial ecological conditions were not exactly the same, we cannot consider the system as chaotic but deterministic.

To be truly chaotic, the plant community may become locked into a changed configuration once human disturbance has ceased, instead of reverting to its original configuration. If a community is disturbed sufficiently, it may shift to a new configuration at which it will remain even if the disturbance is stopped.

The case of ancient and recent forests has been largely investigated by the research group of Martin Hermy in Belgium [38]. It appeared that plant species composition strongly differs between ancient forests (i.e. forests that have continuously existed since many centuries) and recent forests (i.e. forests that have (re-)established on formerly cultivated land), even when biotope conditions were similar. Again it was difficult to prove that ecological conditions (i.e. habitat quality) were exactly the same, but there clearly emerged a functional species group called 'ancient forest species', grouping species which were strongly linked to ancient (but not unmanaged!) forests. The most interesting result was that those species were limited by their dispersal capacity rather than by their ecological requirements. Hence, at a local scale, the major controls on the succession were not the ecological conditions (i.e. the environmental filter) but the local species pool and landscape patterns (e.g. forest fragmentation and patch connectivity). Despite similar initial ecological conditions, the outcome of the succession differs. This conclusion tends to confirm that some plant communities may exist in multiple stable states, and these communities may be confused with non-equilibrium assemblages. This is an important question for conservation and land management, since actually we cannot assume that all communities subjected to human disturbance will return to their original configuration once the disturbance is ameliorated [35]. One may object that dispersal limitation is time-limited. But once again, this is a question of time-scale. For many ant-dispersed forest herbs, covering a distance of one kilometre may take more than a millennium! Studying the vegetation of former Roman villas, I reported that the true forest herb Wood Anemone (*Anemone nemorosa*), despite the presence of a huge seed source at 100 m and suitable habitat conditions, had not yet colonised the archaeological site [10]. This indicated irreversible after-effects of past human disturbances on the outcome of succession, on a historical time-scale, as suggested by Dupouey *et al.* [14]. However, irreversibility could be simply due to the limits of our patience [30].

Following the chaos theory, systems that can provide several equilibrium states have been labelled 'intransitive systems'. Such a system may stay in one or another equilibrium state, but never in both. Only an exogenous disturbance may shift the system from one equilibrium state to another [21]. The critical state after which the succession can follow one or another trajectory has been called 'bifurcation point' [21]. From bifurcation points emerge different solutions. Which one of these solutions occurs is a matter of probability. Thus, determinism and probabilism are not unconciliable but complementary. The existence of bifurcations confers a historical dimension upon the system dynamics [30].

Conversely, apparent chaotic systems are unidirectional, that means that only one final stage is possible, but the trajectories to reach it are multiple.

A good example is the plant colonisation process in abandoned chalk quarries of northern France [8]. I have demonstrated that the earliest successional stage was characterized by strong competition between species belonging to very different phytosociological classes: *Stellarietea mediae, Onopordetea acanthii, Festuco-Brometea erecti, Agrostio-Arrhenatheretea elatioris, Trifolio-Geranietea sanguinei*, *Rhamno-Prunetea spinosae*. The following stages depended on the outcome of the competition, the latter being largely influenced by biotic factors. If the former quarry was, at least temporarily, converted into a pasture, species from the *Agrostio-Arrhenatheretea elatioris* and *Festuco-Brometea erecti* classes became dominant, with cows and sheep respectively. If the quarry was completely abandoned, the outcome depended on the environment. For quarries located near to or within a village, species from the *Onopordetea acanthii* class were dominant. For those located in an open-field landscape, the succession promoted species from the *Festuco-Brometea erecti* class, where rabbits were abundant, and species from the *Trifolio-Geranietea sanguinei* class where rabbits were absent. However, whatever the mid-successional stages, they lead to the same endpoint, consisting in woodland dominated by beech, after an interval of at least a century.

These results are consistent with the '*Initial floristic composition hypothesis*' [15] or the '*Pre-emptive initial floristic model*' [41], both considering that any local succession depends upon which species arrive first in the place and pre-empt the course of succession. But biotic disturbances also play a major role in the succession. Thus, species replacement is not necessarily orderly and directional at each time of the succession, but individualistic and rather unpredictable. However, some species are superior competitors and come to predominate in the climax community. This convergence toward the climax may be delayed when a species present early in the succession inhibit the establishment of later-successional species, in accordance with the '*inhibition model*' [6]. This is particularly significant when an allelopathic species dominates an early-succession stage. Species replacement is thus blocked by the inhibiting resident until it dies. In our example, *Hieracium pilosella* L. often formed monospecific carpets by exhibiting allelopathy, and prevented other herbaceous species from establishing. When autotoxicity occurred, the succession was released from suppression. On the other hand, the convergence may be speeded up when 'nomadic species' (i.e. mid- or late-successional species able to behave as pioneer species under certain conditions) start the succession, in accordance with the '*tolerance model*' [6]. For example, in quarries located near a forest or within a bocage landscape, the pioneer species are often bird-dispersed tree and shrub species, which directly initiate the succession dynamics.

Within the framework of chaos theory, I propose to consider the climax as a 'strange attractor'. At the beginning, the succession is non-directional: a system can follow all trajectories. But in the long term, the only possible trajectories are attractors; others are transient. An attractor is stable. That means that even if a disturbance occurs and strays off the trajectory, the latter rapidly return back to the attractor [21, 34]. The attractor is not only one possible trajectory of a system; it is the trajectory toward which all others converge. As long as the start point is in the vicinity of the attractor, the following points will converge toward this attractor [21]. Thus, the system does not necessarily reach a stationary state but is structurally stable, which means that its behaviour is not influenced by a small variation of a given parameter [33].

I conclude that chaos occurs when the species pool is important, with many species sharing the same habitat and niche. Disturbance promotes chaos by preventing or delaying competitive exclusion. Hence, chaos is expected mainly in the earliest stages of a succession. The more severe the disturbance, the longer the after-effects (i.e. time to go back to the climax) and thus, more important the chaotic behaviour of the system. For certain systems, there may exist a critical threshold beyond which a disturbance may cause a shift from a climax (i.e. an attractor) to another. This shift may be irreversible on a historical time-scale. Finally, because a completely undisturbed ecosystem is unlikely, chaos may be the rule for vegetation dynamics: only the proportion of determinism, and thus of stochasticity, is variable.

Does stochasticity exist in plant community successions?

If communities are non-equilibrium assemblages of species, their composition should change over time and thus, one cannot observe a single steady state. The '*random colonization model*' [28] suggests that species arrival and replacement occur completely randomly, with no interspecific interactions. Succession can thus move in any direction. I believe that if stochasticity does exist, it is an exception. Much more often it may be an observation bias, either in time or space.

To my knowledge there is no evidence that plant community succession may be completely random in the real world. This has been suggested for ecosystems like islands, for example, but stochasticity emerged from numerical simulations – not from long-term field monitoring [13]. Indeed, islands can be viewed as special kinds of traps that catch species that are able to disperse there and establish successfully. The unknown is which species may disperse to the island, since there is no species pool at the beginning of the colonisation process. The compatibility between species' autecology and island habitat conditions is theoretically the unique requisite for a species to establish. A succession can occur only if a sufficient number of species is available. Hence, I consider that in the special case of island systems, the dynamics is conditioned by the size of the species pool and may self-perpetuate in an early- or mid-successional state as long as the number of species is restricted. The immigration rate controls the succession trajectory and outcome. Let us go back to our previous example of old chalk quarries (see above). If we shorten the succession over the 50 first years, we would observe almost as many states as studied cases and thus conclude that the system is completely stochastic. The time limitation would have prevented us from observing the deterministic part of the process, which is the outcome of the interspecific competition when a number of species must pass both the environmental and the functional filters to remain established.

Random assemblage processes have also been reported in relation to colonisation of forest gaps in tropical forests [3]. It has been demonstrated that due to the stochastic availability of gaps and limited recruitment of juveniles, gaps were filled mostly by chance occupants rather than by best-adapted species, slowing down competitive exclusion. Hence, within-gap succession was completely random. However, the approach was limited to a number of gaps, i.e. different patches of the same mosaic. Even if ecological conditions are the same (i.e. the environmental filter is uniform), the vegetation component (i.e. the functional filter) may be heterogeneous due to the lack of disturbance synchronicity over the whole area. In such a situation, it is tempting to conclude that part of a deterministic whole is stochastic. Remember Egler's moral fable (in [22]): to study a herd of cattle, a zootechnician who slices each animal into equal pieces of one pound, and then randomly samples a number of pieces, will definitely conclude that the individual 'cattle' do not exist.

I finally conclude that stochasticity in a dynamic system is controlled by the species pool, including species richness and dispersal ability: the richer the species pool, the less important the stochastic processes. Stochasticity may be an artefact associated either with our limited time-scale for monitoring long-term succession, or with our limited space-scale to assess broad systems.

What consequences for phytosociology and conservation ecology?

Phytosociology, the science of plant communities, is strongly associated with the determinism paradigm. Plant associations are abstractions of statistically repetitive plant species assemblages repeatedly observed in the field [9]. Species assemblages are thus highly predictable, and plant community successions as well. This predictability has been formalized by de Foucault [17] through the 'physical law' and the 'synchorological law'. The physical law postulates that *for a given species pool, one association should correspond to one and only one original set of ecological conditions, and vice-versa.* The synchorological law states that *a given association may be found within a well-defined geographical area and, reciprocally, that a given geographical area supported only a limited number of associations.* Hence, once two of the three components – habitat conditions, plant community, and geographical area – are known, the third could theoretically be predicted. If we now admit that chaos is the rule in vegetation dynamics, part of ecological succession is in contravention of phytosociological laws and thus becomes unpredictable. In a first approach, I think that there are two simple ways to quantify the proportion of stochasticity for a dynamically-defined syntaxon (i.e. a syntaxon which groups relevés of even-aged plant communities representing a same successional stage for a given ecosystem). First, the number of accidental species, traditionally defined as the species occurring in less than 5% of relevés, usually increases as the species assemblage is increasingly random. Second, the histogram showing the number (or percent) of species among frequency classes (decreasing from V to +): a reversed J-shape indicates determinism (i.e. the major part of the species shows a high constancy), while another shape may be a measure of the part of stochasticity. For example, a J-shape indicates a high proportion of stochasticity.

As a succession is progressing toward an attractor, the number of accidental species is expected to decrease and the constancy histogram to tend toward a perfect reverse-J shape, indicating that the succession becomes more and more deterministic. I consider that among successional states, species assemblages should not be described as 'plant associations' as long as their constancy histogram would not show this critical reversed-J shape. Otherwise, one may assign to species assemblages a deterministic character although they are stochastic. Once this critical state reached, the main challenge for phytosociology is to recognize discrete stages along the succession, i.e. plant communities that significantly differ by their species composition along the time sequence. One way to do this is to find the catastrophe points, which correspond to dramatic changes (or accelerations) in plant species turnover [11]. For example, a sudden increase of beta-diversity between two consecutive stages has been found to be a good indicator for a plant community change.

The chaotic nature of many plant community successions has also important consequences for conservation ecology, i.e. the subdiscipline focusing on conservation or restoration of rare and/or valuable plant species and communities.

If the target plant community is a climax (e.g. an old-growth forest), conservation mainly consists of avoiding severe exogenous disturbances. One should remind that there is a probability for the succession, once reinitialized, to shift to another steady state instead of reverting to its original configuration (*true chaotic system*) [35].

In many cases, target plant communities are successional stages that are more or less unstable (e.g. heathland or grassland communities within forest biomes). Restoration of such plant communities (e.g. managing a forest to obtain a grassland) may be difficult since there are many roads leading to the mature state (climax) and which one of the roads will be followed by the succession may be matter of chance. Hence, the target plant community may not be on the road! As a succession usually becomes more and more deterministic as it is approaching climax, the closer to climax a target community, the higher the probability for successful restoration (*apparent chaotic system*).

Moreover, it is necessary to focus not only on the local scale (i.e. the scale at which the target plant community must be restored) but on a broader scale, taking into account both the regional pool of species and landscape attributes. One should remember that a given plant community has established because species from the regional pool have successfully passed both environmental and functional local filters. But if the regional species pool has changed (e.g. arrival of exotic invasive species), new players can participate in the species assemblage game and secondarily modify local filters. For example, the alien *Prunus serotina* Ehrh. has been found to have become the main control on species assemblages and forest succession in several European forests [4]. A set of particular life traits enables it to invade forest gaps, out-compete native tree species and undergo secondarily spread, so that it induces dramatic changes in the forest functioning. Another regional factor controlling local dynamics is landscape structure. To have a chance to pass local filters, a species must reach them! That means that landscape structure must allow species fluxes (i.e. diaspore dispersal), so that suitable species can immigrate into the plant community under restoration. In other words, for a given target plant community, the connectivity between the source patches (i.e. parts of the landscape where the target plant community is already present and from which seeds are dispersed) and the sink patch (i.e. the landscape part where the target community must be restored) must be important. Note that between-patch connectivity also controls local extinction processes by influencing metapopulation dynamics [23]. For example, the colonisation by true forest herb species of recent forest patches scattered in an agricultural landscape has been found to be impaired by the low connectivity with ancient forest, the latter acting as a seed source [25].

Finally, it is a great challenge to understand the exact role of disturbances in a dynamic system. Many plant communities are disturbance-maintained, which means that their persistence is strongly associated with a given disturbance regime that is part of their evolutionary history and, as such, should be regarded as part of normal ecosystem functioning rather than an external disturbance [2]. For example, in Europe some 'close-to-nature' silvicultural systems have been elaborated to mimic natural disturbances (i.e. gap dynamics). Such systems have been applied to forests which were previously managed as traditional coppice-with-standards, to restore the forests' natural state. My colleagues and I have recently demonstrated that such a conversion failed to restore an original forest flora but negatively impacted the 'true forest species', which are rather species adapted to the traditional management. We concluded that current plant communities were disturbance-maintained and that a dramatic change of silvicultural system (i.e. disturbance regime) deflected the succession toward ruderal plant communities [12]. This suggests that there is no general rule about the use of disturbance in conservation ecology and that an accurate understanding of succession drivers is needed before managing plant communities for biodiversity purposes. Because chaos confers the system dynamics a historical dimension, an extensive knowledge of past and current disturbances that have shaped plant communities is required.

REFERENCES

1. Braun-Blanquet J., 1964 - Pflanzenzociologie Grundzüge der Vegetationskunde. 3rd édition. Springer, Wien.
2. Briske D.D., Fuhlendorf S.D., Smeins F.E., 2003 - Vegetation dynamics on rangelands: a critique of the current paradigms. J. Appl. Ecol., **40**: 601-614.
3. Brokaw N., Busing R.T., 2000 - Niche versus chance and tree diversity in forest gaps. Tr. Ecol. Evol., **15**: 183-187.
4. Chabrerie O., Hoeblich H., Decocq G., in press - Déterminisme et conséquences écologiques de la dynamique invasive du cerisier tardif (*Prunus serotina* Ehrh.) sur les communautés végétales de la forêt de Compiègne. Acta Bot. Gallica, **153**.
5. Clements F.E., 1916 - Plant succession: an analysis of the development of vegetation. Carnegie Institute, Washington D.C.
6. Connell J.H., Slatyer R.O., 1977 - Mechanisms of succession in natural communities and their role in community stability and organization. Am. Nat., **111**: 1119-1144.
7. Curtis J.T., 1959 - The vegetation of Wisconsin : an ordination of plant communities. University of Wisconsin Press, Madison.
8. Decocq G., 1999 - La dynamique de recolonisation végétale des anciennes carrières de craie phosphatée du nord de la France: analyse phytosociologique et systémique. Belg. J. Bot., **132**: 77-94.

9. Decocq G., 2001 - Importance et limites de l'observation en phytosociologie. In: Viret J. (ed.), L'observation dans les sciences, pp. 181-191. Editions du CTHS, Paris.
10. Decocq G., 2003 - Phytosociologie, syndynamique et archéologie du paysage. Contrib. Bot., **38**: 13-21.
11. Decocq G., 2005 - Mécanismes de la dynamique naturelle des végétations forestières de montagne: essai de synthèse et conséquences en phytosociologie. Acta Bot. Gallica, **152**: 581-594.
12. Decocq G., Aubert M., Alard D., Dupont F., Wattez-Franger A., de Foucault B., Delelis-Dusollier A., Bardat J., 2004 - Plant diversity in a managed temperate forest: understory response to two silvicultural systems. J. Appl. Ecol., **41**: 1065-1079.
13. Drake J.A., 1990 - Communities as assembled structures: do rules govern pattern? Tr. Ecol. Evol., **5**: 159-164.
14. Dupouey J.L., Dambrine E., Laffite J.D., Moares C., 2002 - Irreversible impact of past land use on forest soils and biodiversity. Ecology, **83**: 2978-2984.
15. Egler F.E., 1954 - Vegetation science concepts. I. Initial floristic composition, a factor in old-field vegetation development. Vegetatio, **14**: 412-417.
16. Falinski J.B., 1986 - Vegetation dynamics in temperate lowland primeval forests. Ecological studies in Bialowieza forest. Geobotany, **8**: 1-537.
17. de Foucault B., 1997 - Nouvelles réflexions sur les lois qualitatives du monde végétal et leur valeur heuristique. Acta Bot. Gallica, **144**: 129-144.
18. Géhu J.M., Rivas-Martinez S., 1981 - Notions fondamentales de phytosociologie. In: Dierschke H. (ed.), Berichte der Internationalen Symposien der Internationalen Vereinigung für Vegetationkunde, pp. 5-33. J. Cramer, Vaduz.
19. Gillet F., de Foucault B., Julve P., 1991 - La phytosociologie synusiale intégrée: objets et concepts. Candollea, **46**: 315-340.
20. Gleason H.A., 1926 - The individualistic concept of the plant association. Bull. Torrey Bot. Club, **53**: 7-26.
21. Gleick J., 1987 - Chaos. The Viking Press, New York.
22. Gounot M., 1969 - Méthodes d'étude quantitative de la végétation. Masson, Paris.
23. Hanski I., Gilpin M., 1997 - Metapopulation biology, ecology, genetics, and evolution. Academic Press, London.
24. Hastings A., Hom C.L., Ellner S., Turchin P., Godfray H.C.J., 1993 - Chaos in ecology: is mother nature a strange attractor? Ann. Rev. Ecol. Syst., **24**: 1-33.
25. Jacquemyn H., Butaye J., Dumortier M., Hermy M., Noël L., 2001 - Effects of age and distance on the composition of mixed deciduous forest fragments in an agricultural landscape. J. Veg. Sci., **12**: 635-642.
26. Kassi J.K., Decocq G., in press - Régénération de la forêt dense semi-décidue dans les stades post-culturaux en forêt classée de Sanaimbo (Côte d'Ivoire). Acta Bot. Gallica, **153**.
27. Krebs C.J., 2001 - Ecology, 5th edition. Benjamin Cummings, San Francisco.
28. Lawton J.H., 1987 - Are there assembly rules for successional communities? In: Gray A.J., Crawlay M.J., Edwards P.J. (eds.), Colonization, succession and stability, pp. 225-244. Blackwell, Oxford.
29. Oldeman R.A.A., 1990 - Forests: elements of silvology. Springer-Verlag, Berlin.
30. Prigogine I., 1993 - Le leggi del caos. Laterza, Roma.
31. Reice S.R., 1994 - Nonequilibrium determinants of biological community structure. Am. Sci., **82**: 424-435.

32. Schaffer W.M., Kot M., 1986 - Chaos in ecological systems: the coals that Newcastle forgot. Tr. Ecol. Evol., **1**: 58-63.
33. Smale S., 1980 - The mathematics of time: essays on dynamical systems, economic processes, and related topics. Springer-Verlag, New York.
34. Stewart I., 1997 - Does God play dice? The new mathematics of Chaos. 2nd edition. Penguin Books, London.
35. Suding K.N., Gross K.L., Houseman G.R., 2004 - Alternative states and positive feedbacks in restoration ecology. Tr. Ecol. Evol., **19**: 46-53.
36. Sutherland J.P., 1990 - Perturbations, resistance and alternative views of the existence of multiple stable points in nature. Am. Nat., **136**: 270-275.
37. Tansley A.G., 1939 - The British Islands and their vegetation. Cambridge University Press, Cambridge.
38. Verheyen K., Hermy M., 2001 - The relative importance of dispersal limitation of vascular plants in secondary forest succession in Muizen forest, Belgium. J. Ecol., **89**: 829-840.
39. Warming J.E., 1896 - Lehrbuch des ökologischen Pflanzengeographie. Gebruder Bornträger, Berlin.
40. Whittaker R.H., 1956 - Vegetation of the Great Smoky Mountains. Ecol. Monogr., **26**: 1-80.
41. Wilson J.B., Gitay S.H., Roxburgh S.H., King W.M., Tangney R.S., 1992 - Egler's concept of 'initial floristic composition' in succession – ecologists citing it don't agree what it means. Oikos, **64**: 591-593.

PHYTOSOCIOLOGIE MODERNE ET CONSERVATION RATIONNELLE DE LA NATURE

Jean-Marie GÉHU
Inter-Phyto, 16 Rue de l'église, **80860 Nouvion (FRANCE)**

Abstract: The paper emphasizes the involvement of modern phytosociology in promoting sensible nature conservation, both from the point of view of the assessment of management plans and for conservation choices and scientific outcomes. It draws attention to the risks associated with a demographic management of plant populations that can conflict with the preservation of the natural equilibrium of ecosystems, in support of enhanced biodiversity.

Introduction

«L'un des problèmes les plus cruciaux de notre époque, l'une des plus grandes responsabilités de notre génération vis à vis de l'avenir est sans conteste la préservation de la biodiversité de la Terre» [16].

Dans la Nature le support de la biodiversité positive et structurée est l'écosystème, ou système de fonctionnement d'une biocénose dans un biotope. A la base de ce fonctionnement se place la végétation qui exprime et conditionne chaque écosystème dans son biotope. La Phytosociologie, science des communautés végétales, s'efforce de décrire chacune des portions du tapis végétal que représentent les groupements végétaux. Elle les décrit et les typologise mais encore en étudie le déterminisme, les causes de développement, d'évolution, de disparition.

L'une des applications les plus importantes et les plus nobles de la Phytosociologie est son apport efficace dans les démarches de la conservation rationnelle de la Nature et en particulier dans l'orientation des choix conservatoires et la gestion préservatrice la plus judicieuse des milieux.

I. La conservation rationnelle de la Nature. Quelques réflexions

Le sujet abordé ici concerne la préservation de la biodiversité dans la Nature, c'est à dire des espèces au sein des biocenoses auxquelles elles participent ou qu'elles structurent. Il s'agit donc de la conservation in situ, dans le cadre de végétations soit naturelles, soit dépendantes à des degrés divers des actions humaines passées ou présentes. Sont donc exclus de ce propos tous les faits de conservation de la biodiversité ex situ mettant en jeu Jardins botaniques, zoo, banques de graines, banques de gènes dont la problématique est différente.

1/ Tout d'abord, en matière de préservation de la biodiversité au sein de la Nature, il paraît essentiel de distinguer la biodiversité négative de la biodiversité positive, structurée, c'est à dire intégrée aux biocenoses naturelles et seminaturelles. La biodiversité négative concerne notamment la pullulation des espèces anthropiques plus ou moins nitrophiles telle qu'il est possible de l'observer autour des terrains de décharge, des dépôts d'ordures, ou encore dans les sites les plus dégradés et anthropisés. Elle ne fait pas normalement partie des objectifs de conservation.

Pourtant les définitions et les indices de diversité usuels ne font pas toujours la différence. Rappelons ainsi que:

- La diversité biologique est un concept traduisant la richesse relative en espèces végétales et animales d'un lieu donné.

- La diversité floristique exprime la richesse et l'abondance en espèces végétales d'un lieu déterminé.

- La diversité alpha (ou intrabiotope) correspond à la richesse en espèces d'une communauté liée à un habitat donné.

- La diversité beta (ou interbiotope) est la diversité spécifique d'un ensemble de communautés reliées entre elles ou organisées le long d'un gradient écologique.

- La diversité gamma (ou géographique) additionne la diversité spécifique de l'ensemble des communautés et habitats d'un vaste territoire.

2/ Il convient ensuite de comprendre que la conservation rationnelle de la Nature n'est pas synonyme de multiplication démographique de quelques populations si précieuses soient-elles au détriment d'un ensemble biocenotique. Favoriser excessivement la prolifération d'une espèce, même initialement rare, peut entraîner l'altération voire la disparition de son milieu de vie, des biocenoses dont elle faisait partie ou qui l'abritaient. On sauve une ou quelques espèces précieuses (pour combien de temps?) mais on détruit la nature et son agencement. L'espèce devient plus importante que son milieu de vie. Ce qui est antinomique et paradoxal. Il se peut que dans certains cas particuliers et sur des espaces limités, il n'existe pas d'autres choix mais alors les objectifs et leurs conséquences doivent être clairement annoncés et compris. Les exemples ne manquent pas. Rappelons seulement la destruction des schorres dans le Golfe du Morbihan par la création de bassins isolés du mouvement de la mer destinés à favoriser le développement de populations de limicoles en raréfaction [17]. Ce procédé à connotation anthropocentrique est curieusement commun aux chasseurs de gibier d'eau à la hutte et aux protecteurs des oiseaux. La recherche d'une gestion globale des milieux naturels doit être préférée à une focalisation sur quelques groupes d'espèces animales.

3/ Les réserves naturelles elles mêmes posent bien des problèmes de rationalisation de la conservation. Mal conçues ou mal gérées, elles peuvent aller à l'encontre du but recherché comme l'avait déjà signalé P. Duvigneaud dès les années 60.

La mise en réserve des sites qui ne l'étaient pas peut en effet avoir des conséquences inattendues et pas nécessairement fastes pour la biodiversité structurée car elle peut entraîner la prolifération d'espèces contenues jusqu'alors, au détriment des équilibres spécifiques des biocenoses. La question est apparue hardue sur le littoral, dans les ilots marins mis en réserve naturelle depuis quelques décennies. L'ensemble des études portant sur ces sites [1, 2, 3, 5, 13, 19] montrent que la pression accrue des oiseaux de mer nicheurs, notamment des espèces prolifiques comme les goelands et les cormorans entraîne une destructuration de la végétation typique originelle de ces sites, ou tout au moins une altération grave nitrophosphatophile des biocenoses. Elle est aussi responsable de l'élimination des espèces aviaires les plus précieuses mais moins compétitives.

«Le problème de la conservation de ces milieux se pose de manière cruciale: peut-on accepter la forte dégradation ou la disparition de compartiments de l'écosystème en privilégiant une espèce ou un groupe d'espèces animales dont la seule présence a parfois suscité la création de telles réserves? Le problème doit être envisagé par une approche globale permettant un bilan objectif du patrimoine et une évaluation de l'état de dégradation du milieu. Cette réflexion nécessaire permettrait de définir un état d'équilibre acceptable entre la pression exercée par les populations d'oiseaux marins et le maintien des phytocenoses caractéristiques des ilots» [3].

Le travail récent effectué par Bioret et Gourmelon [4] de «cartographie dynamique de la végétation terrestre des ilots marins en réserve naturelle» effectué par les auteurs sur les côtes atlantiques et méditerranéennes de France contribue de façon significative à cette réflexion.

Le même problème de gestion des populations aviaires excédentaires dans les milieux humides précieux se pose dans les dépressions interdunaires (ou pannes) ou encore dans les landes tourbeuses oligotrophes.

Dans tous ces cas un suivi phytosociologique attentif, précis régulièrement répétitif doit permettre d'alerter et d'intervenir judicieusement en temps voulu [11].

Les déséquilibres démographiques ne sont évidemment pas les seuls problèmes des réserves naturelles dont bien peu peuvent, tout au moins en Europe, se prévaloir d'une absence de gestion sans préjudice et sans perte de biodiversité ne serait-ce qu'en raison des phénomènes de dynamisme naturel de la végétation si forts en milieu seminaturel mais également perceptibles bien que plus lents dans des milieux considérés comme naturels, tels les liserés littoraux ou les limites sommitales extra sylvatiques. Les réserves intégrales et les réserves gérées doivent faire l'objet de distinction claire dans leurs objectifs et conséquences.

II. La Phytosociologie moderne. Apport à la conservation

L'une des propriétés les plus remarquables de la Phytosociologie moderne est d'analyser très finement, le plus complètement possible les diverses composantes du tapis végétal que constituent les communautés végétales [10]. Ces communautés végétales font l'objet d'analyse floristique très détaillée dont la valeur est objectivée par le caractère de répétitivité exprimé en tableaux phtosociologiques. Chacune de ces communautés définies dans un cadre spatial homogène est conditionnée par divers facteurs d'ordre écologique, mésologique et biotique, d'ordre dynamique, géographique, historique. Ce déterminisme des associations végétales est étudié très précisément par la phytosociologie dans ses aspects synécologiques, syndynamiques, synchorologiques... de telle sorte qu'il devient possible de définir les conditions de vie, de disparition, de survie, de maintien, d'extension, d'évolution des communautés végétales constitutives d'un paysage [14, 15, 16, 18].

Cette connaissance est évidemment le meilleur support possible de définition des plans de gestion conservatoire au sein des réserves naturelles comme à l'extérieur [6]. La conservation rationnelle des écosystèmes précieux, support de la biodiversité positive, passe par cette application des données qu'apporte l'étude phytosociologique des sites.

Les lois de mieux en mieux connues de la Phytosociologie [7], de la synécologie, de la syndynamique autorisent de plus en plus les prévisions dans ces domaines et permettent l'élaboration de modèles prédictifs de plus en plus fiables, selon tel ou tel scénario.

Les gestionnaires conservateurs de la Nature en ont de plus en plus conscience, qui n'acceptent plus de se contenter d'une simple approche physionomique et d'une gestion purement empirique des milieux précieux.

L'attention doit aussi être attirée sur l'utilité particulière de la Phytosociologie dans l'objectivation du choix des sites à préserver ou à ériger en réserve naturelle. La statistique floristique des tableaux détaillés d'association et spécialement le nombre d'espèces présentes (résumé dans le chiffre spécifique moyen du tableau) permet d'orienter le choix vers des individus de groupement les plus représentatifs et les plus riches. Il en va de même pour la symphytosociologie dont les tableaux permettent d'objectiver le choix des meilleurs complexes de biotopes (écocomplexes, synassociations, géosigmassociations…) [12, 20].

Parmi de nombreux autres intérêts de la Phytosociologie dans la conservation de la nature, l'analyse floristique attentive des communautés permet d'éviter les pièges inhérents à une simple observation physionomique. En effet l'érosion floristique des milieux se manifeste généralement bien avant que change leur aspect général. La «désaturation» floristique d'une communauté précède souvent de longtemps sa «destructuration». Bien avant le début de l'effondrement structural visible d'une communauté, l'hémorragie floristique qui la précède peut être irréversible. D'où l'importance primordiale des suivis scientifiques réguliers des réserves naturelles ou de tout site précieux selon la technique de la Phytosociologie moderne Braun-Blanqueto-Tuxenienne.

Conclusion

La conservation rationnelle de la nature passe nécessairement par la préservation et la gestion conservatoire des écosystèmes, supports de la biodiversité, tant végétale qu'animale. De ce point de vue, la conservation de la nature dans la globalité de sa diversité, n'est pas synonyme de gestion démographique des populations animales ou végétales qui, dans bien des cas, va à l'encontre de la préservation des équilibres naturels (exemple en démographie aviaire sur les littoraux).

La végétation exprime et conditionne les fonctionnements écosystémiques dans chaque biotope particulier.

La Phytosociologie moderne, ou science des groupements végétaux, est un outil particulièrement précieux dans la conservation raisonnée de la nature qui nécessite la connaissance précise des objets à préserver, tout comme celle de leur déterminisme écologique et dynamique, base indispensable des interventions gestionnaires conservatrices, hormis les rares cas de milieux naturels dans nos régions.

La Phytosociologie actuelle (dite Braun-Blanqueto-Tuxenienne) par son affinement méthodologique permet la description précise de la nature statistiquement répétitive des communautés végétales. L'originalité territoriale (synendémisme) des combinaisons floristiques décrites, permet une plus juste appréciation de la valeur du génofonds spécifique d'une région (attirant même l'attention sur la singularité de populations non définies nomenclaturalement).

A condition d'être effectuée dans une rigoureuse homogénéité floristique, structurale, écologique, dynamique, la délimitation de chaque communauté est porteuse d'informations précises sur son déterminisme. La connaissance des causalités d'un groupement végétal permet de définir au mieux les modalités de sa gestion conservatoire, notamment dans les milieux seminaturels, les plus fréquents en Europe.

La connaissance de la répartition des communautés (synchorologie) renseigne sur leur préciosité et leur rareté.

La dimension paysagère de la Phytosociologie actuelle, qu'elle soit sériale ou caténale, permet d'étendre à l'ensemble d'un paysage régional les avantages de la connaissance et de la gestion conservatoire de la Phytosociologie classique.

BIBLIOGRAPHIE

1. Bioret F., 1992 - Influence des oiseaux marins sur la végétation des ilots bretons. C.R. du G.I.S.O.M. Brest: 26-31.
2. Bioret F., Bouzillé J-B., Godeau M., 1988 - Exemples de gradients de transformation de la végétation de quelques ilots de deux archipels armoricains. Influence de zoo-populations. Colloque Phytosociologique, **15**: 509-531.
3. Bioret F., Géhu J-M., 1996 - Banalisation floristique et phytocoenotique d'un ilot marin soumis à la surfréquentation par les oiseaux marins nicheurs: l'île des Landes. Colloque Phytosociologique, **24**: 89-109.
4. Bioret F., Gourmelon F., 2004 - Cartographie dynamique de la végétation terrestre des ilots marins en réserve naturelle. Braun-Blanquetia, **37**: 1-31.
5. Bioret F., Leray G., 1995 - Exemple d'altération des phytocoenoses de systèmes microinsulaires soumis à l'influence des colonies d'oiseaux marins: Les ilots de la baie de la Baule. Bull. Soc. Bot. Centre Ouest, **26**: 111-126.
6. Delpech R., Géhu J.-M., 1987 - Intérêt de la Phytosociologie actuelle pour la typologie, l'évaluation et la gestion des écosystèmes. 4ème Colloque A.F.I.E, La gestion des écosystèmes, Bordeaux: 39-52.
7. Foucault B. de, 1986 - Petit manuel d'initiation à la Phytosociologie sigmatiste. Soc. Linnéenne Nord France, Mem. Amiens, 1, 51 pp.

8. Géhu J.-M., 1979 - La phytosociologie comme outil de définition des secteurs côtiers à préserver. Les côtes atlantiques d'Europe. C.N.E.X.O. Colloque 9, Brest: 155-163.
9. Géhu J.-M., 1980 - La Phytosociologie appliquée à l'aménagement et à la conservation. Ecologie appliquée: indicateur biologique et technique d'études. A.F.I.E., Grenoble: 221-231.
10. Géhu J.-M., 1982 - La Phytosociologie d'aujourd'hui. Méthodes et orientations. Notizario Soc. Ital. Fitosociologia, **16**: 1-16.
11. Géhu J.-M., 1984 - De la nécessité de la gestion et du suivi scientifique des écosystèmes seminaturels. Inventaire et gestion des milieux naturels. Institut Européen d'Écologie, Metz.
12. Géhu J.-M., 1987 - Des complexes de groupements végétaux à la Phytosociologie paysagère contemporaine. Informatore Botanico Italiano, **18**: 53-83.
13. Géhu J.-M., 1991 - Vicissitudes du tapis végétal d'une île bretonne (Cézembre) sous l'effet du dérèglement des pressions humaines et animales. Bull. Soc. Bot. Centre Ouest, **22**: 53-76.
14. Géhu J.-M., 1998 - Epistémologie de la typologie phytosociologique de la végétation. Itinera Geobotanica, **11**: 65-83.
15. Géhu J.-M., 2000 - Principes et critères synsystématiques de structuration des données de la Phytosociologie. Colloque Phytosociologique, **27**: 693-708.
16. Géhu J.-M., 2001 - L'intérêt de la typologie phytosociologique pour l'évaluation et la gestion des milieux naturels. Valoracion y gestion de espacios naturales, Jaen: 15-27.
17. Géhu J.-M., Bioret F., 1992 - Etude synécologique et phytocoenotique des communautés à salicornes des vases salées du littoral breton. CR. Session «Halophytes bretons». Bull. Soc. Bot. Centre Ouest, **23**: 347-419.
18. Géhu J.-M., Rivas-Martinez S., 1981 - Notions fondamentales de Phytosociologie. Berichte Intern. Symp. Intern. Verein. Vegetationskunde «Syntaxonomie» Rinteln: 5-33.
19. Paradis G., Lorenzoni C., 1996 - Impact des oiseaux marins nicheurs sur la dynamique de la végétation de quelques ilots satellites de la Corse. Colloque Phytosociologique, **24**: 393-431.
20. Theurillat J.-P., 1992 - Etude et cartographie du paysage végétal (symphytocoenologie) dans la région d'Aletsch. Matériaux pour le levé géobotanique de la Suisse, Lausanne, **68**: 1-324.

ANATOMY OF THE ENDANGERED PLANT *GLAUCIUM FLAVUM* Cr., OCCURRING ON THE ROMANIAN BLACK SEA LITTORAL

Rodica BERCU*, Marius FĂGĂRAŞ, Loreley Dana JIANU
„Ovidius" University, Faculty of Natural and Agricultural Science, Department of Biology, Mamaia Str. 124, **900527 Constanţa (ROMANIA)**;
e-mail: rodicabercu@yahoo.com

Abstract: The paper deals with the anatomy of *Glaucium flavum* Cr., an endangered species growing in Romania on the shores of the Black Sea. The anatomical features of the root, stem, leaf and fruit pericarp have been described and illustrated in detail using a light microscope. The root possesses a typical secondary structure due to its perennial character. The stem exhibits a primary structure (eustele organization with open collateral vascular bundles). The midrib of both types of leaves is closed and collateral. The strengthening of the blades is given by collenchyma cells. Of particular note are the stem, petiole of the basal leaf and pericarp vascular bundles which are collateral, as in the stem vascular bundles; also the presence of non-articulated laticifers in the stem, pinnatipartite blade and pericarp. The xeromorphic features of this plant are due to its occurrence in dry sandy habitats.

Introduction

Glaucium flavum Cr. (Fam. *Papaveraceae*), known as yellow horned-poppy, is an unusual-looking annual, biennual or perennial plant with glaucous leaves that are infolded and ruffled along their margins. In spring and summer its branched stems bear yellow poppy-like flowers that give rise to characteristic, elongate siliquiform capsules. The flowers are yellow with slight tints of orange [13]. The name 'horned-poppy' derives from its very long, swollen and pointed capsules, sometimes bearing horn-like protrusions. The plant contains glaucine, with some medicinal properties, namely it relieves coughs and lowers blood pressure. In ancient times the oil extracted from the seeds was used in lanterns. The plant grows well in areas of low or moderate rainfall, and blooms throughout spring and summer.

Previous investigations have studied the floral anatomy [5] and stem structure of this plant [3] and more recently cyto-morphological aspects (ultrastructure) of the laticifers of the stem and seeds [6, 11, 12]. The detailed anatomical characteristics of *Glaucium flavum* as a whole have not yet been documented.

Material and Methods

Samples were collected from individual plants occurring on the Black Sea littoral at Constanza. Small pieces of root, stem, leaves and capsules were fixed in FAA (Formalin aceto-alcohol), sectioned, subsequently clarified with chloral hydrate and stained with carmine alum and iodine-green. Observations were performed with a BIOROM-T bright field microscope, equipped with TOPICA-6001A video camera connected to a computer.

Results and Discussions

Cross sections of the root exhibit a secondary structure with periderm, cambium and stele (Fig. 1a-d), as most dicotyledonous roots [1, 2, 4, 14, 15]. Phloem fibres are absent. The centrally located secondary xylem elements found in the root of *Glaucium flavum* consist of radially arranged xylem vessels, located in the ground parenchyma. The continuity of xylem is interrupted by primary pith rays (Fig. 1d).

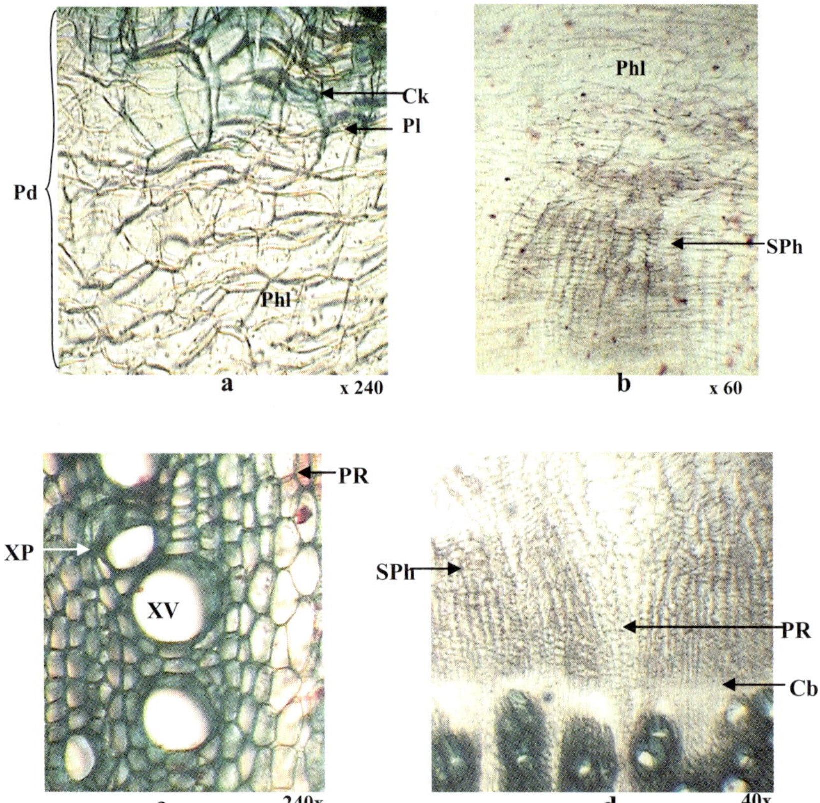

Fig. 1. Cross section of the root showing structure of the periderm (a), the secondary phloem (b), secondary xylem (c) and a portion of the stele (d): Cb, cambium; Ck, cork; Pd, periderm; Pl, phellogen; Phl, phelloderm; PR, pith ray; SPh, secondary phloem; SX, secondary xylem; XP, xylem parenchyma; XV, xylem vessel.

Transverse sections of the stem exhibit epidermis, cortex and stele. The epidermis on its outer surface is covered by cuticle and wax layers interrupted by stomata. They are located below the surface of the epidermis and possess large substomatic cavities (Fig. 3). Epidermal hairs are absent. Below the epidermis is a cortex layer of compactly arranged parenchymatous cells, followed by a layer of sclerenchymatous cells. As previously reported by Esau [7] and Batanouny [3], the latter consists of the same non-articulated laticifers (Figs. 2a, 3) with yellow latex, containing alkaloids such as found in *Papaver somniferum, Nelumbo nucifera* and *Nerium oleander* [6, 8, 10, 12].

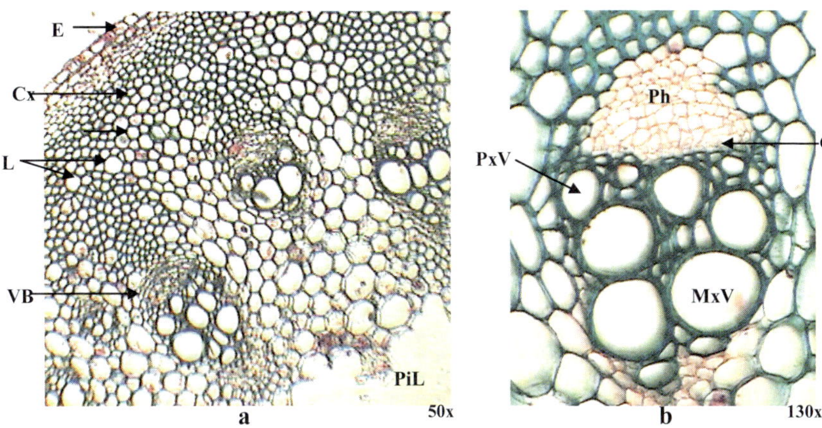

Fig. 2: Cross section of the stem. Portion of the stem (a). Detail of a vascular bundle (b): Cb, cambium; Cx, cortex; E, epidermis; L, laticifers; MxV, metaxylem vessel; Ph, phloem; PiL, pith lacuna; PxV, protoxylem vessel; VB, vascular bundle.

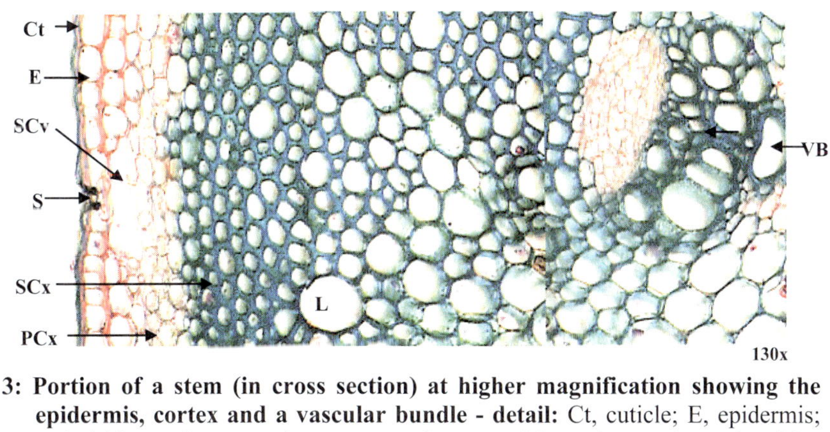

Fig. 3: Portion of a stem (in cross section) at higher magnification showing the epidermis, cortex and a vascular bundle - detail: Ct, cuticle; E, epidermis; L, laticifers; PCx, parenchyma cortex; S, stoma; SCv, substomatic cavity; SCx, sclerenchyma cortex; VB, vascular bundle.

The vascular system of *Glaucium flavum*, mentioned for members of *Papaveraceae* family by Esau [7] and Fahn [9], is represented by a eustele, each vascular bundle being protected by a sclerenchyma sheath [3]. The cross section of the vascular bundle discloses that sclerenchyma cells appear mostly towards the upper side of the vascular bundles (Figs. 2b, 3). The pith is made up of a large number of rounded cells, which are loosely arranged, enclosing towards the centre a large pith lacuna (Fig. 2a).

In cross section the basal leaf exhibits an upper epidermis, the mesophyll and the lower epidermis with irregularly arranged cells. Both epidermides are covered by cuticle and wax layers. The continuity of the lower epidermal cells is interrupted by the presence of stomata and many-celled protective hairs. Some of the hairs are short and pointed, others are long with rounded tips (Fig. 4c, d). The mesophyll is heterogeneous (Fig. 4a). The midrib vascular system is represented by two closed collateral bundles unprotected by bundle sheaths (Fig. 4b).

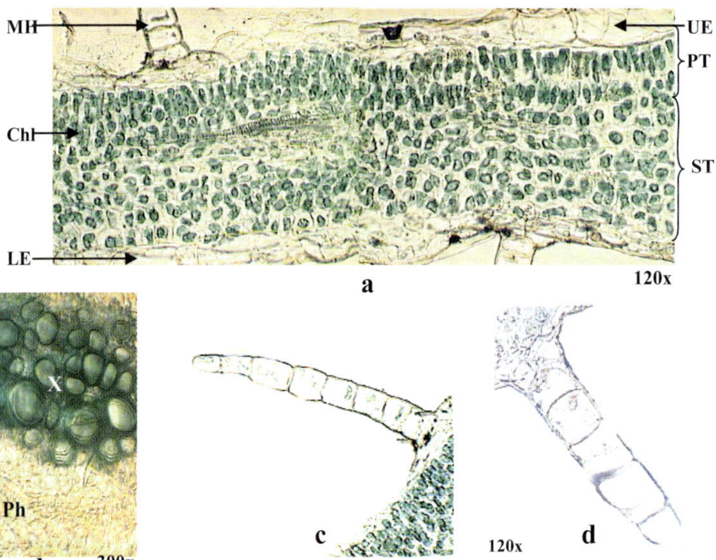

Fig. 4: Cross section of the basal leaf. Portion with mesophyll (a) and a vascular bundle of the midrib (b). Hairs of the upper (c) and lower epidermis (d): Chl, chloroplasts; LE- lower epidermis, MH- many-celled hair, Ph- phloem, PT- palisade tissue, ST- spongy tissue, UE- upper epidermis, X- xylem.

Cross section of the petiole of the basal leaf reveals a plane and a convex surface narrowing towards its edges. Externally, the epidermis is composed of a single layer of compactly arranged thick-walled waxy cells. Stomata occur as well. Below the epidermis 3-4 layers of chlorenchyma cells are present, followed by the usual parenchyma tissue, containing the stele (Fig. 5a). Collateral bundles distributed in a circle form the stele and consist of few phloem elements and normally developed xylem (Fig. 5b). Collenchyma cells are also present. In some places many-celled hairs (protective hairs) occur. They are shorter (3- to 4-celled) and thicker than those of the blade (Fig. 5c).

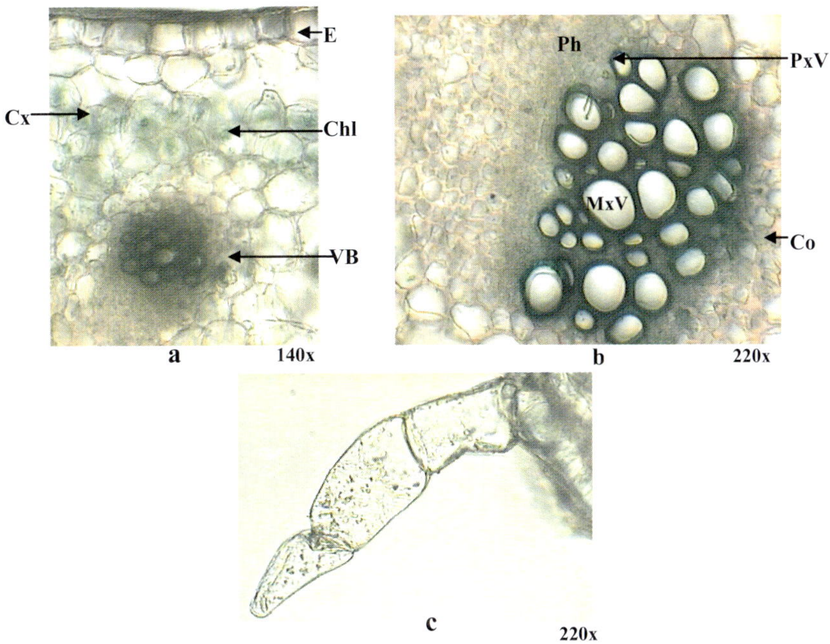

Fig. 5: Cross section through petiole of the basal leaf. Portion with epidermis, cortex and a small vascular bundle (a). Detail of a vascular bundle (b). Hair of the petiole (c): Chl, chloroplasts; Co, collenchyma; Cx, cortex; E, epidermis; Ph, phloem; Mx, metaxylem vessel; Px, protoxylem vessel; VB, vascular bundle.

Transverse sections of the sessile (amplexicaul) pinnatipartite leaf blade show the single layered upper and lower epidermis. The upper epidermis forms a crest just above the midrib, whereas the lower one is just slightly prominent. The mesophyll is differentiated into palisade and spongy tissues (Fig. 6a), containing three closed collateral vascular bundles (in the centre) forming the midrib (first order vein) protected by collenchyma and surrounded by a compact mass of ground tissue (Fig. 6b). The midrib second order veins consist equally of three closed collateral vascular bundles, but they are less developed than those of the main one (Fig. 7 - in medallion). Notable is the presence of laticifers situated around the vascular bundles of the midrib mostly of the second order vein (Fig. 7). Many-celled protective hairs are present on the lower and upper epidermis. The secondary veins are poorly developed, consisting of few xylem and phloem elements protected by a parenchyma sheath (Fig. 6a).

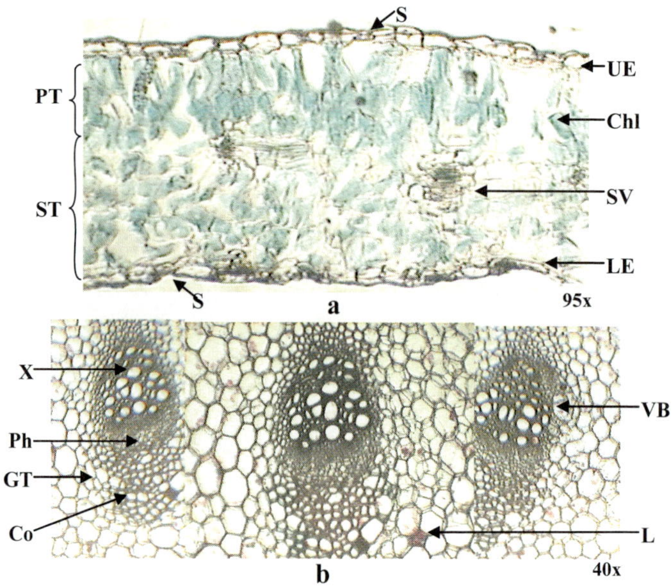

Fig. 6: Cross section of the pinnatipartite leaf blade of the stem. Portion with mesophyll (a). The vascular bundles of the midrib - I order vein (b): Chl, chloroplasts; Co, collenchyma; GT, ground tissue; L, laticifers; LE, lower epidermis; PT, palisade tissue; S, stoma; ST, spongy tissue; SV, secondary vein; UE, upper epidermis; Ph, phloem; VB, vascular bundle; X, xylem.

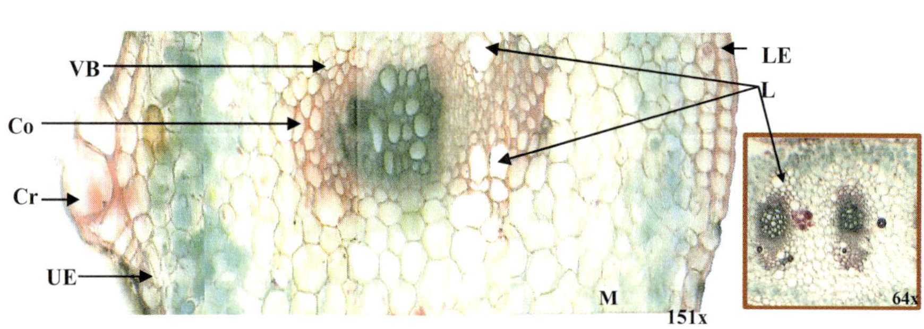

Fig. 7: Cross section of the pinnatipartite blade. Portion with the central vascular bundles of the midrib (II order vain) and the other two vascular bundles (in medallion): Co, collenchyma; Cr, crest; L, laticifers; LE, lower epidermis; M, mesophyll; PT, palisade tissue; UE, upper epidermis; VB, vascular bundle.

The pericarp of the elongate capsules, in transverse section is circular and somewhat sinuous due to sclerenchymatous crests and to the two large ovarian cavities. The outermost layer of cells is the external epidermis, consisting of small regularly arranged cells. Its continuity is interrupted by the presence of crests and stomata. Below the epidermis is the cortex, differentiated into two regions. The external region consists of 3-4 layers of chlorophyllaceous cells whereas the inner zone has 10-12 layers of sclerenchymatous cells, with a mechanical role, followed by a number of radially arranged vascular bundles like the stele of the stem (Fig. 8a, b). The vascular bundles are embedded in a basic parenchyma consisting of starch grains (Fig. 8b). In the median portion, corresponding to the carpel suture line, the sclerenchyma tissue and the vascular bundles are cup-shaped (Fig. 8a). The ovarian cavities are lined by a layer of regularly arranged small cells forming the inner epidermis (Fig. 8a, b). The abundance of laticifers in the sclerenchyma mass of the pericarp (Fig. 8a, b) is notable. The laticifer structure in seedlings of *Glaucium flavum* has previously been reported by Nessler [11].

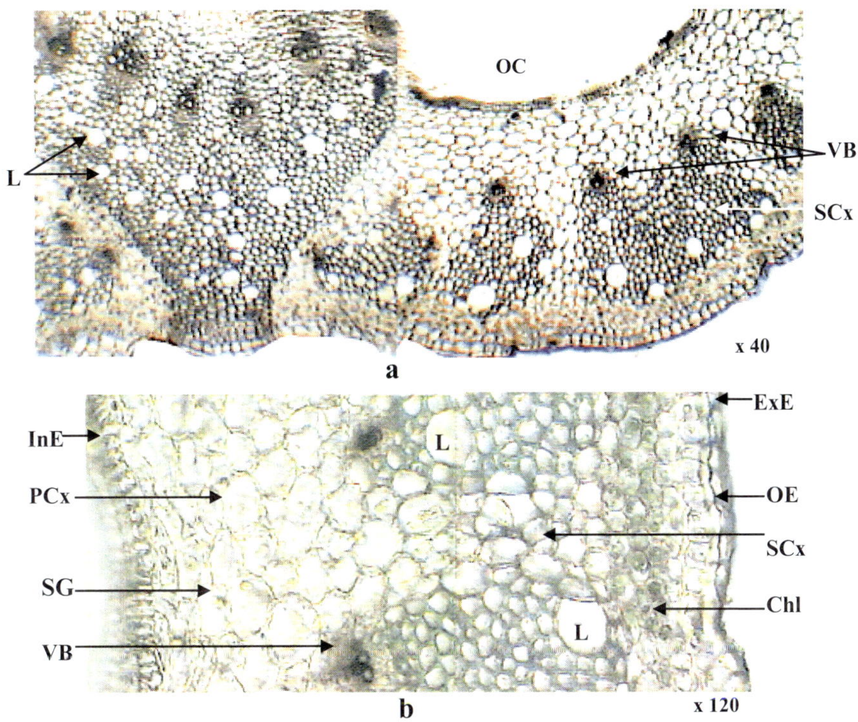

Fig. 8: Cross section of the capsule pericarp. Portion of the pericarp at lower (a) and higher (b) magnification: Chl, chloroplasts; ExE, external epidermis; InE, inner epidermis; L, laticifers; OC, ovarian cavity; OE, outer epidermis; PCx, parenchyma cortex; SB, seed box; SCx, sclerenchyma cortex; VB- vascular bundle.

Conclusions

The present results have revealed that the root possesses a typical secondary structure according to the perennial character of the plant. The stems exhibit primary structures (eustele type). There are several differences between the two types of leaves characteristic for the yellow horned-poppy plant. The many-celled protective hairs occur on stem and both basal and stem leaves.

The strengthening tissue of the leaves is represented by collenchyma, mostly located on both sides of the midrib vascular bundles. The petiole of the basal leaves and the pericarp of capsule are collateral, such as in the stem vascular bundles. Of particular note is the presence of non-articulated laticifers filled with latex, containing alkaloids (glaucine) in the stem, pinnatipartite blade and pericarp.

The epidermal cells of the above-ground vegetative organs are covered by a thick cuticle mixed with wax. In between the epidermal cells sunken stomata are present. This is an anatomical adaption to sandy, xerophilous habitats.

REFERENCES

1. Andrei M., 1978 - Anatomia plantelor. Ed. Did. şi Ped., Bucureşti.
2. Andrei M., Predan G.M.I., 2003 - Practicum de morfologia şi anatomia plantelor. Ed. Ştiinţelor Agricole, Bucureşti.
3. Batanouny K.H., 1992 - Plant Anatomy. University Press, Cairo.
4. Bavaru A., Bercu R., 2002 - Morfologia şi anatomia plantelor. Ed. Ex Ponto, Constanţa.
5. Brancroft A., Dawson H., 1934 - Studies in floral anatomy II. The floral anatomy of *Glaucium flavum* with reference to that of other members of the Papaveraceae. Proc. Linn. Soc. **1**: 175-224.
6. Dickinson P.B., Fairbairn J., 1975 - The ultrastructure of the Alkaloidal Vesicles of *Papaver somniferum* latex. Ann. Bot., **39**: 707-712.
7. Esau K., 1977 - Anatomy of Seed Plants, 2nd edition. Ed. John Wiley & Sons Inc, New York.
8. Esau K., Kosakai H., 1975 - Laticifers in *Nelumbo nucifera* Gaetn. Distribution and structure. Ann. Bot., **39**: 713-719.
9. Fahn A., 1990 - Plant Anatomy, 4th edition. Ed. Pergamon Press, New York.
10. Mahlberg P.G., 1959 - Karyokinesis in the non-articulated laticifer of *Nerium oleander* L. Phytomorph., **9**: 110-118.
11. Nessler C.L., 1992 - Ultrastructure of laticifers in seedlings of *Glaucium flavum* (*Papaveraceae*). Can. J. Bot., **60**: 561-567.
12. Reynold E.S., 1963 - The use of lead citrate at high pH as an electron-opaque stain in electronmicroscopy. J. Cell Biol., **17** (1): 208.
13. Săvulescu T. (ed.), 1955 - Flora R.P.R., 3rd volume. Ed. Acad R.P.R., Bucureşti.
14. Şerbănescu-Jitariu G., Toma C., 1980 - Morfologia şi anatomia plantelor. Ed. Did. şi Ped., Bucureşti.
15. Tarnavschi T.I., Şerbănescu-Jitariu G., Rădulescu-Mitroiu N., Rădulescu D., 1974 - Practicum de morfologie şi anatomie vegetală. Tipografia Universităţii, Bucureşti.

STEPPE–FOREST SUCCESSION AS A CONSERVATION PROBLEM

Yakiv DIDUKH
M.G. Kholodny Institute of Botany, 2 Tereshchenkivska Str.,
01601 Kyiv (UKRAINE); e-mail: didukh@botany.kiev.ua

Abstract: Other conditions being suitable, soil moisture supply is the most important factor determining whether forest or steppe develops on a particular area of land. Supply is limited by dryness, climate aridity, seasonal distribution of precipitation, chemical composition of soil and bed-rock, and economic activity. The speed and direction of changes in steppe vegetation in Mikhailivska Tzilina Nature Reserve were determined through long-term study and analysis of vegetation. Characteristics of forest and steppe ecosystems are given. The main differences in their organization, function, cycling of biogenic elements, and transformation of energy are highlighted. A difference in energy potential drives natural succession from steppe to forest. To conserve steppe, the policy of absolute non-intervention on reserves should be replaced by active management to halt this natural succession.

Introduction

Debate about the ecological balance between forest and steppe has played a key role in forming the Eastern European geobotanical school and has determined much of its character. Each researcher has had their own explanation as to why steppe lacked forest cover, based on personal observations, and tended to disregard or discount patterns found in the observations of other researchers. Summarizing a long debate, Lavrenko [8] concluded that each explanation had some value and that all should be considered together as a complex. To do that, however, forest and steppe ecosystems and their function needed more detailed comparison, particularly on nature reserves.

That work was carried out in the second half of the 20th century, on steppe reserves in Ukraine, with detailed vegetation maps made every 10 years on all reserves (Bilyk, Tkachenko), so that four vegetation "snapshots" exist for each. They provide monitoring information and indicate that vegetation has changed significantly under reserve conditions. In particular, the steppe component of the forest-steppe zone has diminished and the forest component has increased. The impacts of mowing, grazing, fire, and reserve policies of absolute non-intervention have been evaluated (Bilyk, Osychnyuk, Tkachenko, Genov, Lysenko). The policy of absolute non-intervention, maintained continuously on certain reserves since its introduction 75 years ago, has led to significant changes in vegetation cover, particularly in the forest-steppe zone, and mowing has proved ineffective in slowing down the succession from steppe to forest. Steppe ecosystem function has been studied by physiologists (Grodzinsky) and ecologists (Pogrebnyak). Detailed studies of steppe (Osychnyuk, Bystrizkaya, Gordezky, Chuprina) and forest (Belgard, Travlejev) ecosystem function and productivity began in the 1970s within the UNESCO Man and Biosphere framework. This research has provided much interesting information, including data on productivity in different types of community.

Materials and Methods

In the 1990s, the ecological technique of "synphyto-indication" was developed at the M.G. Kholodny Institute of Botany [4, 6] to establish the limiting impact of certain ecological factors, and the character and behaviour of succession as a result of reserve conditions, and to find possible means of slowing down that succession. The technique incorporates scales of ecological amplitudes of plant species with respect to key factors: moisture (Hd), acidity (Rc), soil richness (Tr), and mineral nitrogen content (Nt). Using the results, ordination matrices were constructed, i.e. indirect ordination methods were used (F2). The working up of geobotanical descriptions, prepared from 1928 onwards with the above methods, enabled researchers to establish type and behaviour of successions, to reveal limiting boundaries of certain ecological factors, and to evaluate correlation between these factors and vegetation change, i.e. to evaluate succession and interchange of ecosystems as a whole. This served as a background for the creation of corresponding models. The methods enabled quantitative evaluation of the degree of steppe ecosystem instability and the mode of their transition into forest.

Lavrenko and Zoz's geobotanical descriptions of "Mikhailivska tzilina" (Sumy region), created in 1927, were the initial material for the study. Their descriptions reflected over-grazed meadow steppes [11]. Later the area became a reserve, and a policy of absolute non-intervention was introduced, with mowing permitted in only a few places. Further geobotanical descriptions were prepared in 1971 by Bilyk and Tkachenko [2], and in 1991 by Tkachenko, Genov and Lysenko [18].

Results and Discussion

The steppe has changed significantly over the 64 years for which records exist (Fig. 1) [19], with clear trends in magnitude and acceleration of "ordination fields" for all selected edaphic parameters. In 1927, the soil moisture "ordination field" had a characteristically narrow index range (8.7–9.2 points), but the following decades were marked by an increase of moisture indexes and an acceleration of succession. Inter-correlated indexes of acidity, soil richness and nitrogen content showed similar changes and were similar to those of forests. The reason for this might be the accumulation of a 15-cm litter layer which could not break down completely under conditions of absolute non-intervention. Steppe communities of the *Festuco-Brometea* class dominated by tussock grasses (*Stipa capillata*, *Festuca valesiaca*, *Carex humilis*) almost disappeared and were replaced at first by the meadow class *Molinio-Arrhenatheretea*, dominated by grasses with short (*Bromopsis inermis*, *Arrhenatherum elatior*, *Poa angustifolia*) or long rootstocks (*Elytrigia repens*, *Calamagrostis epigeios*). Later, nitrogen-loving plants (*Urtica dioica*, *Cirsium arvense*, *Galium aparine*, *Glechoma hederacea*), shrubs (*Chamaecytisus ruthenica*, *Prunus spinosa*), and trees (*Ulmus suberosa*, *U. laevis*, *Populus tremula*) appeared. The area covered by tree and shrub communities doubled every ten years [15, 16, 17]. Replacement was somewhat slower on mown plots, but even mowing could not preserve classic steppe.

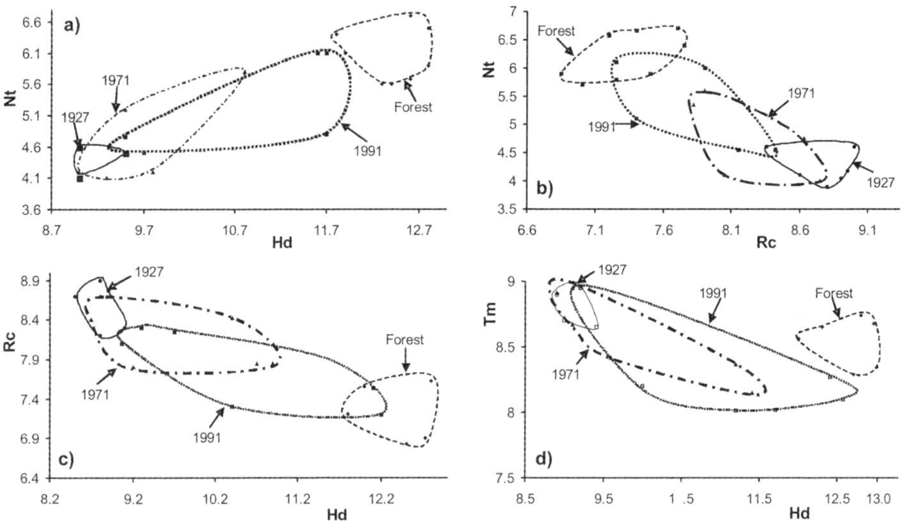

Fig. 1: Changes of ecological characteristics of phytocoenoses of "Mikhailivska tsilina" meadow steppe under the influence of reserve conditions from 1927 to 1991 in relation to forest communities (Nt - mineral nitrogen; Hd – soil moisture; Rc – soil acidity; Tm – thermal regime) [19].

The general tendency along the gradient as one advances from the Atlantic into Eurasia is that xerothermic vegetation becomes predominant, occupies an increasingly larger area, and forms steppe and desert zones. Also xerothermic vegetation spreads to gently sloping watersheds from steep southern slopes, and in the arid Asian regions xerothermic vegetation spreads on to northern slopes descending to river bottom-lands. Walter [20] and Odum [12] consider drought, climate aridity, seasonal distribution of precipitation, and soil water supply to be factors limiting forest–steppe ratio. Probably, soil water supply limits microbial breakdown of litter and circulation of biogenic elements. Presence of calcium ($CaCO_3$) and magnesium ($MgCO_3$) carbonates (limestone, chalk, gypsum, or loess) on which particular soils are formed of high pH value (pH>6.7), that is chernozems or rendzinas, is the key factor influencing water supply of soils and, consequently, the formation of steppe vegetation.

In Western Europe, close to the Atlantic, abundant precipitation falls and evaporates in conditions of increased direct radiation on the steep south slopes to a degree higher than on the level or northern slopes, influencing the formation of xerothermic vegetation. Thus, a hydrothermic regime limiting processes of breakdown and circulation of biogenic elements is a major factor influencing distribution of steppe and forest communities.

When considering forest communities, a definite type of forest should be kept in mind, namely broad-leaved forest of typical tree species formation. Defining characteristics of these forests is necessary, as there are more xerophytic tree species which could in theory grow under such climatic and soil conditions (*Quercus pubescens*, species of *Pinus*, *Juniperus*, etc.). Although certain tree species occupy even more extreme ecotopes on the scale of xerophytism and aridity than steppes, they do not grow in steppes. Under arid conditions, trees are known to form open woodland rather than forest. Open woodlands are absent in the forest-steppe zone but occur in the Mediterranean region and on high mountains. The basic problem of correlation between forest and steppe vegetation is that forest and steppe communities are of incompatible, absolutely different structure and type of functioning. This problem is overcome in open woodlands.

Analysis of phytomass on a zonal profile [9] revealed an abrupt decrease of indexes on the boundaries between forest and steppe, namely from 260 t/ha to 1.5 t/ha, i.e. a 170 times decrease. Furthermore, the annual gain (production) decreases from 5.6 to 1.5 t/ha, i.e. a 3.7 time decrease. Such a discontinuity can be explained if open woodland is not a possibility: the absence of an intermediate which would have enabled gradual transition.

Open woodland is an ecosystem characterized by low stand density (0.2–0.5) growing as separate trees or clusters leading to a well-developed herb layer (up to 80%) including steppe grasses. Crimean open woodland is a good example [5]. Such communities have poor resistance to human interference. Their stands do not renew under conditions of intensive grazing and felling, but a grass canopy forms well. That is why a millennium of human interference has led to reduction of open woodland and replacement by steppe.

It is also of interest to consider characteristics of organization of forest and steppe ecosystems. Development of forest ecosystems is based upon the formation of stable resistant phytocoenoses with high energy status, and in which the leading role of one or several principal trees can be traced. From the very first, succession in forest has a certain trend and its outcome is easily predicted. The impact of the tree layer is so strong that it significantly transforms indexes of the external factors and forms a phytogenic environment influencing species content, phenology of the development of each species, population formation, moisture, warmth, and light regimes, development of consumers and microflora, character of soil processes, circulation of biogenic elements, and energy flows. Thus, 2.5% of rainfall is lost through surface drainage, 5–15% evaporates from the canopy, 72% is used and transpired by plants, and only 15% gets into the soil. Also, forest phytocoenosis is in certain isolation with respect to the surrounding vegetation cover. For instance, the annuals that compose 12% of regional floras are very scarce in forests. The Therophytes/Phanerophytes ratio is within the range 1:2.5–7.6, whereas the Phanerophytes/Hemicryptophytes ratio is 0.22–0.49. The spectrum of main families of the forest flora is qualitatively different from region to region, and includes *Orchidaceae, Asteraceae, Rosaceae, Poaceae, Lamiaceae, Ranunculaceae, Scrophulariaceae, Apiaceae, Caryophyllaceae* and *Fabaceae*.

Functioning of forest ecosystems differs qualitatively from that of grassland ecosystems. In forests, most plant material is accumulated in lignified above-ground (54–83%) organs. Rather less is accumulated in the underground organs (15–35%), i.e. is stored as long vegetating components of the autotrophic. Only 2–15% is accumulated in leaves, which change annually in deciduous forests. Thus the annual gain (production) constitutes 3–6% (7–12 t/ha), out of which 25–27% is a green mass and 40–57% is timber. Dead components constitute 50–70 t/ha (i.e. 1:5, 2–8%) of all plant material, whereas dead material varies within 2–5:1 (20–50 t/ha), as it decomposes slowly [1]. Grass-undershrub and moss cover constitutes from 0.3 to 10% of all biomass [14]. Grass cover species adapt to the phytogenic environment. Shading conditioned by upper layers, i.e. a phytogenic environment, is the main limiting factor defining phenology, type of reproduction, and development of populations. Microclimate is determined to a considerable degree by the phytogenic environment. Consequently dead material decomposes slowly, over 2–2.5 years (e.g. *Carpinus betulus* – 1.5 year, *Acer* – 2 years, *Quercus robur* – 2.5 year) [20]. Humification and chemical decomposition occurs up to the stage of mineral compounds, which are either absorbed by the root system of plants or washed out into the deeper illuvial soil horizons. Thus forest ecosystem development follows the path of accumulation of energy in biomass, i.e. is based on increase of energy potential. The following regularity can be traced down in the phytocoenosis-soil system: the less soil capacity is able to concentrate humus (energy) the larger quantity of energy is stored in biomass, providing the stability of ecosystems in general.

Steppe ecosystems are qualitatively different. The biological moiety is unbalanced and consequently changes constantly. Successions follow the Markov chains type and every subsequent stage might develop in several ways. Therefore, it is either nearly impossible to predict the final stage of a succession or a final stage is beyond the boundaries of a given steppe system. Steppe ecosystems are extremely sensitive to external influences. The sensitivity shows up in exchange of biotic constituents. For instance, the floristic spectrum includes the families *Asteraceae*, *Poaceae*, *Fabaceae*, *Lamiaceae*, *Rosaceae*, *Caryophyllaceae*, *Brassicaceae*, *Scrophulariaceae*, *Ranunculaceae* and *Apiaceae*, and is similar to the spectra of Forest-Steppe and Steppe regional floras. The interrelation of life-forms in steppe ecosystems differs from that in forests: Phanerophytes – Hemicryptophytes ratio is 0.005-0.16:1; Phanerophytes – Terophytes ratio is 0.38–0.94:1, i.e. less than 1 [17].

Species adapt not so much to the phytogenic environment as to the external factors. On the whole, caespitose grasses which are characteristic of steppes are resistant to unfavourable water supply conditions in summer, low winter temperatures, grazing by ungulates, and burning [10]. Their high resistance is due to intercalary growth and location of winter buds close to the soil surface. Plants endure the deficiency of soil moisture, especially in the second half of the summer, by decreasing transpiration intensity dependent on moisture level. Transpiration is decreased by closing stomata and by rolling leaves even during the daytime, while forest species are incapable of doing this. Transpiration in *Festuca* and *Stipa* steppe grass communities constitutes 78–99 ml during the growing season. For comparison, in tree communities transpiration constitutes 200–300 ml during the growing season. Meanwhile, steppe species can change osmotic pressure over a wide range depending on external conditions. Such flexibility enables steppe plants to adapt to sub-lethal humidity deficiency and to use water resources inaccessible to trees [9]. For instance, *Stipa lessingiana* changes osmotic pressure from 26 to 52 atm., *Koeleria cristata* up to 35–80 atm., whereas *Fraxinus excelsior* can only change osmotic pressure in the range 37–39 atm., and *Quercus pubescens* 13–25 atm. [20].

Meadow steppe phytomass constitutes 1.5 t/ha of dry matter (3.2–4.2 t/ha). Of this, 16% accounts for above-ground parts and 84% accounts for underground parts. Thus, the ratio of above-ground to underground parts is 1:5 or up to 1:8. The annual gain (production) totals 140–150% from the preceding year. Thus nearly half (44%) of phytomass separates annually (fall off). Underground parts makes up 64% of the annual fall off, and above-ground parts complete the remaining 36% (60% of all above-ground parts) [3, 7, 10, 20]. The ratio of dead tissue to fresh phytomass is 1:1.25, and that of dead tissue to production is 1:0.8-0.9. Such ratios indicate fast (9–11 months) decomposition of litter [1]. Unlike in forests, readily soluble compounds of organic remnants decompose to final gaseous compounds faster under conditions of precipitation deficiency and certain mechanical and chemical properties of soil (namely carbonate-rich chernozems formed on loess). Furthermore, compounds are incorporated into humus, which is coagulated in the soil up to 2 m deep. Thus on average, humus comprises 7.8% (5–12%) of soil weight. Gradual decomposition of humus is partly a condition of high pH levels (pH>6.5). In summary, the principal energetic potential of steppe ecosystems is accumulated in soil (underground parts of plants included), and ecosystems function in the manner favouring replenishment of energy. Breaking of the biogeochemical cycle and loss of energy which cannot be accumulated in biomass through soil are the driving force of steppe successions [21]. Energy accumulation is prevented by humidity deficiency and the activity of consumers (grazing of ungulates and feeding of gnawing animals) and saprophytes, causing a considerable outflow of energy. Such interrelations are considered as co-evolutionary and coadaptive; without which steppe would not exist.

Forest and steppe ecosystems differ in the extent and speed of cycling of matter. A powerful phytogenic environment, vertical differentiation of phytomass, and diversity of niche types are characteristic of forests. Cycling of matter proceeds via three cycles characterized by different rates: one-year – annual (C_1)

fall off of herbage and leaves (2–10%), several years long (C_2) – fall off and decomposition of branches and litter (11–15%), and perennial (C_3) – timber decomposition (80–87%). The balance of small and large cycles correlates as $C_1<C_2$. A 1-year cycle prevails in steppe. The cycling balance is $C_1>C_2$=3.5-4.3:1 [13].

Conclusions

Comparison of energy potentials of forest and steppe (annual gain + fall off + soil) leads to conclusion that the potential is higher in forest ecosystems. On one hand, this difference is caused by external factors (deficiency of precipitation and soil humification in the steppe zone). On the other hand, such a difference defines direction and rapidity of the development of steppe ecosystems. The latter regularity has been reflected using synphyto-indication methods and ordination matrixes assembly. Energy potential, its magnitude and type of organization (in the above-ground or underground biomass, and soil) is the driving force of successions, defining their direction and characteristics. The 60-year old policy on reserves where there was absolute non-intervention has led to formation of communities which are progenitors of forests. To conserve steppe, reserve policies of absolute non-intervention should be abandoned. An indirect influence by means of external impact should be used as we cannot directly affect the natural climate specificity and ecotopes regime. Scientific experiments on steppe land should be diversified (grazing, burning of litter, etc.) to identify more effective mechanisms of slowing down succession and preserving biological diversity in steppe ecosystems in general.

REFERENCES

1. Bazilevich N.I., 1993 - Biological productivity of north Eurasian ecosystems. M., Nauka. (in Russian).
2. Bilyk H.I., Tkachenko V.S., 1972 - Modern status of vegetation cover of "Mikhailivska tzilina' reserve (Sumy region). Ukr. Bot. J., **29** (6): 696-701. (in Ukrainian).
3. Bystricka T.L., Osychnyuk V.V., 1975 - Soils and primary biological productivity of Priazov steppes. M., Nauka. (in Russian).
4. Didukh Y.P., 1990 - Methodological approaches to the problems of ecological factors phytoindication. Ukr. Bot. J., **47** (6): 5-12. (in Ukrainian).
5. Didukh Y.P., 1992 - Vegetation cover of Mountain Crimea. K., Naukova dumka. (in Russian).
6. Didukh Y.P., Plyuta P.H., 1994 - Ecological factors phytoindication. K., Naukova Dumka. (in Ukrainian).
7. Gordeckyi A.V., Bystrycka T.L., Osychnyuk V.V., 1978 - Energy and ecochemical characteristics of steppe ecosystems autotrophic components. In: Gordeckyi A.V. (ed.), Natural feeding resources of the USSR and their application, pp. 116-132. M., Nauka. (in Russian).

8. Lavrenko E.M., 1940 - Steppes of the USSR. In: Shyshkin B.K. (ed.). Vegetation of the USSR, pp. 1 – 265. 2nd volume. M.; L.: AS USSR Publishers. (in Russian).
9. Lavrenko E.M., Andreev V.N., Leontjev V.L., 1955 - Productivity profile of overground part of the USSR natural vegetation cover from tundra to deserts. Ukr. Bot. J., **40** (3): 415-419. (in Russian).
10. Lavrenko E.M., Karamysheva Z.V., Nikulina E.M., 1991 - Steppes of the Eurasia. L., Nauka. (in Russian).
11. Lavrenko E.M., Zoz I.G., 1928 - Vegetation of Mikhailivsky farm stud (former Kapnista) of Sumy region. Protection of natural sites. Kharkiv, 3b, 2: 23-36. (in Ukrainian).
12. Odum E.P., 1983 - Basic ecology. 2nd volume. M. Mir. (in Russian).
13. Pogrebniak P.S., 1969 - Interrelation of substance conversion cycles in steppe and forest landscapes and coenoses. In: Golubets M.A. (ed.), Landscape geochemistry, pp. 5-24. (in Ukrainian).
14. Tereshenkova I.A., Samiliak S.I., Schastnaja L.S., Rastvorova O.G., Tsyplenkov V.P., 1974 - Intake of organic matter into forest soil with fall off and its decomposition. LSU scientific notes, Biological Sciences Series, **3** (367): 169-194. (in Russian).
15. Tkachenko V.S., 1984 - About nature of meadow steppe of "Mikhailivska tzilina" reserve and prognosis of its further development in reserve conditions. Ukr. Bot. J., **69** (4): 448-457. (in Russian).
16. Tkachenko V.S., 2004 - Phytocoenotic monitoring of successions in the Ukrainian nature steppe reserve. Kyiv. Phytosociocentre. (in Ukrainian).
17. Tkachenko V.S., Didukh Y.P., Genov A.P., Duduka I.O., Vasser S.P., Bojko M.F., 1998 - Ukrainian nature steppe reserve. Plants. Kyiv. Phytosociocentre. 280 pp. (in Ukrainian).
18. Tkachenko V.S., Genov A.P., Lysenko G.M., 1993 - Structure of "Mikhailivska tzilina" reserve steppe vegetation based on 1991 large-scale mapping. Ukr. Bot. J., **50** (4): 5-15. (in Ukrainian).
19. Tkachenko V.S., Lysenko G.M., Vakal A.P., 1993 - Changes in "Mikhailivska tzilina" meadow steppe ecotopes (Sumy region) during succession. Ukr. Bot. J., **50** (3): 50-56. (in Ukrainian).
20. Walter H., 1968 - Die Vegetation der Erde in öko-physiologischer Betrachtung. Bd. II. Die gemäßigen und arktischen Zonen. VEB Gustav Fischer Verlag, Jena.
21. Zherikhin V.V., 1993 - Grass bioms nature and history. In: Karamysheva Z.V. (ed.), Steppes of Eurasia: problems of preservation and renewal, pp. 29-49. L-M. (in Russian).

CLONAL GROWTH MODES IN PLANT COMMUNITIES ALONG A STRESS GRADIENT IN THE CENTRAL APENNINES, ITALY

Roberto CANULLO[1*], Giandiego CAMPETELLA[1], Melinda HALASSY[2], Ladislav MUCINA[3]

[1] Dept. of Botany & Ecology, University of Camerino, Via Pontoni 5, **62032 Camerino, MC (ITALY)**; email: roberto.canullo@unicam.it

[2] Institute of Ecology & Botany of the Hungarian Academy of Science, **2163 Vácrátót (HUNGARY)**

[3] Dept. of Plant Sciences, University of the Free State, QwaQwa Campus, P.B. 339, **9300 Bloemfontein (SOUTH AFRICA)**

Abstract: A simplification from species to functional groups using the concept of clonality is particularly attractive for predictive modelling of vegetation processes and preparing guidelines for nature conservation. This important functional trait based on a modular structure including resource-acquiring units (ramets, *feeding sites*) and spacers, has been studied in three plant communities (xeric grassland, mesic grassland, and beech forest) under different levels of environmental stress (related to soil moisture and fertility) in the Montagna di Torricchio Nature Reserve near Camerino, Central Apennines, Italy. The study sought to reveal patterns of clonal growth modes (CGMs) in the three plant community types, and to test a series of hypotheses on the importance of selected CGMs along the stress gradient. Clonality was shown to have different importance in the grassland communities, due to differences in the importance of various CGMs (representing syndromes of clonal traits). Below-ground positioning of CGOs, shorter spacers, higher multiplication potential, permanent physical connection between ramets, large bud bank, and increased importance of bud protection were frequently found in water-stressed xeric grasslands, suggesting the adaptive value of these clonal traits. The major differences between grassland communities were due to the dominant CGMs: turf graminoids (with an effective way of protecting growth meristems in dense tussocks) dominated xeric grasslands, whereas rhizomatous graminoids (typical of competitive resource-rich habitats) dominated mesic grasslands. The beech forest had fewer clonal species (67%) and lower CGM diversity. Based on the assumption that different environments promote different selection pressures, the tests revealed the following results: (1) Plants with clonal organs below ground have significantly higher cover values in stressed habitats. (2) Species with short spacers are more frequent in less favourable environments, and their importance is almost ten times higher in the xeric grassland than in the forest (71% to 7.6%). (3) The number of species able to produce numerous ramets is highest in the most stressed habitat. (4) The number of species with a potential for long-lasting connection between ramets increases towards stressed environments. In contrast to our expectations, the mesic grasslands (occupying the central position along the studied stress gradient) have the highest number of species with storage organs. (6) In stressed habitats, species with forms of bud protection were the most frequent.

Introduction

Plant species respond individually to their environment, and in order to better understand this behaviour in the context of complex natural systems, researchers have suggested examining this in terms of functional species groups with similar adaptations, and hence similar responses to environmental gradients [10, 28]. Increasing demands for a predictive ecology and its application in guidelines for nature conservation have further stimulated the interest in functional plant ecology. Life-history traits, believed to carry functional messages, offer an effective tool for this purpose [26]. The crucial step in classifying plants into functional types involves selecting a list of key traits held to be informative for predictive theories [46]. Following Harper's concept, clonality itself is a trait [46] based on a modular structure [48] including resource-acquiring units (ramets, *feeding sites*) and spacers [1, 29].

Klimeš *et al.* [20] developed a classification system of Clonal Growth Modes (CGM) that can help reveal the relationship between the types of clonal growth and the functions that are attributed to clonality at higher levels of organization. Most vascular plants of Central Europe were classified into 21 (later extended) hierarchically related categories based on a combination of criteria, including origin and placement of Clonal Growth Organs (CGOs), storage functions, and spacer length and longevity. The differences in the pattern of distribution of clonal plant traits in different habitats can raise important questions about the adaptation of certain traits to special environments, and hence about the underlying mechanisms of functional processes.

Recognition of the importance of spatio-temporal scales [4, 18, 33] and of the interconnection of pattern and process in community ecology [45], calls for a re-evaluation of the role that clonality plays in community organization, functioning and maintenance of diversity level [15, 19]. In fact, clonal plant growth leads to the formation of intricate hierarchical spatial structures [7, 14, 30], of vital importance in creating vegetation patchiness at smaller scales.

In this paper, using the CGM classification system developed by Klimeš *et al.* [20], we compare the occurrence of various CGMs in selected plant communities in the Central Apennines. The selected communities form a ecocline ranging from high-stress (dry grasslands) to low-stress (mesic grasslands and forests) communities. We suppose that different environments impose different selection pressures by favouring certain mechanisms of clonal growth, with consequent differences in shaping CGM spectra. This information may prove useful in better understanding mechanisms of species coexistence, diversity level and dynamics processes in these secondary plant communities.

The following clonal traits, assumed to be of adaptive importance in stressed environments (low nutrient content and low water retention capacity) were targeted: (*a*) position of the connection between the mother and daughter ramets with respect to the soil surface (above- or below-ground); (*b*) spacer length (shorter or longer than 10 cm; excluding categories without a rhizome or stolon); (*c*) the possibility for multiplication (frequent – numerous ramets produced every year; infrequent – ramets produced in some years only); (*d*) longevity of connection between ramets (shorter or longer than 2 years); (*e*) the presence of storage (specialized organs, e.g. tubers, bulbs, or storage in organs other than those operating for clonal growth); (*f*) bud protection by specialized leaves.

We postulate that under more stressed (xeric) conditions:

(H1) connections between the mother and daughter ramets are more likely to be found below-ground than above-ground

The soil environment affords some degree of protection to susceptible (often short-lived) connections, in contrast to the stress-intensive above-ground environment, where the chance of desiccation is higher.

(H2) spacers will be shorter than they are in mesic conditions

Stressed environments bottleneck production of biomass and increase energy expenditure costs. Short spacers also imply low transport costs [16].

(H3) the frequency of multiplication (formation of CGOs) is lower than it is in mesic conditions

Less frequent formation of ramets (and CGOs, for that matter) is an energy and matter saving strategy. Lack of excessive spreading (by formation of new ramets) results in a safe-site effect, which is basically an expression of lowering the risk of extinction or damage. These features might be controlled by allometric relations [22].

(H4) connections between ramets will be more persistent than they are in mesic conditions

Because of the adversity of the stressed environments, biomass and stored energy are worth conserving, hence preservation of the present status is the preferable strategy. Long-living (active) connections assure quick regeneration, thus serving as a buffer from damage that may happen [17, 30, 34].

(H5) the presence of storage organs is more frequent than it is in mesic conditions

We limit ourselves here only to grasslands since the stressful period in forests is qualitatively different from that in the grasslands. The growth rhythm of forest geophytes (bulbous or rhizomatous plants) is controlled to a large extent by the regime of light penetration to the under-storey [36]. In grasslands, the stress relates to decreased water availability (hence also nutrient lack). The storage organs are a mean of dispersal over time (designed to assist the plant in overcoming adverse time periods). Since the dry grasslands suffer higher water discharge, the plants may experience extreme drought conditions [41].

(H6) bud protection by specialised leaves will be more important than it is in mesic conditions

Protection of meristems in environments experiencing temporary stress is of vital importance for maintaining the regeneration pool [30].

Materials and Methods

Study area

The study area is the Montagna di Torricchio Nature Reserve near Camerino, in the Central Apennines (Italy) – an area of 317 ha strictly protected since 1970. The Torricchio Reserve spans altitudes between 820 and 1491 m, and is situated on two slopes of the Val di Tazza, divided by a deep valley running SW-NE. The mean annual precipitation here is c. 1250 mm, and the mean annual temperature about 11° C. The area is dominated by Jurassic-Cretaceous calcareous rocks [5]. Soils on these calcimorphic substrates show very little taxonomic differentiation. Poorly developed, shallow and skeletal soils on steep slopes are a result of erosion associated with the presence of rocky outcrops [42].

Previous syntaxonomical studies [8, 24, 32, 42] distinguished two major grassland groups in Torricchio: **dry grasslands** (*Centaureo bracteatae–Brometum erecti* Biondi *et al.* 1986 on soft marly substrate, *Seslerio nitidae–Brometum erecti* Bruno & Covarelli 1968 and *Asperulo purpureae–Brometum erecti* Biondi & Ballelli 1981 on hard rocks called "scaglia rosata") and **mesic grasslands** (*Campanulo glomeratae–Cynosuretum cristati* Ubaldi 1979 on the valley-bottom, and *Brizo mediae–Brometum erecti* Biondi & Ballelli 1982 in the semi-mesophilous high-altitude areas). These grasslands have been surveyed in 34 and 31 relevés respectively. We have also included five relevés from the **beech forest** (*Polysticho–Fagetum* Feoli & Lagonegro 1982) in our analyses. Basic information on the character of the communities studied and the data sources are given in Tab. 1.

Table 1: Geomorphological and pedological characterisation of plant communities present in the "Torricchio Mountain" Nature Reserve and their categorisation as vegetation complexes based on the most important stress factors.

Vegetation complexes	Slope	Moisture status	$A_{0\,(cm)}$	Soil depth	Erosion	No. of relevés	Sources
Xeric grassland							
Centaureo bractatae-Brometum erecti	24°	xeric	0-10	shallow	strong	18	[32]
Seslerio nitidae-Brometum erecti	35°	extremely xeric	0-10	very shallow	strong	7	[32]
Asperulo purpurae-Brometum erecti	17°	xeric	0-4	shallow	medium	9	[32]
Mesic grassland							
Brizo medie-Brometum erecti	5°	semi-mesic	0-6	medium	weak	10	[32]
Campanulo glomaretae-Cynosuretum cristati	7°	mesic	0-7	deep	very weak	21 (8 + 13)	[9] + [32]
Beech forest							
Polysticho-Fagetum	26°	mesic	0-15	deep	weak	5	[4]

These three plant communities form a natural coenocline differentiated by levels of ecological stress, here defined as temporary deficiency of water in soils linked to low nutrient content (*e.g.* organic matter). The dry grasslands and forests range over the driest and most mesic extremes of the coenocline, respectively.

Data collection

In the classification system used, the CGMs are defined on the basis of a combination of the criteria of CGO origin (stem, root, other), CGO initial and final position (above-ground *vs*. below-ground), presence of special storage organs (tuber and bulbs), and the length and longevity of spacers between ramets [20] (Tab. 2). Most species in the study region were classified in clonal categories using the CLO-PLA2 database of Klimeš & Klimešová [21]. Of the 271 species here, about 75% were found in the database. We revised the classification by Klimeš & Klimešová (l.c.) with direct field observations for the dominant species of each of the 12 CGM categories. The remaining taxa (not featured by the Klimeš & Klimešová database) were classified into the CGMs on the basis of specimens collected (Appendix) using the same criteria as Klimeš *et al.* [20]. Nomenclature follows Pignatti [31].

Data analyses

The habitat groups were compared on the basis of both presence-absence and cover data. Species cover was estimated on the basis of Braun-Blanquet's scale [2] in all studies that served as data sources. Prior to statistical analysis, the cover codes were converted into a mean percentage scale [43]. In the classification system of Klimeš *et al.* [20], a number of plants can have more than one single type of clonal growth; only the dominant (most important) one was considered in the analyses. Counting species by combining several modes of clonal growth separately for each type (as they were different species) did not influence the results for frequency data. Statistical comparisons between the habitats targeted the participation of CGMs and clonal traits by using the non-parametric Kruskal-Wallis test (after percentage standardization and arcsine transformation for cover data [38]). Diversity was calculated for each sampling unit based on the frequencies of species belonging to each clonal growth category, using the Shannon-Wiener diversity index. A *t*-test allowed the diversities of the three habitat groups to be compared [25].

Results

Patterns of clonality in plant communities

Xeric and mesic grasslands contain 88% and 84% of clonal plant species, respectively. Species with some forms of clonal growth were more frequent here than in the forest – 67% (Fig. 1). The differences were more pronounced when cover was considered as the basis for the comparisons. The mean total cover of clonal plants was as high as 128% and 126% in the xeric and mesic grasslands, respectively, while in the beech forest only 58% of the total cover was accounted for by clonal plants. The mean total cover of all species (incl. non-clonal ones) was the highest in the forest (180%), due to the effect of summarizing all the vegetation layers. The total cover of xeric and mesic grasslands was 140% and 136%, respectively (Table 3).

The forest is differentiated from both grassland types significantly (*t* test; $p<0.001$; Shannon-Wiener index) by diversity of CGMs. The forest supports a lower number of species belonging to a restricted number of CGMs, while in the grasslands a higher number of clonal plants form a wider spectrum of CGMs.

Table 2: Characteristics of Clonal Growth Modes (CGO) that formed the basis of clonal traits with possible adaptive functions to stressed environment [20].

Clonal Growth Mode	CGO origin	CGO position	Spacer length	Multiplication	Integration longevity	Storage	Bud bank	Bud protection	Spread
1. Trifolium pratense	root	below	-	infrequent	long	-	-	-	rare
2. Alliaria petiolata	root	below	short	infrequent	long	-	large	-	rare
3. Rumex acetosella	root	below	short	frequent	long	-	large	-	fast
4. Ranunculus ficaria	root	below	-	infrequent	short	special	small	-	rare
5. Lycopodium annotinum	stem	above	long	frequent	long	-	large	-	fast
6. Festuca ovina	stem	below	short	frequent	long	-	large	special	slow
7. Rumex obstusifolius	stem	below	short	frequent	long	-	large	-	slow
8. Rumex alpinus	stem	below	long	frequent	long	-	large	-	slow
9. Dactylis glomerata	stem	below	short	frequent	long	-	large	-	fast
10. Aegopodium podagraria	stem	below	long	frequent	long	-	large	-	fast
11. Fragaria vesca	stem	above	long	frequent	short	-	large	-	fast
12. Caltha palustris	stem	below	short	frequent	short	-	small	-	slow
13. Asperula odorata	stem	below	long	frequent	short	-	small	-	fast
14. Calystegia sepium	stem	below	long	frequent	short	special	large	-	fast
15. Lycopus europaeus	stem	below	short	frequent	short	special	large	-	rare
16. Corydalis solida	stem	below	-	-	short	special	small	-	rare
17. Corydalis cava	stem	below	-	infrequent	long	special	small	-	rare
18. Galanthus nivalis	stem	below	-	-	short	non CGO-	small	special	rare
19. Ornithogalum gussonei	stem	below	-	frequent	short	non CGO	small	special	rare
20. Tulipa sylvestris	stem	below	short	-	short	non CGO	small	special	rare
21. Polygonum viviparum	special	above	-	frequent	short	non CGO	small	special	rare
22. Cardamine pratensis	special	above	-	frequent	short	non CGO	small	-	rare
23. Dentaria bulbifera	special	above	-	frequent	short	non CGO	small	-	rare
24. Aldrovanda vesiculosa	special	below	-	frequent	short	non CGO	small	-	rare
25. Lemna gibba	special	below	-	frequent	short	non CGO	small	-	rare
26. Elodea canadensis	special	below	-	frequent	short	non CGO	small	-	rare
27. Botomus umbellatus	special	below	-	frequent	short	non CGO	small	-	rare

Table 3: Relative importance of clonal traits in the three habitat types. Presence-absence and cover data standardised by the total number of species and the total cover for each relevés (mean ±SE) are given in italics. The results of nonparametric Kruskal-Wallis tests performed on arcsine transformed data, are indicated for the respective median values (n=5 for the beech forest, 31 for mesic and 34 for xeric pastures). The mean cover of clonal plant and the mean total cover are based on the original data (in bold).

	Number of species			Chi-square	Prob. value	**Cover (%)**			Chi-square	Prob. value
	Xeric grasslands	**Mesic grasslands**	**Forest**			**Xeric grasslands**	**Mesic grasslands**	**Forest**		
CGO position										
above-ground CGO	6.48 *(6.56±0.64)*	4.44 *(4.53±0.44)*	0 *(1.24±0.77)*	14.55	0.00	7.03 *(4.94±1.52)*	9.03 *(4.13±1.72)*	0 *(1.63±1.58)*	5.58	0.06
below-ground CGO	81.60 *(82.11±1.20)*	80.64 *(79.73±1.34)*	70.83 *(65.80±4.44)*	11.28	0.00	75.53 *(89.96±1.80)*	68.44 *(85.27±1.75)*	34.15 *(29.64±3.65)*	18.93	0.00
Spacers length										
long spacers	21.70 *(23.73±1.26)*	22.85 *(23.07±1.03)*	27.58 *(29.12±3.03)*	3.29	0.19	16.97 *(17.19±2.93)*	22.20 *(19.45±2.66)*	22.85 *(17.17±2.09)*	2.55	0.28
short spacers	48.47 *(47.62±1.09)*	42.85 *(41.96±1.49)*	25 *(30.67±4.00)*	15.46	0.00	60.34 *(71.04±3.17)*	52.64 *(62.30±3.14)*	11.83 *(7.62±2.92)*	17.90	0.00
Multiplication ability										
frequent multiplication	63.48 *(64.84±1.93)*	60.52 *(61.70±1.45)*	58.62 *(57.72±2.98)*	2.65	0.27	66.95 *(84.38±1.59)*	63.97 *(78.14±2.37)*	29.73 *(23.07±2.99)*	17.69	0.00
infrequent multiplication	18.51 *(17.94±1.17)*	14.63 *(15.07±0.84)*	3.57 *(3.29±0.88)*	14.86	0.00	13.94 *(7.36±0.91)*	10.61 *(5.19±0.89)*	3.71 *(3.95±3.66)*	8.22	0.02
Connection permanency										
Long-lasting connection	70.71 *(71.38±1.21)*	66.66 *(67.32±1.40)*	41.66 *(41.13±3.96)*	17.34	0.00	72.11 *(86.221±1.91)*	59.87 *(75.42±2.26)*	26.26 *(17.44±3.31)*	27.58	0.00

	Number of species			Chi-square	Prob. value	Cover (%)			Chi-square	Prob. value
	Xeric grasslands	**Mesic grasslands**	**Forest**			**Xeric grasslands**	**Mesic grasslands**	**Forest**		
short-lasting connection time	17.31 *(17.29±0.63)*	17.64 *(16.94±1.04)*	25 *(25.92±2.87)*	7.31	0.03	13.79 *(8.68±1.17)*	19.16 *(13.97±2.28)*	20.38 *(13.84±2.52)*	7.43	0.02
Storage organs										
specialised organs	2.98 *(3.72±0.64)*	5 *(6.05±0.64)*	4.16 *(3.98±1.14)*	7.74	0.02	5.83 *(2.42±0.85)*	4.71 *(4.13±2.02)*	7.56 *(2.62±1.49)*	0.08	0.96
nonCGO storage	3.64 *(4.23±0.46)*	4 *(4.59±0.68)*	5.26 *(4.81±1.44)*	0.35	0.84	3.65 *(1.03±0.35)*	6.21 *(2.41±0.62)*	3.71 *(5.53±3.66)*	10.25	0.01
Size of bud bank										
large bud bank	59.00 *(59.94±1.43)*	59.10 *(58.13±1.33)*	37.50 *(38.55±3.42)*	12.81	0.00	66.78 *(80.94±2.00)*	60.47 *(73.68±2.46)*	18.51 *(13.54±2.53)*	18.71	0.00
small bud bank	18.39 *(17.58±0.79)*	15.55 *(16.95±1.36)*	26.31 *(27.96±3.57)*	8.16	0.02	13.99 *(8.72±1.17)*	14.73 *(11.53±2.30)*	24.38 *(17.69±4.08)*	5.06	0.08
Bud protection										
with specialised leaves	14.28 *(14.62±0.76)*	10 *(10.68±1.02)*	7.14 *(6.05±1.54)*	15.06	0.00	48.79 *(52.51±3.74)*	27.23 *(21.42±3.46)*	7.56 *(2.98±1.78)*	30.25	0.00
Vegetative spreading										
fast spread	27.54 *(28.99±1.52)*	28.88 *(29.56±1.35)*	31.03 *(26.52±3.38)*	0.61	0.74	27.70 *(26.93±3.37)*	43.49 *(47.50±4.30)*	22.42 *(17.69±3.41)*	16.52	0.00
insignificant spread	59.59 *(57.90±1.38)*	55.55 *(54.02±1.80)*	33.33 *(38.44±5.32)*	10.24	0.01	59.30 *(65.95±3.85)*	39.94 *(38.80±3.71)*	15.96 *(11.60±4.39)*	28.56	0.00
rare spread	2.27 *(1.78±0.26)*	0 *(0.68±0.25)*	2.63 *(2.07±0.88)*	10.36	0.01	3.43 *(2.02±0.86)*	0 *(3.10±2.03)*	2.73 *(1.72±1.56)*	6.26	0.04
Clonal plants	**31.09±1.15**	**28.16±1.31**	**20.00±2.45**			**128.28±5.08**	**126.18±7.46**	**58.40±10.58**		
Total	**35.09±1.27**	**33.52±1.54**	**29.60±2.29**			**136.00±6.02**	**140.69±7.82**	**180.90±17.19**		

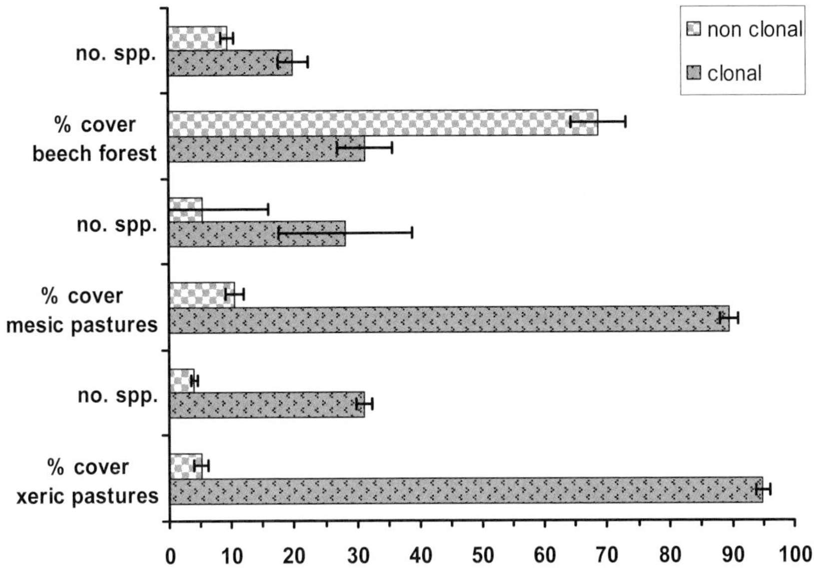

Fig. 1: The participation of clonal and non-clonal plants in the three vegetation complexes. Bars and whiskers represent mean±SE of the number of species and percentage cover (standardised data) for each category.

Species capable of fast vegetative spread did not show any preference for a community, but their abundance was significantly higher in mesic grasslands than in the other two communities. Plants with poor vegetative spread (<0.05 m per year) dominated the vegetation of xeric grasslands (when both frequency and cover data are considered) and were found to be less important in mesic habitats (Fig. 2).

The spectrum of CGMs varied between the communities, in particular between the group of grasslands and the beech forest (Fig. 3). In both grasslands, the top five CGMs (contributing more than 80 % in cover) were identical, though they differed in the order of importance.

The *Festuca ovina* CGM contributed to 66% of vegetation cover in xeric grassland, whereas the other CGMs played only a minor role in this community. Mesic grasslands were dominated by the *Dactylis glomerata* CGM (40%), followed by *Festuca ovina* CGM (26%). Lower cover values characterised the forest, with the *Asperula odorata*, *Aegopodium podagraria* and *Corydalis cava* CGMs as the more important ones (with 16.5%, 10% and 8%, respectively).

Our forest differs from the grasslands studied primarily in the layer of (non-clonal) trees. By considering the cover of the trees we have contributed to diminishing the importance of clonal species in quantitative (cover) terms in the forest. At the same time, the tree foliage creates a level of stress for the understorey species through the effect of shade. Traits found to be characteristic for the beech forest (showing higher frequency and/or abundance than in the grasslands) were the short permanence of connection between ramets, and the presence of specialized storage organs. The former trait occurred especially in the *Asperula odorata* CGM (represented by *Galium odoratum* and *Cardamine bulbifera* – plants with short-lived below-ground plagiotropic stems) and in some geophytes. Species having specialized storage organs were geophytes, with the spring ephemeroid *Corydalis cava* dominating the herb-layer. However, we acknowledge that their importance might have been underestimated because of the timing of vegetation sampling (when most of the above-ground foliage of *Corydalis cava* had already withered).

The grasslands showed high mutual similarity. Although differences regarding the relative importance of certain CGMs were found between them, these mainly resulted from the higher cover of clonal plants present in the xeric grasslands (Fig. 2). Almost all traits could be explained by one or two highest ranked CGMs, for example, the *Festuca ovina* CGM in the xeric grasslands, and *Dactylis glomerata* CGM joined by the *Festuca ovina* type in the mesic grasslands. Species attaining undisputed dominant status in the xeric grasslands were *Bromus erectus*, a matrix grass of the *Centaureo bracteatae–Brometum erecti*, and *Sesleria nitida* in the *Seslerio nitidae–Brometum erecti*. In the mesic grassland, *Festuca circummediterranea, Arrhenatherum elatius, Dactylis glomerata* and *Lolium perenne* contributed to the dominance of *Dactylis glomerata* CGM, while *Bromus erectus, Cynosurus cristatus* and *Koeleria splendens* were the most important representatives of the *Festuca ovina* CGM.

The most striking difference between the two grassland types was found in the "bud protection by specialized leaves" trait (a characteristic trait of many species in xeric grasslands), and in the "capability for vegetative spread" (prevalent in mesic habitats). In fact, the dominant CGMs for the two grassland types had almost identical growth characteristics, except for a few traits, including bud protection and vegetative spread. Plant species belonging to the *Festuca ovina* CGM possess specialised leaves that protect buds, whereas vegetative spread in the *Dactylis glomerata* CGM can be fast, covering several metres per year. The presence of the *Aegopodium podagraria* CGM in the mesic grasslands, marked by species capable of fast vegetative spread (*Galium verum* and *Lathyrus pratensis*) was also conspicuous (Fig.2).

Position of clonal organs

The majority of CGMs in our data have below-ground CGOs, and thus not much information can be gained from species frequency data alone. The cover data for above-ground CGOs is also restricted, and shows no difference between habitats. Species with below-ground CGOs have significantly higher cover values ($p<0.001$) in the grasslands than in the forest. The relative importance of this habit (in terms of total cover of species) also differed significantly between the xeric (90%) and mesic grasslands (85%; $p<0.05$). The difference was partly due to species having root-derived CGOs (CGMs 1–4 in the system of Klimes *et al.* 1997) and those with an extensive perennial root system.

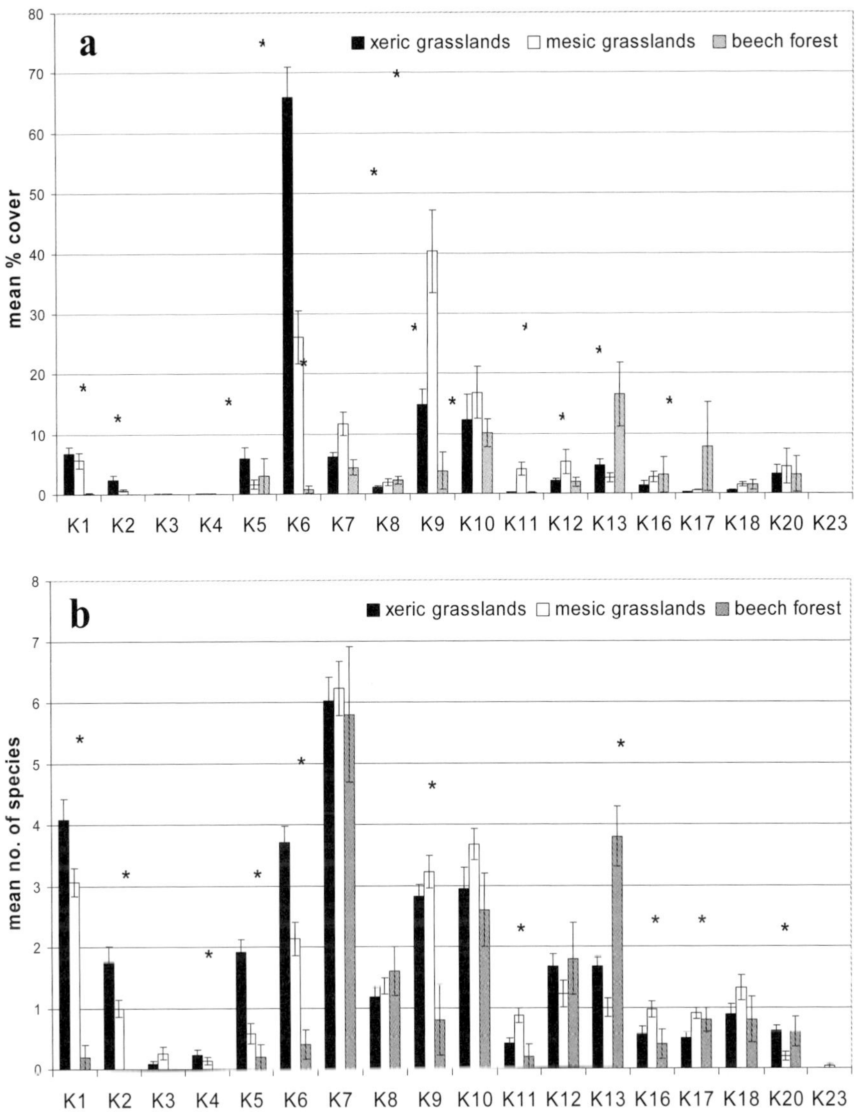

Fig. 2: The importance of clonal growth modes in xeric grasslands, mesic grasslands, and beech forest based on mean cover (a) and mean species number (b). Codes on the x axis indicate clonal categories according to Table 2. Standard errors and significant differences at $p<0.05$ based on the nonparametric Kruskal-Wallis test are indicated. (n=5 for the beech forest, 31 for mesic and 34 for xeric pastures).

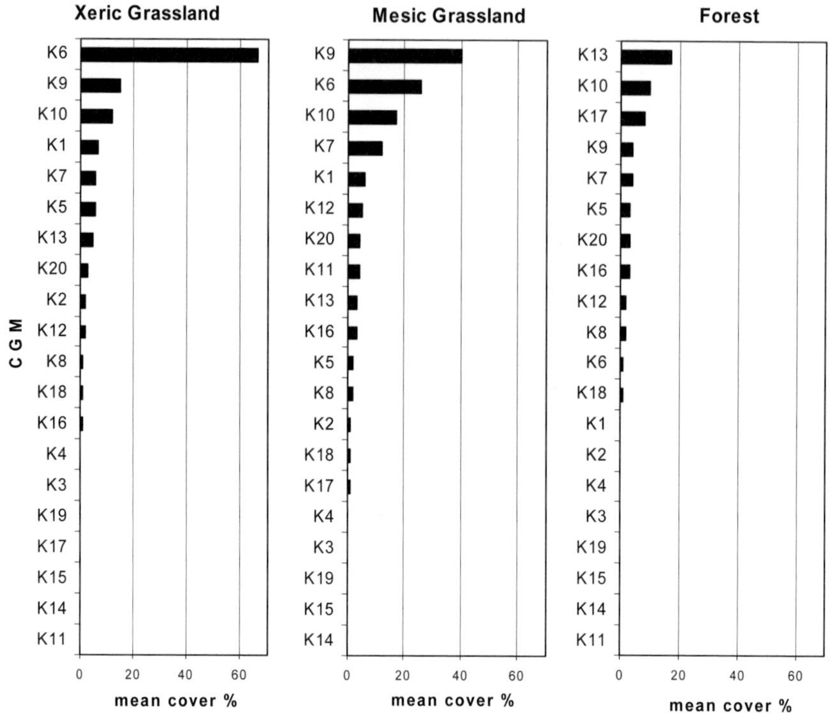

H (Shannon)	2.25	2.22	1.86
Clonal cover (%)	94.90	89.39	31.28
Non clonal cover (%)	5.10	10.61	68.72

Fig. 3: The abundance-dominance of clonal growth mode (CGO) in the three vegetation complexes based on the standardised mean cover values. Numbers indicate clonal categories according to Table 2.

Spacer length

The number of species with the capacity to develop long spacers did not differ among the three communities studied. Species with short spacers became more frequent (and important in terms of cover) in less favourable habitats. The importance of species with short spacers was almost ten times higher in xeric grasslands than in the forest (71% to 7.6%).

Multiplication ability (and bud bank)

Species producing numerous ramets every year were frequent in all three communities, but their importance was the highest in the xeric grasslands (expressed in terms of cover: 84%, as compared to 78% for mesic grasslands and 23% for the forest).

A large bud bank is a pre-requisite for an advantageous strategy under unfavourable conditions, for it allows flexibility in growth response, and thus in multiplication ability. We have not submitted this bud bank as a clonal trait to direct testing, because of its high level of correlation with the trait of multiplication ability [11].

Connection permanency

CGMs characterised by long-lasting connections between ramets are encountered more frequently in the grassland communities than in the forest. The xeric grasslands had significantly higher number of species characterised by long-lasting connections than the mesic ones (71% and 67%, respectively). Clonal plants with long-lasting connections contributed 86% of the total cover in xeric and 75% in mesic grasslands. In the beech forest, on the contrary, the number of species with short-lasting connections is higher than in grasslands.

Storage organs

The presence of storage organs and the abundance of species with this feature were supposed to differ between the studied communities, as storage could be a useful strategy for surviving periods unfavourable to plants. In contrast to our expectations, the only significant difference found was that the mesic grasslands hosted more species with CGOs with this storage function, compared to the total number of species, than the other two habitats. The beech forest was expected to differ from the grasslands, because of the widespread occurrence of spring-flowering geophytes in this community. However, only a few geophytes were found in the herb layer of the forest, and all had low cover values. This can be ascribed to the late date of vegetation sampling, carried out in summer when most of the geophytes had already withdrawn into the subterranean stage. At the same time, species with such specialised storage organs were present in the grasslands as well (*e.g.* orchids, *Ranunculus bulbosus*, *Eranthis hyemalis*).

Bud protection

The relative number of species with bud protection tended to increase as the habitat changed from xeric grasslands to mesic grasslands, and to forest. The cover data was one order of magnitude higher in the grasslands than in the forest. Species with bud protection provided twice the cover in xeric grasslands than they did in the mesic (53% and 21%, respectively).

Discussion

The major approaches applied to study plant clonality consider (*i*) detailed demographic studies of a limited number of species under a restricted variety of environmental conditions (experiments not conducted in the field, or observations of a small number of habitats; see [44]) that allow for studying correspondence between clonal behaviour and habitat characteristics; (*ii*) comparative morphological studies of a large number of species from large geographic areas at higher taxonomic levels in the search for evolutionary trends [27]; and (*iii*) spatially explicit simulation techniques [40] that help to determine the potential adaptive value of certain growth patterns in different environments. Our study largely follows the first approach while assuming some elements of the second. The large number of species (271) included in the analysis did not allow a detailed demographic study, and the evaluation of Clonal Growth Modes in the field could not be carried out for all species at this stage. However, including such a high number of species and analysing clonal traits in habitats that are part of an important successional context in the Central Apennines, has afforded interesting results on the adaptive value of certain growth forms and on their role in spatial patterns and processes of secondary vegetation.

The beech forest is a community on its own, obviously very different from the grasslands, in terms of both clonal plant occurrence and CGM diversity. Clonal plants were found to play only a minor role in the forest, and the morphological diversity of clonal species was lower here than in the grasslands. At the same time, the clonal growth mode-abundance curve was rather even, showing the equalised importance of various modes. Traits important in this community are linked to geophyte life form, considered to be very important in deciduous temperate forests [36].

Structure and processes of grassland ecosystems are usually determined by a few keystone species; in our case, these were found to be long-lived, clone-forming graminoids in mesic and high alpine grasslands [7, 12]. We found the mono-dominance of *Festuca ovina* CGM in xeric grasslands and the dominance of *Dactylis glomerata* CGM along with *Festuca ovina* CGM in mesic grasslands. The two CGMs differ only in a few characteristics. *Festuca ovina* CGM includes turf graminoids with long-lived, below-ground stems formed above-ground. Buds are protected by specialized leaves in this type, and young ramets start to photosynthesize immediately after their initiation. The *Dactylis glomerata* CGM has long-lived, below-ground plagiotropic stems formed below-ground. Species belonging to this type tend to have additional types of CGOs and can show secondary thickening. Vegetative spread can be fast and cover several metres per year [20]. The differences between the two CGMs and the role that they play in the two communities bring to mind the responses of modular plants to mesic and tundra environments [3, 47]. Callaghan [3] found that abiotic control in tundra vegetation was associated with deterministic growth (cushion and tussock formation) and weak competitive ability. Cushion form is efficient in buffering extreme conditions (low temperatures and drought), and the outer ring of dead modules in the tussock growth form provide protection and nutrients for young modules. The *Festuca ovina* CGM represents this "protective strategy" adaptive to harsh environments of exposure, extreme temperatures, and mineral soils. Mesic grasslands, on the contrary, are relatively resource-rich, but the closed vegetation results in a strongly competitive environment. Clonal perennial plants are characterized by higher plasticity of modular constructions that allows foraging for resources and the avoidance of interspecific competition in mesic habitats [3]. The *Dactylis glomerata* CGM capacity for intensive lateral spread and the presence of additional growth modes indicate a "competitive strategy" adaptive to environments under phytocoenotic control (i.e. competition).

Below-ground position, short spacers, the capacity for frequent multiplication, the maintenance of physical connection between ramets, large bud bank, and bud protection prevailed in xeric grasslands compared to mesic grasslands, accord with our expectations. Plants showing potential for fast lateral spread were more important in mesic grasslands, whereas species characterized by insignificant spread abilities dominated xeric grasslands. Storage organs played a more important role in mesic grasslands, in contrast to our expectations. This can be explained by the fact that many different organs can serve as storage sinks in clonal plants [41], but very few species were characterised by special storage organs (such as bulbs and corms) in the CGM system used [20].

The correspondence between clonal traits and ecological characteristics is a sign of adaptive behaviour [13, 20, 39]. Tightly packed modules were reported to be advantageous in open habitats [3]. Most traits associated with xeric grasslands (short spacers, potential of frequent multiplication, prolonged period of physical connection between ramets) can be interpreted as mechanisms that determine tight packing of modules.

The mechanism of fast lateral spread was more important in the mesic grasslands, whereas species characterized by insignificant spread abilities dominated the xeric grasslands. The clonal "spread ability" can be adopted as a measure of "plant mobility". Sammul *et al.* [35] introduced "ramet turnover speed" (a plant demographic measure) and found an increase of importance of this clonal feature in fertilized grassland communities – an interesting finding, which they related to species-richness depletion. In our study we found higher species richness in xeric grasslands – where low-mobility is more common – than in the other two communities. Klimeš [19] looked into plant turnover of seasonally dry, species-rich grasslands. He found that the mobility of a group of species characterized by potential extensive clonal growth was not much higher than clonal plants with poor clonal growth. More importantly, he concluded that low plant mobility does not mean high species richness – a fact indirectly supported by our data as well. However, even if plant mobility is not directly linked with diversity, a high level of different growth forms may promote coexistence in species-rich grasslands.

Clonal plants show high morphological plasticity in terms of spacer length and branching intensity as well as in changing resource acquisition strategies [6]. Moreover, it is difficult to judge whether the occurrence of various traits is a result of adaptations to a particular habitat or whether it reflects evolutionary processes of the past [23], or possibly both. The system of Clonal Growth Modes [20] does not provide for testing how individual life-history traits respond to ecological factors. However, since it assumes the adaptive value of a group of traits, such as the type and location of the organs of clonal growth, it offers an effective exploratory tool for searching out evolutionary and macro-ecological patterns in multi-species systems.

These aspects draw our attention to the importance of and need for specific field studies targeting how clonal traits interact with community structure and functions, in order to formulate more appropriate guidelines for systems conservation and restoration [37].

Acknowledgements: This work was partially supported by the Regione Marche PTRAP financial plan. We thank Domenico Lucarini for technical support and Sanyi Bartha for helpful comments on an earlier version of the manuscript. M.H. thanks the Hungarian Scholarship Board of the Ministry of Education. L.M. & M.H. thank the staff of the Department of Botany and Ecology of the University of Camerino for their hospitality and logistical support.

REFERENCES

1. Bell A.D., 1984 - Dynamic morphology: a contribution to plant population ecology. In: Dirzo R., Sarukhán J. (eds.), Perspectives on Plant Population Ecology, pp. 48-65. Sinauer Associates, Sunderland, MD.
2. Braun-Blanquet J., 1964 - Pflanzensoziologie. Grundzüge der Vegetationskunde. 3rd ed. Springer, Wien.
3. Callaghan T.V., 1988 - Physiological and demographic implications of modular construction in cold environment. In: Davy A.J., Hutchings M.J., Watkinson A.R. (eds.), Plant Population Ecology, pp. 111-135. Blackwell, Oxford.
4. Campetella G., 1999 - Struttura spaziale e scala del sottobosco in aree permanenti di studio. PhD thesis, Dept. of Botany & Ecology, Univ. of Camerino, Italy.
5. Deiana G., Pieruccini U., 1976 - Geologia e geomorfologia della Montagna di Torricchio. La Riserva naturale di Torricchio, **1**: 27-76.
6. de Kroons H., Hutchings M.J., 1995 - Morphological plasticity in clonal plants: the foraging concept reconsidered. J. Ecol., **83**: 143-152.
7. Falińska K., 1991 - Plant demography in vegetation succession. Kluwer, Dordrecht.
8. Francalancia C., 1976 - Carta della vegetazione della Riserva naturale di Torricchio. La Riserva naturale di Torricchio, **1**: 77-99.
9. Francalancia C., Galli P., Paradisi L., 1995 - Variazioni nella composizione floristica dei prati a *Cynosurus cristatus* L. delle alte valli di Tazza e di Fematre (Appennino marchigiano) in rapporto alle pratiche colturali. Fitosociologia, **29**: 89-93.
10. Gitay H., Noble I.R., 1997 - What are functional types and how should we seek them? In: Smith T.M., Shugart H.H., Woodward F.I. (eds.), Plant Functional Types: their relevance to ecosystem properties and global change, pp. 3-19. Chambridge University Press, Chambridge, UK.
11. Geber M.A., Watson M.A., de Kroon H., 1997 - Organ preformation, development, and resource allocation in perennials. In: Bazzaz F. & Grace J. (eds.), Plant Resource Allocation, pp. 113-141. Academic Press, San Diego.
12. Grabherr G., 1989 - On community structure in high alpine grasslands. Vegetatio, **83**: 223-227.
13. Grime J.P., 1979 - Plant strategies and vegetation processes. J. Wiley & Sons, Chichester.
14. Herben T., Hara T., Marshall C., Soukoupova L., 1994 - Plant clonality: biology and diversity. Folia Geobot. Phytotax., **29**: 113-122.
15. Herben T., Krahulec F., Hadincova V., Pechackova S., 1997 - Is a grassland community composed of coexisting species with low and high spatial mobility? Folia Geobot. Phytotax., **29**: 459-468.

16. Hutchings M.J., Slade A.J., 1998 - Morphological plasticity, foraging and integration in clonal perennial. In: Davy A.J., Hutchings M.J., Watchinson A.R. (eds.), Plant Population Ecology, pp. 83-109. Blackwell, Oxford.
17. Jonsdottir I.S., Watson M.A., 1997 - Extensive physiological integration: an adaptive trait in resource-poor environments?. In: de Kroon H., van Groenendael J. (eds.), The ecology and evolution of clonal plants, pp. 109-136. Backhuys Publ., Leiden.
18. Juhász-Nagy P., 1992 - Scaling problems almost everywhere: an introduction. Abstr. Bot., **16**: 1-5.
19. Klimeš L., 1999 - Small-scale plant mobility in a species-rich grassland. J. Veg. Sci., **10**: 209-281.
20. Klimeš L., Klimešová J., Hendriks R., van Groenendael J., 1997 - Clonal plant architectures: a comparative analysis of form and function. In: de Kroon H., van Groenendael J. (eds.), The Ecology and Evolution of Clonal Plants, pp. 1-29. Backhuys Publ., Leiden.
21. Klimeš L., Klimešová J., 1999a - CLO-PLA2 - A database of clonal plants in central Europe. Plant Ecology, **141**: 9-19.
22. Klimeš L., Klimešová J., 1999b - Root sprouting in *Rumex acetosella* under different nutrient levels. Plant Ecology, **141**:33-39.
23. Krahulec F., 1994 - Clonal behaviour in closely related plants. Fol. Geobot. Phytotax., **29**: 277-289.
24. Kwiatkowski W., Venanzoni R., 1994 - Carta dei suoli della Riserva naturale di Torricchio (Appennino Centrale). La Riserva naturale di Torricchio, **9**: 15-21.
25. Magurran A.E., 1988 - Ecological diversity and its measurement. Croom Helm, London.
26. McIntyre S., Díaz S., Lavorel S., Cramer W., 1999 - Plant functional types and disturbance dynamics - Introduction. J. Veg. Sci., **10**: 604- 608.
27. Mogie M., Hutchings M.J., 1990 - Phylogeny, ontogeny and clonal growth in vascular plants. In: van Groenendael J.M., de Kroon H. (eds.), Clonal growth in plants: regulation and function, pp. 3-22. SPB Acad. Publ., The Hague.
28. Nygaard B., Ejrnaes R., 2004 - A new approach to functional interpretation of vegetation data. J. Veg. Sci., **15**: 49-56.
29. Oborny B., 1994 - Spacer length in clonal plants and the efficiency of resource capture in heterogeneous environments: a Monte Carlo simulation. Fol. Geobot. Phytotax., **29**: 139-158.
30. Oborny B., Bartha S., 1995 - Clonality in plant communities - an overview. Abstr. Bot., **19**: 115-127.
31. Pignatti S., 1982 - Flora d'Italia. Edagricole, Bologna.
32. Pinzi M., 1995 - I pascoli della Riserva Naturale di Torricchio. M.Sc. thesis, Dept. of Botany & Ecology, Univ. of Camerino, Italy.
33. Podani J., Czárán T., Bartha S., 1993 - Pattern, area and diversity: the importance of spatial scale in species assemblages. Abstr. Bot., **17**: 37-51.
34. Price E.A.C., Marshall C., 1999 - Clonal plants and environmental heterogeneity. Plant Ecol., **141**: 3-7.
35. Sammul M., Kull K., Tamm A., 2003 - Clonal growth in a species-rich grassland: results of a 20-year fertilization experiment. Folia Geobot. Phytotax., **38**: 1-20.
36. Shorina N.I., Smirnova O.W., 1985 - The population biology of ephemeroids. In: White J. (ed.), The Population Structure of Vegetation. Handbook of vegetation science, **3**: 225-240. W. Junk, Dordrecht.

37. Sluis W.J., 2002 - Patterns of species richness and composition in re-created grassland. Restoration Ecology, **10**: 677-684.
38. Sokal R.R., Rohlf F.J., 1981 - Biometry. W.H. Freeman & Co., San Francisco.
39. Stuefer J.F., 1996 - Potential and limitations of current concepts regarding the response of clonal plants to environmental heterogeneity. Vegetatio, **127**: 55-70.
40. Sutherland W.J., Stillman R.A., 1990 - Clonal growth: insights from models. In: van Groenendael J., de Kroon H. (eds.), Clonal growth in plants: regulation and function, pp. 95-112. SPB Acad. Publ., The Hague.
41. Suzuki J.I., Stuefer J.F., 1999 - On the ecological and evolutionary significance of storage in clonal plants. Plant Species Biology, **14**: 11-17.
42. Venanzoni R., Kwiatkowski W., 1995 - Analisi integrata del paesaggio in un settore dell'Appennino Centrale (Riserva naturale Montagna di Torricchio). Coll. Phytosociol., **24**: 187-201.
43. van der Maarel E., 1979 - Transformation of cover-abundance values in phytosociology and its effects on community similarity. Vegetatio, **39**: 97-114.
44. van Groenendal J., de Kroon H., 1990 - Preface. In: van Groenendael J. & de Kroon H. (eds.), Clonal growth in plants: regulation and function, pp. vii-ix. SPB Acad. Publ., The Hague.
45. Watt A.S., 1947 - Pattern and process in the plant community. J. Ecol., **35**: 1-22.
46. Weiher E., van der Werf A., Thompson K., Roderick M., Garnier E., Eriksson O., 1999 - Challenging Theophrastus: a common core list of plant traits for functional ecology. J. Veg. Sci., **10**: 609-620.
47. Wilhalm T., 1995 - A comparative study of clonal fragmentation in tussock-forming grasses. Abstr. Bot., **19**: 51-60.
48. White J., 1979 - The plant as a metapopulation. Ann. Rev. Ecol. Sist., **10**: 109-145.

Appendix

Species	Clonality I.	II.	III.	Species	Clonality I.	II.	III.
Acer obtusatum	N			*Festuca curvula*	6		
Adenostyles australis	7			*Festuca inops*	6		
Alyssum minus	N			*Galium lucidum*	13		
Anchusa barrelieri	N			*Genista januensis*	N		
Arabis collina	7			*Helianthemum oelandicum*	5		
Armeria canescens	7			*Helichrysum italicum*	7		
Asperula purpurea	13	7		*Helleborus bocconei*	8		
Asphodelus albus	10			*Inula montana*	10		
Astragalus depressus	2			*Knautia purpurea*	7		
Astragalus sempervirens	10			*Leontodon cichoraceus*	12	17a	
Avenula praetutiana	6			*Leontodon crispus*	12		
Bupleurum baldense	N			*Linum bienne*	N		
Calepina irregularis	N			*Medicago lupulina*	2	N	
Campanula apennina	7			*Narcissus poeticus*	18		
Cardamine greca	N			*Ostrya carpinifolia*	N		
Cardamine heptaphylla	13			*Phleum ambiguum*	6		
Carex macrolepis	7			*Polygala major*	N		
Carlina corymbosa	7			*Polygala nicaeensis*	N		
Centaurea ambigua	7			*Primula veris*	8		

Species	Clonality I.	Clonality II.	Clonality III.	Species	Clonality I.	Clonality II.	Clonality III.
Cerastium ligusticum	N			*Ranunculus millefoliatus*	16	4	
Chamaecytisus hirsutus	N			*Rubia peregrina*	13		
Cirsium arvense	N			*Rubus ulmifolius*	N	14	
Colchicum lusitanum	16	16a		*Sanguisorba minor*	2		
Coronilla minima	N			*Saxifraga bulbifera*	18	19	23
Crepis lacera	N			*Senecio apenninus*	12		
Crepis vesicaria	N			*Sesleria nitida*	6	9	
Crupina vulgaris	N			*Silene italica*	9	2	
Cyclamen repandum	17			*Solenanthus apenninus*	N		
Cynoglossum magellense	N			*Stachys tymphaea*	N		
Cytisus sessilifolius	5	N		*Thlaspi alliaceum*	N		
Daphne laureola	N			*Thymus longicaulis*	5		
Dianthus sylvestris	7			*Tragopogon samaritani*	N		
Digitalis micrantha	8			*Valeriana tuberosa*	17a		
Eryngium amethystinum	1	2		*Veronica orsiniana*	7		
Erysium pseudorhaeticum	N			*Vicia onobrychioides*	10		
Euphorbia gasparrini	10			*Viola eugeniae*	13		
Festuca circunmediterranea	9						

RELATIONSHIP BETWEEN PHENOLOGY AND ABOVE-GROUND PHYTOMASS IN A GRASSLAND COMMUNITY IN CENTRAL ITALY

Andrea CATORCI*, **Renata GATTI, Alessandra VITANZI**
Dipartimento di Botanica ed Ecologia, UNICAM, Via Pontoni 5,
62032 Camerino - MC (ITALY); e-mail: andrea.catorci@unicam.it

Abstract: We analysed the relationship between phenology, phytomass temporal pattern and productivity dynamics in two grassland plots in the Marche Region (Central Italy). The plots were placed in grasslands belonging to the same syntaxon, previously defined as *Brizo mediae-Brometum erecti* using the phytosociological approach. The two sites were chosen at the lowest and the highest altitude within the distribution area of this syntaxon. The analysis showed a clear relationship between the observed variables and other ecological variables, such as climate. This information should be useful to plan future management strategy in the study area.

Introduction

Montane grasslands cover about 8% of the territory of the Marche Region (Central Italy). They are one of the most important ecosystems for biodiversity conservation, both at regional and national scale. Indeed, they show high levels of floristic diversity (25% of the regional flora), as well as high species richness of invertebrates and vertebrates; for instance, 13 nesting bird species out of 18 protected by the European Bird Directive (79/409/CEE) were found in Marche montane grasslands.

Montane grasslands have been the most important resource for the local economy for thousands of years. They were used largely for cattle forage and breeding, in some places for the whole year, in other places only recurrently (transhumance) [8]. Due to its long history, pasture deeply transformed the pristine ecosystems and created a functional ecological equilibrium. The strong decrease of pasture, which occurred in the second half of the 20th century, is going to modify this equilibrium, and the consequences are still unknown. Therefore, long-term management plans for grasslands, based on scientific data and inferences, are now necessary to match different needs: preserving the high species and community diversity, maintaining the ecological processes that allow the existence of a such high diversity [1, 4], and improving the zootechnical use [16, 18].

Different types of vegetation analysis can contribute to the management plans. Among these analyses, the study of plant community phenology [6] and its relationship with productivity can be a powerful insight in the whole ecosystem dynamics.

Materials and Methods

We analysed a semi-mesophilous community type ("syntaxon" in the phytosociological approach), previously defined as *Brizo mediae-Brometum erecti* Bruno in Bruno *et* Covarelli 1968 corr. Biondi *et* Ballelli 1982 [11]. This plant association belongs to the class *Festuco-Brometea*. From the whole floristic data set (Table 1), two relevés were chosen to locate the study sites. Two permanent plots ("Prati di Gagliole", "Monte Pennino") of 3x3 m were established at 880 m a.s.l. and 1460 m a.s.l. respectively; these elevations represent the lower and the upper limits of the syntaxon altitudinal range [3]. The plots were surrounded by rigid fences to exclude grazing animals.

Inside each plot, from the beginning of May up to the beginning of October, the vegetation in a 1x1 m square was mown every 15 days. In addition, from the beginning of June the plant re-growth inside the previously mown square was clipped again. This approach allowed us to quantify both the above-ground phytomass and its ability to re-grow after consumption [10, 13, 17].

Phenological analysis was carried out following the method of Puppi [20]. According to this, the phenological phases have to be surveyed along a linear transect or inside a fixed area, on each individual of all the species that are present in the sample. Phenological surveys were done every 11–12 days from the beginning of May up to the beginning of August, and after that every 30 days up until the end of October. Phenophases were classified using the system of Puppi [20], modified according to other authors [12, 14, 15]. A number indicates each phenophase as shown in Table 2.

Species were grouped by their life form, as follows: geophytes (G), caespitose hemicryptophytes (H caesp), scapose hemicryptophytes (H scap), rosette hemicryptophytes (H ros), therophytes (T) and chamaephytes (C) [19]. The last two groups were merged in a single group named "others".

Results

The association *Brizo mediae–Brometum erecti* is a semi-mesophilous grassland dominated by *Bromus erectus*. Its typical habitats are gentle slopes on calcareous massifs of the Central Apennines, mostly of northern exposure and on Lithic Hapludoll soils [5]. Its altitudinal range spans 800–1400 m; this range was included by Biondi *et al.* [2] in two bioclimatic belts, the upper meso-temperate, the lower supra-temperate. These grasslands host many transgressive species from the *Molinio-Arrhenateretea* class, which are associated with mesophilous species of the *Festuco-Brometea* class [3, 9]. The *Brizo mediae–Brometum erecti* grasslands are characterized by a dense turf, so they are largely used for cattle and sheep grazing. Where the terrain is not too uneven, the turf is periodically mown. Both study sites are used for cattle grazing. Stock density is higher at Prati di Gagliole than at Monte Pennino. Mowing is not carried out at any one site.

Table 1: ***Brizo mediae-Brometum erecti*** **Bruno in Bruno** ***et*** **Covarelli 1968 corr. Biondi** ***et*** **Ballelli 1982:** 1 - "Prati di Gagliole"; 2 - "Monte Pennino".

Releves no.	1	2
Altitude (m)	880	1460
Aspect	N	W
Slope (degrees)	10	4
Total vegetation cover (%)	100	100
Area sampled (m^2)	150	150
Charact. and diff. of *Brizo mediae-Brometum erecti*		
Leontodon cichoraceus (Ten.) Sanguin.	+	1
Briza media L.	+	1
Filipendula vulgaris Moench	+	+
Centaurea triumfetti All. subsp. aligera (Gugler) Dostál	+	+
Betonica officinalis L.	+	+
Plantago lanceolata L. var. sphaerostachya Mert. et W.D.J. Koch	.	+
Charact. and diff. of *Phleo ambigui-Bromion erecti*		
Muscari neglectum Guss. ex Ten.	1	+
Trifolium montanum L. subsp. rupestre (Ten.) Nyman	+	2
Avenula praetutiana (Parl.) Pignatti	+	+
Centaurea ambigua Guss. ambigua	+	.
Phleum ambiguum Ten.	+	.
Koeleria lobata (M. Bieb.) Roem. et Schult.	.	1
Potentilla rigoana Th. Wolf	.	+
Charact. of *Brometalia erecti* and *Festuco-Brometea*		
Bromus erectus Huds.	4	3
Festuca stricta Host subsp. sulcata (Hack.) Patzke	2	2
Sanguisorba minor Scop. subsp. balearica (Bour. ex Nym.) Muñoz G. et C. Navarro	1	1
Thymus longicaulis C. Presl	+	2
Cerastium arvense L. subsp. suffruticosum (L.) Nyman	1	+
Ranunculus millefoliatus Vahl	1	+
Salvia pratensis L. subsp. pratensis	1	+
Valeriana tuberosa L.	+	1
Hieracium pilosella L. (s.l.)	+	1
Carex caryophyllea Latourr.	+	1
Armeria canescens (Host) Boiss.	+	1
Onobrychis viciifolia Scop.	+	+
Saxifraga bulbifera L.	+	+
Anthyllis vulneraria L. subsp. polyphylla (DC.) Nyman	+	+
Orchis mascula (L.) L. subsp. mascula	+	+
Koeleria cristata (L.) Roem. et Schult.	+	+
Trifolium ochroleucum Huds.	+	+
Aira caryophyllea L. subsp. caryophyllea	+	+
Helianthemum nummularium (L.) Mill. subsp. obscurum (Celak.) Holub	1	.
Linum bienne Mill.	1	.
Ranunculus bulbosus L. subsp. aleae (Willk.) Rouy et Foucaud	1	.

Trifolium incarnatum L. subsp. molinerii (Hornem.) Syme	1	.
Orchis ustulata L.	+	.
Ranunculus illyricus L.	+	.
Thymus praecox Opiz subsp. polytrichus (A. Kern. ex Borbás) Ronniger	+	.
Dianthus carthusianorum L. subsp. carthusianorum	+	.
Linum catharticum L. subsp. catharticum	+	.
Plantago lanceolata L. var. lanceolata	+	.
Festuca laevigata Gaudin subsp. laevigata	.	2
Ranunculus bulbosus L. subsp. bulbosus	.	+
Asperula cynanchica L.	.	+
Hippocrepis comosa L.	.	+
Rumex acetosella L. subsp. angiocarpus (Murb.) Murb.	.	+
Transgr. of *Molinio-Arrhenatheretea*		
Anthoxanthum odoratum L.	1	2
Trifolium pratense L. subsp. pratense	1	2
Knautia purpurea (Vill.) Borbás	1	1
Lotus corniculatus L.	1	1
Cynosurus cristatus L.	1	+
Rhinanthus minor L.	+	2
Euphrasia stricta D. Wolff ex J.F. Lehm. (incl. E. pectinata Ten.)	+	+
Plantago media L.	+	+
Allium vineale L.	+	+
Vicia sativa L. subsp. nigra (L.) Ehrh.	1	.
Rhinanthus alectorolophus (Scop.) Pollich subsp. alectorolophus	+	.
Bromus hordeaceus L. subsp. hordeaceus	1	.
Festuca rubra L.	1	1
Festuca circummediterranea Patzke	.	1
Rumex acetosa L.	.	+
Poa molinerii Balb.	.	+
Trifolium dubium Sibth.	.	+
Narcissus poeticus L. subsp. poeticus	.	+
Transgr. of *Nardetea strictae*		
Luzula campestris (L.) DC.	+	1
Viola eugeniae Parl.	.	+
Other species		
Galium corrudifolium Vill.	1	+
Trinia glauca (L.) Dumort. subsp. carniolica (A. Kern. ex Janch.) H. Wolff	+	1
Trifolium campestre Schreb.	+	+
Bupleurum baldense Turra	+	+
Senecio provincialis (L.) Druce	+	+
Veronica arvensis L.	+	+
Asphodelus macrocarpus Parl.	1	.
Trifolium striatum L. subsp. striatum	1	.
Cynosurus echinatus L.	+	.
Cerastium ligusticum Viv.	.	2

Table 2: Phenological phases.

1	Early visible growth
2	Completely extended leaves
3	Appearance and rising of the culms
4	Presence of flower gems
5	Flowering in progress
6	Flowering term
7	Immature fruits
8	Mature fruits and seeds
9	Initial leaf yellowing
10	Total leaf yellowing
11	Appearance of new green leaves

Figure 1 represents the biological spectra of the phytosociological relevés (Table 1). The most relevant difference between the sites is the higher abundance of therophytes at Prati di Gagliole (20% of the local flora) than at Monte Pennino (13%). Therefore, in the first site the group H caesp and the group H scap were under-represented in comparison to the second site. We argue that this difference is due to the lower elevation of Prati di Gagliole (800 m is the lower altitudinal range of the *Brizo mediae–Brometum erecti* association) and higher stock density – and consequent trampling – which can affect community structure and composition [13].

Maximum phytomass in the plots reached 740 g/m^2 at Prati di Gagliole at the end of May, and 520 g/m^2 at Monte Pennino in mid-June (Fig. 2). The delay between the two plots was about 20 days. Seasonal dynamics of phytomass showed a similar pattern for both plots. The highest curve slope was observed in May. At the beginning of May, which is the usual starting point of the growing season, the phytomass at Prati di Gagliole was already greater (400 g/m^2). At Monte Pennino the corresponding data are missing due to the presence of snow. The first mowing was done the day after complete snow melt, on May 17th, and the phytomass was 100 g/m^2. This observation suggests that at lower elevation (Prati di Gagliole) new phytomass production began earlier than at higher elevation (Monte Pennino), where in mid-May the vegetation was still dormant.

Regarding the vegetation re-growth after mowing, in both plots the phytomass increased highly in July and negligibly in August. A slight increase of re-growth was observed in September and October (Fig. 2).

The phenological pattern of the community as a whole was described considering all the individuals of all the species encountered in each plot on the survey day (Figs. 3 and 4). At Prati di Gagliole, the peak of the flowering phases (no. 4, 5, and 6, indicating the start of flowering, full flowering and the end of flowering) – represented by the highest number of individuals in those phases – was recorded on May 24th (76% of individuals), June 4th (81% of individuals) and June 15th (78%). At Monte Pennino, the peak of the flowering phases was recorded on June 7th (69%), June 18th (81%, maximum number of individuals in phase 5) and June 29th (62%).

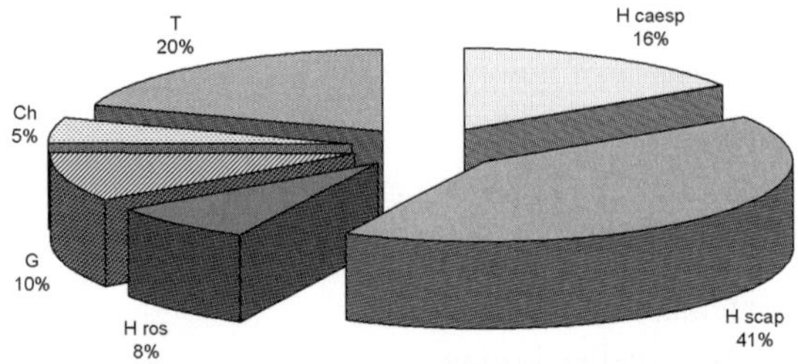

A

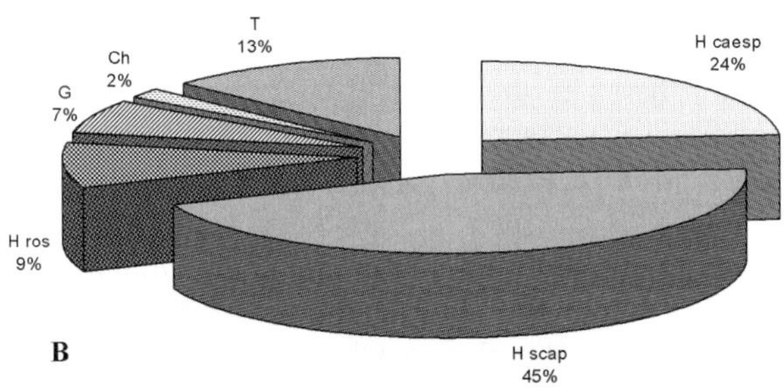

B

Fig. 1: Biological spectrum of the grassland samples in Table 1: A. "Prati di Gagliole"; **B.** "Monte Pennino". G - Geophyte; H caesp - Hemicryptophyte caespitose; H scap - Hemicryptophyte scapose; H ros - Hemicryptophyte rosulate; Ch - Chamaephyte; T - Therophyte.

At Prati di Gagliole, almost all individuals were producing seeds (phases no. 7 and 8) in July, and one third were already yellowing (phase no. 9). At the beginning of August, almost 80% of individuals were completely dry (phase 10). At Monte Pennino, 65% of individuals were producing seeds in July, but 30% were still in flower and at the end of flowering; in August, only 15% of individuals were dry (phase 10).

At Prati di Gagliole, 50% of individuals were in vegetative re-growth (state 11) at the beginning of October, while at Monte Pennino 95% of individuals were in this phase, most having been still fully vegetative for the whole previous month.

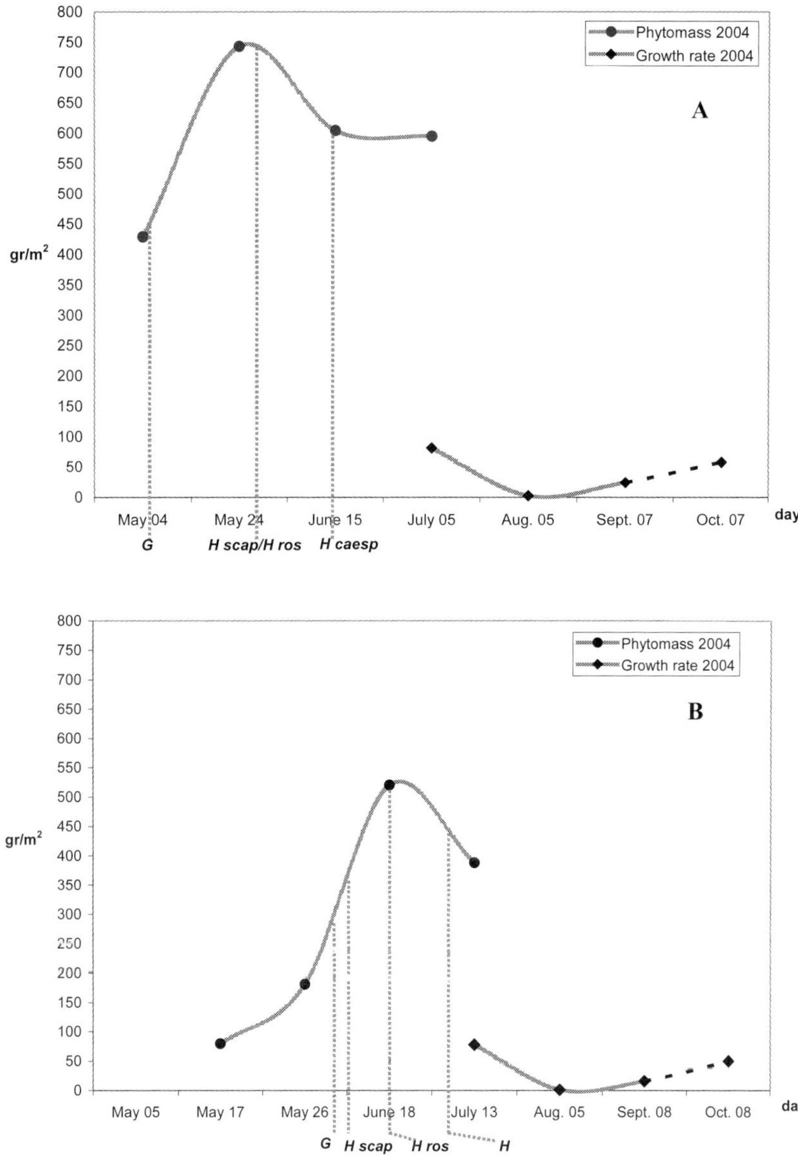

Fig. 2: Seasonal change in phytomass production (—●—) and plant re-growth rates (—◆—) in *Brizo mediae-Brometum erecti* during 2004. A - "Prati di Gagliole"; **B** - "Monte Pennino". Vertical lines correspond to the timing of the flowering (phenological phase 5) in different plant life forms. For abbreviations see Fig. 1.

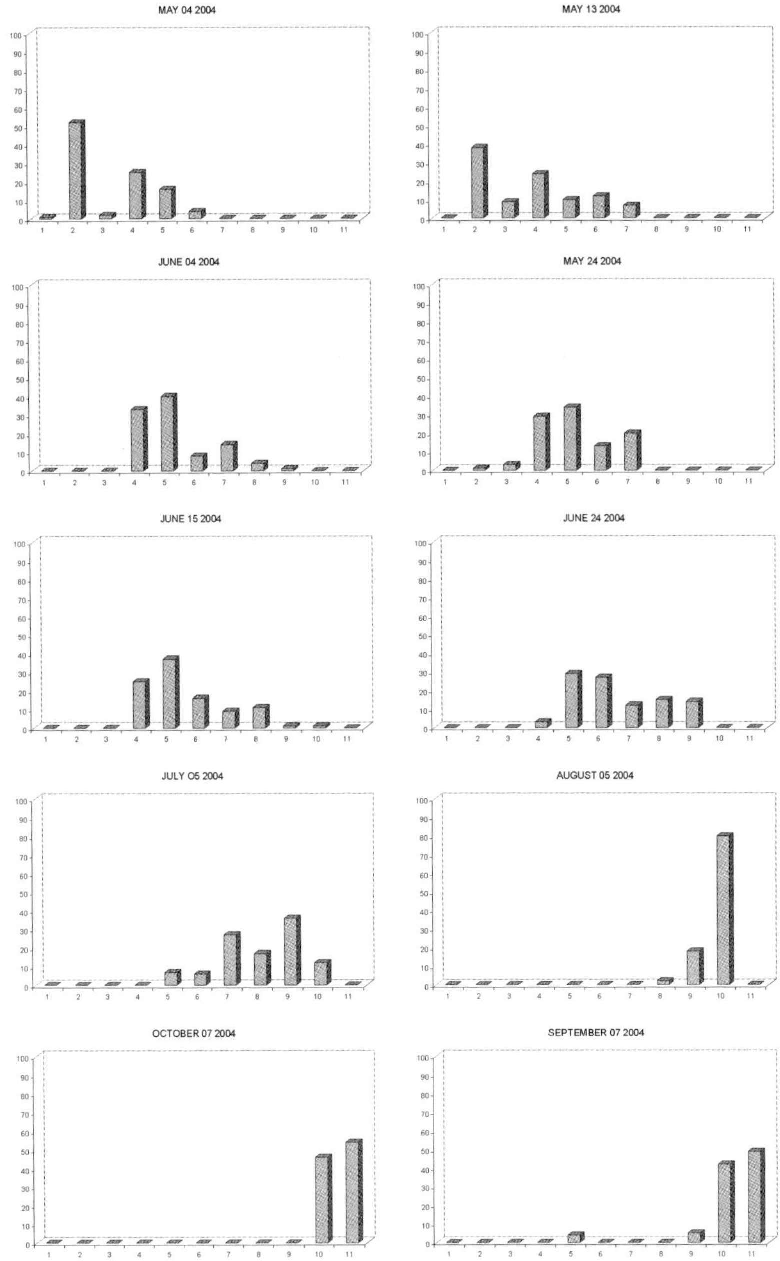

Fig. 3: Seasonal phenologic dynamics of the *Brizo mediae-Brometum erecti* in "Prati di Gagliole". The x axis represents the phenophases, whereas the y axis represents proportion (%) from the total number of individuals.

Table 3: Phenogramme of the "Prati di Gagliole" biological groups (phenological phase in rows as in Table 2). The number in each cell represents the percentage of individuals in each phenological phase. The shaded cells correspond to the highest percentage in each day.

Brizo mediae-Brometum erecti

Geophyte

	MAY 04	MAY 13	MAY 24	JUNE 04	JUNE 15	JUNE 24	JULY 05	AUGUST 05	SEPT. 07	OCTOBER 07
1										
2	13	17								
3					33					
4	12		10		**67**	2				
5	**54**	5		8		23	22		28	
6	21	**78**	10				17			
7			**80**	**84**			5			
8						**59**				
9				8		6	**39**	**85**		
10							17	15	22	**65**
11									**50**	35

Hemicryptophyte cespitose

	MAY 04	MAY 13	MAY 24	JUNE 04	JUNE 15	JUNE 24	JULY 05	AUGUST 05	SEPT. 07	OCTOBER 07
1										
2	**54**	22								
3		6								
4	43	**54**	**60**	**64**	23	2				
5	3	16	13	12	**59**	15				
6		2	20	6	8	**48**	8			
7			7	10	6	14	8			
8				6	4	8	11			
9						10	**63**	5		
10							10	**95**	18	46
11									**82**	**54**

Hemicryptophyte scapose

	MAY 04	MAY 13	MAY 24	JUNE 04	JUNE 15	JUNE 24	JULY 05	AUGUST 05	SEPT. 07	OCTOBER 07
1	2									
2	**48**	**35**	2							
3		3	7	8		1				
4	37	30	26	19	24	3				
5	10	14	**37**	**52**	**27**	**43**	11		2	
6	3	7	12	11	17	14	3			
7		11	16	5	23	15	**47**			
8				5	9	14	30	4		
9						10	8	22	14	
10							1	**74**	**75**	47
11									9	**53**

Hemicryptophyte rosulate

	MAY 04	MAY 13	MAY 24	JUNE 04	JUNE 15	JUNE 24	JULY 05	AUGUST 05	SEPT. 07	OCTOBER 07
1										
2										
3	**60**	40								
4	40	**60**	**50**							
5			**50**	**100**						
6					**100**					
7										
8										
9						**100**				
10							**100**	**100**		
11									**100**	**100**

	Other									
	MAY 04	MAY 13	MAY 24	JUNE 04	JUNE 15	JUNE 24	JULY 05	AUGUST 05	SEPT. 07	OCTOBER 07
1										
2	**97**	**59**								
3		22								
4	3		17	29	22					
5			**66**	**56**	**37**	39				
6		6	17	5	**37**	**54**				
7		8		10	2	7	**75**			
8		5			2		25			
9								8		
10								**92**	48	**52**
11									**52**	48

Table 4: Phenogramme of the "Monte Pennino" biological groups (interpretation as in Table 2).

	Brizo mediae-Brometum erecti									
	Geophyte									
	MAY 04	MAY 17	MAY 26	JUNE 07	JUNE 18	JUNE 29	JULY 13	AUGUST 05	SEPT. 08	OCTOBER 10
1										
2		**50**	**65**	5						
3		19								
4		19	16	10						
5		12	19	**52**	**35**	13	17		**62**	
6				14	**35**	**87**	**33**			
7					30		17			
8							**33**	20		
9								**40**		
10								**40**	38	
11										**100**

Hemicryptophyte cespitose

	MAY 04	MAY 17	MAY 26	JUNE 07	JUNE 18	JUNE 29	JULY 13	AUGUST 05	SEPT. 08	OCTOBER 10
1										
2		**47**	29	31						
3		12			7					
4		41	**66**	**61**	**46**	5				
5			5	3	**42**	**50**	30			
6				5	3	22	15			
7					2	14	**27**	6		
8						9	26	32		
9							2	**45**	10	
10								17	**57**	
11									33	**100**

Hemicryptophyte scapose

	MAY 04	MAY 17	MAY 26	JUNE 07	JUNE 18	JUNE 29	JULY 13	AUGUST 05	SEPT. 08	OCTOBER 10
1										
2		**57**	**54**	25						
3				4	3					
4		34	18	22	13	8				
5		9	26	**45**	10	17	5			
6			2	4	**43**	19	2			
7					30	**56**	**48**	8		
8							37	33	20	
9							2	**43**	**50**	
10								16	20	3
11									10	**97**

Hemicryptophyte rosulate

	MAY 04	MAY 17	MAY 26	JUNE 07	JUNE 18	JUNE 29	JULY 13	AUGUST 05	SEPT. 08	OCTOBER 10
1										
2		**87**	**50**	**55**	21					
3		9	25							
4		4	25	36	7	11				
5				9	**64**	18	18	4		
6					8	15	14	4		
7						**56**	7	22		
8							**61**	**41**		
9								22	33	
10								7		15
11									**67**	**85**

Other

	MAY 04	MAY 17	MAY 26	JUNE 07	JUNE 18	JUNE 29	JULY 13	AUGUST 05	SEPT. 08	OCTOBER 10
1										
2		**67**	**46**	12	27					
3		33	15	24	27					
4			39	**35**	**37**					
5				29	9	28	11			
6						**68**	32	4		
7						4	**54**	23		
8							3	23	40	
9								**38**	**60**	
10								12		
11										**100**

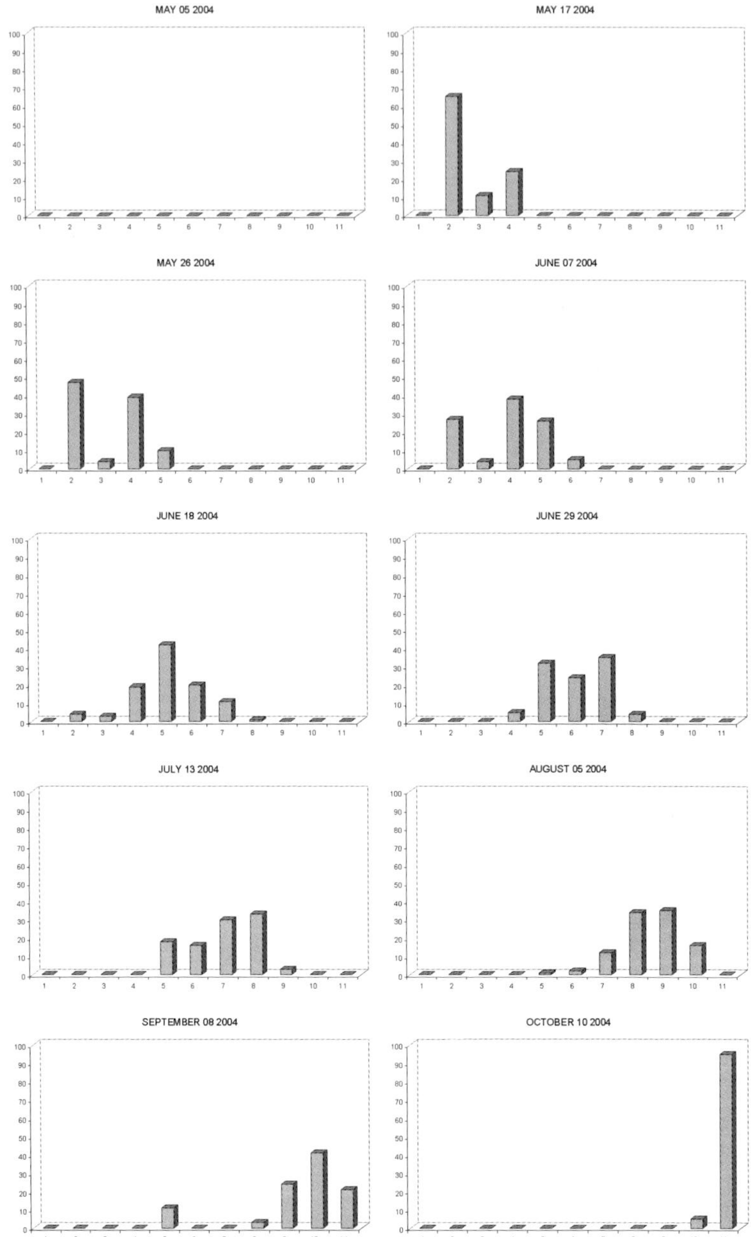

Fig. 4: Seasonal phenologic dynamics of the *Brizo mediae-Brometum erecti* in "Monte Pennino" (x and y axes as in Fig. 3).

In synthesis, the phenological pattern (especially the flowering peak) had 12–15 days delay at Monte Pennino compared to Prati di Gagliole. Moreover, at Monte Pennino plants persisted in a vegetative state during summer, while at Prati di Gagliole they suffered a period of desiccation (more than one month).

A further insight in phenological pattern was carried out taking into account species life forms. For each species group a diagram was realized, as shown in Tables 3 and 4.

The flowering peak of G happened to be earlier than that of the community as a whole in both study sites. The flowering peak of H caesp corresponded to the late period of flowering phases of the community (May 24th – June 15th at Prati di Gagliole, June 18-29th at Monte Pennino), while for H scap and H ros the flowering peak corresponded to the early period of flowering phases of the community. Finally, the group of "others" had the flowering peak in the same moment as the community as a whole.

At Prati di Gagliole, the flowering peak of G (especially *Muscari neglectum* and *Anemone hortensis*) was reached between May 3-13th; three other peaks were on June 24th (*Phleum ambiguum*), July 5th (*Allium vineale*), and September 7th (*Scilla autumnalis*). The flowering peak of H caesp was between May 24th and June 15th, with *Festuca stricta* subsp. *sulcata* the first flowering species, followed by *Bromus erectus*, which is the dominant species in the community. Regarding all the other groups (H scap, H ros and others), the flowering peak was reached between 24 May and 15 June, and involved mostly the following species: *Linum bienne*, *Lathyrus pratensis*, *Centaurea triumfetti* subsp. *aligera*, *Cerastium arvense* subsp. *suffruticosum* (H scap); *Hieracium pilosella* (H ros); *Trifolium stellatum*, *Trifolium campestre*, *Cynosurus echinatus* (and others).

At Monte Pennino, the flowering peak of G (mostly *Muscari neglectum*, *Allium vineale*, *Ornithogalum umbellatum*) was reached between June 7–29th; on September 8th there was also a peak of *Scilla autumnalis*. The H caesp and H scap showed the same flowering pattern, with a peak between June 7–29th. In the H caesp group, four species follow one another in the flowering peak: *Poa alpina, Anthoxanthum odoratum, Festuca stricta* subsp. *sulcata*, and *Bromus erectus*. In the H scap group, *Potentilla cinerea*, *Cerastium arvense* ssp. *suffruticosum*, and *Stachys recta* are the most relevant flowering species. The group H ros showed a shorter flowering peak, concentrated between June 18–29th and involving *Leontodon cichoraceus*, *Plantago media* and *Hieracium pilosella*. Finally, for the group of other species we registered the flowering peak around June 29th (*Aira caryophyllea*, *Arenaria serpillifolia*, *Medicago lupulina*, *Euphrasia stricta*), but this is only an approximation, due to the fact that this heterogeneous group included both annuals (T) and perennials with different life forms

A further insight into phenological dynamics was accomplished by focusing on three grass species, which are the most relevant elements of the plant community due to their abundance: *Anthoxanthum odoratum*, *Festuca stricta* subsp. *sulcata* and *Bromus erectus* (Fig. 5). Even if there is a 15-day time-lag between the plots, in both of them the first species to flower was *Anthoxanthum odoratum*; the second *Festuca stricta* subsp. *sulcata* (phase 2 at the middle of May, phase 4 at the end of May in both plots; full flowering at the beginning of June in the first plot, in mid-June in the second); and the last *Bromus erectus* (fully flowering in mid-June in the first plot, at the end of June and the beginning of July in the second).

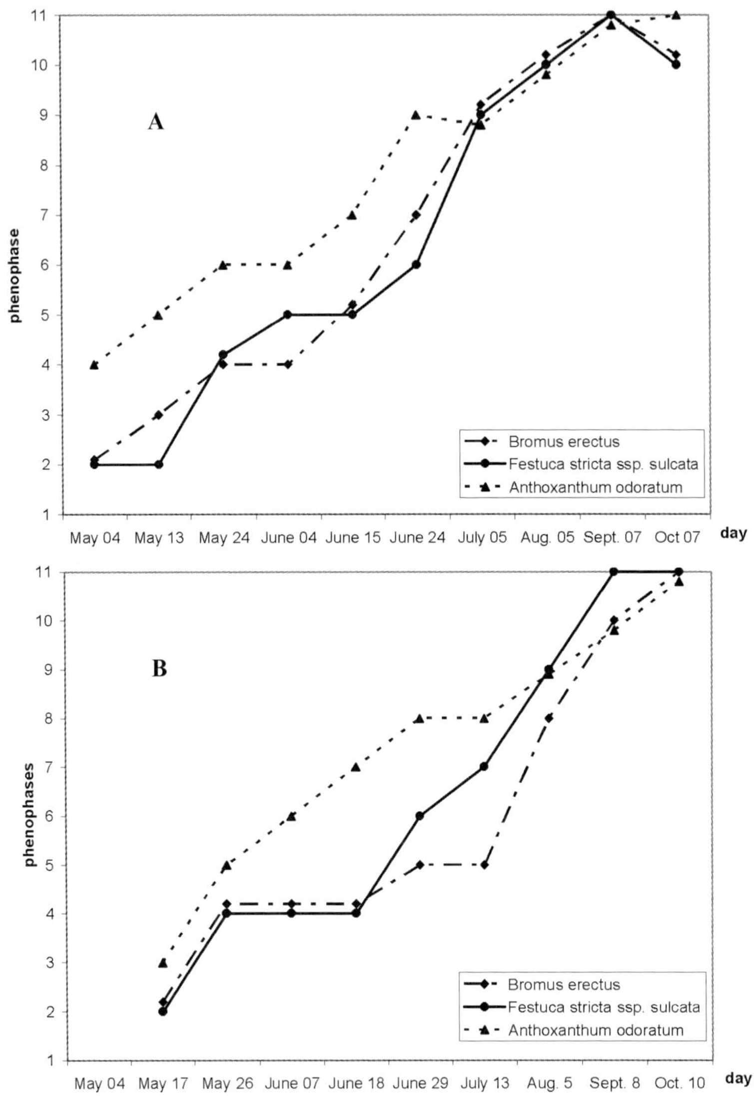

Fig. 5: Seasonal phenologic dynamics of *Bromus erectus* (‑◆‑), *Festuca stricta* ssp. *sulcata* (—●—) and *Anthoxanthum odoratum* (‑▲‑). A: "Prati di Gagliole"; **B:** "Monte Pennino".

Discussion

The phenological dynamics in the two study areas showed a parallel pattern, with a time-lag of 15 days at Monte Pennino (higher elevation) compared to Prati di Gagliole (lower elevation). The flowering peak (phase 5) showed the same between-group sequence in both the areas, starting with the geophytes (4 May at Prati di Gagliole, 7 June at Monte Pennino), followed by scapose hemicryptophytes (June 4th and June 7th, respectively), rosette hemicryptophytes (June 4-15th and June 18th, respectively), caespitose hemicryptophytes (June 15th and June 29th, respectively). As regards the "others" group, there was no definite flowering peak.

The phenological dynamics of the community as a whole are consistent with its structural dynamics. Indeed, a low turf characterizes the first weeks of the growing season; over the turf the geophytes can stand up with their flowering stalks, their growth possible due to the nutrients stored in the underground organs. Then, geophytes are not bound to the start of the nitrogen-fixing bacterial activity, which in turn is related to temperature increase. After that, while sward height is increasing and is going to reach its final height (25–35 cm), the scapose and rosette hemicryptophytes raise their stalks over the sward and produce flowers. This behaviour probably arises from the need of the plants to complete the strongest energetic effort (flowering and fruiting) before the sward starts to shade their leaves. Finally, the caespitose hemicryptophytes (mostly grasses in this community) raise their stalks over the sward, to improve pollen and seed dispersal by the wind.

This pattern seems to be confirmed by the phenological dynamics of the most relevant grasses – *Anthoxantum odoratum*, *Bromus erectus* and *Festuca stricta* subsp. *sulcata*. The time-lag between their phases seems to be related to their size; indeed, the sequence of flowering peak is *Anthoxantum odoratum*, the shortest grass (average height 10–15 cm), followed by *Festuca stricta* subsp. *sulcata* (20–30 cm), and finally by *Bromus erectus* (40–60 cm).

The time-lag in the phytomass accumulation peak between the two study sites is parallel to the time-lag in the flowering peak. The range of the phytomass maximum agrees with the flowering peak in rosette hemicryptophytes and scapose hemicryptophytes; conversely, the flowering peak of caespitose hemicryptophytes is slightly delayed. This feature could be explained by considering that the phytomass of such life groups is already completely formed at buds appearance, and that in these grasses the leaf dry weight is much more elevated than that of the culms and flowers.

Conclusions

The present work confirms the close relationship between the timing of flowering and phytomass maximum in grasslands [13]. A good knowledge of phenological phases is important for a better understanding of ecological trials and rational grassland management.

In fact, when flower heads begin to appear in the sward, a change in plant life-cycles occurs as reproduction takes over and resources are diverted to production of flowers and seeds in preference to vegetative growth [13]. Production of plant biomass therefore decreases as photosynthetic capabilities are reduced through a combination of leaf senescence, shading and loss of soil moisture. In addition, this fall in growth rate is accompanied by a loss of nutritional quality, because the stalk in its reproductive phase contains a much fibrous stem and seed-head material, and much smaller amounts of the more digestible green leaves and developing shoots, which contain soluble sugar and proteins [13].

The present work has allowed to approach a syntaxonomic structural model and to correlate ecological and productive aspects in a practical way; therefore it can represent a good starting point for the management definition of Apennine grasslands. Grassland utilization can represent a difficult compromise for the farmer in trying to achieve a balanced but thorough removal of material, so that as much of the phytomass is removed without reducing its potential for future growth, and without damaging the structural and floristic characteristics of the syntaxon. Future data collection and analyses will focus on the theoretical equilibrium between plant growth, community productivity and phytomass removal by grazing animals.

REFERENCES

1. Biondi E., 2001 - Paesaggio vegetale e potenzialità pastorali. In: Atti 36° Simposio Internazionale di Zootecnia "Prodotti di origine animale: qualità e valorizzazione del territorio", 1, pp. 5-22. Greppi & Enne, Portonovo (Ancona).
2. Biondi E., Baldoni M.A., Talamonti M.C., 1995 - Il fitoclima delle Marche. In: Atti Conv. "Salvaguardia e gestione dei beni ambientali nelle Marche", pp. 21-70. Tipolit. Trifogli, Ancona.
3. Biondi E., Ballelli S., Allegrezza M., Zuccarello V., 1995 - La vegetazione dell'ordine *Brometalia erecti* Br.-Bl. 1936 nell'Appennino (Italia). Fitosociologia, **30**: 3-45.
4. Bonanomi G., Allegrezza M., 2004 - Effetti della colonizzazione di *Brachypodium rupestre* (Host) Roemer et Schultes sulla diversità di alcune fitocenosi erbacee dell'Appennino centrale. Fitosociologia, **41** (2): 51-69.
5. Calandra R., 1995 - I suoli della montagna marchigiana. In: Atti Conv. "Salvaguardia e gestione dei beni ambientali nelle Marche", pp. 117-123. Tipolit. Trifogli, Ancona.
6. Caniglia G., Cappai A., 1990 - Ipotesi applicative della sinfenologia in cenosi prative. Informatore Botanico Italiano, **20**: 723-730.
7. Caporali F., 1995 - Agroecosistemi. In: Pignatti S. (ed.), Ecologia vegetale, pp. 383-434. UTET, Torino.
8. Castagnari G., 2002 - Lineamenti storico-ambientali. In: G. Fortunati (ed.), Aspetti geobotanici e lineamenti storico-ambientali dell'Alto Esino (Appennino umbro-marchiginao). Provincia di Ancona – Assessorato alla Tutela dell'Ambiente. Tip. La Nuova Stampa, Camerino.

9. Catorci A., Orsomando E., 2001 - Note illustrative della Carta della vegetazione del Foglio Nocera Umbra (N. 312 – Carta d'Italia I.G.M. – 1:50.000). Braun-Blanquetia, **23**: 1-94.
10. Catorci A., Fortunati G., Gatti R., Pieruccini P., Scocco P., 2003 - The grazing agri-ecosystem of mounts Rogedano-Puro (Marches Region, Central Italy): preliminary results of a multidisciplinary study. Contrib. Bot., **38** (2): 31-47.
11. Catorci A., Gatti R., Ballelli S., in press - Studio fitosociologico della vegetazione delle praterie montane dell'Appennino centrale. In: Catorci A. & Gatti R. (eds.), Le praterie montane dell'Appennino maceratese (Appennino umbro-maechigiano - Italia centrale). Braun-Blanquetia.
12. Cenci C.A., 1983 - Metodi di rilevamento fenologico in Graminee foraggere. Quaderni di Bioritmica 1. PMA, Palombi, Roma.
13. Crofts A., Jefferson R.G., 1999 - The Lowland Grassland Management Handbook (2nd edition). English Nature/The Wildlife Trusts, Peterborough.
14. Dierschke H., 1972 - Zur Aufnahme und Darstellung phänologischer Erscheinungen in Pflanzengesellschaften. In: Grundfragen und Method in der Pflanzensoziologie, Ber. Int. Symp. der Int. Verein für Veg. Junk Verlag, Den Haag.
15. French F., Sauer G.H., 1974 - Phenological studies and modeling in grasslands. In: Lieth E. (ed.), Phenology and Seasonality Modeling, pp. 227-236. Springer Verlag, New York.
16. Gatti R., Catorci A., 2005 - Contributo alla caratterizzazione dei pascoli alto-collinari dell'Appennino umbro-marchigiano ai fini zootecnici (Prati di Gagliole e Monti Rogedano-Puro). Tip. Arti grafiche Gentile, Fabriano.
17. Gratani L., Rossi A., Crescente M.F., Frattaroli A.R., 1999 - Ecologia dei pascoli di Campo Imperatore (Gran Sasso d'Italia) e carta della biomassa vegetale. In: Ricerche di geobotanica ed ecologia vegetale di Campo Imperatore (Gran Sasso d'Italia). Braun-Blanquetia, **16**: 227-247.
18. Hodgoson J., Illius A.W. (eds.), 1998 - The Ecology and Managemente of Grazing Systems. CAB International, Wallingford.
19. Pignatti S., 1982 - Flora d'Italia. Edagricole, Bologna.
20. Puppi A., Branzi G., 1989 - Rilevamenti fenologici su piante della flora spontanea. Quaderni metodologici CNR-IPRA (Roma), **12**: 9-38.
21. Sheath G.W., Clark D.A., 1998 - Management of Grazing Systems: temperate pastures. In: Hodgoson J. & Illius A.W. (eds.), The Ecology and Management of Grazing Systems, pp. 301-323. CAB International, Wallingford.

EVALUATION OF THE ITALIAN APENNINE ECOSYSTEMS WITH RESPECT TO ANATOMICAL AND ETHOLOGICAL CHARACTERISTICS OF THE ROE DEER

Paola SCOCCO[1*], Daniele SPARVOLI[2], Andrea CATORCI[2]
[1] Dipartimento di Scienze Veterinarie, UNICAM, Via Circonvallazione 93-95, **62024 Matelica (ITALY)**; e-mail: paola.scocco@unicam.it
[2] Dipartimento di Botanica ed Ecologia, UNICAM, Via Pontoni 5, **62032 Camerino (ITALY)**

Abstract: Based on a previous landscape synphytosociological analysis, the present work intends to test the degree of suitability for the roe deer (*Capreolus capreolus*) of the different Apennine habitats, relating the ecological and structural characteristics of the forest Apennine ecosystems (bioclimate, floristic structure and composition, food availability) with the specific anatomical characteristics of the digestive apparatus and alimentary habits of this ungulate. Although the roe deer is a ruminant, it must be considered a grazing rather than a browsing animal that carefully selects what it ingests, choosing plants or plant parts rich in cellular sap of high protein content, and avoiding, as far as possible, plants with a high content of raw fibre. The data in hand seem to point out a range of suitability for the roe deer of the hilly and low-mountain wooded habitats, among which, nevertheless, remarkable differences in ecological and environmental level of suitability have been recorded. One of the principal limiting factors is the relationship between the availability of trophic resources present at the beginning of the autumn, and the length of the winter period (corresponding to the phase of vegetative dormancy). Also of notable importance are vegetation patches, types of forest management and the different floristic and structural characteristics of the ecotonal belts.

Introduction

The present landscape of central Italy shows different levels of human influence that are strictly related to the landform characteristics of the area. In fact, while in the Apennine zone strong natural elements are still preserved, since the higher ground is occupied by extensive pastures and the slopes by woods and shrubby areas, within the steeper hilly sector the woods alternate with agricultural areas, small uncultivated surfaces and ancient rural settlements, often fortified, which dominate the narrow valleys below [4]. Thus this is an ecologically very diverse landscape, in which numerous ecosystems with different floristic and structural characteristics alternate with one another.

Classification of these ecosystems and mapping them represent important tasks for realizing a hierarchical system of ecosystem representation, since this type of analysis can offer a rational basis for many problems inherent in land management, sustainable development and nature conservation [8, 20]. A proposal to create a hierarchy for the ecosystems for the Italian territory has been adopted by Blasi *et al.* [1], who affirm that ecosystems are generally identifiable on the basis of their ecological homogeneity, in relationship to the surrounding territory. The need to establish a hierarchical scheme of ecosystem representation for the identification of the relationships between fauna and vegetation has also been demonstrated by Brandmayer *et al.* [2].

The methodology of landscape hierarchy lies at the basis of the publishing of the Map of the Ecosystems of the Province of Macerata (Central Italy), produced in support of animal-hunting planning with GIS [5], and mainly referring to population management of roe deer, since this species is re-colonizing the Apennines of central Italy.

The roe deer has a predominantly nocturnal life, but in winter it intensifies diurnal activity in the search for food. It can also be seen at dawn and sunset when grazing activity is more intense, with grazing periods of 2–6 hours distributed in 10–11 forays [7, 12]. The roe deer habitually frequents the edge of woods or clearings, above all after strong rains because of its perturbation by water that falls from leafy branches after storms [11].

Under optimal environmental conditions density should not exceed 8–12 units per 100 ha [12], even if in some zones of strong influx the density can be 20–25 head per 100 hectares [16, 17]. This is rarely compatible with the forestry and/or agricultural activities of the territory, nor with the maintenance of plant biodiversity and the structure of the forest ecosystems.

Concerning the ingestive system, the roe deer has a short and pointed face which helps it select plants or parts of them with great precision, and it is also helped by the presence of an arched line on the incisor teeth [14]. The roe deer possesses large salivary glands, about 25% greater than those of cattle (in proportion to their relative dimensions) [10]; this is linked to the fact that it directly carries part of the cellular sap of the plants (sugars and triglycerides) to the abomasum (glandular stomach) through the oesophageal groove; in this way it directly exploits the protein portion contained in the plants eaten.

The anatomical peculiarities of the roe deer's digestive apparatus make it more demanding in respect of diet. It is considered a concentrated selector [3, 10] and poorly adaptable to diets low in proteinaceous substances such as raw fibres, preferring leaves, buds and wild fruits.

In this work, the data present in the Map of the Ecosystems of the Province of Macerata and the territorial hierarchical reading have been correlated to anatomical and ethological considerations related to the roe deer, with the purpose of cartographically identifying the areas mostly suitable for this ungulate.

Material and Methods

The Map of the Ecosystems of the Province of Macerata (scale 1:100000) represents the distribution of 33 ecosystems, arranged in 19 subsystems (Geosigmeta); these characterize the nine landscape systems (Macrogeosigmeta) present in the territory of the Province of Macerata [5].

Each ecosystem was defined on the basis of: Prevailing structure (wood, pasture, fields under cultivation); Species that defines the vegetation physiognomy; Phytosociological characterization; Vegetation series and dynamic tendencies; Floristic and structural aspects of the prevailing phytocoenosis; Edge; Trophic characterization and productivity.

The suitability of the territory for roe deer has been tested starting from the data contained in this account. To this end, every ecosystem has been valued through the compilation of the card proposed by Simonetta [19] and based on the following environmental parameters referring to roe deer requirements: Altitude (evaluation range 0–3); Exposure (evaluation range 0–4); Morphology (evaluation range 0–5); Soil (evaluation range 0–8); Water endowment (evaluation range 0–5); Forest cover (evaluation range 0–30) (especially valued in reference to the food availability in the winter period); Presence of clearings (evaluation range 0–14); Contact with other habitats (evaluation range 0–14); Human activity (evaluation range 0–10) and Predators (evaluation range -5–0).

Results and Discussion

By means of the compilation of cards we attributed a total value to each ecosystem. Table 1 shows example evaluation cards referring to different sample ecosystems.

Table 1: Example of evaluation cards relative to some different ecosystems.

Environmental parameters	Wood of *Ostrya carpinifolia*	Tilled land	Wood of *Fagus sylvatica*	High Apenninic grassland	Wood of *Salix alba*	Wood of *Quercus pubescens* and *Quercus cerris*
Altitude	3	3	0	0	3	3
Aspect	3	2	0	1	1	2
Morphology	1	2	1	2	4	4
Soil	7	4	7	3	5	6
Water endowment	1	2	1	0	5	1
Forest cover	26	0	15	0	22	28
Presence of clearings	12	0	4	0	4	14
Contact with other habitats	10	0	10	8	10	10
Anthropization	9	5	10	10	7	8
Predators	-1	0	-1	-1	0	-1
Total value	**71**	**18**	**47**	**23**	**61**	**75**

Analyzing the data obtained we subdivided the ecosystems into three suitability classes: according to Simonetta [19] we assign the value ranging from 0 to 40 to the Class defined as Scarce or Null, the value ranging from 41 to 70 to the

Class defined as Poor, and the value ranging from 71 to over 85 to the Class defined as Good–very good.

In particular we define as:

– Good – Very good, habitat that guarantees trophic resources for the whole year and numerous sites for shelter; these two factors can allow an increased density and homogeneous distribution of individuals. According to the model of Pulliam [18], in this habitat source populations can develop; population that are under favourable environmental conditions and in which the number born is higher than those that die.

– Poor, habitat that indicates trophic resources available only for a limited period of the year or a significantly small number of shelter sites; in this case the presence of roe deer is possible, also at good densities, only in some seasons or in some phases of the daily activity of the animal, that can use a part of the habitat as a feeding site. According to the model of Pulliam [18], in this habitat only sink populations can develop, which without an influx of individuals coming from source areas would run into decline with an increased possibility of extinction. The presence of corridors (characterizing to landscape staircase) connecting these habitat with those in which the source populations live (good–very good habitat) is of fundamental importance for the presence of roe deer.

– Scarce or Null, habitat that does not indicate available trophic resources nor a large enough number of shelter sites; these two factors do not allow the installation of stable populations but only the presence of few isolated and/or occasional individuals. According to the model of Pulliam [18], this habitat represents a link area in which there is no possibility for development and maintenance of a roe deer population, also because the eventually fit area shows an elevated fragmentation index [8], with an almost total absence of corridors connecting between different patches.

This subdivision is essentially founded on the availability of food and the presence of suitable shelter, which represent two of the three essential parameters for the attribution of the characteristics of eco-field to a habitat [6]; the third parameter is represented by the index of reproduction.

In Table 2 the attribution to each class of suitability of the ecosystems analyzed is shown; assigning a different colour to each suitability class and transferring the data on the Map of the Ecosystems of the Province of Macerata, we obtained a derived map showing the distribution of the three suitability classes in the area studied (Fig. 1).

In particular, the ecosystems mostly suitable for roe deer are hilly woods of deciduous trees combined with a well-structured and rich shrub layer, with coppice forest management, above all if surrounded or adjacent to pastures or agricultural areas rich in natural nuclei (hedges, small woods, grassy uncultivated areas) [9]. In fact, in these ecosystems is recorded the maximum potential productivity of edible forage both inside the forest structure and in the ecotonal band. Besides, the phase of vegetative dormancy is relatively brief (90–100 days), with irregular presence of snow on the soil [15].

Table 2: Ecosystems classification according to their vocational classes.

MEDITERRANEAN REGION			
Bedrock system of conglomerate and sands			
Subsystem of reliefs with meso-mediterranean bioclimate			
	Good-Very good	Poor	Scarce or null
Hilly agro-ecosystem with few natural nuclei			X
Bedrock system of clay and sands			
Subsystem of reliefs with meso-mediterranean bioclimate			
	Good-Very good	Poor	Scarce or null
Hilly agro-ecosystem with few natural nuclei			X
System of alluvial sediments			
Subsystem of riverbeds			
	Good-Very good	Poor	Scarce or null
Ecosystem of riparian forests		X	
Subsystem of current and recent alluvial terraces			
	Good-Very good	Poor	Scarce or null
agro-ecosystem on plain with few natural nuclei			X
TEMPERATE REGION			
Bedrock system of limestones			
Subsystem of reliefs with low hilly bioclimate			
	Good-Very good	Poor	Scarce or null
Forest ecosystem of thermo-xerophilous oak woods	X		
Piedmont agro-ecosystem with some natural nuclei			X
Subsystem of reliefs with high-hilly bioclimate			
	Good-Very good	Poor	Scarce or null
Forest ecosystem of thermophilous oak woods	X		
Piedmont agro-ecosystem with many natural nuclei			X
Forest ecosystem of hilly hop hornbeam woods	X		
Ecosystem of secondary grasslands on hills		X	
Ecosystem of mostly rocky areas			X
Forest ecosystem of holm–oak woods on hills		X	
Subsystem of reliefs with low-montane bioclimate			
	Good-Very good	Poor	Scarce or null
Forest ecosystem of mesophilous oak woods	X		
Ecosystem of secondary sub-acidophilous grasslands		X	
Forest ecosystem of thermophilous beech woods		X	
Ecosystem of secondary low-montane grasslands			X
Subsystem of Karst-tectonic basins			
	Good-Very good	Poor	Scarce or null
Ecosystem of wet meadows and fens			X
Subsystem of reliefs with high-montane bioclimate			
	Good-Very good	Poor	Scarce or null
Forest ecosystem of mesophilous beech woods		X	
Ecosystem of secondary high-montane grasslands			X
Subsystem of reliefs with subalpine/alpine bioclimate			
	Good-Very good	Poor	Scarce or null
Ecosystem of primary grasslands			X

Subsystem of slopes with subalpine bioclimate			
	Good-Very good	Poor	Scarce or null
Ecosystem of rocky slopes			X
Bedrock system of marly limestones			
Subsystem of reliefs with high-hilly bioclimate			
	Good-Very good	Poor	Scarce or null
Forest ecosystem of semimesophilous woods with hop hornbeam and oaks	X		
Hilly agro-ecosystem with many natural nuclei		X	
Bedrock system of sandstones			
Subsystem of reliefs with high-hilly bioclimate			
	Good-Very good	Poor	Scarce or null
Forest ecosystem of sub-acidophilous chestnut woods	X		
Bedrock system of clay and sandstones			
Subsystem of reliefs with low-hilly bioclimate			
	Good-Very good	Poor	Scarce or null
Forest ecosystem of thermophilous woods with hop hornbeam and Turkey oak	X		
Hilly agro-ecosystem with some natural nuclei		X	
Subsystem of reliefs with high-hilly bioclimate			
	Good-Very good	Poor	Scarce or null
Forest ecosystem of semi–mesophilous woods with hop hornbeam and Turkey oak	X		
Hilly agro-ecosystem with many natural nuclei		X	
Bedrock system of clay and sands			
Subsystem of mostly sandy reliefs with low-hilly bioclimate			
	Good-Very good	Poor	Scarce or null
Hilly agro-ecosystem with few natural nuclei			X
Subsystem of mostly clayey reliefs with low-hilly bioclimate			
	Good-Very good	Poor	Scarce or null
Hilly agro-ecosystem with few natural nuclei			X
System of alluvial sediments			
Subsystem of riverbeds			
	Good-Very good	Poor	Scarce or null
Ecosystem of riparian forests		X	
Subsystem of current and recent alluvial terraces			
	Good-Very good	Poor	Scarce or null
agro-ecosystem on plains with few natural nuclei			X
Subsystem of old alluvial terraces			
	Good-Very good	Poor	Scarce or null
agro-ecosystem on plains with few natural nuclei			X

From a phytosociological point of view, these formations mainly belong to the alliance *Carpinion orientalis*, and to the associations *Scutellario columnae-Ostryetum carpinifoliae*, *Aceri obtusati-Quercetum cerridis*, *Peucedano cervariae Quercetum pubescentis*, *Roso sempervirentis-Quercetum pubescentis*, *Carici sylvaticae-Quercetum cerridis* and *Cyclamino hederifolii-Castaneetum sativae*.

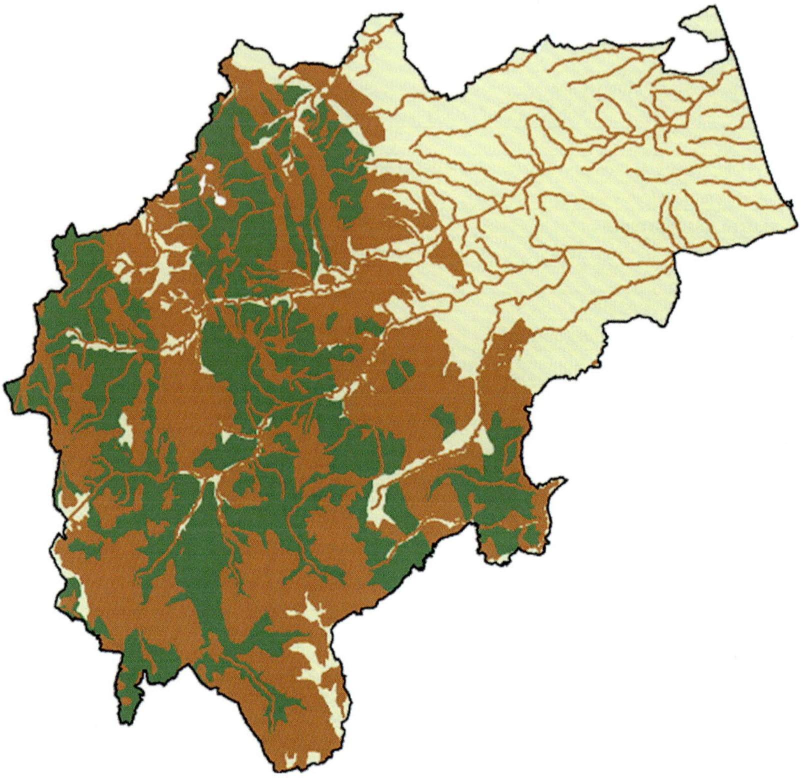

Fig. 1: Derived map showing the territory distribution of the three vocation classes: green = good-very good; brown = poor; yellow = scarce or null.

Conclusion

This study represents the subsequent phase of the analysis of the landscape through the application of a hierarchical scheme and, currently, the preliminary phase for appliance searches on the roe deer carrying capacity of the ecosystems present in the territory of the Province of Macerata.

Other research are in progress in order to:

– Examine, at a greater scale (e.g. 1:10.000), some sample areas to evaluate the relationships between landscape models and the animal community, and then develop predictive models.

– Verify the predictability of appropriate environmental models, for some kinds of particular management interest (i.e. roe deer), by means of field data.

– Verify the density of the roe deer population in sample areas.

These studies can therefore have important consequences in the evaluation of: 1) the economic and environmental sustainability of the numerical increase of the population of roe deer (agricultural problems and traffic safety); 2) the degree of tourist attraction of the territory.

Acknowledgements: The authors thank Mrs. Sara Tulczyjew for the linguistic revision.

REFERENCES

1. Blasi C., Carranza M., Frondoni R., Rosati L., 2000 - Ecosystem classification and mapping: a proposal for Italian landscape. Appl. Veg. Sci., **3**: 233-242.
2. Brandmayer P., Cagnin M., Mignozzi T., Scalercio S., Pizzolotto R., 1997 - Misura efficace della biodiversità animale in ambienti mediterranei e sue applicazioni. S.It.E. Atti, **18**: 581-586.
3. Catorci A., Fortunati G., Gatti R., Pieruccini P., Scocco P., 2003 - The grazing agri-ecosystem of mounts Rogedano-Puro (Marches region, central Italy): preliminary results of a multidisciplinary study. Contrib. Bot., **38** (2): 31-47.
4. Catorci A., Gatti R., 2004 - La vegetazione forestale della Provincia di Macerata. In: Provincia Macerata (ed.), Alberi custodi del tempo, pp. 13-47. Biemmegraf, Macerata.
5. Catorci A., Perna P., Sparvoli D., Vitanzi A., 2005 - Carta degli ecosistemi della Provincia di Macerata ai fini faunistico-venatori. Bieffe, Recanati.
6. Farina A., 2004 - Lezioni di ecologia. Ed. UTET Libreria, Torino.
7. Ferloni M., Simonetta A. M., 1998 - Cervidi. In: Greentime (ed.), Principi e tecniche di gestione faunistico-venatoria, pp. 225-250. Grafiche Zanini, Bologna.
8. Forman R., Godron M., 1986 - Landscape ecology. John Wiley & Sons, New York.
9. Fratoni F., Corrado G., 1984 - Gli interventi selvicolturali in un'azienda faunistica estensiva dell'Appennino umbro. Atti V Congr. Allevamento di selvaggina a scopo alimentare, pp. 117-124. Regione dell'Umbria.
10. Hofmann R.R., 1984 - L'adattamento dell'apparato digerente nei cervi. Atti V Congr. Allevamento di selvaggina a scopo alimentare, pp. 81-104. Regione dell'Umbria.
11. Kerschagl W., 1952 - Rehwildkunde. Ed. Hubertusverlag Richter Springer, Vienna.
12. Ladini F., 1998 - Il capriolo. Ed. Tassotti, Bassano del Grappa.
13. Lewis L.D., 1998 - Alimentazione e allevamento del cavallo. Ed. EMSI, Roma.
14. Lucifero M., Giorgetti A., 1987 - Lo sfruttamento degli ungulati selvatici in ambiente collinare appenninico: possibilità e prospettive. Atti IX Conv. Gruppo di studio per allevamenti di selvaggina, pp. 149-175. Regione dell'Umbria.
15. Orsomando E., Catorci A., Pitzalis M., Raponi M., 2000 - The phytoclimate of Umbria. Parlatorea, **4**: 5-24.
16. Perco F., Perco D., 1979 - Il capriolo. Ed. Carso, Trieste.
17. Perco F., 1987 - Ungulati. Ed. Lorenzini, Udine.
18. Pulliam H.R., 1988 - Sources-sinks and population regulation. American Naturalist, **132**. 652-661.
19. Simonetta A.M., 1998 - Vocazioni faunistiche, metodi di valutazione della capacità attuale e potenziale in rapporto a possibili miglioramenti. In: Greentime (ed.), Principi e tecniche di gestione faunistico-venatoria, pp. 25-41. Grafiche Zanini, Bologna.
20. Taiton N.M., Morris C.D., Hardy M.B., 1996 - Complexity and stability in grazing systems. In : Hodgson J. & Illius A.W. (eds.), The Ecology and Management of Grazing System, pp. 275-299. CAB International. New York.

V

CONSERVATION PLANNING AND MANAGEMENT

PLANECO - PLANNING IN ECOLOGICAL NETWORKS

Piergiorgio BELLAGAMBA
Facoltà di Architettura, Universita' degli Studi di Camerino, Colle dell'Annunziata, **Ascoli Piceno (ITALY)**; e-mail: piergiorgio.bellagamba@unicam.it

Abstract: PLANECO is a research project founded by the Italian Ministry of University and Scientific Research in the years 1998–2000. It involves five universities: Camerino, L'Aquila, Pescara, Isernia and Roma 3. Among its goals is the definition of planning criteria for a territory considered as an ecological system, with particular attention to environmental continuity, and including protected areas that are not isolated but linked as components of an ecological network.

Introduction

In Italy the interconnection between the natural and cultural systems, together with the overlay of a national ecological network, interacts at all levels of planning and land use transformation.

The Planeco Project aims to contribute to the solution of this problem in three ways:

1- With discussion of the planning role, useful to individualize, conserve and restore environmental continuity;

2- With directions to orient the land use within protected areas, considered as core areas of an ecological network;

3- With new possible forms of urban development, to allow the connective functionality of suburban areas.

In the first case, 'dynamic indicators' will be used to determine the present situation and possible scenarios. Among these indicators are settlement dispersion, infrastructure density and permeability, and spatial and quantitative relationship between natural and urban land use.

In the second, the new consideration of a ramified zone structure of protected areas takes the place of the old consideration of a centralised zone structure. This allows a better environmental continuity between the inside and the outside of the parks, and supports an eco-centric planning process to be tested in areas of special environmental protection.

In the third, town-planning principles like linear urban development are reviewed because they fragment environmental continuity and threaten the quality of nature.

Results

One of the first results of the Planeco Project is the identification of the elements connected to environmental continuity on a national scale. CORINE Land Cover data have been used to elaborate the National Bio-permeability Map. This provides evidence for the geography of land use to which the different levels of nature and the most relevant national environmental systems can be linked. An uncertain role is played by the agricultural areas, as further data are necessary to elaborate their specific 'fragmentation indicators' (e.g. related to geometry, texture and conductivity). The analysis of the Planeco research at the moment is therefore based mainly on environmental systems and land use. In the preliminary recognition of bio-permeability conditions in Italy, the analysis focuses on three macro-categories based on land use features: hydro-morphology; agriculture and forestry; barriers.

The Planeco Project has produced the following outputs through different instruments of territorial interpretation:

- National map of bio-permeability (Fig. 1), environmental continuity (Fig. 2), and complex infrastructural and urban barriers;
- Central Apennines map of the environmental systems and landscape units (Fig. 3), environmental fragmentation of infra-structural (IF indicator) and urban (UF indicator) barriers (Fig. 4), and fragmentation tendency;
- Central Apennines map of zoological biodiversity (Fig. 5), the relationship between potential species, habitats and fragmentation tendency, and the relationship between environmental fragmentation and municipal planning (Fig. 6).

Discussion

Scientific knowledge in the field of biogeography and landscape ecology has assured that environmental fragmentation causes decrease in biodiversity. It is important to find the responsibilities of territorial planning in this phenomenon and the possibilities of the territorial sciences for producing new methods of environmental analysis.

A methodology to assess the state of an ecological network has to take into account a certain number of criteria that should not be confined to species' habitats, but should be applied to the larger context of landscape. The latter is becoming a fundamental topic in studies on nature conservation. In areas of ancient human usage it represents the idiosyncrasy of the cultural elements within an ecological network. The wider landscape, therefore, is the appropriate context in ecological network development. Cultural landscapes, applied landscape ecology and environmental policy can assist in this process. A common objective of European policies is to reinforce the ecological sustainability of the environment. This should be considered within the framework of sustainable development, which retains both natural and cultural aspects. Many studies, however, lack the latter. It is hence proposed to use the wider public policy context, including sustainable development, in ecological network development.

Another common objective of European policies is to conserve biological and landscape diversity. In order to achieve this, it is necessary to move beyond ecology and consider human influence on the ecological structure in terms of socio-economic needs and pressures on the territory. The European policies, accordingly, have to be supported by national, regional and local policies dealing with physical planning and having thus a strong cultural input.

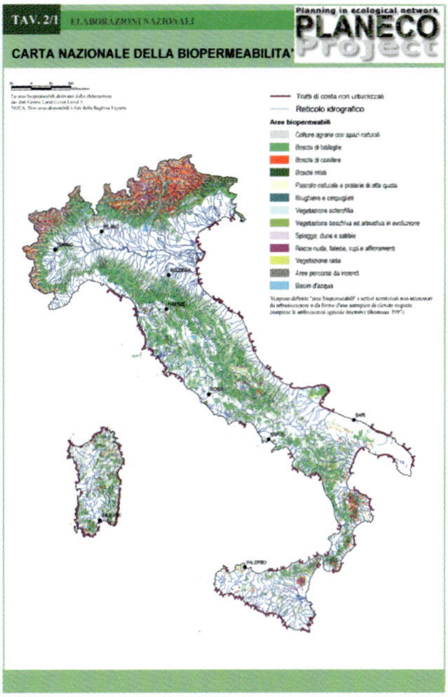

Fig. 1: National map of bio-permeability.

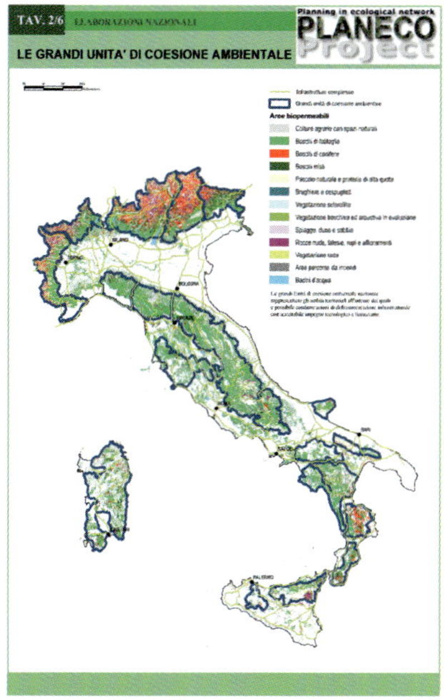

Fig. 2: The major units of environmental continuity.

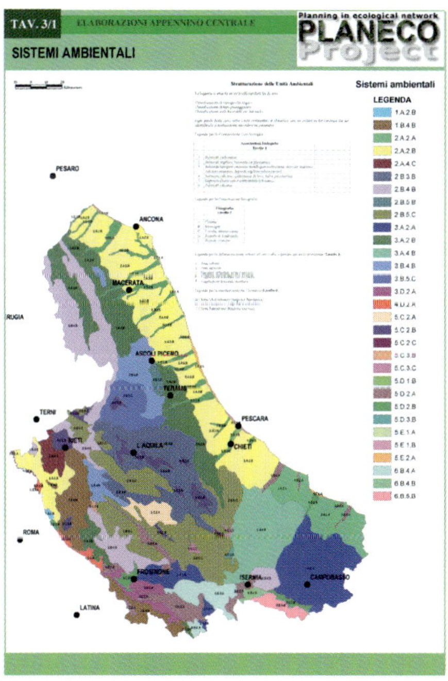

Fig. 3: Environmental systems and landscape units.

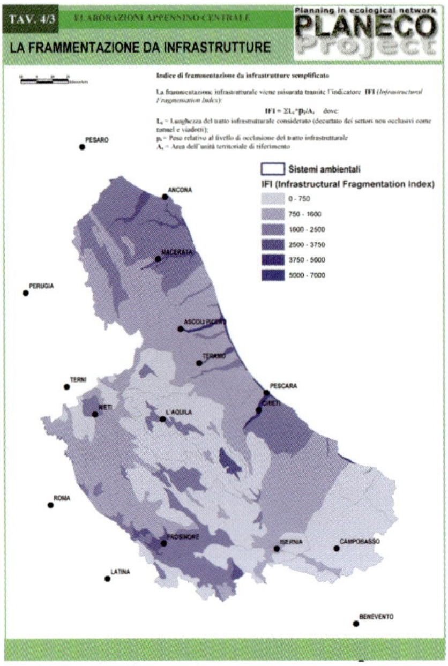

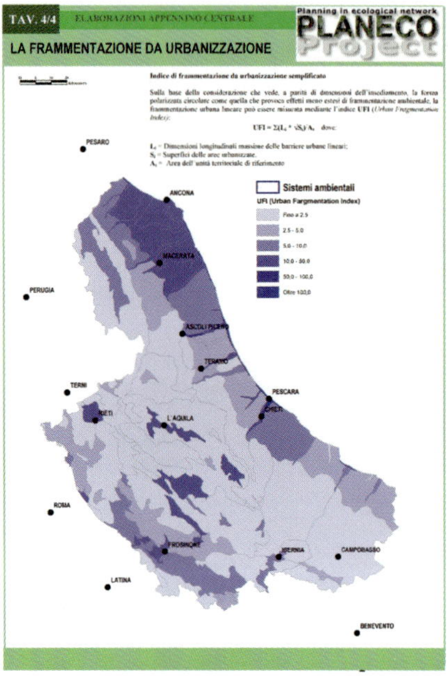

Fig. 4: Environmental fragmentation of infrastructural (left) and urban (right) barriers.

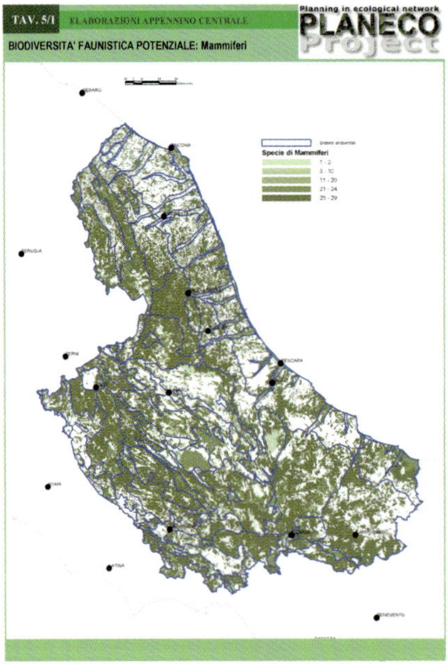

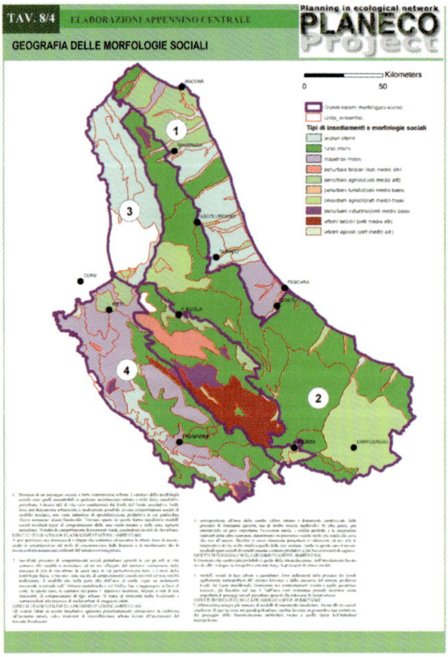

Fig. 5: Potential zoological diversity. **Fig. 6: Geography of social morphologies.**

In Italy, however, physical planning has met a series of limitations that have led to advance the concept of 'de-anthropization of the plan'. The concept arises from the consideration that in the country any practical and implementation action on the de-fragmentation of natural habitats can be planned only at municipality level. But at this level environmental continuity is invisible, while it is more evident at national and regional level. This is therefore the scale to use for theoretical inputs and actions.

Past research shows how ecological networks are the results of science based on nature conservation, landscape ecology, land use and environmental planning. This indicates how human activities, education and awareness here play a relevant role. Accordingly, the development of ecological networks in Europe and, particularly, in Italy should be carried out incorporating cultural aspects into the features of an ecological network and, in turn, considering the interaction between its natural and cultural components. Socio-economic trends and functions of the site should be taken into account, linking ecology to environmental economics. Then, the integration of ecological networks into the planning system becomes imperative.

Conclusions

The wider landscape context should be used in the development of ecological networks, including human processes that have shaped the landscape and its natural and cultural components. The wider public policy context can be addressed in parallel, with the goal of balancing nature conservation with sustainable development. This assumption emphasizes how ecological networks, in areas of ancient human occupation like Italy, are located between nature and culture. For this reason the planning of future landscapes in these areas should imply not only ecological and natural aspects, but also cultural, social, economic and political considerations.

REFERENCES

1. Adams L.W., Leedy D.L. (eds.), 1991 - Wildlife conservation in metropolitan environment. Proceedings of a National Symposium on Urban Wildlife, Cedar Rapids, Iowa.
2. Ahern J., 1994 - Greenways as ecological networks in rural areas. In: Cook E.A. & Van Lier H.N. (eds.), Landscape planning and ecological networks, pp. 159-178. Elsevier, Amsterdam.
3. Arnofi S., Filpa A., 2000 - L'ambiente nel piano comunale. Guida all'eco-management nel PRG. Il Sole 24 Ore, Milano.
4. Bellagamba P., Cannata G., Olivieri M., 1993 - Piano del sistema dei Parchi della Regione Lazio. Università LUISS, Roma.
5. Bellagamba P. (in gruppo ASTAC), 1999 - Studi PIT Regione Marche. Problematiche di distrettualizzazione, Ancona.

6. Cook E.A., Van Lier H.N., 1994 - Landscape planning and ecological networks. Elsevier, Amsterdam.
7. Di Ludovico D., Properzi P., Romano B., Tamburini G., 2000 - The "natural" town. Methodology hypothesis for defragmentation of the urban organism. World Conference "The Human Being and the City", Università di Napoli "Federico II", Napoli.
8. Filpa A. (ed.), 2000 - Il Parco Nazionale dei Monti Sibillini nel sistema dell'Appennino. Sala Ed., Ascoli Piceno.
9. Flink C.A., Searns R.M., 1993 - Greenways. A guide to planning, design and development. Island Press, Washington DC.
10. Forman T.T., 1996 - Land Mosaic. Cambridge University Press, Cambridge.
11. Gambino R., 1994 - I parchi naturali europei. NIS Ed., Roma.
12. Gambino R., Negrini G., Peano A., 1997 - Parchi e territorio in Europa: nuovi orientamenti per la pianificazione dello sviluppo sostenibile. Atti XVIII Conferenza Italiana di Scienze Regionali Europa e Mediterraneo, Siracusa.
13. Malcevschi S., Bisogni L.G., Gariboldi A., 1996 - Reti ecologiche e interventi di miglioramento ambientale. Il Verde, Milano.
14. Peano A., 1997 - Parchi naturali in Europa. Il CED-PPN. Urbanistica dossier, n. 7, Roma.
15. Properzi P., Romano B., Tamburini G., 1998 - La continuità ambientale nella pianificazione dei territori naturali. Atti XIX Conferenza AISRE, L'Aquila.
16. Romano B., 1999 - La continuità ambientale nella pianificazione. Urbanistica, **112**: 156-160.
17. Romano B., Tamburini G., 2000 - Continuità ambientale. Andromeda Ed., Roma.

THE CARTOGRAPHY OF LANDSCAPE UNITS IN SÃO BENTO DO SAPUCAÍ – SERRA DA MANTIQUEIRA (SÃO PAULO, BRAZIL)

Marcello MARTINELLI
Departamento de Geografia, Universidade de São Paulo, Av. Gen. Cavalcante de Albuquerque 960, **05638-010 São Paulo (BRAZIL)**; e-mail: cartotem@ig.com.br

Abstract: This paper sets out our reflections on the reality of relations between nature and society, which will integrate landscapes and their dynamics. The cartography of landscape units takes as its methodological basis a proposition in line with this philosophy and establishes a combined series of analyses, which contemplate structures, functions and dynamics considered in scientific terms, with a bearing on nature and society. Arguing for a synthesis, we can construct the map of landscape units for São Bento do Sapucaí County, which shows groups of spatial elements characterized by groups of attributes. The organization of the legend will explain the methodological position adopted for the study, as well as the reasoning by which the map of landscape units was achieved.

Introduction

Studies of ***landscape*** are not a novelty. As we work with the units of landscape, a series of questions emerges, which considers the relationship between nature and society.

We can take into account reflections about this relationship back to Antiquity. In the pre-Socratic era (>5th century BC), human beings integrated nature into the totality of all that existed. It was the *Physis*, to which belonged the sky, the land, stones, plants, animals and men in all their labours and actions, and the gods. With Plato and Aristotle (4th–3rd centuries BC), an evaluation took place of man and his ideas. The concept of nature as non-human was set down in writing at the critical moment for Greek democracy. However, the great statement of opposition to nature came through Judaic-Christian influence. For Jewish and Christian people, man possessed privilege before nature. God created man, in His own image and likeness.

René Descartes (1596–1650) reconciled this antagonism to establish thoughts which we still have today. Nature is seen as a resource, and its knowledge is useful in our lives [22]. During the era of the 18th century Enlightenment, with the full affirmation of capitalism, comprehension of reality came from within itself, abandoning religious dogmas. In the 19th century, industrial capitalist civilization crystallized a world vision with an objective nature external to man [34]. Thus, the human sciences separated from natural sciences. It was ecology and, mainly, the ecological movements themselves that promoted and created some new and more integrated concepts of the world, pointing to new approaches for present day scientific studies [22]. With the General System Theory of Bertalanffy [7], scientists took back the concept of integrated reality. Ecology was the first scientific field to absorb this, and Tansley proposed in 1935 the concept of the ***ecosystem***, which was consistent with it.

The first scientist to work with this concept was Sotchava [49]. He created a method to study Physical Geography – the ***geosystem***. He presented his conceptualization as: "geosystem is the expression of the natural phenomenon, or, in other words, the ecological potential of certain space where there is biological exploitation, which can be influenced by social and economic factors, in the structural and spatial features, but, without having necessarily, according to dynamic procedures, an internal homogeneity". Later, Bertrand proposed the study of ***landscape***, possible only within the bounds of Global Physical Geography. "Landscape is, in a space, the result of the physical, biological and human elements in dynamic combinations, which reacting one over another produces the landscape as a unique set in a continual evolution" [8]. In the taxonomy of landscape, this author selects the ***geosystem*** unit as –"the scale where there is a larger part of the interrelation phenomenon between the landscape elements, involving dialectic conditions, and they are of main interest to the geographer". It is the human spatial-temporal scale [8].

Another field of research linked to the study of landscape is ***Landscape Ecology***, suggested by Troll, as the science of landscape structure and function [52]. In landscape, from ecological studies emerges the concept of ***Land unit***, a portion of land ecologically homogeneous, at a certain level of scale [57]. Tricart's ***Eco-dynamics*** is another integrated proposal: landscape study made through its dynamic behaviour. He established landscape eco-dynamic units, pointing out three major categories: very stable, intergrading and strongly unstable media [51]. Not much later, the suggestion emerged of ***Eco-geography***; Tricart and Kilian formulated it, designed for natural planning. The authors devised methods for the natural environment and offered global knowledge resources, as well as their dynamics. In this sense, we may observe the unequal degree of appreciation of the ecosystems in which society articulates its actions [50]. The ***Gaia Theory***, elaborated by James Lovelock in the 1960s, is another holistic proposal. This theory considers Earth and life as a complex and indissoluble system. Rocks, water and sky do not have life, but they are very much integral to our life processes.

At the present moment in the history of human society, the capitalist mode of production, today dominant worldwide, has instituted nature in opposition to man and viewed objectively. Nature is an inexhaustible resource for society, in spite of many disastrous consequences for the environment. In the 1960s–1970s came an increased awareness about nature and its exhaustible resources and, therefore, for a tendency to initiate development which would become unsustainable, due to uncontrolled economic growth.

In historical and dialectic materialism, in search of integration, the relation between man and nature is achieved by means of a mediation operated by the workers, members of a society, predisposed to certain historically defined social relations. Meanwhile, current industrial capitalism has converted the worker into an object, separating him from his products, their means of production and his own nature. What society produces from nature starts to take part in human history, in its space and time [31].

Even with minor manifestations emerging at the end of the 19th century, at the peak of the Second Industrial Revolution, with its industries, the environmental question exploded after World War II, awakening to new considerations about reality. Society is now conscious of the finite character of natural resources. Therefore, society is beginning to question scientific and technological advances conceived mainly in previous eras. In the determinist consideration of Traditional Geography, nature imposed conditions and obstacles on the development of society. The Critical Geography of the 1970s opened horizons for a social transformation where knowledge would be its own artisan [34]. With a growing incorporation of science, technology and information into this field, in progress since the 19th century and with a major increase in the 20th, the natural background has transformed itself within the milieu of technical-scientific-information.

At the present time, the consciousness about the environmental debate is evolving with ever more impulse. Today, we can also count upon a strong input from the non-governmental organizations that have emerged since the 1960s [14, 44, 45, 46].

Landscape

It seems that the aesthetics of landscape emerged in China, at the time of the first centuries of the Christian era. The concept of Landscape expands during the Renaissance period where the expression of landscape was through outdoor oil paintings. Landscape was perceived as integral to beauty, harmony and amenity [6]. Therefore, in a whole world vision outside of theological considerations, the West found a new interest in nature through landscape painting. Even so, people began to look at nature from other perspectives, finding delight in its beauty. At the same time, nature also became an object of study.

In geography, landscape appeared from the 18th century, demonstrating visual expression of the physiognomy of a certain area. Alexander von Humboldt (1769–1859) exalted landscape in his Nature Note books, published at the beginning of the 19th century. Landscape is an image. at the end of the 19th century, photography would transform landscape painting, available to just a few people, into a printed image reproduced for the benefit of many. Landscape does not show only what nature has made. It incorporates too the consequences of culture, which at the same time compounds it – therefore, it is not out of social context in its historic movement. Landscape and its values integrate the whole environment – we can read into it values, representations and symbols [5, 16, 47].

Landscape is the physiognomic expression of the geographical space that we can observe; it constitutes the scenery of our existence. Therefore, its concept is impregnated with cultural and ideological connotations. Several components participate in the moment of its capture. Perception of landscape interacts not only with vision, but also with smell, flavour, sound and texture, besides the life experience of each person crystallized by his or her culture, thoughts and feelings. This improves the imagination too.

Morandi & Gil [35] emphasize that landscapes are unmatched spectacles. A landscape image does not have the capacity to reveal certain aspects, only by direct contact: "the sound, the temperature, the smell, the texture, the taste, the movement, at last, its real dimension". In their observations they assert that, "the landscape has the power of waking up the feelings and the most profound emotions of humans beings: ecstasies, curiosity, tenderness, solitude, auto-confidence, fear, inspiration, courage, joy, melancholy, tranquillity, excitation, greatness, smallness".

The panoramic appreciation of landscape has, as a first visual impact, a frontal view from the ground, providing a sequence of silhouettes in vertical parallel planes that follow into the distance. The arrangement of these planes organizes the visualization of the landscape in a scalar succession that reduces in direction to the horizon, at the same time that it interposes outlines masking part of the spatial ensemble that remains behind [13, 25, 27]. We may perceive the same landscape in distinct forms. Therefore, we can have a landscape for the geographer, the tourist, the conservationist, the planner and the ecologist [20].

For the geographer, since the end of the 19th century, landscape has been an objective study, and it is possible to describe and analyze it, therefore, without considering the aesthetics and the form that we perceive or interpret. The landscape of the tourist is the screening of a portion of space as a spectacle. A photographic record may be proof that somebody was in that place. The landscape of the conservationist is a land for preservation as a testimony to interactions between environment and human actions. For the planner, the landscape is a field, which will receive management, due to its value in a prospective vision. The ecologist sees the landscape as an organizational level of ecological systems, where they occur, and how they are controlled by a certain number of processes. Recently, the European Landscape Convention, held in Florence in 2000, defined landscape as: "a territory portion as it is perceived by people, where the characteristics result from natural and/or human actions and also from its interrelations" [20].

The study area

São Bento do Sapucaí County, with a total area of 279 km^2 and 10,355 inhabitants, is located in the south-western portion of the Serra da Mantiqueira highlands in the São Paulo state of Brazil, on the Minas Gerais state frontier. The county is situated between longitudes 45°35' and 45°45' W. and latitudes 22°35' and 22°45' S., and consequently located completely in the tropical zone. Its foundation occurred at the end of the 19th century, with a little church erected at São Bento. Its initial activities were related to dairy cattle breeding. Today tourism is becoming the county's greatest economic activity [48].

A picture, as an inked drawing of a field sketch, may present the whole landscape of the study area (Fig. 1).

Fig. 1: The landscape of the São Bento do Sapucaí county as viewed from a relief height (Alto Do Campestre – 1910 m) located in the far west side, looking to the far east sector.

Delimitation of landscape units

The identification of landscape units in the county, the object of our study, has brought into focus knowledge of diverse components in an integrated way. They are the geological structure, lithology, relief, climate, soil, vegetation, land use and all other manifestations of human activity throughout history, as well as the varying analyses of interrelations, which model a multiple reality [41, 42]. These orientations accord with the methodological positions adopted by the literature linked to the study. They consider landscapes from the scale of wide-ranging surfaces, to that of small-ranging areas, expressing the unity of each place, reflecting the natural and cultural history of a territory at a specific moment in time. In addition to this holistic vision, there is also a continuous dynamism viewed as unique for each place. The adoption of a work scale will certainly have implications for the decision for a conscious taxonomy.

We used studies carried out by various authors in different scientific disciplines on the territory of São Paulo state and other parts of Brazil, as the foundation to recognize in the county of São Bento do Sapucaí the proposition of the space-time scalar level of ***geosystem***. This systemic level is in accordance with the landscape hierarchical assertion of Bertrand. A refined study at more detailed space-temporal levels is welcome. Our work will not research the level of these subdivisions [8].

Therefore, we are able to see that the fundamental methodological question for the discernment of the study of units of landscape is the spatial-temporal scale. Its appropriateness must be compatible with the category of study, the interested part of reality compatible with the resolution of the phenomena studied, which needs certain time for its organization with consequent manifestation of spatial characteristics. However, these phenomena not only occur consecutively, but also simultaneously. In space, we may consider the coexistence of diverse time-scales. For that reason, the spatial, and concomitant temporal, categories of interest would be inserted into a certain level of the hierarchical sequence of spatial-temporal scales. These can cause variation in the order and in the respective relation of degrees of biological or social character, each corresponding to the suggestion of a cartographic scale for its material representation [6, 8, 10, 11, 12, 18, 19, 21, 23, 24, 26, 44, 47].

The reasoning for the identification of the units of a landscape for its subsequent cartography follows a methodological flow of reference. This sequence starts from awareness about the questions asked of the environment, which must be directed towards nature as much as society, remembering that it is from our own society that the problems emerge significantly [33, 36]. Since we are aware of this basic point of view, in beginning construction of this reasoning, we shall have to consider as fundamental, the geo-structural and litho-geomorphologic knowledge, at the dynamic level of reality. In this sense, São Bento do Sapucaí County is in the Atlantic morpho-structural complex; a crystalline structure caused by a tectonic process, the Paiol Grande fault.

It is possible to recognize in the territory, as lithological units, the gneisses and varied migmatites of the lower Proterozoic, found almost everywhere, with metadiorite intercalations of the lower Proterozoic too, but present more in the western part. Near the flood plains of the main drainage axis, the Sapucaí-Mirim River and its larger tributaries, and the Serrano and Paiol Grande rivers accumulated Holocene alluvial sediments.

The large morpho-structural complex mentioned above, genetically represented by the middle and upper Precambrian folding zone is indicative of inherent relief patterns, with large altitude variation and parallel ridges and valleys. Enclosed within it, at a second level of taxonomic order of terrestrial relief, is the Planalto Atlântico morpho-sculptural complex [43]. Thus, São Bento do Sapucaí County is inserted in the Planalto and Serra da Mantiqueira relief unit within the Planalto Atlântico morpho-sculptural complex, with specific geotectonic, structural and lithological characteristics. Locally, this complex exhibits a block of relief built up of varied lithology, reaching altitudes of 1,700–2,000 m, starting at the local base level of 850 m [43].

The forms of the relief result predominantly from denudation processes, presenting high hills with sharp summits and convex crests, with enclosed valleys, creating steep hillsides, sometimes with slopes of over 60% gradient. This morpho-sculptural ensemble exhibits a dendritic drainage pattern. Meanwhile, when it adapts itself to faults, folds and characteristic lithological contacts of regional structural directions, it shows a rectangular drainage pattern. In its entirety, the county constitutes an irradiation sector of the regional drainage networks. In particular, the main basin axis of the Sapucaí-Mirim River cuts transversely across the directions of the structure, forming a series of defiles, visually evident in the landscape.

In spite of the development of soil cover, as Cambisol, Lithosol, Podzol types, this area exhibits many rocky outcrops that perceptibly mark the singular characteristics of the landscape [1, 43]. The county's climate articulates itself at the zone controlled by equatorial and tropical air masses. It is an area with raised participation of tropical and maritime air mass with flows coming from the east that give stability to the weather in winter.

Since the plateau block reaches altitudes of over 1,700 m, this constitutes an altitudinal factor mitigating the local climate, which has an annual average temperature of 13° C. In the winter period, temperatures fall below 10° C, while in summer they oscillate around 16° C. Due to São Bento do Sapucaí being situated on the other side of the Serra da Mantiqueira mountain range, in the orographic rainfall shadow, more intense on the escarpments, it displays a lower rainfall, remaining at the annual average of 1,670 mm. In winter, there is a consequent reduction of precipitation: from 200 to 400 mm. The passage of cold fronts promotes the occurrence of bad weather; however, when these arrive they are weakened [32].

From the accurate research compiled by Diniz and Furlan [15] about the vegetation of a park in the adjacent county of Campos do Jordão, we can make some similar considerations of the geobotanical framework of the study area. Large transformations made by human activities in the region, during the period of coffee cultivation, affected this county. Coffee cultivation started near this locality and spread to the neighbouring state of Minas Gerais at the end of the 19th century. Due to exhaustion of the soil and lack of slave labour, the coffee plantations were transformed into pastures for the extensive breeding of dairy livestock. Later, in the last half of the 19th century, the county had some industrial development. Today São Bento do Sapucaí is a centre of tourist attraction that takes advantages of cooler weather, besides having the infrastructure and satisfactory access roads for visitors. In this respect, the primeval vegetation has been deeply degraded, due to deforestation with burning to prepare the soil for pastures. Even so, there were significant replacement of autochthonic components by allochthonic, and the spread of cosmopolitan species arriving from other Brazilian ecosystems [17].

Thus the study area presents, as phyto-geographical units, Subtropical Forest of *Araucaria angustifolia* and *Podocarpus lambertii* relicts, so-called araucaria woods, and patches of Subtropical Leafy Forest clumps and, on inter-fluvial ridgeline summits, high grassland clearings. The Subtropical Leafy Forest clumps are on the ridge alignments, relief slopes and alluvial bottomlands. In addition, there are different secondary formations, both forest and meadow, in areas occupied by human activities over several economic cycles that were part of the region's social history, besides patches of reforestation with *Pinus elliotti* [15, 56].

The layout of the potential vegetation, as the basic information for the proposed cartography, is very complex. Meanwhile, based on the studies of Ab'Saber & Bernardes [1], Ab'Saber [2, 3, 4], Diniz & Furlan [15], Troppmair [53] and Viadana [55], we have solid foundations. Therefore, we can confirm that in the region in which São Bento do Sapucaí is located, the primeval vegetation coverage was represented by mixed woodland (Subtropical Leafy Forest with *Araucaria angustifolia*), araucaria woods (Subtropical Forest with *Araucaria angustifolia* and *Podocarpus lambertii*) and high grassland (Subtropical Grass-wood Meadow).

According to these authors, the rise of the Serra da Mantiqueira, which has occurred since the Cenozoic (Tertiary), has elevated the land above 1,100 m altitude. This elevation has allowed the region to be in a colder climatic zone – from 5° to 6° C above to the actual temperature – suitable for the establishment of an ecological refuge, which today exhibits high grasslands alternating with patches of araucaria woods. Consequently, from these assertions regarding the dynamic tendencies, we could say that, at some restricted high grassland areas over 1,600 m, as well as in the remaining elevated patches of the Subtropical Leafy Forest associated to the Subtropical Forest of *Araucaria angustifolia* and *Podocarpus lambertii* relicts, we could see something like a primary fluctuation [54].

The next step of our study will consider the space produced by human society. It is utilized territory, where land use and soil coverage, both urban and rural, have their greatest expression. In this stage, we shall evaluate carefully the actual vegetation and also its dynamic tendencies [17, 37, 40]. Therefore, in the county's grazing areas, once forest, but today almost completely abandoned, processes of secondary succession are frequently present. On the lower slopes, lower mounds and lowlands, parts very densely occupied by rural people with some fields near to the main rivers and streams, we can see an anthropogenic fluctuation.

As a possible approach to the degree of naturalness of the vegetation, we will consider the synanthropization level of the phytocoenoses, which it is possible to access through landscape observation. This process consists in the gradual substitution of natural vegetation by the anthropogenic, introduced by the action of human activities in the production of his social space. Therefore, we will see in the landscapes various stages of these alterations, which are going from parcels with remained primary vegetation to those, which are intensively cultivated or urbanized [39, 40].

At this point, we may establish an analytical cartography for the county; showing on a map all these integrated aspects studied (Fig. 2).

Cartography of landscape units

This is the last stage of our study. We shall consider the synthetic reasoning that will be established on the map, and landscape units that are patches, i.e.

synthetic units, involving groups of spatial unitary elements characterized by groups of attributes [28, 30].

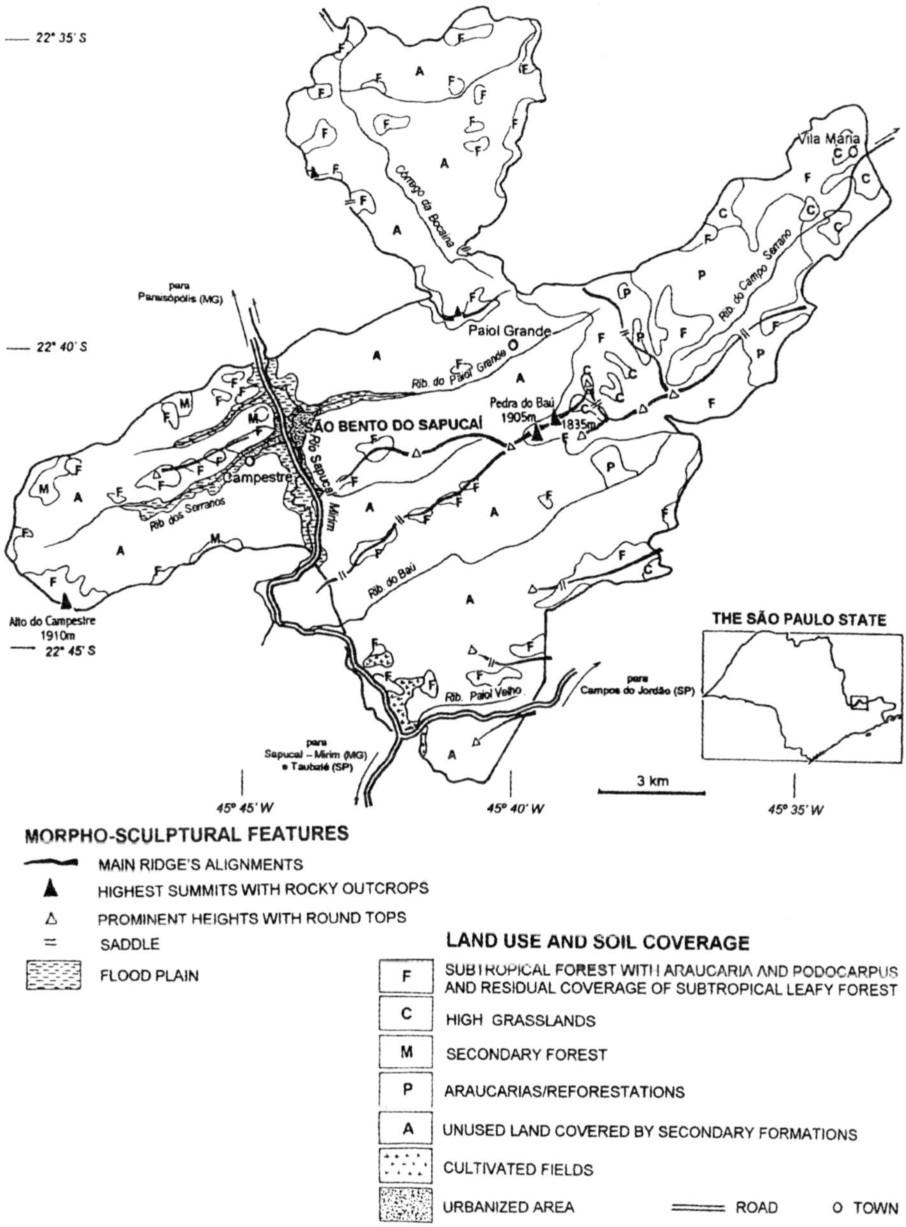

Fig. 2: Morpho-sculptural features and land use of the São Bento do Sapucaí County.

We conceive this cartography as an environmental synthetic cartography, once the landscape units themselves behave as ***geosystems***, and consequently, they have morphology, functionality and behaviour. We may define the morphology by vertical structures, characterized by physiognomy, mass, and energy; and horizontal, viewed through combinations of vertical structures that vary in time. Functionality corresponds to the ensemble of transformations due to the intervention of various factors. This behaviour matches to the set of changes in the internal state that intervenes historically [6, 18].

The organization of the legend will give the appropriate interpretation of the methodological position taken in this work, as well as of the reasoning developed to search the map of landscape units. In this sense, it would be ideal to adopt the organization of the legend in a taxonomic order that considers several levels of categorization. In a first level would be the large sets of landscape units scaled from the most natural to the most artificial, as produced by human society. In a second level, we would consider the arrangement of geomorphologic complexes anchored in a solid structural origin. In the third level, it would be the final unfolding individualizing the landscape units themselves [37, 38].

However, since the county is an area of small extent (279 km^2), that legend organization would place the whole county in a unique scheme for the upper levels. Consequently, we shall judge only one categorization level – that in which the landscape is a unit – however, it is organized from the most natural to the most artificial.

Even so, if the communication by map is essentially monosemic, the polisemy of the figurative image will be welcome, ensuring satisfactory results in the cartographic message received by the user. Thus, for every legend's verbal speech in its latest organization level, it will be possible to associate a visual speech that will refer to certain identifying defining characteristic of every landscape unit that is on the map (Fig. 3).

According to all our assertions here expressed, we propose the following map legend:

1. Mountainous relief summits characterized by rocky outcrops of exotic form that stand out from the steep upward slopes covered by Subtropical Forest with *Araucaria angustifolia* and *Podocarpus lambertii* and well-preserved high grassland clearings.
2. High mountain summits and ridge lines with residual coverage of Subtropical Leafy Forest.
3. Upper zones of relief and drainage head areas with Subtropical Forest of *Araucaria angustifolia* and *Podocarpus lambertii* and/or reforestation and high grassland clearings.
4. Upper and middle upward slope zones with some cultivated areas, active and abandoned grazing fields of secondary succession.
5. Middle and lower slopes outlining the bottom of valleys occupied by farms, gardens and holiday camps with moderately intensive land use.
6. Alluvial bottomlands of the main drainage axes with some cultivated fields.
7. Urbanized area.

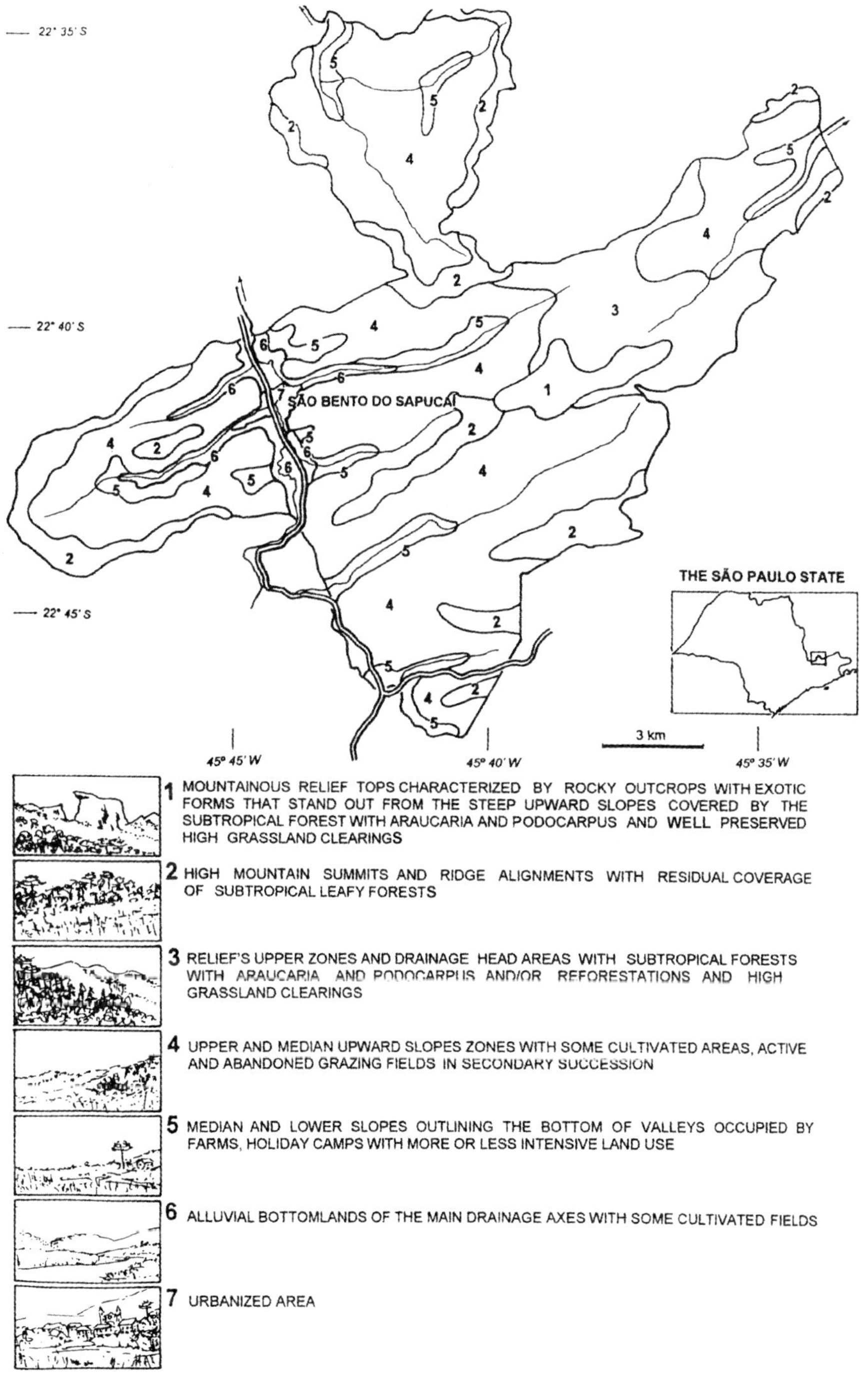

Fig. 3: The landscape units of the São Bento do Sapucai County.

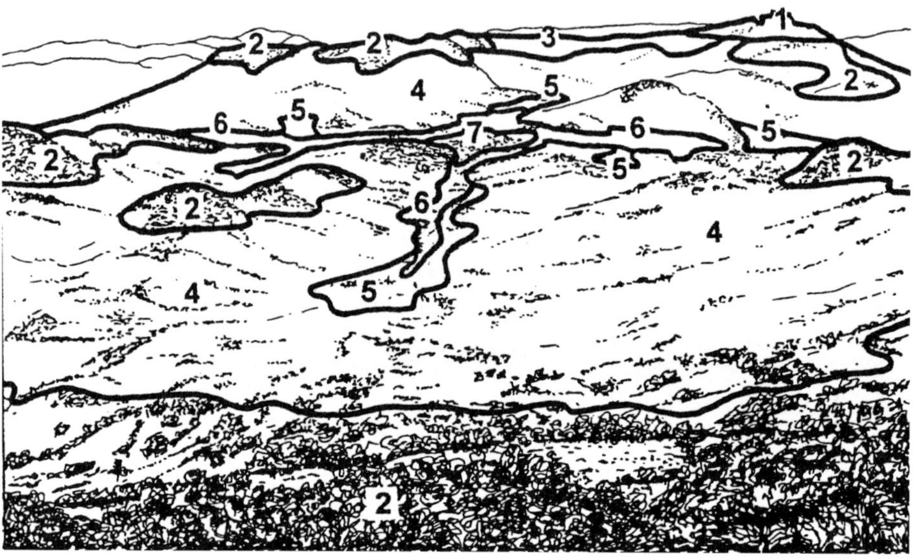

Fig. 4: Panoramic sight of the landscape units arrangement in the São Bento do Sapucaí County (legend as in Fig. 3).

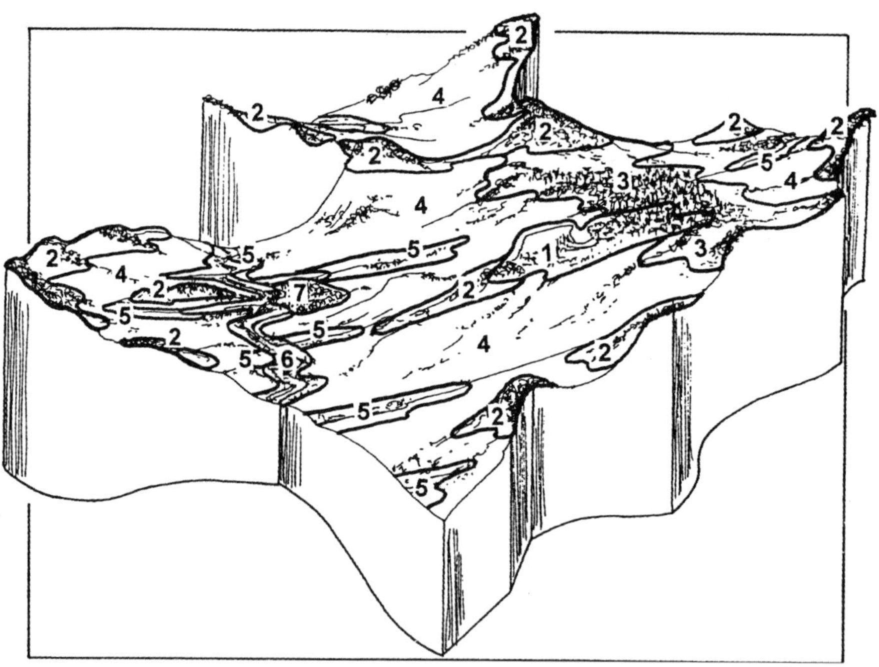

Fig. 5: Block-diagram of the landscape units arrangement in the São Bento do Sapucaí County (legend as in Fig. 3).

To increase the cartographic visualization power of the map of landscape units, we may stimulate the creativity and the artistic drawing to compound a complementary global landscape view, as an oblique air photo view or as a block-diagram sketch, both with the landscape units incorporated over them.

These two representations may facilitate tourists' visits to the county, ensuring them an easy interpretation of the potentialities of the landscape units seen on the ground (Figs. 4 and 5) [9, 29].

REFERENCES

1. Ab'Saber A.N., Bernardes N., 1958 - Vale do Paraíba, Serra da Mantiqueira e arredores de São Paulo. CNG, Rio de Janeiro.
2. Ab'Saber A.N., 1967 - Domínios morfoclimáticos e províncias fitogeográficas do Brasil. Orientação, **3**: 45-48.
3. Ab'Saber A.N., 1977 - América do Sul. Domínios naturais há 13.000 – 18.000 anos. Primeira aproximação. IGEOG, São Paulo.
4. Ab'Saber A.N., 2003 - Os domínios de natureza no Brasil. Potencialidades paisagísticas. Ateliê Editorial, Cotia.
5. Alves T., 2001 - Paisagens – em busca do lugar perdido. Finisterra, **36**: 67-74.
6. Beroutchachvilli N., Bertrand G., 1978 - Le géosystème ou système territoriel naturel. Révue Géographique des Pyrénées et du Sud-Ouest, **49** (2): 167-180.
7. Bertalanffy L. van, 1968 - General system theory: foundations, development and application. George Braziller, New York.
8. Bertrand G., 1968 - Paysage et géographie physique globale. Équisse méthodologique. Revue Géographique des Pyrénées et du Sud-Ouest, **39** (3): 249-272.
9. Castro I.E., 2002 - Paisagem e turismo. De estética, nostalgia e política. In: Yázigi E. (ed.), Turismo e paisagem. Contexto, São Paulo.
10. Castro I.E., 1995 - O problema da escala. In: Castro I.E. (ed.), Geografia: conceitos e temas. Bertrand Brasil, Rio de Janeiro.
11. Coniat D., 1985 - Analyse de la cartographie géopolitique et géostratégique. (Memoire de maîtrise) - Université de Paris I, Paris.
12. Cruz O., 1985 - A escala temporal-espacial nos estudos dos processos geomorfológicos erosivos atuais. Geomorfologia, **33**: 1-12.
13. Cullot M., 1986 - Points de vue sur la perception des paysages. L'Espace Géographique, **15** (3): 211 217.
14. Cunha S.B., Guerra A.J.T. (eds.), 2003 - A questão ambiental. Diferentes abordagens. Bertrand Brasil, Rio de Janeiro.
15. Diniz A., Furlan S.A., 1998 - Relações entre classificações fitogeográficas, fitosociologia, cartografia, escalas e modificações sócio-culturais no Parque Estadual de Campos do Jordão (SP). Revista do Departamento de Geografia, **12**: 123-162.
16. Evaso A.S., 1999 - A refuncionalização do espaço. Experimental, **6**: 33-54.
17. Falinski J.B., 1975 - Anthropogenic changes of the vegetation of Poland. Phytocoenosis, **4** (2): 97-116.

18. Ferreira M.C., 1997 - Mapeamento de unidades de paisagem com Sistemas de Informação Geográfica: alguns pressupostos fundamentais. Geografia, **22** (1): 23-36.
19. Géhu J.-M., 1991 - L'analyse symphytosociologique et géosymphytosociologique de l'espace. Théorie et méthodologie. Colloques Phytosociologiques, **17**: 11- 46.
20. Godart M.-F., Deconinck M., 2003 - Les paysages: celui de l'écologue, du géographe, du touriste. In: Phytosociology at landscape ecology service, Abstracts, pp. 22. Universitatea Babeş-Bolyai, Cluj-Napoca.
21. Gomes H., 1983 - Espaço/Tempo em geografia. Boletim Goiano de Geografia, **1-3** (3): 105-134.
22. Gonçalves C.W.P., 1990 - Os (des)caminhos do meio ambiente. Contexto, São Paulo.
23. Henning M.D., 1983 - Toward an understanding of scale and its relevance to cartographic communication. The Cartographic Journal, **20** (2): 119-120.
24. Joly F., 1976 - La cartographie. PUF, Paris.
25. Lacoste Y., 1977 - A quoi sert le paysage? Qu'est-ce qu'un beau paysage?. Herodote, **7**: 3-41.
26. Lacoste Y., 1976 - La géographie, ça sert, d'abord, à faire la guerre. Maspero, Paris.
27. Lecoeur C., 1987 - Le paysage comme cadre physique. Herodote, **41**: 45-50.
28. Martinelli M., 1999 - La cartographie environnementale: une cartographie de synthèse. Phytocoenosis, **11**: 123-130.
29. Martinelli M., 2001 - Cartografia do turismo e imaginário. In: Rodrigues A.B. (ed.), Turismo rural: práticas e perspectivas. Contexto, São Paulo.
30. Martinelli M., Pedrotti F., 2001 - A cartografia das unidades de paisagem: questões metodológicas. Revista do Departamento de Geografia, **14**: 39-46.
31. Marx K., Engels F., 1979 - The German ideology. International Publishers, New York.
32. Monteiro C.A.F., 1973 - A dinâmica climática e as chuvas no estado de São Paulo. Estudo geográfico sob forma de atlas. IGEOG, São Paulo.
33. Moraes A.C.R., 1990 - Bases epistemológicas da questão ambiental: o método. Orientação, **8**: 43-48.
34. Moraes A.C.R., 1981 - Geografia. Pequena história crítica. Hucitec, São Paulo.
35. Morandi S., Gil I.C., 2000 - Espaço e turismo. Copidart, São Paulo.
36. Moreira R., 1986 - Sobre a educação ambiental. Boletim Campo-grandense de Geografia, **1**: 1-10.
37. Orsomando E. (ed.), 1998 - Carta del paesaggio vegetale del Consorzio della Bonificazione Umbra. Bacino idrografico dei fiumi Topino e Marroggia (scala 1:100.000). XXVIIIème Colloque Phytosociologique, Camerino.
38. Orsomando E. (ed.), 2000 - Carta delle unità ambientali-paesaggistiche dell'Umbria, scala 1:100.000. Università di Camerino, Camerino.
39. Pedrotti F., 1998 - Le serie di vegetazione nella cartografia ambientale. Genio rurale, **61** (1): 15-17.
40. Pedrotti F. (ed.), 1997 - Le unità ambientali de Parco Nazionale dello Stelvio. L'Uomo e l'Ambiente, **28**: 1-104.
41. Pinto-Correia T. (ed.), 2001 - Identificação de unidades de paisagem: metodologia aplicada a Portugal Continental. Finisterra, **36**: 195-206.
42. Rodriguez J.M.M., Silva E.V., 2002 - A classificação das paisagens a partir de uma visão geossistêmica. Mercator. Revista de Geografia da UGC, **1** (1): 95-112.

43. Ross J.L.S., Moroz I.C., 1997 - Mapa geomorfológico do estado de São Paulo. Vol 1. IPT/FAPESP, São Paulo.
44. Santos M., 2002 - A natureza do espaço. Técnica e tempo. Razão e emoção. Edusp, São Paulo.
45. Santos M., Silveira M.L., 2001 - O Brasil. Território e sociedade no início do século XXI. Editora Record, Rio de Janeiro.
46. Santos M., 1991 - Pensando o espaço do homem. Hucitec, São Paulo.
47. Santos M., 1994 - Técnica, espaço, tempo. Globalização e meio técnico-científico-informacional. Hucitec, São Paulo.
48. Sirgado J.R., 2001 - Espaço turístico e desenvolvimento no Cone Leste Paulista. In: Rodrigues A.B. (ed.), Turismo rural: práticas e perspectivas. Contexto, São Paulo.
49. Sotchava V.B., 1976 - The study of geosystems. Special issue for the 23rd International Geographic Congress, **51**: 3-40.
50. Tricart J., Kilian J., 1979 - L'éco-géographie et l'aménagement du milieu naturel. Maspero, Paris.
51. Tricart J., 1977 - Ecodinâmica. IBGE/SUPREN, Rio de Janeiro.
52. Troll C., 1950 - Die geographische Landschaft und ihre Forschung. Studium Generale, **3**: 4-5.
53. Troppmair H., 1969 - A cobertura vegetal primitiva do estado de São Paulo. Biogeografia, **1**: 1-11.
54. Tuxen R., 1951 - Die heutige potentielle natüralische Vegetation als Gegenstand der Vegetations Kartierung. Pflanzensoziölogie, **13**: 5-12.
55. Viadana A.G., 2002 - A teoria dos refúgios florestais aplicada ao estado de São Paulo. Edição do Autor, Rio Claro.
56. Victor M.A.M., 1975 - Cem anos de devastação. OESP, São Paulo.
57. Zonneveld. I.S., 1989 - The land unit – a fundamental concept in landscape ecology and its applications. Landscape Ecology, **3** (2): 67-86.

ENVIRONMENTAL MANAGEMENT FOR BIODIVERSITY CONSERVATION

Massimo SARGOLINI
Facoltà di Architettura, Università di Camerino, Viale della Rimembranza,
63100 Ascoli Piceno (ITALY); e-mail: massimo.sargolini@unicam.it

Abstract: What is the role of environmental management in biodiversity conservation? Why is urban management always linked more to environmental management, and environmental management linked more to urban planning?

I attempt to answer this question through some planning experiences in protected areas, where you can find the importance of environmental planning in the processes of natural and cultural resource conservation (Alpi Apuane Regional Park, Monti Sibillini National Park, Cilento Vallo di Diano National Park).

Traditionally, there has been a separation between environment policies and urban, territorial and regional policies. On the international stage, these relations are changing; separateness is giving way to integration or convergence.

As a key element of individual and social well-being and quality of life, the environment plays an important part in human fulfilment and in reinforcement of European identity. It has an important public interest role in the cultural, ecological and social fields, and constitutes a resource favourable to economic activities, particularly tourism. Unfortunately, developments in agriculture, forestry, industrial and mineral production techniques and in regional planning, town planning, transport, infrastructure, tourism and recreation have often damaged the environment and landscapes, or obliterated their distinctiveness.

Environmental policies are becoming territorialized, in the sense that they have to address the problems, threats, needs and expectations of the development of the regions or territorial contexts in which they are situated. At the same time territorial policies are induced or forced to recognize an important role for environmental policies and in particular those for protected areas.

Introduction

The policies for parks and protected areas have assumed in the few last decades, in Europe as well, a growing economic, territorial and cultural importance [3]. The Italian case is symbolic: several thousand or so protected areas now cover more than 11% of national territory (the European average is 14%). The 152 national or regional parks (30 years ago there were fewer than ten) cover 25% of national territory in all, with 32% of Italy's population (Fig. 1). This quantitative growth has been accompanied by a considerable increase in economic impact and the role that the experiences of parks and protected nature play in the social and cultural spheres [2].

In research on park planning, many diverse fields of expertise (geology, biology, architecture, history, archaeology, ecology, etc.) are trying out approaches that are increasingly of a trans-disciplinary and trans-scale nature, seeking new methods for working together, starting from concrete problems of management. In various European countries, there are diversified experiments of participation, co-operation and collaboration among institutions [1]. In these cases, there are many exchanges of experiences between urban and environmental planning (Fig. 2).

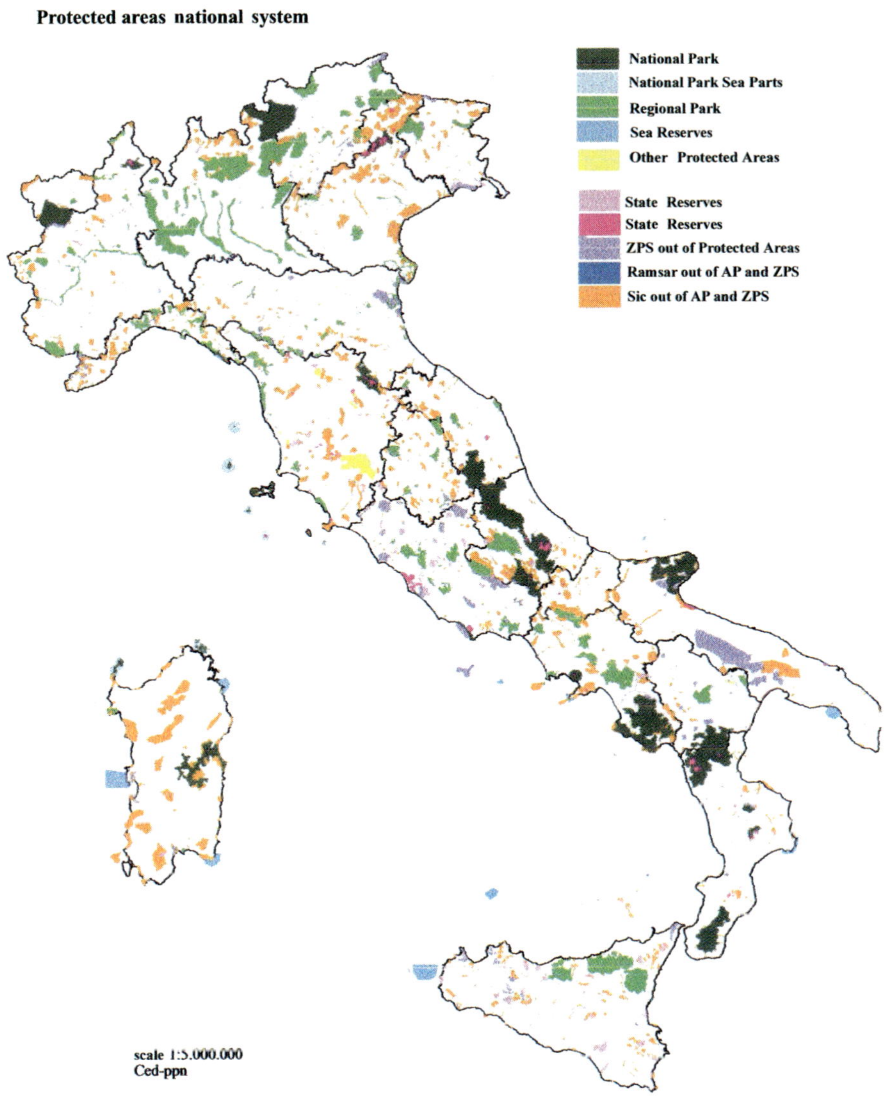

Fig. 1: Protected areas (CED-PPN, Politecnico di Torino).

Fig. 2: Nivolet - Gran Paradiso National Park.

The Italian landscape alternates between absolute and universally acknowledged values (cities/centres, and historical landscapes, parks and protected natural areas) with diffuse and indifferent negligence for most of the territory, a place where most of people live and whose organization for safety, at least against disasters that are recurrent, natural or directly or indirectly induced by anthropogenic factors, does not even attract any concern [5].

A territory whose government is shared by diverse institutions which combine actions and interventions aimed at multiple purposes in a disorderly manner (admittedly *per se* not censurable), without any efficient and effective co-ordination, and in fact with all but latent conflict which multiplies norms, darkens the real responsibilities each one holds, and actually reduces the certainty of rights for all.

In some cases, most of the territory is occupied by common property that includes pastures, forests, irrigation systems and other natural or man-made resources (Fig. 3). Many of those 'traditional' commons were managed effectively and survived for centuries [4]. Due to important changes in agriculture and industrial development, many have lost their economic importance. Moreover, many European governments of the 18th–19th centuries attempted to abolish commons in favour of private property. Nevertheless, in several cases they are still relevant for the local population and maintain a crucial ecological and cultural significance.

Fig. 3: Agricultural landscape at the border of Sibillini National Park.

Three Italian study-cases

In the planning activities of three Italian parks (Apuan Alps Regional Park; Cilento and Vallo di Diano National Park; Monti Sibillini National Park), it is very important to link biodiversity conservation with environmental planning.

The Apuan Alps Regional Park possesses a number of specific features [6]:

– the wealth and variety of its natural resources and the uniqueness of its landscape (relationship between the coastal belt and the peculiar Alpine-type geomorphological configurations);

– a particularly strong relationship between the problems and prospects of the dominant economic activities (above all those based on the extraction of marble, historically and culturally rooted in the Apuan Alps) and those of landscape and environmental protection;

– a strong exposure to pressures deriving from the coastal belt with its great development of tourist facilities and widely urbanized pattern.

The noteworthy diversification of the values and problems to be considered for purposes of Park management introduces a number of profiles of interpretation:

– geological and geo-morphological, characterized by the singular geological nature of the Apuan complex which makes it stand out as a recognisable unit with respect to the configuration of the Apennine chain, to the historic, landscape and cultural importance of the marble resources, to the interest of the karst phenomena and of the subsurface hydrology, and to the hydro-geological risk widespread especially in certain parts of the Apuan territory;

– ecological and vegetational, characterized by nine groups of environmental units (outstanding natural non-forest ridge or upper slope areas, degraded forest areas, areas with spontaneous woods in upland plains, areas with spontaneous woods in the lower plain of mixed and variable composition, areas of chestnut woods, artificially wooded areas, areas with water resources and wet woods, and cultivated areas). An environmental unit is understood to be a relatively homogeneous area from the macro-lithological and bioclimatic aspect (vegetation belt), characterized by a number of interacting vegetation patches, by anthropogenic action and by morphology;

– historical settlement, characterized by two settlement patterns: a permanent one linked to the Mediterranean world and to the Etruscan civilisation, and a semi-nomadic, pastoral one, linked to immigrant waves of Celtic, Apuan and Ligurian civilization. Recent developments have profoundly modified the overall settlement pattern;

– landscape perceptive, characterized by various landscape sectors: the 'heartland' consisting of the main ridge and landscapes, dominated by the most natural components of greatest iconic value; the mountain complex dominated by the historical interaction between man and nature; and the outer ring or "cornice" from which the overall image of the Park can be appreciated, comparing it with the other surrounding spaces.

The Plan pursues the objectives of conservation and sustainable development by carrying out the following functions:

1) it establishes rules and regulations governing the activities and measures of rehabilitation, improvement or transformation permitted in the protected territory, preventing them from being prejudicial to the various sites and resources;

2) it co-ordinates and guides the actions and the programmes of measures for which the various public and private bodies are responsible, within the territory of the Park and, as far as possible, also in the surrounding area of influence; and

3) it motivates protection and action choices, their margins of negotiability, the conditions and the modalities of dialogue and confrontation among the various institutional bodies involved in managing the territory concerned.

The Apuan Alps Park seems to make evident a problem which in recent decades has increasingly involved the management of the majority of the European nature parks, i.e. that of relations with their social, economic and territorial context. The Plan expresses an open development strategy vis-à-vis the economic and territorial context, integrated in a plurality of convergent policies, agreed and supported by all the institutional bodies and local players. It is subdivided into various lines of action:

A, the management of natural resources, for the safeguarding of natural, landscape and environmental values, and the active conservation and enhancement of the ecosystems that define the structure and the overall image of the Park and its various parts;

B, the enhancement of the historical-cultural heritage, for the safeguarding and active conservation of the cultural values and of the individual resources that define the quality of Apuan territory and the well-knit system of local identities;

C, agricultural-forestry-grazing improvement, for the maintenance, development and upgrade of productive and management techniques and practices;

D, the management of extractive activities, with the promotion of forms of knowledge, programming and discipline for a sensible economic use of resources;

E, urban planning and infrastructure reorganisation for the upgrading of the settlements and of the networks of infrastructures and services, rehabilitation of run-down or abandoned areas, and the sensible use and enhancement of the building and urban planning assets; and

F, the promotion of tourism and the social enjoyment of the Park, with actions aimed at fostering and guiding the development of tourism and the use of the Park for recreational, sporting, educational and cultural purposes.

The Cilento and Vallo di Diano National Park occupies an area of 178,300 ha (almost half the large province of Salerno, south of Naples, five times the average size of European parks) and involves 80 municipalities and eight mountain communities [7].

In 1998 UNESCO recognized the Park area for its value as a cultural landscape of world-wide importance. UNESCO has stressed that the Cilento is not a mere container of individual natural or cultural resources, biotopes or monuments, but a complex system of landscapes characterized by the balanced association of the anthropogenic world with the natural world, such as has become consolidated in the course of time. This recognition confers a special significance on the goals (established by National Law 394/91) of the conservation of natural and landscape resources.

The national law attributes different tasks and purposes to the various Park management instruments (Park Plan, Long-term Economic and Social Plan, Regulations). The Cilento Park Plan pursues the following objectives:

a) to consolidate the Park's image through the inclusion of the Cilento area in the inter-regional and European ecological, productive, economic and socio-cultural networks;

b) to regulate and provide incentives for traditional and productive activities, safeguarding natural and cultural assets and consolidating local economic systems;

c) to safeguard and promote local identities and all those components that enhance the 'identity' of the area with respect to its context;

d) to facilitate the development and self-organisation of local systems; and

e) to avoid usages that consume and degrade the resources of the area.

The necessary coexistence of conservation and innovation, with a view to sustainable development, requires:

- the joint consideration of the economic, ecological, social and cultural dimensions, in consultation with the local community and taking care of relations with the territory considered globally;
- the enhancement of specific identities and diversities;
- the development of local specializations able to connect with the network economies of the context;
- the productive interaction of services and investments, between internal and external local systems.

The Plan expresses the following strategic lines of approach:

A, conserving ecosystem diversity and functional aspects with measures aimed at:

- setting up ecological networks connecting the Park to surrounding natural spaces;
- improving agriculture and forestry;
- upgrading the systems of monitoring bio-coenoses and health balance;
- increasing scientific research.

B, endogenous development and reduction of internal imbalances of socio-economic type with measures aimed at:

- upgrading cropping techniques;
- converting non-sustainable activities;
- reorienting the building industry and handicraft activities;
- recovering existing building assets;
- strengthening the self-organizing capacities of the local systems.

C, development of sustainable tourism and of appropriate forms of social use (recreational, cultural and educational pursuits) of the Park and of its resources, with policies and measures aimed at:

- providing incentives for a balanced spread of visitor flow;
- improving relations between seaside/marine tourism and inland tourism;
- increasing and upgrading hospitality and diffuse accommodation facilities;
- stimulating a better knowledge and more adequate use of natural and cultural resources;
- developing interpretative and formative activities and social communication at an international level.

D, improvement of the quality of settlements, with policies and measures aimed at:

- improving housing conditions and accessibility of the territory;
- trying out new settlement models in equilibrium with the environment;
- providing incentives for landscape restoration, and the rehabilitation and re-use of historic centres and cultural assets;
- trying out innovative forms of bio-architecture and energy saving.

The Plan also lays down specific procedural guidelines for the various environmental systems and subsystems, and in particular for:

- protecting the natural hydro-geomorphological evolution of the mountain slopes;
- protecting water resources;
- safeguarding the surviving natural areas of the coastal belt;
- safeguarding fluvial belts;
- safeguarding and enhancing natural floristic and faunal features, and geological sites of stratigraphic, palaeo-environmental, palaeo-biological, structural and geo-morphological interest;
- conservation and upgrading of wooded areas;
- safeguarding of agricultural-forestry-grazing activities;
- maintenance and conservation of sectors of the rural landscape, contexts of historic centres, sites of archaeological interest, panoramic views and roads;
- upgrading and improvement of areas and elements of specific historical, artistic, cultural and archaeological interest;
- imposing control measures on roads and transport;
- upgrading systems of equipment and services;
- controlling the practices of hunting and fishing.

The Monti Sibillini National Park covers an area of 70,000 ha, and involves 18 municipalities, five mountain communities, three provinces (Ascoli Piceno, Macerata and Perugia) and two regions (Marche and Umbria) (Fig. 4). There are 13,200 inhabitants in the Park area [8].

The Plan integrates objectives of general management, already recognized by national law, with the following basic choices expressed by the Park Statute (Resolution no. 111 of 21/08/1997):

- recognition of the role of the Monti Sibillini Park in the Apennine system of natural spaces and in ecological and environmental networks represented at various levels, from European to local;
- enhancement of the historical, cultural and artistic assets, also through achieving a direct relationship with the cultures of the local communities;
- providing incentives for policies of promotion and of active defence, as the complement to the protection policies, and the assumption of precise social commitments by the Park Authority.

Therefore, on the basis of these key choices, the following Plan objectives can be identified:

1) consolidating the Park's image, strengthening its role in the Apennine system;

2) identifying forms of regulation and incentives for traditional activities and for activities of practical use, guaranteeing both conservation and development;

3) enhancing local identities, resources, landscapes and local cultures, also through the inclusion of the diversified local realities in the spheres of utilisation, trading and production;

4) ensuring a rigorous limit to the consumption of essential resources (such as land and water) and effective maintenance of the landscape and environmental heritage.

Fig. 4: Ambro's Mountains in the Sibillini National Park.

The Plan expresses the necessity to act simultaneously along the following strategic lines:

A, management of natural assets, with measures aimed at:

- conserving habitats of significant interest;
- safeguarding and making improved use of water resources;
- safeguarding and recovering land resources;
- safeguarding and increasing the networks of ecological connection and the environmental matrices;
- reintroducing and repopulating extinct and endangered species of particular natural history importance;
- implementing a system of controlling the effectiveness of the above measures.

B, risk control and prevention, with the assistance of the competent bodies already called in, by means of measures aimed at:

- mitigating the hydraulic risk;
- mitigating the hydro-geological and geo-morphological risk;
- reducing the risk of earthquakes;
- upgrading the hydro-meteorological risk forecasting systems.

C, improvement of the agricultural-forestry-grazing territory, guaranteeing effective maintenance of the ecological and landscape heritage, through measures aimed at:

- rationalising the management of agricultural activities;
- disciplining forestry management;
- co-ordinating and guiding local sheep-rearing and animal husbandry;
- promoting and upgrading agro-tourism and rural tourism.

D, rehabilitation and enhancement of local heritage, specific resources and local cultures through measures aimed at:

- upgrading municipal implementation and general urban planning instruments;
- conserving and upgrading scattered nuclei and agglomerates of historical and architectural value;
- conserving and upgrading isolated cultural assets;
- reinstating the network of the most significant historical walks;
- limiting settlement, productive and infrastructure development;
- reorganizing and upgrading spaces and artefacts for the benefit of tourism.

E, development of tourism and social enjoyment, through measures aimed at:

- improving the network of accommodation facilities and structures;
- mitigating environmental and landscape impacts caused by pedestrian and vehicular routes;
- upgrading spaces and artefacts for improved use by tourism.

F, improvement of the landscape and local identities through measures aimed at:

- improving the overall image of the Monti Sibillini;
- enhancing the symbolic images and the individual landscapes of the Monti Sibillini;
- enhancing local cultures and specific features.

The Plan also introduces an ensemble of co-ordinated rules and regulations for:

- zones having different levels of protection;
- landscape units;
- different categories of resources and activities;
- improvement projects and programmes.

The zones having different levels of protection are (in accordance with Law 394/91):

A, Class A reserves, where the natural environment has to be conserved in its absolute integrity;

B, general and specific reserves, where ecosystem functions are fostered, as well as conservation of the cultural landscape resources present, also by reducing disturbance factors;

C, zones of protection, in which agriculture-forestry-grazing activities are encouraged, in keeping with traditional usages, or introducing methods of biological agriculture; and

D, zones of economic and social promotion, which concern the urbanized areas, those to be urbanized and degraded zones, even if only partly. In these zones all activities are permitted which are compatible with the institutional aims of the Park and are intended to improve the social and cultural life of the local communities and the optimal enjoyment of the Park by visitors.

REFERENCES

1. Gambino R., 1994 - I parchi naturali europei. Dal piano alla gestione. La Nuova Italia Scientifica, Roma.
2. Gambino R., 2003 - AP Il sistema delle aree protette. Alinea Editrice, Firenze.
3. Niccolini F., 2004 - L'azienda turistica sostenibile. Prospettive culturali e assetti di network. Edizioni ETS, Pisa.
4. Sargolini M., 2003 - Le carbonaie. Un progetto per Cessapalombo. PROCAM – Ministero per l'ambiente, Università di Camerino.
5. Sargolini M., Pedrotti F., Gafta D., 2000 - Manuale per la zonizzazione dei parchi nazionali. Documento di sintesi. Ministero dell'Ambiente, Roma.
6. Parco regionale Alpi Apuane, 2000 - Piano per il Parco. Relazione. Seravezza.
7. Parco nazionale del Cilento Vallo di Diano, 2000 - Piano per il Parco. Relazione. Vallo della Lucania.
8. Parco nazionale dei Monti Sibillini, 2001 - Piano per il Parco. Relazione. Visso.

THE MODENA BOTANIC GARDEN: PLANT CONSERVATION AND HABITAT MANAGEMENT STRATEGIES

Carlo Del PRETE*, **Daniele DALLAI, Elisabetta SGARBI, Luigi MAFFETTONE**
Botanic Garden, University of Modena and Reggio Emilia,
Department of Paleobiology and Botanic Garden, Viale Caduti in Guerra 127
41100 Modena (ITALY); e-mail: delprete.carlo@unimore.it

Abstract: The Botanic Garden of Modena (Northern Italy) was founded in 1758 by Duke Francesco III d'Este. Its original functions of "Hortus simplicium" evolved over more than two centuries of its academic history, until its present structure. Today, like the majority of the University Botanic Gardens in Italy, the Modena Botanic Garden is involved in several projects for plant conservation and ecosystem management that are compliant with the Global Strategy for Plant Conservation. Even if its original location in the centre of the town does not permit enlargement or structural modifications, it does, however, greatly help the education activities carried out in the Garden, while the conservation activities are developed outside the Garden by means of numerous co-operations with Parks, natural reserves and other administrative bodies or institutions.

The Botanic Garden of Modena recently started various *in situ/ex situ* integrated conservation projects with Parks and other land-managing institutions. Even if such projects involve only a limited number of Endangered and/or Vulnerable species/populations, some interesting and noticeable results have been achieved. Plantlets (e.g. indigenous Orchidaceae) are often obtained by the way of non-conventional techniques such as *in vitro* culture.

Co-operation with local Museums has provided interesting sources for future action: the collections that they preserve are mostly related to circumscribed local areas or well-defined periods, so they are the source of fundamental information about past situations.

Investigation on flora and vegetation and the monitoring of plants and plant communities, carried out in protected areas, cooperating with the National Forestry Service, is a very important source of knowledge to improve conservation and verify the effectiveness of conservation actions.

1. Introduction

Over recent years most of the University Botanic Gardens in Italy have been involved in projects and activities regarding plant conservation and ecosystem management [4, 18, 19, 20]. Today these activities are carried out in implementation of the Global Strategy for Plant Conservation [21, 22, 23, 41]. Taxonomists, plant ecologists, technicians, gardeners and other professionals with different skills from all branches of Plant Science, are involved with University Botanic Gardens. This provides a good opportunity for co-operation with National and Regional Parks, Nature Reserves and other local institutions and official bodies as regards terms of land management, sharing of knowledge and establishing guidelines for conservation activities. The guidelines consist in providing vegetation and floristic information on protected areas, plants and habitat preservation, landscape and biotope management, and *in situ* and *ex situ* integrated plant conservation.

2. Modena Botanic Garden Actions

The Botanic Garden of Modena (Northern Italy) is a historical Botanic Garden: it was founded in 1758 by the Duke of Modena, Francesco III d'Este, with the original function of *Hortus simplicium*. Its functions have been enlarged throughout two centuries of academic history [5, 10, 13, 14, 15, 17, 25, 39, 40].

Modena Botanic Garden's original location in the centre of the town does not allow enlargement of the area or any structural modification. However, its special location greatly encourages its role as a scientific and cultural centre, and an integral part of the "local network of Museological System" [2] – which groups different types of museums within the local area – supporting the educational activities carried out inside the Garden [8, 9, 11, 12]. Instead, conservation activities are often developed outside the Garden by means of extensive co-operation with National and Regional Parks, Nature Reserves and other institutions, as reported below.

Intense industrial expansion is a common feature in the history of the Emilia – Romagna region where the Garden is located, whereas significant environmental protection actions have only recently been embarked upon. In the region, at least 15 plant species are now considered Extinct (provisional data, IUCN source), 49 species are Endangered, and over 150 are considered Vulnerable or Rare. This entails an arduous task for the region's four University Botanic Gardens (Bologna, Ferrara, Modena, Parma), which are the scientific institutions most qualified to offer the required know-how on ecosystems and plant conservation. University Botanic Gardens are also supported by certain local Alpine or Thematic Gardens, generally owned and managed by local institutions.

In this context, the Botanic Garden of Modena has recently begun a number of different conservation projects with the co-operation (sometimes with financial support) of public administrative bodies and associations that handle environmental policies.

2.1 Vegetation map production

Within the general framework of regional vegetation mapping of protected areas, several phytosociological maps have been drawn up, i.e. vegetation maps of the Regional Park of the Northern Apennines [34, 35], Regional Park of "Corno alle Scale" [37], Regional Park of the "Sassi di Rocca Malatina" [38], and the geotourist map of the Natural Reserve of the "Salse di Nirano" [3] (Fig. 1a).

2.1.1. The Regional Park of the Northern Apennines

The Regional Park of the Northern Apennines territory covers a wide area defined by the watershed ridge of the Tuscany and Emilia regions, and at the border between the provinces of Bologna and Reggio Emilia and the massif of Mt.

Cimone (Fig. 1b). Of its total surface of approximately 15,000 ha, 9,000 ha are covered by the Park – all of which is situated at altitudes of over 1,000 meters – and the remaining 6,000 belong to the pre-Park area. The phytosociological vegetation map of the entire area, on 1:25000 scale, was realized by Tomaselli *et al.* [34, 35]. A monographic survey of flora and vegetation has been published too, which provides a comprehensive overview of botanical knowledge of the area considered and guidelines for plant and habitat conservation [36].

2.1.2. The Regional Park of "Corno alle Scale"

The Regional Park of "Corno alle Scale" is very important from a natural history point of view, since it is characterized by a rich variety of plant and animal species, geological and mineralogical features, and beautiful landscapes (Figs. 1c, 1d, 1e). The Park can be divided into two parts: first, it contains habitats typical of the Apennine chain, which are extremely rich in vegetation, flora, and fauna; second, the high peaks, such as Corno alle Scale, La Nuda and Mt. Gennaio, feature typical Alpine habitats.

2.1.3. The Regional Park of the "Sassi di Rocca Malatina"

The protected area of the Regional Park "Sassi di Rocca Malatina" extends from the bottom of the Panaro Valley, along the right bank of the river, safeguarding an important hilly landscape which surrounds the pinnacles of the Sassi (Fig. 1f) where spectacular arenaceous cliffs (used by peregrine falcons as a nesting area) dominate the surrounding territory.

2.1.4. The Natural Reserve of the "Salse di Nirano"

The Natural Reserve of the "Salse di Nirano" is situated in the Modena Apennines, covering an area of 207 ha, with altitudinal ranges of 140–308 m. The reserve was constituted in 1982 with the aim of preserving an area characterized by a geological phenomenon known as "salse". These mud volcanoes originate from a mixture of hydrocarbon deposits, salt water and surface soil; as the salt water rises it melts or dissolves the clay and, on reaching the surface, forms characteristic volcano-like mud cones (Fig. 1g). The "Salse di Nirano" area is the most important of its kind in Italy and the mud represents a very special habitat, allowing only the growth of halophytic species.

2.2 Managing biotopes and projects for *in situ/ex situ* plant conservation

In the Po Valley, wetlands are considered important environments: rivers, bogs, marshes and channels are at present the only relict habitat where many endangered hygrophyte species still occur (Figs. 1h, 1i, 1l). Today, the wetlands mainly depend on water management, land reclamation and irrigation activities. They represent more critical factors for these particular environments than pollution or other human activities. For this reason, co-operation with land-reclamation bodies has become necessary to implement good conservation policies.

Fig. 1: a. Vegetation Maps and other publications on protected areas of the Regional landscape; b. The watershed ridge of the Tuscany and Emilia Romagna Regions, massif of Mt. Cimone; c. *Viola cavillieri*; d. *Geranium argenteum*; e. Wood of *Fagus sylvatica* in the Regional Park of "Corno alle Scale"; f. Landscape of the Regional Park of Sassi di Roccamalatina; 1g. Characteristic volcano-like mud cones in the natural reserve "Salse di Nirano"; h. A typical wetland in the Po Valley (Emilia Romagna Region); i. *Senecio paludosus*, an endangered hygrophytes species; l. *Trapa natans*, an aquatic plant growing in a relic habitat (between Modena and Ferrara Provinces).

The "Consorzio di Bonifica Burana, Leo, Scoltenna e Panaro" operates in the territory of Modena, dealing in land reclamation, and is in charge of all the existing waters in the territory, with regards to agricultural uses and safety issues. For several years this Consortium has been working with the Botanic Garden of Modena on conservation projects dedicated to preserving and increasing populations of certain hygrophyte plants and managing their habitats, from the dual perspective of nature conservation and correct use.

2.2.1. Marsilea quadrifolia L. project

Marsilea quadrifolia L. (Marsileaceae) is an aquatic fern (Fig. 2a) which, until recent times, had been considered extinct in the Po Valley. It was rediscovered in a ditch near Mirandola (Modena Province). Given the importance of this find, a co-operative agreement has been signed by the Botanic Garden and the local Consortium. The species was monitored *in situ* and in the Botanic Garden (*ex situ*), where some specimens were cultivated in an experimental ditch (Fig. 2b), created in the Garden as an example of a typical rural biotope threatened with extinction [6, 7]. Some local schools were also involved in this project through educational activities.

In this co-operative agreement subscribed to and binding between Botanic Garden and "Consorzio of Burana, Leo, Scoltenna e Panaro", the most relevant aims are not only to save and increase the *Marsilea* population, but also to overcome traditional conflicts with this kind of institution, which pursues social and economic goals rather than those of nature conservation. One current project is particularly significant: a practical booklet describing the threatened aquatic plant species has been printed and distributed and explained to Consortium operators in order to carry out an accurate census of their local sites of occurrence. In conclusion, by involving the bodies appointed to manage resources in an innovative, articulate plan of action (with an unexpected outcome) rather than taking an inflexible ideological stance against them, Modena Botanic Garden has obtained positive results in the conservation of plant biodiversity.

2.2.2. Saving the Italian Viola pumila population

Another experiment in *ex situ/in situ* conservation, carried out in the Botanic Garden of Modena, is linked to saving a small surviving population of *Viola pumila* Chaix (Fig. 2c) in the Province of Reggio Emilia (Northern Italy). This site is one of the very few localities formerly known in Italy for this species [27], and the only one recently confirmed. The Botanic Garden promoted a programme to preserve the habitat and to increase the population, through an agreement with the corporate body in charge of local land management. *Viola pumila* seeds were cultivated in the Botanic Garden in both *in vivo* and *in vitro* cultures (Fig. 2d) to induce germination and development of a large stock of plantlets. *In vitro* experiments proved to be effective in obtaining a large number of seedlings, much more than *in vivo* conditions. The plantlets were then transferred from *in vitro* culture to soil in pots, following which they showed a high percentage of rooting. Finally, the well-developed plants were re-transplanted into the place where the seeds were collected [26, 33]. The operation, which involved pupils from local schools during the first *in situ* reintroduction, was successful (Fig. 2e) and the *Viola pumila* population in Italy can actually be considered safe.

2.2.3. Increasing rare and endangered orchid populations from seed

Another conservation project, begun at the Botanic Garden in 1996, concerns wild terrestrial orchids and involves *in vitro* sowing and propagation techniques. *Ex situ* orchid propagation was carried out through *in vitro* asymbiotic seed germination, as this technique is widely recognized to be the most appropriate strategy to conserve plant genetic diversity [24, 28]. During these years methods for growing many different orchid species were standardized and optimised, so that a large number of plantlets can be obtained (Fig. 2f), [29, 30]. After culture *in vitro*, plantlets of some species of orchid were potted up in loam successfully and placed outdoors in the Botanic Garden (Figs. 2g and 2h). Plants of *Serapias vomeracea* subsp. *laxiflora*, obtained by means of *in vitro* culture, were cultivated in the Botanic Garden and finally reintroduced into their natural habitat (Fig. 2i) [31, 32].

2.3 Collaborations with Museums: flora from the Modena territory in herbaria

Co-operation with local Museums has produced interesting sources for future actions within conservation projects: their collections are mostly related to circumscribed local areas or well-defined periods, so they are the source of fundamental information about past situations. Some years ago, for these purposes, a census of the local flora collected in herbaria at the Modena Botanic Garden was carried out. The data obtained, but not yet published, actually constitute a considerable database to be added to the local flora study project.

In another significant study regarding the reorganization of the "F. Minghelli" Natural History Museum (Pavullo nel Frignano, Modena, Northern Apennines, Italy), two unpublished herbaria have been discovered, one of which, arranged by D. Riva, contains over 1400 specimens, while the other consists of about 2500 specimens collected by V. Mori [16].

2.4 Montecristo Project (National Park of the Tuscan Archipelago)

Studies conducted on flora and vegetation of protected areas constitute a very important source of knowledge to verify the effectiveness of conservation actions. An interesting project, carried out in cooperation with the National Forestry Service, regards Montecristo Isle, in the National Park of Tuscan Archipelago (Italy). Investigation and monitoring of plants and plant communities living on the island are conducted with an integrated approach to evaluate the effectiveness of 25 years of total protection.

Fig. 2: a. *Marsilea quadrifolia* growing in a ditch near Mirandola (Modena Province); b. The artificial ditch created into the area of Botanic Garden of Modena where *Marsilea quadrifolia* grows *ex situ*; c. *Viola pumila*; d. *V. pumila* plantlets growing in *in vitro* conditions; e. A moment of the reintroduction of *V. pumila* near the original sites (Rolo Valleys, Reggio Emilia Province); f. *In vitro* cultures of *Serapis vomeracea* ssp. *laxiflora* plantlets; g. Some orchid plants after the transplanting in pots; h. *Serapias vomeracea* ssp. *laxiflora* in the Botanic Garden; i. The natural area into the Reserve "Tomboli di Cecina", where orchid plants were reintroduced; l. A school during a visit at the Botanic Garden of Modena; m. An experience of vegetables cultivation by a local school; n. Some drawings realized during didactic activities in the Botanic Garden; o. A didactic activity about the ancient trees growing in the Garden; p. Musical event in the Garden.

2.5 Educational programs

In accordance with the Botanic Garden Conservation Strategy [1], the Botanic Garden of Modena includes a flexible education program among its priorities. The aim is to create an understanding and awareness of the needs for conservation activities at both local and international levels. Enterprises devoted to the diffusion of botanical knowledge and culture are well-established practise.

The Garden is visited every year by thousands of people, with numerous activities dedicated to all levels of schools (Figs. 2l, 2m and 2n), cultural associations (Fig. 2o) and other groups – conferences, exhibitions, courses on gardening and botany, Garden Club activities, floral decoration, play activities, music and theatre in the Garden (Fig. 2p), etc.

In Italy today, there is a specific law oriented towards supporting projects aimed at the diffusion of scientific culture (L.6/2000): the Botanic Garden of Modena, like several other University Gardens, sustains its education programmes mainly by means of this law and specific projects, along with local Institutions or Administrative Bodies.

These activities are always under-funded, which causes considerable difficulties for the University Botanic Gardens when promoting education programmes outside their academic activities.

Remarks

University Botanic Gardens, together with other similar institutions, have been appointed to:

– carry out *ex-situ* plant conservation,

– promote botanical knowledge at all levels,

– preserve floral heritage, architecture and tradition,

– endorse any action oriented towards fostering respect and love for plants among the public and, in particular, among young people.

Furthermore, thanks to their scientific capabilities, these institutions can and must became focal points for local Administrations, National and Regional Parks, Natural Reserves and other administrative bodies or institutions, in programming and developing environmental interventions, proposing and applying the right strategies in order to preserve the biodiversity of plants, plant communities and habitats.

Acknowledgements: The authors are grateful to all people, institutions and administrative bodies for their successful cooperation and endorsement of Botanic Garden's actions.
We acknowledge all the staff of the Modena Botanic Garden, the Cartographic Department of the Emilia Romagna Regional Council, Modena Provincial Council, Government National Forestry Service, Regional Park of Northern Apennines, Regional Park of Corno alle Scale, Regional Park of Sassi di Rocca Malatina, Nature Reserve of Salse di Nirano, Land Reclamation Consortium of Burana, Leo, Scoltenna & Panaro, Voluntary Nature Rangers of Modena, and other land-managing institutions producing vegetation and floral information on protected areas in the Province.

REFERENCES

1. AA.VV., 1995 - Orti Botanici e Strategia della Conservazione. BGCI & Orto Botanico Dipartimento di Scienze Botaniche, Università di Pisa. [Original version: The Botanic Gardens Conservation Strategy. WWF, IUCN-BGCS, Gland and Richmond, 1989].
2. AA.V.V., 2001 - Musei della Provincia di Modena. Provincia di Modena, Modena.
3. Barozzini E., Bertogna I., Castaldini D., Cosmin C., Dallai D., Del Prete C., Gorgoni C., Ilies D., Sala L., Valdati J., 2004 - Riserva Naturale Regionale delle Salse di Nirano. Carta turistico-ambientale.
4. Bedini G., Del Prete C., Dallai D., 1998 - Trends in the evolution of the Italian Botanic Gardens Network. In: Abstracts, pp. 4. 5th Intern. Bot. Gard. Cons. Congr. Kirstembosch.
5. Bertolani Marchetti D., 1992 - L'Orto Botanico dell'Università di Modena. In: Raimondo F.M. (ed.), Orti Botanici, Giardini Alpini, Arboreti Italiani, pp. 119-122. Edizioni Grifo, Palermo.
6. Bonafede F., Dallai D., Del Prete C., 1995 - Ritrovamento di *Marsilea quadrifolia* L. nella Pianura Padana meridionale e prospettive per la sua conservazione in situ. Giorn. Bot. Ital., **129** (2): 97.
7. Bonafede F., Dallai D., Del Prete C., Maffettone L., 2003 - *Marsilea quadrifolia* L. in Emilia Romagna: distribuzione, ecologia e problematiche di conservazione integrata *in situ/ex situ*. Atti Soc. Nat. Mat. Modena, **133**: 183-211.
8. Dallai D., 1993 - Current guidelines for exhibitions of the Modena Botanical Garden collections. Giorn. Bot. Ital., **127** (3): 672.
9. Dallai D., 1996 - Gli Orti Botanici per la conservazione della biodiversità e la diffusione della conoscenza naturalistica. In: Atti Convegno "Biodiversità, Conservazione e Sviluppo", pp. 71-77. Provincia di Modena.
10. Dallai D., Antonini E., Del Prete C., 1998 - Ruolo storico dell'Orto Botanico di Modena nella diffusione e coltivazione delle piante ornamentali. Museol. Sci., **14** (1): 271-282.
11. Dallai D., Del Prete C., 1994 - Il Riassetto museologico delle serre storiche dell'Orto Botanico di Modena. Primi risultati e obiettivi futuri. Giorn. Bot. Ital., **128**: 408.
12. Dallai D., Del Prete C., 1996 - Esperienze di didattica pre-universitaria all'Orto Botanico dell'Università di Modena. Atti Riunione Scient. "Didattica pre-universitaria negli Orti Botanici". Inf. Bot. Ital., **28**: 118-122.
13. Dallai D., Garbari F., 1998 - L'Orto Botanico dell'Università di Modena nella ricerca e nella didattica biologico-naturalistica. Atti Conv."Giornata di Studi in Ricordo di Daria Bertolani Marchetti". Deput. St. Patr. Ant. Prov. Modenesi, Biblioteca, n.s. **150**: 83-98.
14. Dallai D., Maffettone L.; Barberini E.; Bosi G., Del Prete C., Accorsi C.A., 2001 - La Xiloteca Storica dell'Orto Botanico dell'Università di Modena e Reggio Emilia: interventi di recupero e valorizzazione museale. Atti Soc. Nat. Mat. Modena, **131** (2000): 35-61.
15. De Toni G.B., 1906 - Il R. Orto Botanico di Modena dal 1772 al 1906. Malpighia, **20**: 272- 283.

16. Florini S., 2004 - Riorganizzazione del Museo Naturalistico "Ferruccio Minghelli" di Pavullo nel Frignano: valutazione conoscitiva preliminare delle collezioni botaniche ed eventuali emergenze. Atti Soc. Nat. Mat. Modena, **134** (2003): 347-349.
17. Franchini F., Del Prete C., Dallai D., 2005 - Emilio Chiovenda a Modena. In: Russo L. (ed.), The succulent plant of Eastern Africa. In: Atti Intern. Symp. "Le piante succulente dell'Africa orientale": storia, esplorazione e ricerca botanica, pp. 157-175. Verbania.
18. Garbari F., 1985 - Il ruolo degli orti botanici nella conservazione della diversità genetica regionale e nell'educazione ambientale. In: Atti Sem. "Le piante e l'uomo", pp. 39-47. Catania.
19. Garbari F., 1990 - Orti Botanici e significato delle collezioni. Museol. Sci., **7** (1-2): 123.
20. Garbari F., Raimondo F.M., 1986 - Botanical Gardens in Italy: their history, scientific role and future. Museol. Sci., **3** (1-2): 57-81.
21. Heywood V., 1992 - Botanic Gardens and the conservation of biodiversity. Atti Conv. Internaz. "I 400 Anni dell'Orto Botanico di Pisa. L'Orto Botanico, il passato chiave per il futuro?". Museol. Sci. **9**: 21-32.
22. IUCN, 1980 - The World Conservation Strategy. IUCN, Gland.
23. IUCN, 1986 - Recommendations passed at "Botanic Gardens and the World Conservation Strategy". IUCN Conservation Monitoring Centre.
24. Johansen B., Rasmussen H., 1992 - *Ex situ* conservation of orchids. Opera Bot., **113**: 43-48.
25. Mor C.G., Di Pietro P., 1975 - Istituto ed Orto Botanico. In: "Storia dell'Università di Modena II" Leo Olschki, pp. 477-479. Firenze.
26. Morselli A., 2004 - Propagazione *in vitro* e *in vivo* di *Viola pumila* Chaix., specie rara e minacciata di estinzione nel territorio emiliano. Atti Soc. Nat. Mat. Modena, **134**: 339-341.
27. Pignatti S., 1982 - Flora d'Italia. Edagricole, Bologna.
28. Ronse A., 1989 - *In vitro* propagation of orchids and nature conservation: possibilities and limitations. Mem Soc.Roy Bot. Belg., **11**: 107-114.
29. Sgarbi E., 2002 - La germinazione *in vitro* delle Orchidee: un approccio metodologico. Atti Soc. Nat. Mat. Modena, **132**: 65-74.
30. Sgarbi E., Ronconi L., Del Prete C., 1999 - Germinazione asimbiotica *in vitro* di Orchidaceae spontanee dell'Italia centrale: un'esperienza di propagazione *ex situ* per la reintroduzione *in situ*. In: 94th Congr. Soc. Bot. Ital., Abstracts, pp. 94.
31. Sgarbi E., Del Prete C., Ronconi L., Dallai D., Perini C., 2001a - Wild Italian orchids: from seed to plant experience in a project for *in situ* reintroduction and *ex situ* conservation. Atti Conv. Planta Europa.
32. Sgarbi E., Del Prete C., Ronconi L., Dallai D., 2001b - Asymbiotic micropropagation of wild Italian Orchids from seed to plant in a project for *in situ* reintroduction. J. Europ. Orch., **33** (1): 395-404.
33. Sgarbi E., Dallai D., Del Prete C., 2004 - Propagazione *in vivo* e *in vitro* di *Viola pumila* Chaix in un programma per la sua conservazione nel territorio emiliano. In: 99th Congr. Soc. Bot. Ital., Abstracts, pp. 179.
34. Tomaselli M., Manzini L., Del Prete C., 1995a - Carta della vegetazione con itinerari naturalistici 1:25.000 del Parco Regionale dell'Alto Appennino Modenese. Foglio Est. Regione Emilia-Romagna: Servizio Cartografico.

35. Tomaselli M., Manzini L., Del Prete C., 1995b - Carta della vegetazione con itinerari naturalistici 1:25.000 del Parco Regionale dell'Alto Appennino Modenese. Foglio Ovest. Regione Emilia-Romagna: Servizio Cartografico.
36. Tomaselli M., Del Prete C., Manzini M.L., 1996 - Parco regionale dell'Alto Appennino Modenese: l'ambiente vegetale. Regione Emilia Romagna Servizio Paesaggio, Parchi e Patrimonio Naturale.
37. Tomaselli M., Manzini L., Del Prete C., Rossi G., 2003 - Carta della vegetazione del Parco Regionale del Corno alla scala 1:15.000. Servizio cartografico della Regione Emilia-Romagna.
38. Tomaselli M., Manzini L., Del Prete C., Spettoli O., 1997 - Carta della vegetazione 1:10.000 del Parco Regionale dei Sassi di Roccamalatina (Provincia di Modena). Regione Emilia-Romagna: Servizio Cartografico.
39. Tretiac M., Dallai D., 1990 - L'erbario lichenologico dell'Università di Modena (Mod). I licheni dell'erbario Baglietto. Not. Soc. Lich. Ital., **3** (suppl. 1): 51-56.
40. Vaccari A., 1928 - Notizie sull'Erbario Vaccari ceduto all'Istituto Botanico della R. Università. Atti e Mem. Accad. Sci. Lett. Arti Modena, **4** (2): 43-48.
41. Wyse Jackson P., 1989 - Resolution passed at the Second International Botanic Gardens Conservation Congress. I.U.C.N. Botanic Gardens Conservation Secretariat, Kew.

FOREST MANAGEMENT IN PROTECTED AREAS IN ITALY

Renzo FELIZIANI
Smilax Association, Department of Botany and Ecology, University of Camerino
Via Paolo Buonamici 6, **63041 Acquasanta Terme, AP (ITALY)**;
e-mail: renzofel@libero.it

Abstract: In Italian protected areas, traditional silviculture has not always given the rightful priority to the objectives of conservation and natural improvement of the forest heritage. In addition, approaches to silvicultural management vary quite significantly. In some cases, for example in the Abruzzo National Park, most of the forests are left to free evolution or at least treated by natural criteria in cutting, thus obtaining significant improvements in population structure and biodiversity; in other cases, for example, the Monti Sibillini National Park, no particular conservation measures have been adopted, and thus currently there is no substantial difference between usage inside and outside the protected area. Of particular importance is the possibility of and need to create A Zones in the Parks (whole reserves) to maintain some significant sections of forest, excluding any anthropogenic intervention, in order to allow conservation of all extant forest associations and thus their species, phytosociological and genetic diversity. Hence, forest assessment projects in protected areas must take into account the economic exigencies of the resident population, but also be attentive to identifying areas of the forest where the complete biological cycle of the wood can happen, and the different associations can be maintained, promoting the adoption of strictly natural silvicultural criteria. In the current conditions in Italy, this is possible only if we enact a policy of renting or, preferably, public acquisition of wooded areas with the highest degree of natural conditions, as well as wooded areas that are degraded today, for which a recovery process is timely.

Introduction

In the Italian protected areas, management choices, particularly those involving silviculture, do not always give rightful priority to the objectives of conservation and naturalistic improvement of the forest heritage; rather, they tend to safeguard the interests of the local population, in particular, rights connected with private property and civic use. This work critically describes the status of Italian forests in terms of their conservation; it analyses the current outlook and considers the opportunities offered by the protected areas for their qualitative and quantitative recovery, providing general indications for correct management of the forest heritage. The study outlines the reasons for forest degradation, examines current problems, explains ecologically possible solutions, and, while it criticizes much current management of forest resources in protected areas, it praises the positive cases, emphasizing the role of parks in the conservation of this resource.

Materials and Methods

Preparation of the present work drew upon the current bibliography on the subject, statistical data extrapolated from the Corpus of Foresters website [9, 12] and from the APAT Yearbook of Environmental Data [2], as well as personal studies and experience. Analysis and critical processing of these data has made it possible to provide a general overview of the situation of the forest heritage in protected areas, and to formulate necessary suggestions for its conservation.

Results

The forest resources in Italy

Italian forests cover 9,857,000 ha [9], a little less than a third of the country's territory. Of this forested land, about 6,860,000 ha comprise old-growth forest, coppices and Mediterranean maquis [12], while the remainder is represented by minor formations, with a low level of tree cover. The degree of cover in Italian forests does not diverge much from the European average, while the structure is of fairly poor quality, because the prevalence of coppice (57%) is conspicuous. Simple coppices alone represent over 40% of the total, while mixed old-growth forests or even broadleaved ones, comprise less than 20% [9]. Forest property in Italy is prevalently private (60%), while much of the public forest property belongs to municipalities (27.4%) and, to a lesser degree, to the national government, or other bodies (12.6%). The Italian forests have a timber mass of over 1 billion cubic meters, which increases by about 30 million cubic meters each year. For purely economic reasons, part of this increase in woody mass (about 9 million cubic meters per year) is used to partially meet national demand for timber. In terms of dynamics of the forest area in the period from 1948 to 2001, there was an increase of 1.238.303 ha, equivalent to an increase of 22% (Fig. 1); while in terms of dynamics of the different types of governance (cutting regime for forest reproduction and renewal), one can observe during the same period a notable decrease in composite coppices, in favour of simple coppices, and a general increase in all old-growth forests (Fig. 2). Analysis of the data and the dynamic tendencies illustrated give strong reason to hope for the recovery of the Italian forest heritage, over-exploited in the recent past. In this context, protected areas become important opportunities for permitting the natural and qualitative/quantitative recovery of this resource.

The forest resources in the National Parks

Forest distribution notably varies in the different protected areas. While some parks are prevalently forested (Foreste Casentinesi, Val Grande), others have little forest vegetation (Mesola woods). In general, the woods and forests (Corine classes that are made up of hardwood forests, conifer forests and mixed woods) in the protected areas are 17.40% of total forest cover (Fig. 3), equivalent to an estimated area of approximately 1,193,640 ha. In addition, one should also consider the woods that lie within the territory of the SICs (sites of European Community Importance) (21.62%) and ZPS's (Zones of Special Protection) (9.03%), as some of them are not included in the previous category of protected areas [5]. If we consider the percentage cover of the different forest types in the country, it appears evident that the protected areas are proportionally much better forested than the remaining territory (Fig. 4).

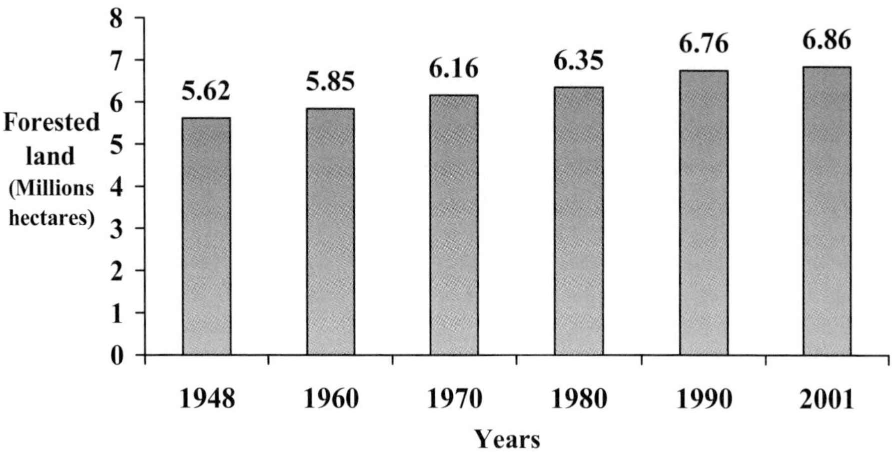

Fig. 1: Dynamics in forested lands – from 1948 to 2001 (data from [2]).

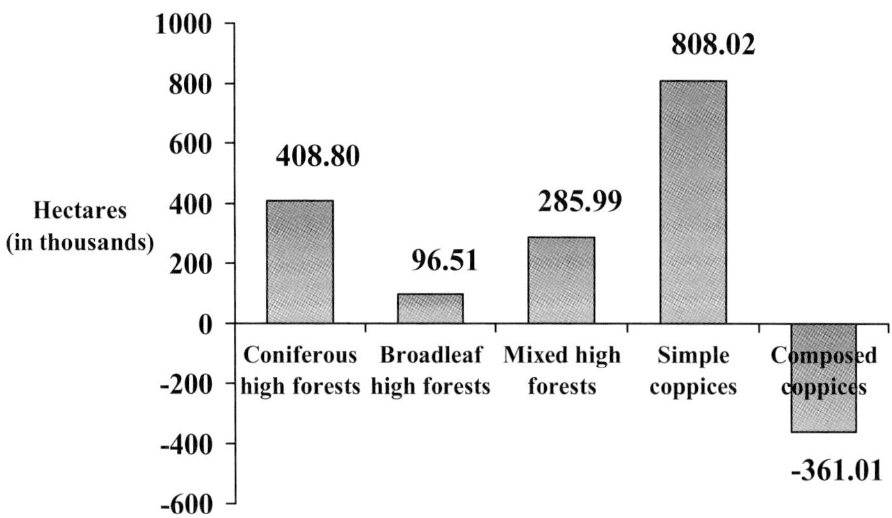

Fig. 2: Dynamics in forested lands by typology – from 1948 to 2001 (data from [2]).

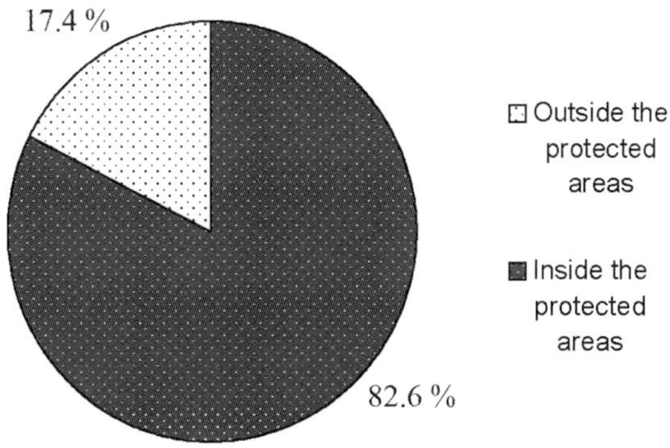

Fig. 3: Distribution of woods and forests (Corine Classes) in the protected areas and in the rest of the territory (data from [4]).

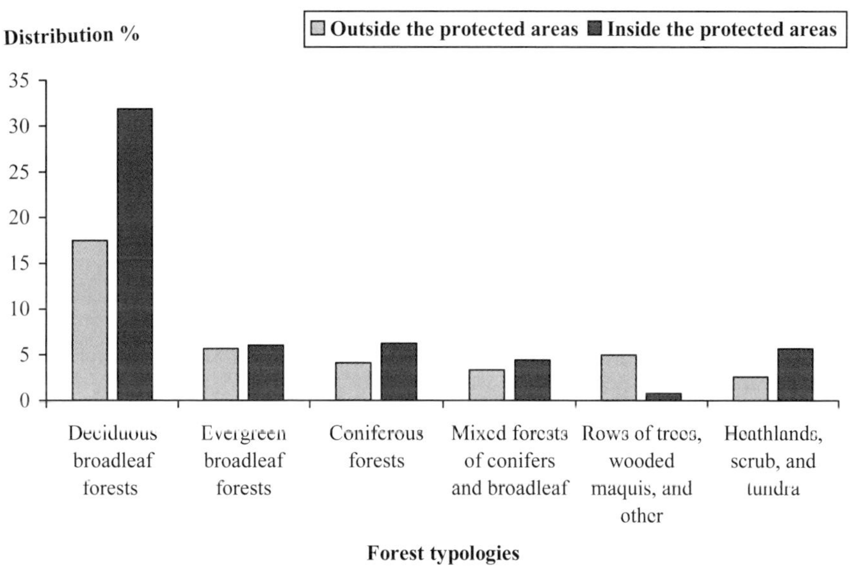

Fig. 4: Typology of forest habitats throughout Italy and in protected areas (data from [2]).

The state of the woods in the protected areas

The current state is the result of intense and in many cases devastating anthropogenic activity, conducted on the forest resource over the course of centuries, and continued to the present day, with different trends and justifications at various times in history (Fig. 5). In particular, during the last century the forest landscape was tremendously disturbed by actions such as the agrarian reform of 1950 (which devastated vast wooded areas), the campaigns to reclaim and improve mountain areas, with road construction and development of tourist facilities (ski slopes, lifts, etc.), reforestation in the Apennines (achieved prevalently with Austrian pine) and in the Alps (especially the establishment of conifers in the beech zone), grazing (which has a negative impact on forest renewal) and forest utilization (trends of wood cutting) [15]. The last activities, conducted in the recent past for exclusively economic reasons, have been the principal causes of the naturalistic degradation of this resource, causing losses to the ecosystem's diversity, complexity and stability. In particular, they have

- ✓ favoured coppicing over old-growth forest governance;
- ✓ transformed mixed woods into monospecific woods;
- ✓ selected some species over others (this action is the principal cause of the fragmentation of mixed woods of beech and silver fir in the Apennines, and of Norway spruce and silver fir in the Alps);
- ✓ simplified the forest's vertical structure;
- ✓ created forest areas composed exclusively of populations of the same age;
- ✓ interrupted the ecological cycle of the woods (absence or scarcity of decomposing wood in Italian forests).

Underway dynamics in the forests of the Italian protected areas

There are no virgin forests in Italy, and by far, most of the old-growth forests are not at all close to natural and climax phase conditions, and thus they are subject to succession. Only in very small strips of forest, generally located within ample forest complexes, does the process of fluctuation dominate. Examples are to be found in the mixed fir-beech stands of the Sasso Fratino Reserve in the Casentinesi Forests National Park [11], the hornbeam woods of the Quarto forest and the beech woods of the Umbra forest in the Gargano National Park [10], some strips of holm oak (ilex) woods in the Gennargentù National Park [3] and some small areas of beech wood in the Abruzzo National Park and other protected areas. Most other old-growth forests are monospecific and even-aged, far from natural conditions, or mixed uneven-aged woods, too often characterized by degenerative problems linked to excessive use for grazing or mistaken thinning out. These forests, along with the multitude of coppices, are mostly in a phase of degeneration and often even in regression. The latter process can be seen in some strongly human-influenced Italian parks, such as the slopes of Monte Vettore in the Monti Sibillini National Park [13], or in some areas of the Quarto woods in the Gargano National Park [10]. However, the plans for utilization of the coppices in many protected areas, which foresee leaving uncut a greater number of 'mother trees', and sometimes require conversion from coppice to old-growth forest, give good reason to hope for a slow but always possible regeneration, as has happened, for example, in the vast areas of beech wood in the Abruzzo National Park [6].

Fig. 5: Evident degradation of the landscape (Bolognola, Monti Sibillini) due to excessive anthropic pressures in the 1950's (photo: V. Marchesoni) [4].

While primary succession is a rare process, observable for example on the lava flows and lapilli deposits in the Vesuvio National Park, secondary succession is a very widespread process in meadows and fields abandoned in recent decades as a result of changed socio-economic conditions (Fig. 6). The vegetation landscape of these areas undergoes radical modifications in species and physiognomy, towards disordered and initial transitional phases that will lead to the development of old-growth forests. This natural expansion of woods is considered by many as an absolutely negative process to be halted by any means, in order to maintain meadows and grazing lands and their productive functions. Others, among them some Park managers, observe this process with great concern and think they should limit its extension in order to avoid the disappearance of some meadow associations and rare species. In reality, this is a natural process that leads to the formation of stable phytocoenoses, and one should certainly not aim to halt it, except in limited areas such as the Fiavé Peatbog biotope in Trentino, where for years now, mowing has been continued in order to maintain the special vegetation of the wet meadows [7]. If this were not done, the area would be transformed into a single marshy thicket of *Salix cinerea* with consequent loss of biodiversity and a general negative effect on the ecological complexity of the marshy, peaty environment, as seen in geobotanical studies conducted in a similar case, the Laghestel di Piné biotope, which is being taken over by an *Alnus glutinosa* wood [16].

Fig. 6: Colonisation by trees of the field crops and pasturelands abandoned in the 1960's at Gaglierto (Monti della Laga and Gran Sasso National Park) (photo S. Evangelisti).

Operations currently conducted in the forest patrimony of the protected areas

In order to give a better overview of the Italian forest heritage, and since it is impossible to indicate all the types of operations currently employed (they vary in different protected areas, different regions and provinces, forms of governance and treatment, etc.), a few brief examples will be given to illustrate what has happened and what is currently being done in some Italian protected areas. We can begin with the Torricchio Natural Reserve, managed and monitored by the University of Camerino (Botany Department), where 86 ha of predominantly beech coppice (with small nuclei of old-growth forest), after centuries of intense forestry and grazing, have been left since 1970 to free evolution under an attentive monitoring program [14]. In the nearby Monti Sibillini, Gran Sasso and Monti della Laga National Parks, which are still waiting for their Plan and Regulations, tree cutting is carried out by the same criteria both inside and outside the Parks. Fortunately, the existing Regional procedure reminds the protection functions of woods, stipulating, for example, that in coppices a certain number of 'mother trees' and trees of undefined age be left uncut, and requiring conversion in aged coppices, etc. This procedure, based as it is on classic silviculture, does not require that areas be left to free evolution, nor does it exclude senseless actions that can prove extremely harmful to the forest ecosystem, such as "cleaning out" the old-growth forests (cutting the underbrush, the new growth, the dead trees, etc.), clear felling in the old chestnut woods, or thinning the already compromised riverside vegetation, etc. Among the positive aspects, Swiss mountain pine has been reintroduced in the Sibillini, and in the Monti della Laga there are plans to create spacious integral reserves to preserve important forests (the Martese wood, the San Gerbone forest and the Valle della Corte fir wood). To give an extreme example, which both in concept and in terms appears contestable in any policy for development of a protected area, consider the socio-economic multi-year Plan of the Cilento National Park, which proposes "restoring" the woods to productive purposes, by cleaning out the underbrush and eliminating dead wood, converting simple coppices into compound ones, and where possible, into old-growth forests. In this manner, the Park hopes to improve the forest patrimony in both quality and quantity, thus permitting "a rational use that is economically remunerative" [1].

The situation in the Casentinesi Forests National Park is certainly different. Here, the Park Authority authorizes forestry activities through the Corps of Foresters, and there are entirely protected forests, such as the Sasso Frattino Natural Reserve, where since 1959 the mixed fir-beech stands have been left to evolve freely. In Sardinia, the situation is extremely difficult in the territory of the Gennargentù National Park, which in fact exists only on paper. In the Asinara Island Park, instead, the situation is well-defined (abandoned by now, characterized by thinned-out Mediterranean maquis with a small nucleus of holm oaks); this is also the case in the Maddalena Archipelago Park. In the Abruzzo National Park, some areas have been set aside as integral reserves where the woods are left to evolve freely. Cutting criteria have been adopted that promote systems of re-conversion towards structures composed of trees of different ages and species. In the level areas, 'hole' cuts have been made to create small clearings, and in the beech woods selective thinning has been carried out to promote associated species. This kind of forest management, made possible through a rental policy, has yielded positive results in terms of increased biodiversity and complexity. In addition, it has provided important indications, for example: 1) the natural importance of leaving at least some rotting or dead trees; 2) tree cutting performed according to natural criteria can even increase the specific faunistic diversity; 3) sparing the most environmentally valuable woods from cutting allows the survival of the most demanding species [17].

Guidelines for ecosustainable management of the forest resources of the National Parks

In this regard, the Ministry of the Environment decided to give the Italian Academy of Forest Sciences responsibility for preparing a management document, which comprises a final report and a manual [8]. This work provides guidelines for regulating forest utilization in the national parks, and articulates specific indications for the various types of woods, according to the different plans foreseen. They call for the introduction of innovations of systemic silviculture, in the process overcoming some limits of traditional (or classical) silviculture. Systemic silviculture is an extensive method in which man intervenes in the forest, bearing in mind the ecosystem's natural evolutionary process. It does not stipulate structures, shifts (the period between one cut and the next) or minimum earnings from the forest in terms of quantity of wood. Instead, it works toward the spontaneous mixing of species, and its main objective is to conserve and increase forest complexity. The guidelines stipulate no action of any kind in zone A, while in zone B they indicate operations of systematic silviculture in the better structured old-growth woods and interventions of re-naturalization in the other forest types. Only zones C and D permit, as an alternative to systematic silviculture and re-naturalization, utilization conducted on the basis of traditional silviculture (Table 1).

Table 1: Forms of management according to the degree of naturality of the forest systems and zoning [7].

	Natural		**Artificial**		
ZONES	**Old-growth forests** (autochthonous species with a complex structure and varied composition)	**Old-growth forests** (autochthonous species with a complex structure and/or simplified composition)	**Coppices**	**Old-growth forests** (Artificial reforestation with autochthonous species)	**Old-growth forests** (Artificial reforestation with exotic species)
A	Natural	Preservation		Artificial	
B	SS	R	R	R	R
C	SS	R	R-(ST)	R - (ST)	R - (ST)
D	SS - (ST)	ST - (R)	ST-(R)	ST - (R)	ST - (R)

Legend: R = renaturalization; SS = systemic silviculture; ST = traditional or classic silviculture

Discussion

Summary considerations on the status of the Italian forest heritage

Briefly, one can affirm that the forest heritage in the Italian protected areas, though generally characterized by a hardly exceptional state of conservation and by a low degree of naturalness, is of great interest from the phytogeographical, phytocoenotic and ecological points of view. In fact, this heritage is composed of woods of elevated biological diversity and with features of considerable conservation value. Today, the data on the average annual increase of the Italian forest heritage, as well as information on the dynamic tendencies currently underway, indicate good cause for hope, but a great deal of work is still required to improve the quality and quantity of this resource, from the natural point of view, and above all, to increase its overall ecological stability. There are notable differences in how forests are managed in the various protected areas, because no one specific body of legislation regulates such activity. In many National Parks (for example, the Monti Sibillini and, the Gran Sasso and Monti della Laga National Parks), since there still is no specific regulation of the Park Entity, there are no substantial differences in wood-cutting on lands inside or those outside the protected area. Other Parks, instead, require an authorization permit from the management Authority and/or a report outlining the plans for forest use (Abruzzo National Park, Foreste Casentinesi, etc.). It should be noticed that in all the protected areas in Italy, there is considerable resistance to the imposition of primarily natural criteria for cutting, given that a great part of the forest heritage is on private property. Thus, forest utilization regulations do not always take into consideration conservation exigencies, which instead should be of first priority in the National Parks. In addition, in the Special Conservation Zones (SIC and ZPS) in Italy, wood-cutting should be done only after an Environmental Impact Evaluation, but this does not happen uniformly across the various Italian Regions.

Conclusions

In conclusion, we propose a series of guidelines on interventions to adopt in order to safeguard the forest heritage of Italian protected areas:

– It is fundamental to leave to free evolution a significant percentage of the forest heritage, in particular all the woods of greater natural value, following (at least in part) the Ministry of the Environment guidelines, which call for this action in all Zone A woods. In order to do so, it is necessary to overcome the strong resistance posed by the legitimate interests of individuals on private property, utilizing for this purpose government reserves, and eventually acquiring the most interesting wooded areas for Park Service ownership, rather than renting them;

– Do not fear the natural expansion of forests (secondary succession) and do not try to stop it, except in limited areas, in order to conserve rare and important environmental resources;

– In silvicultural choices, always promote systemic over classic silviculture, because it aids re-naturalization of the woods and hence increases overall complexity, stability and diversity. In particular, it is necessary to promote cultivation cuts that optimize the conservation function of the forest habitat, for example, 'hole' cutting, which creates small clearings in the woods, or sparing dead or very old trees, even if they have fallen to the ground. In addition, targeted interventions for conifer reforestation are required in order to direct their re-naturalization, promoting the insertion of native hardwoods;

In silvicultural operations, it is always necessary to respect the underbrush, the renewal of the present and rare species, and hence never permit, much less finance, so-called forest 'cleaning' interventions;

– In conclusion, we strongly urge that the European Council's recommendation to leave dead or rotting wood in the forest be respected, thus enabling the forest ecosystem to complete its natural cycle. In this regard, an acceptable value would be a limit of 30-40 cubic meters/hectares of dead wood, at least in zones A and B of national parks.

It is hoped that there will soon be a reference procedure to unify cutting trends in the various Protected Areas, and that authorization permits will be required from the Park Entity for any and all utilization, to be issued only after an on-site survey, and with instructions for cutting criteria. National Corps of Foresters staff could fulfill this role, since they have specific training and historical experience in this sector. In the current situation in Italy, in order actually to allow the realization of these actions, it is always necessary to respect private property and encourage and/or compensate appropriately those who will have to take responsibility for these measures.

REFERENCES

1. AA.VV., 2000 - Misura 7. Sviluppo, ruralità, commercializzazione prodotti tipici, agriturismo in Piano pluriennale socio economico. Parco Nazionale del Cilento e Vallo di Diano, Vallo di Lucania.
2. APAT, 2003 - Annuario dei dati ambientali. I Vol. Agenzia per la protezione dell'Ambiente e per i servizi tecnici. Roma.
3. Arrigoni P.V., Di Tommaso P.L., 1991 - La vegetazione delle montagne calcaree della Sardegna centro-orientale. Boll. Soc. Sarda Sc. Nat., **28**: 201-310.
4. Ballelli S., Lucarini D., Pedrotti F., 2005 - Catalogo dell'erbario dei Monti Sibillini di Vittorio Marchesoni. Braun-Blanquetia, **38**: 1-259.
5. Boitani L., Falcucci A., Maiorano L., Montemaggiori A., 2002 - Rete ecologica nazionale. Il ruolo delle Aree Protette nella Conservazione dei Vertebrati. Dip. B.A.U., Università La Sapienza - Dir. Conservazione della Natura, Ministero dell'Ambiente e della Tutela del Territorio - Istituto di Ecologia Applicata. Roma.
6. Canullo R., Pedrotti F., 1993 - The cartographic representation of the dynamical tendencies in the vegetation: a case study from the Abruzzo National Park, Italy. Oecologia Montana, **2**: 13-18.
7. Canullo R., Pedrotti F., Venanzoni R., 1994 - La torbiera di Fiavé. In: Pedrotti F. (ed.), Guida all'escursione della Società Italiana di Fitosociologia in Trentino, pp. 78-110. Dipartimento Botanica Ecologia, Camerino.
8. Ciancio O., Corona P., Marchetti M., Nocentini S., 2001 - Linee guida per la gestione ecosostenibile delle risorse forestali e pastorali nei Parchi Nazionali. Ministero dell'Ambiente, Servizio Conservazione della Natura - Accademia Italiana Scienze Forestali, Firenze.
9. EUROSTAT, 1998 - http//www.corpoforestale.it/wai/Archivio/statistiche/index.html.
10. Falinski J.B., Pedrotti F., 1990 - The vegetation and dynamical tendencies in the vegetation of Bosco Quarto, Promontorio del Gargano, Italy (maps in scale 1:10000 with comment). Braun-Blanquetia, **5**: 1-31.
11. Hofmann A., 1965 - L'abieti-faggeto di Sasso Frattino e i suoi aspetti fitosociologici. Arch. Bot. Biogeogr. Ital., **9** (4): 148-162.
12. ISTAT, 1997/1998 - http//www.corpoforestale.it/wai/Archivio/statistiche/index.html.

13. Pedrotti F., 1996 - Il territorio forestale dei Parchi: situazione attuale ed orientamenti gestionali. In: Parco Nazionale delle Foreste Casentinesi (ed.), Atti convegno di studio "Pianificazione e gestione nelle aree protette, pp. 1-23. Poppi.
14. Pedrotti F., 1997 - La Riserva Naturale di Torricchio. In: Fermanelli A.(ed.), Atti "Prima conferenza regionale sulle aree protette, pp. 80-89. Ancona.
15. Pedrotti F., 2002 - Lo stato dei boschi dell'Appennino. In Cavazza C. (ed.), l'Appennino dal passato al futuro. I cento anni della Società Emiliana Pro Montibus et Sylvis, pp. 41-48. Società Emiliana Pro Montibus et Sylvis. Bologna.
16. Pedrotti F., 2004 - Ricerche geobotaniche al Laghestel di Piné (1967-2001). Braun-Blanquetia, **35**: 1-51.
17. Sulli C., Bernoni M., 1993 - Primi riscontri ai criteri di gestione delle risorse forestali adottati nel Parco Nazionale d'Abruzzo. Coll. Phytosoc., **21**: 373-378.

TRADITIONAL MANAGEMENT OF THE RURAL AREAS IN WALLONIA (BELGIUM)

Nathalie FEREMANS, Marie-Françoise GODART, Mireille DECONINCK
Université Libre de Bruxelles, Institute of Environment Management and Regional Planning (IGEAT), CP 130/02, Av. F. Roosevelt 50, **1050 Brussels (BELGIUM)**;
e-mail: mfgodart@ulb.ac.be

Abstract: The study presented is based on an investigation conducted among rural participants concerned with nature and landscape. They belong to the scientific world, administration, the social and political world, and local management such as farmers or foresters. Information provided by this investigation has been supplemented by a bibliographical survey, while the evolution of biodiversity and the rural areas in the past is being characterized by an historical approach. These various methods show the spatial continuity of rural areas. This continuity also exists in the way in which the various functions recognized for rural areas are taken into account. The integration of nature, landscape and environment by the producer can be remunerated by the society, in order to compensate for the loss of earnings due to practices more respectful for the environment. In certain extreme cases, one can consider that the function of production is additional while the service to the community is the principal activity. Various levels of nature's integration in production functions of the rural areas imply compensations adapted to the effort provided and to collective et patrimonial interest. Most of the actions of interest for landscape and biodiversity result in the support of ecotones, particularly rich in species diversity. Their maintenance or restoration makes it possible to materialize continuity of the rural areas, to recreate a mosaic of landscapes and to support the special environments.

Introduction

Rural areas, in the sense we use here, are the ‘countryside’, as opposed to the city. They consist of farmland (farms, fields, meadows, pathways, etc.), forests (woods, paths, etc.), animals, patches of native vegetation, roads, villages, and all the boundaries between these elements – waysides and road verges, field margins, wetlands, rivers, patches of wasteland, etc. Here is a place where people live, work and walk. It is also a place where nature and pleasant landscapes find their own niche.

Rural areas, their landscapes and structure, the activities undertaken there, and the nature which still exists there, are the heritage of long natural evolution and human history of which we are custodians and in which we are participants at the same time: they are still evolving to create spaces, landscapes, the nature of tomorrow.

The responsibility for the ‘participants’ in rural areas, people who live, work and travel there, is thus important. Their activities testify to their respect for the beauty, the complexity of that for which they are responsible. It also testifies to their will to live and work the land according to their needs and those of their contemporaries, and in the way in which they prepare tomorrow’s landscapes.

The research reported here sought out rural areas and their managers through time and in the Wallonia of today. A better knowledge of rural areas, and their bonds with nature and landscapes, allows us to apprehend both its diversity and its continuity in time and space. A better knowledge of rural participants also invites us to understand their specialities and the technical actions they carry out. Approaching these technical aspects enables us to suggest concrete answers as to how to manage a semi-natural space or how to combine production, nature and landscape in Wallonia today.

We propose to present here some of the results of this research, by focusing on aspects related to nature conservation.

Methods

In order to understand the complex relationships between nature, landscape and the management of rural areas, several investigations were conducted. A bibliographical review of historical aspects made it possible to put into perspective the appearance of biodiversity and rural areas, within their evolution through the ages. A bibliographical review of technical and legislative aspects was also carried out. Lastly, a qualitative survey was carried out among participants involved in the 'traditional' management of rural areas. Those participants were selected from the scientific world, administration, the social and political community, field managers, etc. In all, 40 people were interviewed. The objective of these conversations was to supplement and direct technical and legislative research, as well as to understand the roles and the relationships of rural participants of Wallonia; and to understand the close relationships existing between rural areas and the human environment.

Results

Forestry and agricultural history of the Walloon rural areas

With the end of the last Ice Age, vegetation became increasingly woody as the climate warmed. From this moment, man probably had an impact on the environment. Since the Neolithic era, human presence has gradually modified, worked and managed the land [10]. The history of agriculture has a rhythm of technical upheavals involving deep socio-cultural and, of course, landscape reshaping. The agricultural revolution of the Neolithic era, with the arrival of slash-and-burn farming, and that of Antiquity, and the use of the curved plough, constitute the first stages of those changes [9]. The agricultural revolution of the Middle Ages, with the appearance of improved ploughs and more effective means of transport, allowed the development of the stalling of animals in winter and the use of their manure. The agricultural system improved thanks to an association between cultivation and breeding. Mowing of meadows goes back to that period.

The first agricultural revolution of modern times, during the 16th century, constitutes an important modification of the farming system. Fallow, until that time used to clean farmland and reconstitute fertility, was replaced by fodder crops: either weeded cultivations (turnips, etc.) or meadows of grasses or leguminous plants. Artificial meadows (leys) thus appeared at this time. This change made it possible to combine the advantages of fallow with an improvement in agricultural productivity. This improvement was at the same time direct (production of meat) and indirect (vegetable production improved thanks to a more significant use of manure).

The second agricultural revolution of modern times, in the 20th century, was largely encouraged by European agricultural policy. The mechanization, which started in the 19th century, continues and involves both machinery and use of agricultural chemicals. These changes have generated an increase in labour productivity (each man can exploit a greater area) and at the same time of yield from the land. Such changes have also induced a very important global increase in yields [5].

Agricultural production, animal husbandry and forest production originally greatly overlapped in space. Multiple uses were probably always assigned to forest. From Antiquity, forest was treated as coppice for the production of firewood, including charcoal. In The Middle Ages, it was a place of numerous common rights: firewood, timber, brushwood, gathering, mowing, extraction of sand, clay or stones, cattle and other pasturage (pigs, oxen), etc. The medieval use of forests was thus polymorphic, comprising timber trees, scrub, meadows, clearings, ponds, marsh, heaths, etc. Its 'wild' character was purely relative, because forest was, at the time, greatly explored and utilized.

During the 18th–19th centuries, forest was the place of wood production for the iron mills and blast furnaces. Since the end of the 19th century, when outlets for charcoal decreased, foresters have sought other products (e.g. mine pit-props and timber). Plantation forest, evergreen and resinous, then gradually replaces coppice [1]. Forest, employed at all times for multiple uses – which have themselves evolved considerably – is in perpetual change. Its evolution over time is not linear: planting and deforestation follow one another according to demographic, industrial and technical pressures. In addition, the multiple uses which it has always provided have involved various conflicts of interest.

These conflicts of use have always existed: variations in population density, competition for resources or space, existed at various times and stages in agricultural development. For a long time, the interest of one or more parties might be in contradiction: wood production is not compatible with agriculture, or even with animals grazing in the forest. Another example is hunting: its main goal is to maintain game densities as high as possible, which is not favourable to the production of wood. One could find many other examples of such conflict.

As a result of such episodes, major elements of our current rural landscape that we often wish to preserve for their aesthetic, heritage or biological character (wood-pasture, coppice, hay-meadows, permanent meadows, etc.) appear or disappear during agricultural history. They are functional elements of an agrarian system at a given time and the expression of that system on the landscape. The concept of 'traditional' agriculture or forestry is therefore quite relative, and refers in general to the 19th century; although this period cannot be considered a result of the evolution of biodiversity, nor as an ideal state to go back to. However, its cultural or heritage aspects constitute values to be protected [6].

Consequences for nature conservation

Such findings raise questions about the objectives of human actions on the landscape from the point of view of nature conservation: which environment to maintain, which species to protect? Even when the conservation objectives for a site are defined, which techniques are needed to achieve the objectives? Choice of management objective results from a decision: it is possible to aim at the greatest possible diversity, or naturalness, or the protection of particularly rare species.

The search for naturalness, in spite of its legitimacy, does not have to divert us from a nature closer to human beings but at the same time, threatened. This reflection takes a particular direction in Wallonia, where naturalness is generally rather weak. On the other hand, the most biologically interesting habitats found in our regions often relate to human intervention [6].

The weight one allots to a species at the time of the evaluation of a habitat depends on its rareness. The more threatened it is, the higher the weight, perhaps with inclusion on the 'Red Lists' of threatened species drawn up by the International Union for the Nature Conservation (IUCN). However, we must also be careful of the desire to impose human will on a nature which appears creative of disorder, lest we to get too close to a botanic or zoological garden, while trying to support such species or such threatened habitats. In addition, the significance granted to a species or a group of species in an environment is not necessarily related to the naturalness of that environment (for example, coppices and grazing forests can be extremely interesting for certain species) or its biodiversity (*Tilio–Acerion* forest of ravines is not very diverse but comprises species that are rare and specific to this habitat).

As for the intervention chosen to support a species, it can be related to biodiversity or degree of naturalness. Thus, by laying out various types of nest-boxes in a natural reserve, it is possible to increase biodiversity, but to the detriment of naturalness. Generally, biodiversity is so important that, even in a temperate climate, we cannot hope to elucidate the individual needs of all the species we wish to preserve. Consequently, one uses species that indicate habitats in danger and rich in different species (e.g. birds, deadwood beetles).

Sylvo-agricultural techniques and nature conservation

With the objective to recreate or maintain a habitat or to support a species (or suite of species), various interventions are possible. In the Walloon region, management related to nature conservation can take two different paths. On the one hand, conservation of open environments often consists of maintaining them, either by mowing or pasture, thanks to the intervention of volunteers, the administration or farmers. On the other hand, the conservation of forested environments is one of limitation of forest interventions, and even the creation of 'integral reserves' in which nature evolves spontaneously. They make it possible to maintain very old trees, indigenous species, etc., by using various techniques such as mowing and grazing, fire and clearing, coppice, wood-pasture, and minimal intervention.

a) Cultivated lands

Mowing was the first management technique used to maintain open habitats in Wallonia with the objective of nature conservation. It was regarded as favourable to the maintenance of great plant and insect diversity [7]. But for about 15 years, pasture has been tried and has been used more and more.

Nowadays, it is generally considered that the best management method for open environments, and the most natural and least detrimental for the insect fauna, is extensive pasturage. Mowing is a much more aggressive technique which causes heavy loss among the insect fauna. In addition, one can consider that no species is specifically bound to mown grass: all these species must originate from a natural environment.

Whether it is or not a traditional practice on the site concerned, everywhere in Europe, people in charge of semi-natural open zones (i.e. natural reserves) increasingly choose extensive pasturage as a management tool. For practical as well as biological reasons, this kind of grazing uses farm animals. Extensive pasturage allows the creation of a mosaic of habitats: zones regularly grazed and maintained very short, zones of refusal (plants that are not consumed), intermediate zones, more trampled places where the ground becomes bare. This heterogeneity favours biodiversity [4].

The choice of species and breed must be based on environmental characteristics such as climate, ground, land organization, and of course the fodder quality of the vegetation. The sheep are preferentially used in dry habitats or steep zones. Equine and bovine species are preferred in association with wet environments. Of course, the herd itself needs careful follow-up. As for the management plan, it must be maturely considered according to the conditions of the site, aims in view, availability of the manager, etc.

b) Forest habitats

Like wood-pasture or pollard trees, traditional coppice and under-storey coppice benefits those forest species associated with clearings or the first stages of recolonization. Currently, in the Walloon region, coppices are in the process of being converted into forest and represent just about 15,000 ha, that is to say 4% of land surface in a regular stand.

The dynamics creating a mosaic of environments (and thus of habitats, able to take in a great number of species) are strongly reduced in production forests. Moreover, it can be maintained only for those forests with a surface large enough to ensure the permanence of each succession stage. The conservation management of our forests must attempt to implement management practices that mimic the effects of disturbance and natural processes, while maintaining or restoring senescent stages (retaining old trees and deadwood). Very old trees, maintained beyond their usual age of exploitation, often carry cavities due to the breaking of a branch, a peck-hole or other injury – used for nest-holes by species of birds, bats, etc. Species which are typically related to senescent forest stages or natural dynamics of the forest must, to develop, have integral natural reserves at their disposal where trees are not exploited at all and deadwood is left in place.

Finally, in addition to integral reserves, there must be some unexploited places near the centre of the exploited part of the forest. Such places consist of some old trees that will never be cut and will age within the forest until their natural death. These small islands constitute links between the various reserves, intended to allow the significant species to move from one to the other. In these reserves, letting the forest evolve spontaneously, without any human intervention, makes it possible to develop a mix of species and its own structure. In fact, the principle of management, even ecological, directs the development of the forest (by choosing which trees to take out or maintain).

c) Evaluation of management

A precise follow-up to interventions is necessary in order to be able, if necessary, to re-orientate the technique chosen according to the objectives defined in the management plan. In the case of pasturage, the species and breed of the animals, the pressure, periodicity and frequency of grazing, staying times and selected conditions (e.g. shelter, disease prevention) must be followed. In the case of mowing, the number, period and frequency of the interventions will be listed. This follow-up of interventions must be associated with a biological monitoring of the site to evaluate their effects correctly [8].

Implication of the traditional managers

The techniques available are, however, feasible only on the basis of the management engineering it is possible to implement. As we can see, if techniques used in nature conservation are not practices of modern forestry or agricultural, they can approach older practices (without being identically copied) or, at the very least, be practised extensively. The participation of farmers and foresters is thus particularly valuable, but unfortunately financial resources to encourage such participation may not be available. Despite this, traditional managers of rural areas are implicated in various ways in the management of natural environments.

a) Open areas

In the Walloon region, as long as the areas of private natural reserves in open areas were on a modest scale, volunteers who came to clear or mow the vegetation could manage them. When areas became too great to be managed only in this way, several solutions were adopted. The volunteers continued generally to participate in clearing, and in herding cattle and sheep. In addition, agreements with farmers could be established. They come to mow mechanically when conditions allow, or are responsible for monitoring the grazing on the site. Collaboration with a farmer is carried out within the framework of a "convention of occupation (of the grounds of the reserve) on a purely temporary and free basis". It is established according to site history. Often, when an interesting site is purchased for a reserve, a farmer has already exploited it – the priority is thus to propose his collaboration. This type of process can be undertaken by nature conservation associations or by the administration in charge, "Nature and Forest Division". The grazing in the reserves, first, and secondly, collaboration with farmers, has clearly improved the existing relationship between field naturalists and farmers. The idea of using animals to manage one's own ground is a step that farmers can easily understand, and of which they approve. Moreover, closeness to a fellow stock breeder creates bonds of neighbourliness and mutual aid. Thus interesting dialogues can take place.

b) Forest areas

The administration in charge of forests was ten years ago renamed "Nature and Forests Division". The expertise of foresters was, at the same time, widened to include nature management. Foresters then became active partners in nature conservation, in open areas as well as in forest, respecting their new powers to create natural reserves. Generally, foresters consider that forests are much less intensified than agricultural land. Drainage, as well as spraying, remains less important and frequent than the intensive practices of open areas. For them, nature's capital is thus better preserved in forest than elsewhere. Consequently, the foresters in charge of nature turned first their attention to nature conservation in open areas, where the fast disappearance of species required more urgent intervention. As for the integral forest reserves, only a few foresters, even among those who are sensitive to nature conservation, find them interesting and are ready to act in this way. Natural reserves in forest, integral or not, remain today the 'poor relations' of the Walloon reserves.

In Wallonia half the forests are private properties and not managed by public administration. The private forest owners feel their priority as producers: they do not like the 'vacuum', and expect their forest to produce trees. Open zones, old trees, non-productive zones are thus for them often difficult to accept. They consider, however, for the most part that their forest management is respectful of nature, environment and landscape. The concept of 'nature' for private owners is thus probably different from that of naturalists, who do not always agree concerning their management practices [2]. One of the great characteristics of private foresters is certainly their personal involvement in the management of their forests. They see it as a richness of heritage. They dedicate their time and energy to manage their forests, and quickly develop a feeling of interference when somebody else gives an opinion concerning their management. They think generally that a forest could not develop fully without human intervention, which is thus to be a major participant, as a manager and a beneficiary of management. Many owners are ready to act in favour of nature, which is compatible with forestry production. But even limited interventions in forest and particularly integral reserves, although important objectives for nature conservation, are not easily conceivable in private forests.

c) Hunting

One can distinguish hunting in the open country from hunting in the forest for their differentiated impact on the structure of the territory. The hunter of the open country seeks to structure space by maintaining or creating hedges, grassy margins or sunken roads. These elements indeed favour game because they provide food and shelter. In the forest, on the other hand, hunters pursue the goal of creating open spaces: clearings, forests tracks, etc. Those are intended to provide food for big game, like stags, which need grassy zones where they are assured a lack of disturbance [3]. The interventions of hunters also favour other wild fauna and flora, in the open country as well as the forest: they maintain useful habitats for many vulnerable species. This is the case, for example, of a butterfly which needs open zones in forest and which finds ideal conditions on forest tracks mown by hunters for game.

Conclusions

The wild species that have been able to survive throughout Wallonia's long history of sylvo-agriculture have been able to find, in the humanized landscape, surroundings allowing their survival. These habitats constitute major elements of our current rural landscape, which we often wish to preserve for their aesthetic, heritage or biological character. They appear or disappear throughout agricultural history, as functional elements of an agrarian system at a given time. The concept of agriculture or 'traditional' forestry is thus quite relative, and refers in general to the 19th century, even if this period does constitute a result of biodiversity evolution, nor an ideal state to find [6].

Nature conservation must thus think out its interventions in terms of objectives such as biodiversity, naturalness – and, finally, the protection of indicator species. Activities that can be carried out in its name, however, depend upon the techniques available and participants willing to take action according to such objectives. Administration, nature conservation associations, farmers, foresters and hunters can find common interest in collaboration in favour of nature conservation, with more or less efficiency depending on individual cases. Increasing the awareness of participants potentially interested in any action in favour of the growth of nature conservation would most probably be profitable: farmers and foresters work in contact with nature, know it in their everyday life, are emotionally attached to it. It is a starting point in favour of a better knowledge of nature's needs, and the means of taking it into account in its management.

REFERENCES

1. Billen C., Gaiardo L., Godart M.F., 1992 - Etude historique de la Forêt d'Anlier. GEVERU ULB. Rapport de recherche, Communauté Française de Belgique.
2. CPDT, 2005 - Gestion territoriale de l'environnement (thème 4) – sous-thème: implication des gestionnaires traditionnels de l'espace rural dans les activités de gestion des milieux naturels et des paysages. Rapport final. ULB & ULG.
3. Cuvelier M., Dierstein A., 2003 - Mise en oeuvre d'une méthode d'aménagement pour une forêt multifonctionnelle: l'intégration de la composante faune sauvage. Forêt wallonne, **63**: 26-31.
4. Delescaille L.M., 2002 - Nature conservation and pastoralism in Wallonia. In: Redecker B., Finck P., Härdtle W., Schröder E. (eds.), Pasture landscapes and nature conservation, pp. 39-52. Springer, Berlin – Heidelberg.

5. Destain M.F., 2002 - De la moissonneuse des Trévires à l'agriculture de précision. Leçon inaugurale, Séance d'ouverture de l'année académique 2002-2003, Faculté des Sciences agronomiques de Gembloux.
6. Feremans N., Godart, M.F. (eds.), 2004 - Gestion de l'espace rural, nature et paysages en Wallonie. Etudes et Documents, 5. Min. Région Wallonne, DGATLP-DGA-DGRNE, CPDT.
7. Goffart P., 1998 - Gestion des milieux, entomofaune et réflexions sur la conservation de la nature. Parcs et Réserves, **53** (3): 12-17.
8. Lecomte J., 2001 - Conservation de la nature, des concepts à l'action. Courrier de l'environnement de l'INRA, **37**: 59-73.
9. Mazoyer M., Roudart L., 1997 - Histoire des agricultures du monde, du néolithique à la crise contemporaine. Ed. Seuil, coll. Points Histoire, Paris.
10. Noirfalise A., 1995 - Les origines et la diffusion de l'agriculture néolithique en Europe. Bull. Rech. Agron. de Gembloux, **30** (4): 373-396.

VI

NATURE CONSERVATION IN PRACTICE

EFFECTS OF IRRIGATION ON SWAMP FORESTS DRAINED BY LIGNITE MINING

Rüdiger WITTIG[1*], Carla MICHELS[2], Carolin WETZSTEIN SUNKE[1]
[1] Chair of Ecology and Geobotany, Johann Wolfgang Goethe-University, Siesmayerstraße 70, **60323 Frankfurt (GERMANY)**;
e-mail: r.wittig@em.uni-frankfurt.de
[2] LÖBF, Dezernat Biomonitoring und Erfolgskontrolle, Castroper Straße 312-314, **45665 Recklinghausen (GERMANY)**

Abstract: In the lignite mining area of the Rhineland (Germany) some wet forests, valuable from the point of view of ecology and nature conservation, have been affected by the drainage caused by the large and deep mining pits. In order to restore these drained forests and to protect others from drainage, an irrigation project was started in 1987. Preliminary investigations had begun in 1982. The aims of the total project were to document the changes in vegetation and soil caused by drainage, to test the suitability of different methods of irrigation in order to restore forests affected by drainage, and to document and evaluate the development of the forests under the influence of irrigation. The drainage of the different *Alno-Ulmion* communities and of the *Carici elongatae-Alnetum*, all typical for undrained areas, leads to *Alnus glutinosa*-forests dominated by *Deschampsia cespitosa*, by *Urtica dioica* or by *Rubus fruticosus* agg. In some cases *Betulo-Quercetum alnetosum* is the final result, in other cases *Alnus glutinosa*-forest with a *Carpinion*-like herb layer. Seepage of water at the edge of the valley has proved to be an adequate method to re-transform drained alder forests into wet alder forests. However, the habitat conditions achieved up to now by irrigation seem to support the development of riparian forest (*Alno-Ulmion*) more than alder swamp (*Carici elongatae-Alnetum*).

Introduction

Lignite is an important source of energy in Germany. Exploitation is by surface mining, and large open pits up to a depth of several hundred metres have been established. These pits cause drainage of their vicinity, which leads to a lowering of the groundwater level. The human water supply of this densely populated area, mainly based on the groundwater stock, may also contribute to the lowering of the water table. Among other biotopes, some wet forests – very valuable from the point of view of ecology and nature conservation – have been affected by this drainage. To document the perturbations caused by drainage and to initiate the most effective restoration, in 1982 a study of the wet forests of the so-called Schwalm-Nette-Gebiet (area of the rivers Schwalm and Nette) was initiated. The field investigations were carried out by order of the company RWE Power. The aims of the project are:

- to document the changes in vegetation and soil caused by drainage (completed);
- to test the suitability of different methods of irrigation in order to restore forests affected by drainage (completed);

• to document and evaluate the development of the forests (vegetation, fauna, soil, water level in the soil) under the influence of irrigation (still ongoing; due to space limitation only vegetation and soil water level will be dealt with in this paper).

In the following these three parts of the project will be treated in separate sections. Each of these sections will first present methods and then the results.

After preliminary investigations (June 1982 until June 1986) the main phase of the project started in July 1986 and ended in July 1991. After the end of the main phase the measurement of the soil water level was continued. The survey of the vegetation transect was repeated in 1995, 2000, 2002, and 2004. In 1995 the flora of the permanent plots was completely documented again, and in 2004 the cover of selected indicator species was reported for the permanent plots. In the following account we shall present some results of these long-term investigations, which, taking the main phase as starting point, by 2004 had been carried out over 18 years. Including the preliminary phase, the investigation period totals 22 years. Such long-term studies of the development of swamp forests and riparian forests definitively are very rare.

Area of investigation

The area of investigation is situated in the Rheinland lignite mining area near the border with the Netherlands and belongs to the town area of Mönchengladbach. It is part of the lowlands of the lower Rhine [1], in the north-western Germany. The area has an Atlantic climate (mild winters and temperate summers with plentiful precipitation). The average precipitation ranges from 700 to 750 mm per year.

The zonal vegetation is represented by mixed oak and beech forests (*Fago-Quercetum* and *Milio-Fagetum*). The characteristic natural vegetation of the valleys is formed by alder swamps (*Carici elongatae-Alnetum*), as well as by riparian communities of *Alno-Ulmion* and by birch-alder-oak forests (*Betulo-Quercetum alnetosum*). In the transition area between these wet forests and the zonal vegetation, not influenced by soil water, *Carpinion* communities form the potential natural vegetation [cf. 8].

Due to the influence of the deep lignite mining pits and perhaps also to the high water demand of the neighbouring cities, between 1963 and 1986 the groundwater table in the area under investigation had been lowered 1–2 m [7]. At a depth of about 1 m the soil of the Mühlenbach valley contains a loamy-clayey layer. This layer is not totally impermeable but prevents soil water from seeping for some time. Thus, in the first months of the year, as well as after periods of rainfall, the soil water level is comparatively high in the valley, in spite of the lowering of the groundwater table. The *Carici elongatae-Alnetum*, however, needs a continuously high soil water level (not deeper than 20 cm) all over the year. This condition was not met in the area of investigation for some years at the beginning of the project.

Vegetation changes caused by drainage

Methods

In order to document the changes caused by drainage, we compared stands which had already been drained for some years with those which were just on the way to being drained and those not up to then affected by drainage. The following methods were used:

- comparative vegetation analyses by phytosociological relevés and tables;
- vegetation mapping;
- studying vegetation transects.

Results

According to our phytosociological relevés and tables [9, 10], the succession caused by drainage can be characterized as follows: the typical undrained forests of the area are represented by *Alno-Ulmion* communities and by *Carici elongatae-Alnetum iridetosum*, *typicum* and *sphagnetosum* (*Alnion glutinosae*). The first is the typical community of the floodplains, of rivulets and small rivers, and occurs on mineral soil or on soil with a thin peat surface. The latter is a swamp forest, occurring in depressions of the valley on a considerable layer of peat. At the edge of the valley a small *Carpinion*-zone exists at some locations. By drainage, both the *Alno-Ulmion* communities and the *Carici elongatae-Alnetum iridetosum* can be transformed into *Alnus glutinosa* forests with dominance of *Deschampsia cespitosa*. The transformation runs via *Alnus glutinosa* forests dominated by *Impatiens noli-tangere* or *Urtica dioica* [3, 6], whereas the *Carici elongatae-Alnetum sphagnetosum* changes via *Alnus glutinosa* forests dominated by ferns (*Athyrium filix-femina*, *Dryopteris carthusiana*) and/or brambles (*Rubus fruticosus* agg.) into a *Betulo-Quercetum alnetosum*. The herb layer of the drained forests finally looks *Carpinion*-like, while the herb layer of those forests which have an intermediate position between undrained and drained shows some phytosociological similarity to *Alno-Ulmion* communities. In many cases the vegetation changes could be identified as results of changes in peat cover, soil-pH and nutrient content of the soil [4, 5].

Suitability of different methods of irrigation

Methods

In the course of the project, four different methods of irrigation were tested:

- conduction of water into the rivulet;
- sprinkling of water onto the forest floor;
- seepage of water at the edge of the valley (Fig. 1);
- enrichment of the groundwater.

Whereas the first three methods focus on the enrichment of the soil water that can be found over a (nearly) water-impenetrable layer of clay at a depth of 40-100 cm in the valley (see Fig. 2), a fourth experiment attempted to raise the groundwater level.

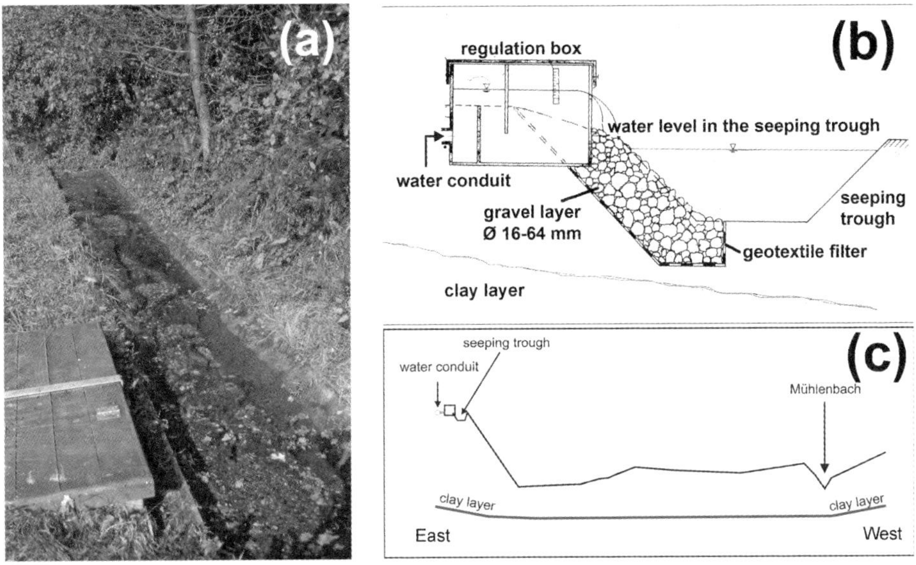

Fig. 1: Photo (a), diagram of a seepage trough (b), and profile of the valley after installation of the seepage troughs (c).

The suitability of these methods of irrigation was assessed by documenting the level of the soil water above the clay layer. In the immediate vicinity of each of the permanent plots (the location of the permanent plots is shown in Fig. 2), we installed a perforated plastic tube (diameter 7.5 cm) down to the level of the clay layer in order to allow the measurement of the soil water level. Additionally, in the area of one of the permanent plots a measuring tube of 2-m length penetrating the clay layer was installed in order to monitor the effects of the planned irrigation on the groundwater level. Between 1986 and 1988 measuring took place weekly; since 1989 twice per month. The irrigation started in May 1988. The soil water level was not only measured in order to evaluate the different irrigation methods, but also served as a basis for the evaluation of the success of the whole project.

Results

The first method shows results only for the rivulet itself, because due to its very loamy bed the rivulet does not contribute to the groundwater. The second method shows effects only in the immediate surrounding of the sprinklers. As forests full of sprinklers do not look very natural, we cannot recommend this method. In contrast to the results obtained in other areas [2], the fourth method showed no effects up to now in the upper Mühlenbach valley. Obviously the third is the best for this area.

By this method the level of the water in the soil is raised even at some distance from the seepage troughs. In many of the permanent plots, the soil water level is still strongly fluctuating (cf. Fig. 3: plot C). At some places, however, the irrigation was successful in rising the ground water level permanently up to less than 20 cm below the surface (cf. Fig. 3: plot N). At a few places the water rose above the surface level and caused the formation of temporary or permanent pools. And there were also some plots where the irrigation showed only little or no effect.

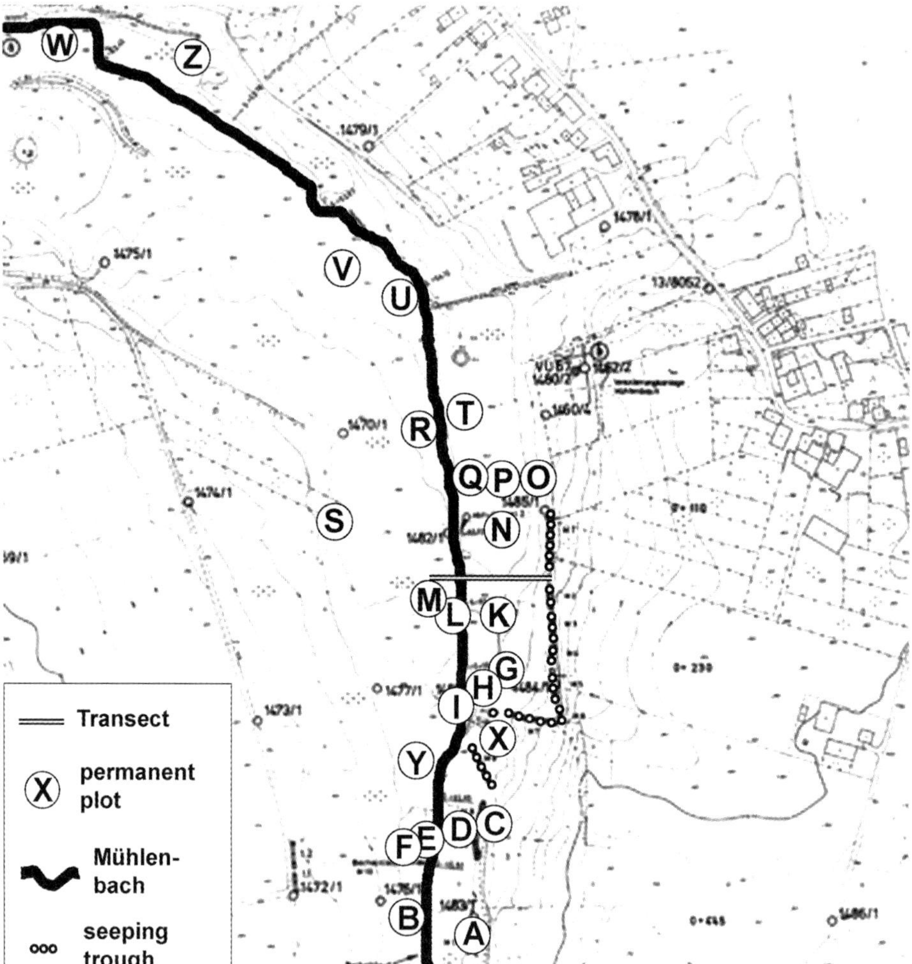

Fig. 2: Location of permanent plots, transect and seepage troughs.

Development of the forests under irrigation
Methods

The vegetation monitoring was carried out on permanent plots installed in representative locations of the area of investigation and along a transect crossing the valley. Additionally, a vegetation map for the whole area was elaborated in the preliminary phase [10]. The units of this vegetation map are documented by relevés. After 20 years of irrigation (*i.e.*, in 2008) we shall develop a new vegetation map.

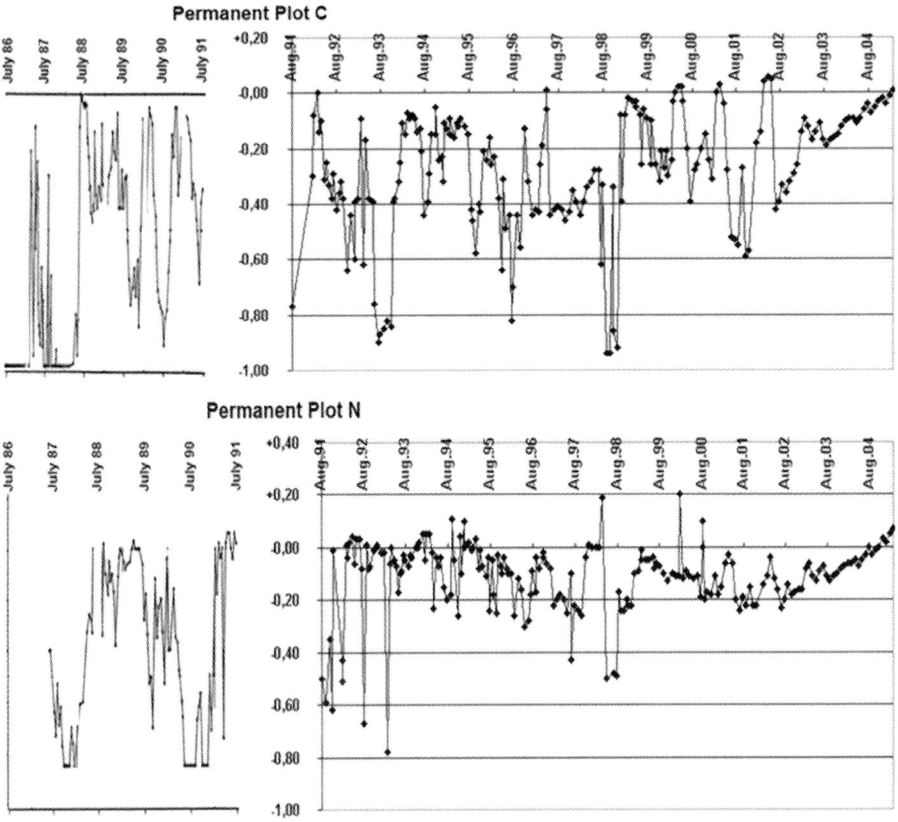

Fig. 3: Soil water level in selected permanent plots from July 1986 to January 2005.

In November 1986, at the end of the vegetation growing period, ten permanent plots were installed in habitats that show ecological homogeneity. The results obtained in the first year showed the necessity of adding more permanent plots, which were established in November 1987. In the upper part of the valley, where ecological conditions are comparatively heterogeneous, small permanent plots were established preferentially, whereas in the lower part only large permanent plots (Fig. 4) were installed. The position of the permanent plots is shown by Fig. 2.

The small plots (A–L) consist of three squares of 1 m^2, each divided into 100 grids of 10 x 10 cm. All species occurring in a permanent plot were recorded separately for each of the grids. The abundance of each species in a permanent plot was estimated as the percentage of the grids occupied by that species related to the total number of grids (*i.e.*, of 300). Within the large plots, five squares of 2 x 2 m each were delimited along one of the diagonals. In each of these five squares the cover of each species was estimated using the Londo scale: <1%, 1–3%, 3–5%, 5–15%, 15–25%, 25–35%, 35–45%, 45–55%, 55–65%, 65–75%, 75–85%, 85–95% and 95–100%. Additionally, a Braun-Blanquet relevé was recorded within each large plot. Up to 1991 the inventory of the permanent plots was taken yearly at the flowering time of *Impatiens noli-tangere*. A further inventory covering all species was compiled in 1995 at the same phenological date. In the year 2004 the coverage of some selected indicator species was documented.

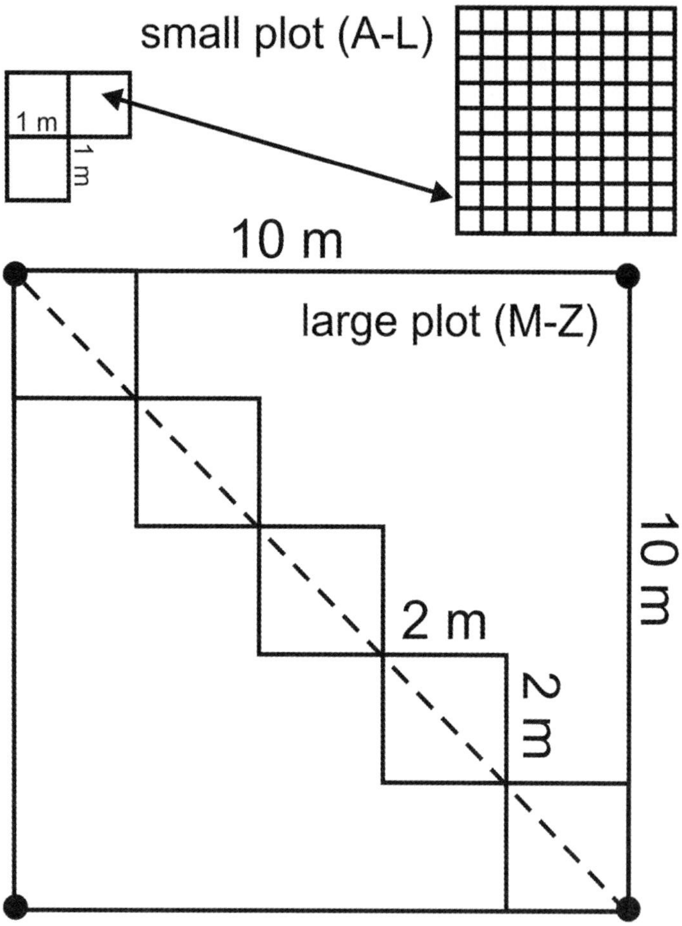

Fig. 4: Design of the permanent plots.

In 1988 a 4 m-wide transect was installed across the Mühlenbach valley (Fig. 2). From 1988 to 1991, as well as in the years 1995, 2000, 2002 and 2004, in each of the transect metres (1 x 4 m), the cover of each species was estimated according to the Braun-Blanquet scale. A second transect was established outside of the study area in the further course of the Mühlenbach valley, in a stand of still natural (*i.e.*, undrained) vegetation. The results of the investigation of this transect were reported by Denz [2].

Results

In this section, the results are presented for four indicator species:

• *Carex elongata*, character species of the *Carici elongatae-Alnetum glutinosae*, which represents the characteristic alder swamp community of the area of investigation;

• *Carex remota*, character species of riparian forests (*Alno-Ulmion*) of rivulets;

• *Circaea lutetiana*, an *Alno-Ulmion* character species, which avoids alder swamps;

• *Urtica dioica*, an indicator of drainage when growing in (former) alder swamps.

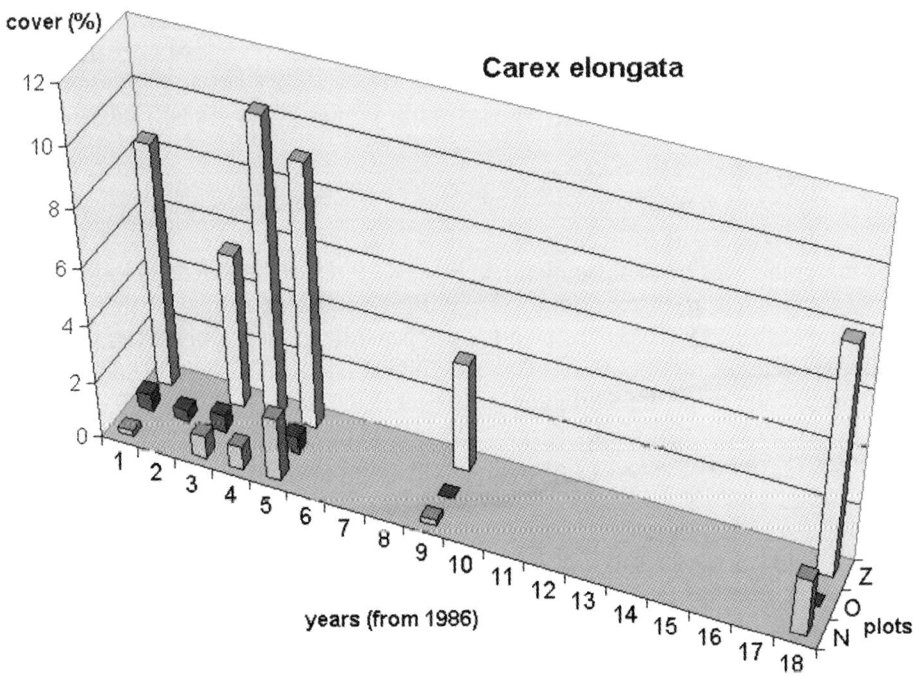

Fig. 5: Development of *Carex elongata* in the permanent plots.

Carex elongata

At the beginning of the irrigation *Carex elongata* was present in three permanent plots (Fig. 5). In one of them (N) its cover has slightly increased. In a second plot (Z) it has maintained its relatively high cover. In a third plot (O), where it was represented with low coverage at the beginning of the project, it has meanwhile totally disappeared.

In the transect area a remarkable decrease occurred from 1988 (at that time occurring within 19 transect metres, in three of them with Braun-Blanquet values of 2 or 3) to 1995 (only 13 transect metres, in all with Braun-Blanquet values of ≤1). From 1995 to 2004 it had increased, and now grows in 18 transect metres, showing a Braun-Blanquet value of 2 within five transect metres (Fig. 6).

Carex remota

At the beginning of the project, *Carex remota* occurred in only one permanent plot (A) and at low frequency. Having totally disappeared in 1989, it has re-established and finally reached its initial level in this permanent plot. In three further permanent plots (D, E, N) it has newly established within the course of the project. In plot E, where it was reported not earlier than in 1995, up to now, it has reached only low frequencies (4), while in D the frequency is comparatively high. In N it reached medium frequencies (7 to 8) for some years but now is of minor importance (Fig. 7).

Along the transect, primarily *Carex remota* was represented only in a very few places with Braun-Blanquet values of <2. Since 2000 a strong increase of the species took place in the former alder swamp region at the eastern edge of the valley (Fig. 6).

Circaea lutetiana

This species is still present in all permanent plots where it was already growing at the start of the irrigation (A, B, F, K, P, Q, and Y). Additionally it has newly established on three plots (C, D, and N) within the course of this project. From one of these (N) it has meanwhile disappeared. On the major part of the plots *Circaea lutetiana* shows a (remarkably) higher frequency than it did at the beginning of the irrigation. Particularly striking is the increase in plot F from 0.3 to 35. On two plots (P and Q) the species has maintained its level, and on two others (A and K) it is today less frequent than at the start of the project. In both of the latter cases from 1987 to 1990 an increase was reported, but this increase was followed by a strong decrease: on A from 50% to 4%, in K from 51% to 1% (Fig. 8). Along the transect, *Circaea lutetiana* has spread continuously and has also increased in coverage (Fig. 6).

Urtica dioica

When the main phase of the project started, *Urtica dioica* was represented in many of the permanent plots (A, B, D, E, K, N, P, Q), in some of them at high frequency or cover. (A: >40%; B and D: >80%; E: >70%; Q: >60%). During the period of investigation this species has conspicuously decreased in all these plots, and from some (E, K, N) it has totally disappeared. Only within two plots (C and F) has it newly re-emerged, but until now it plays only a minor role (Fig. 9).

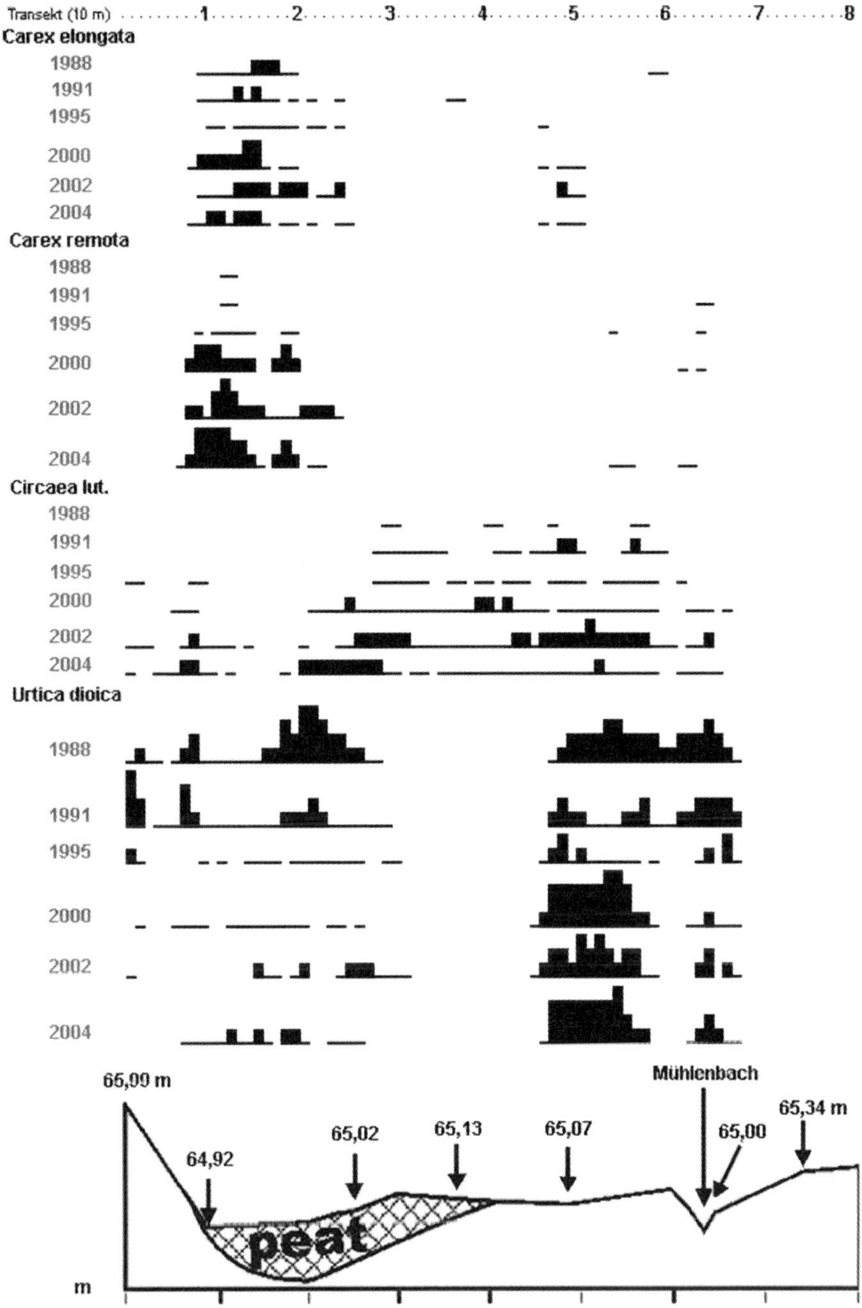

Fig. 6: Development of *Carex elongata*, *Carex remota*, *Circaea lutetiana* and *Urtica dioica* in the transect area.

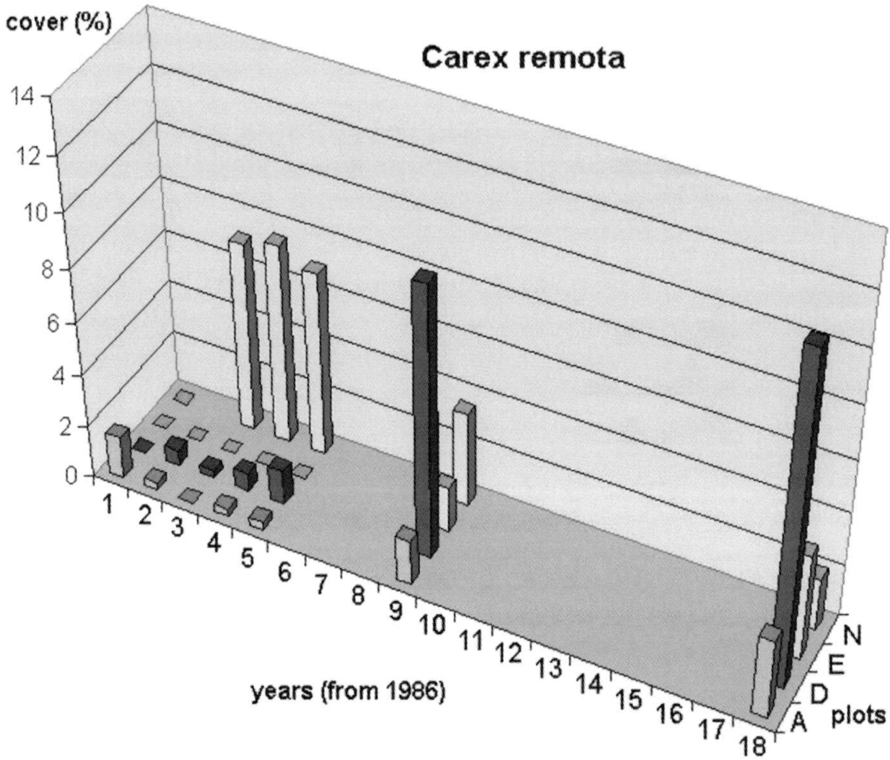

Fig. 7: Development of *Carex remota* in the permanent plots.

Along the transect *Urtica dioica* was, with the exception of the medium part, very strongly represented in 1986. Up until 1995 it decreased remarkably, but then a strong increase was reported for the western part of the transect. Here, in some transect meters, the Braun-Blanquet values were higher in the year 2000 than in 1988. Considering the number of transect metres occupied by the species, in the eastern edge of the valley, the decrease has continued up till now. In some metres, however, the Braun-Blanquet values have increased from 2000 to 2004. Obviously there is no clear tendency for its development in the eastern part of the valley.

Without any doubt, however, *Urtica dioica* today plays a remarkably minor role in the eastern part of the valley as compared to at the start of the irrigation. In the central part of the transect, no occurrences of *Urtica dioica* were reported during the whole project phase (Fig. 6).

Discussion

Regarding both the development of the soil water level and the development of the flora, there is no uniform tendency over the total area of investigation, as the different plots have developed differently. Locations where a continuous raise of the soil water level up to 20 cm or less below the surface was achieved (*e.g.*, plot N) show a positive development of their flora, in particular an increase or at least a maintenance of the frequency or cover of species typical of alder swamps. The parallel increase of *Carex remota*, mainly regarded as character species of riparian forests (*Alno-Ulmion*), does not represent a negative trend because this species is also typical of spring habitats. Thus, it can be kept as a differentiating species of alder swamps with springs. Another sign of the positive development of such plots is the strong decrease of disturbance indicators, like *Urtica dioica*. Also, where the water level was raised above the soil surface, all indicators of disturbance have disappeared, but in these plots no or only few species of alder swamps have returned. Plots still showing a strong fluctuation of soil water level do not, or only to a much lower degree, show such positive trends. In some plots (*e.g.*, C), *Urtica dioica* has even newly emerged during the course of the project.

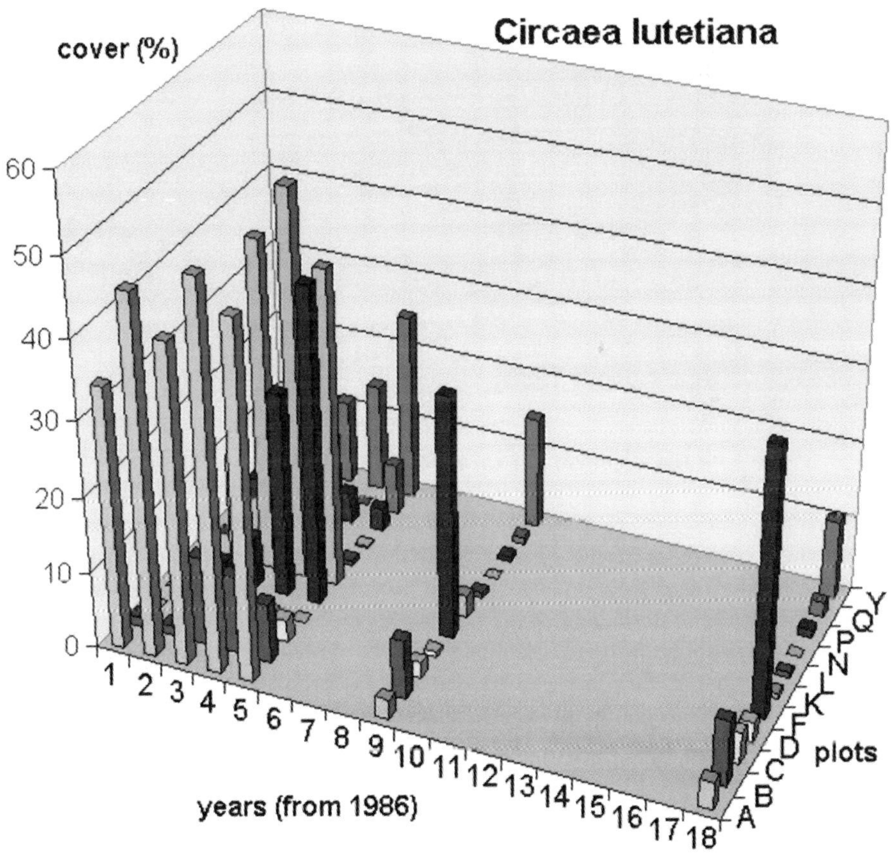

Fig. 8: Development of *Circaea lutetiana* in the permanent plots.

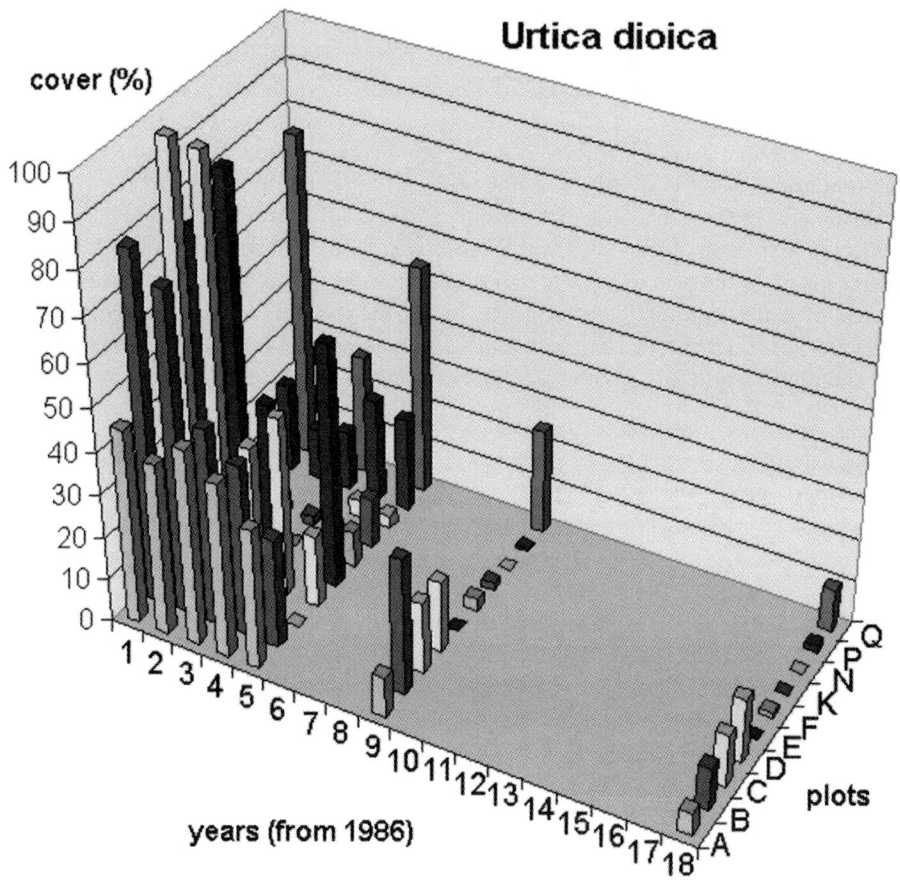

Fig. 9: Development of *Urtica dioica* in the permanent plots.

The estimation of the results from the permanent plots given above is supported by the results obtained along the transect. Here a positive development has occurred at the eastern edge of the valley. This area is situated in the immediate vicinity of the seepage troughs, thus permanently showing a high level of the soil water.

In spite of the comparatively positive development of the eastern transect area, the restoration of a typical alder swamp has not been achieved up to now. This is demonstrated by the enormous increase of *Circaea lutetiana*, which strongly avoids alder swamps, but prefers riparian forests. The increase of *Carex remota*, too, can be regarded as an indication of development in the direction of a riparian forest. As *Carex remota* prefers spring habitats, its increase may be related to the infiltration of water from the seepage troughs.

Conclusions

The present habitat conditions seem to support the development of a riparian forest more than that of an alder swamp. However, the results obtained up to now do not totally exclude the possibility of regeneration of an alder swamp. But even if it should not be possible to regenerate an alder swamp, the irrigation project was successful considering the following issue: a riparian forest does not represent the former natural vegetation of the total area, but nevertheless is typical for some parts of the valley. In any case, it is more natural and, from the point of view of nature conservation, has a higher value than the *Urtica dioica-* and *Rubus fruticosus-Alnus glutinosa* forests, which had developed under the influence of drainage before the beginning of the irrigation project. We have learned that, by using the seepage trough method for irrigation, it is possible to restore a wetland forest, but not necessarily the intended one.

Acknowledgements: We thank the RWE Power company (former Rheinbraun) for financing the project, Dr. Oswald (RWE Power) for the soil water data, and Dick Byer for correcting the English.

REFERENCES

1. Dahmen F.W., Kierchner G.-J., Schwann H., Wendebourg F., Westphal W., Wolff-Straub R., 1973 - Landschafts- und Einrichtungsplan Naturpark Schwalm-Nette. Beitr. Landesentwickl, **30**: 226 S.
2. Denz O., 2002 - Regenerationspotenzial entwässerter Bruch- und Erlen-Eschenauenwälder. Floristische und pflanzensoziologische Aspekte. Naturschutz und Landschaftsplanung, **34**: 247-251.
3. Dinter W., 1982 - Waldgesellschaften der Niederrheinischen Sandplatten. Dissertationes Botanicae 64. J. Cramer, Vaduz.
4. Kazda M., Verbücheln G., Brans S., Luwe, M., 1991 - Mapping of vegetation and soil changes in an alder carr affected by a decrease of water-table height. Phytocoenosis **3** (N.S.), Suppl. Cartographiae Geobotanicae, **2**: 243-250.
5. Kazda M., Verbücheln G., Luwe M., Brans S., 1992 - Auswirkungen von Grundwasserabsenkungen auf Erlenbruchwälder am Niederrhein. Natur und Landschaft, **67** (6): 283-287.
6. Rödel D., 1987 - Vegetationsentwicklung und Grundwasserabsenkungen, dargestellt am Beispiel des Fuhrberger Feldes in Niedersachsen. Landschaftsentwicklung und Umweltforschung, Schr.r. Fachber. Landschaftsentw. Techn. Univ. Berlin, Sonderheft **S1**.
7. Stadt Mönchengladbach, 1987 - Braunkohle und Sümpfung. Natur, Landschaft, Ökologie. Auswirkungen des Braunkohlentagebaus auf die Stadt Mönchengladbach. 3., völlig neu bearb. Aufl., Oberstadtdirektor Stadt Mönchengladbach.
8. Trautmann W., 1973 - Vegetationskarte der Bundesrepublik Deutschland 1:200 000 - Potentielle natürliche Vegetation - Blatt CC 5502 Köln. Schr.r. Vegetationskde., **6**: 1-172.

9. Verbücheln G., Krechel R., Wittig R., 1990 - Die erlenreichen Waldgesellschaften der Schwalm-Nette-Platten und ihrer Randgebiete. Tuexenia, **10**: 419-432.
10. Wittig R., 1983 - Vegetation, Gefäßpflanzenflora, Schutzwürdigkeit aus botanischer Sicht, Gefährdung und Pflege. In: Wittig R., Volpers T., Heinen W., Mühlenbachtal (eds.), Flora, Fauna, Schutzwürdigkeit, Pflege, Entwicklung. Schutzwürdige Biotope in Mönchengladbach, **5**: 1-38.

AUTOCHTHONY AND CONSERVATION OF A RELICT POPULATION OF *ABIES ALBA* MILLER IN THE APUAN ALPS REGIONAL PARK (TUSCANY – ITALY)

Alessia AMORFINI[1], Antonio BARTELLETTI[1], Gordon CAVALLONI[1], Gianfranco GENOVESI[1], Emanuele GUAZZI[1*], Isabella RONCHIERI[1], Paolo PIOVANI[2]

[1] Parco Regionale delle Alpi Apuane, v.le Stazione 82, **54100 Massa (ITALY)**;
e-mail: eguazzi@parcapuane.it

[2] Dipartimento di Scienze Ambientali dell'Università di Parma,
Parco Area delle Scienze 11/A, **43100 Parma (ITALY)**

Abstract: A small and relict population of Silver Fir (*Abies alba* Miller) in the Apuan Alps (Tuscany – Italy) has been confirmed and immediately analyzed in order to evaluate its level of autochthony. Some 19th–20th century floristic evidence, along with an extraordinary 17th century cartographic document, testify to the presence of Silver Fir in the same place. Results obtained from chloroplast genome analysis have been compared with Alpine and Apennine population data in order to identify any genetic relationship. The phylogenetic tree obtained shows a strong affinity between the Apuan populations and those in the Apennines, confirming with the historical floristic documents the autochthony of the Apuan fir population. Conservation activities have consisted, *in situ*, in improving vegetation habitats, through thinning surrounding woods, which will allow natural regeneration. *Ex situ*, near the Forest Garden of Camporgiano (Lucca), saplings have been obtained from seeds of the autochthonous Silver Fir population; later, these were planted out close to the relict population, so constituting a new experimental conservation nucleus. Thus a new nature conservation activity for a relict habitat, of priority on the 92/43/CEE Directive ("Apennine beech forests with *Abies alba* and beech forests with *Abies nebrodensis*"), has been initiated.

Localization

A small relict population of Silver Fir *(Abies alba* Miller) in the Apuan Alps (Tuscany – Italy) is located in the northern sector of this mountain chain, at c. 1500 m above sea level, on the northern slope of Contrario Mountain (Fig. 1), within the valley of Orto di Donna-Serenaia [2]. This population consists of 20 plants (approximately) of different ages, growing in a rather poor vegetative condition (Fig. 2).

Autochthony analysis

Before taking any conservation measures, much research has been carried out to check the level of autochthony of the Silver Fir in this part of the Apuan Alps (exotic specimens of Silver fir having been much employed for reforestation in the Apuan Alps). In the end, confirmation of the autochthony of the population on M. Contrario has been based on the following evidence:

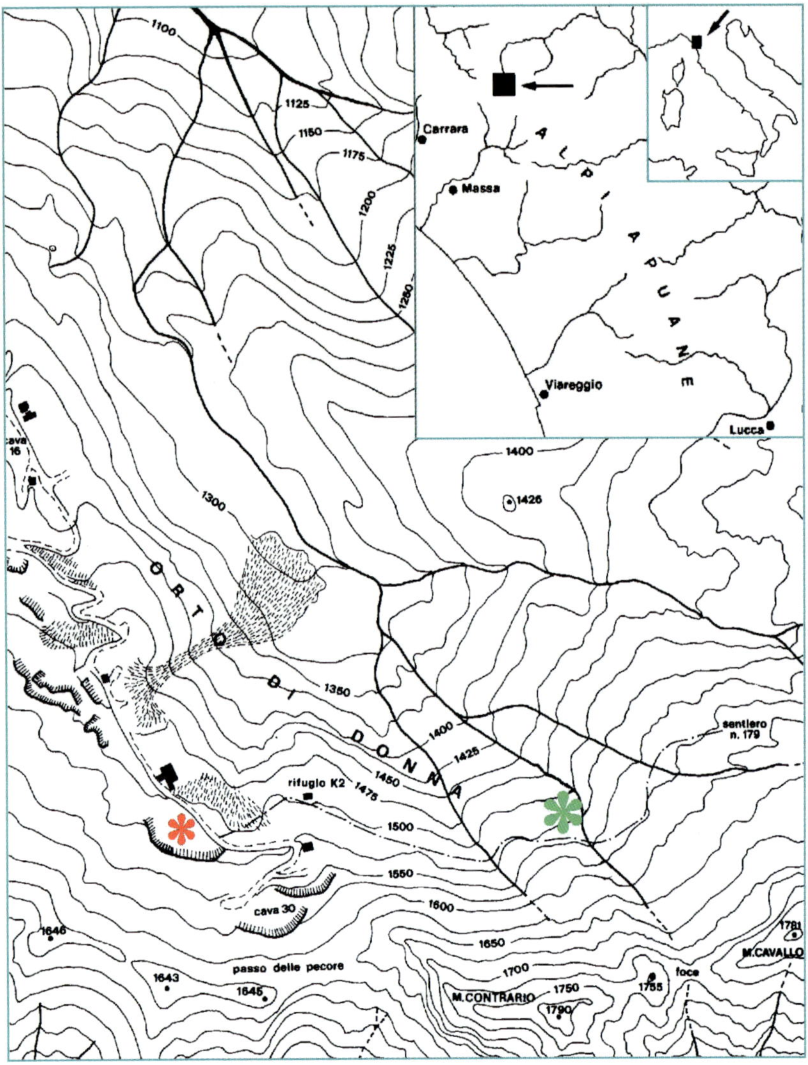

Stazione relitta - Relict population

0 250 m

Nucleo di nuovo impianto - New positioning nucleus

Fig. 1: Localization of *Abies alba* autochthonous population in the Apuan Alps.

Fig. 2: *Abies alba* in the Apuan Alps (M. Pisanino (1947 m) in the background).

a) The available floristic reports of *Abies alba* in the Apuan Alps by Bertoloni [6], De Stefani [4] and Longo [5], are all about the Orto di Donna-Serenaia valley, the same locality as the illustrated population;

b) Herbarium *exsiccatae* confirm this indication of the same Apuan valley; particularly remarkable is the *exsiccata* shown in Fig. 3, collected in 1812 by Antonio Bertoloni in this same area;

c) An extraordinary 1667 map (Fig. 4a) testifies that the actual population of *Abies alba* almost coincides (Fig. 4b) with a spot called "*Piella*", an ancient local popular word meaning silver fir [1]. This was already documented by Bertoloni [3] for the same area: "*incolae hujus loci dicunt Piella"* (Fig. 4b);

d) Genetic analysis — based on some highly repeated and hyper-variable fields of the chloroplast genome [7] — underlines a strong likeness between the population in the Apuan Alps and two autochthonous populations from the nearby Apennines (La Nuda and Monte Nero); the geographical distribution of the genetic variability is illustrated on the phylogenetic tree shown in Fig. 5a; the results confirm the similarity of geographically close populations along postglacial migration routes from southern to northern Italy (Fig. 5b).

e) The relict population grows today in a steep and barely accessible place; moreover, the random distribution of the plants and their different ages exclude any anthropogenic origin.

Fig. 3: ***Abies alba*** **individual collected in the Apuan Alps by Antonio Bertoloni and conserved in the Herbarium of Bologna Botanical Garden.**

Conservation

The conservation measures taken to protect the relict population of *Abies alba* in the Apuan Alps, *in situ* and *ex situ*, are:

Fig. 4: a) A 17th century drawing of the "Orto di Donna-Serenaia" valley, showing the position of the locality "Piella" (=silver fir) at label n. 13; b) A shot of the same valley with overlapping numbers showing localities of the previous drawing. The green star shows the position of the relict population; the red one shows the position of the new silver fir nucleus.

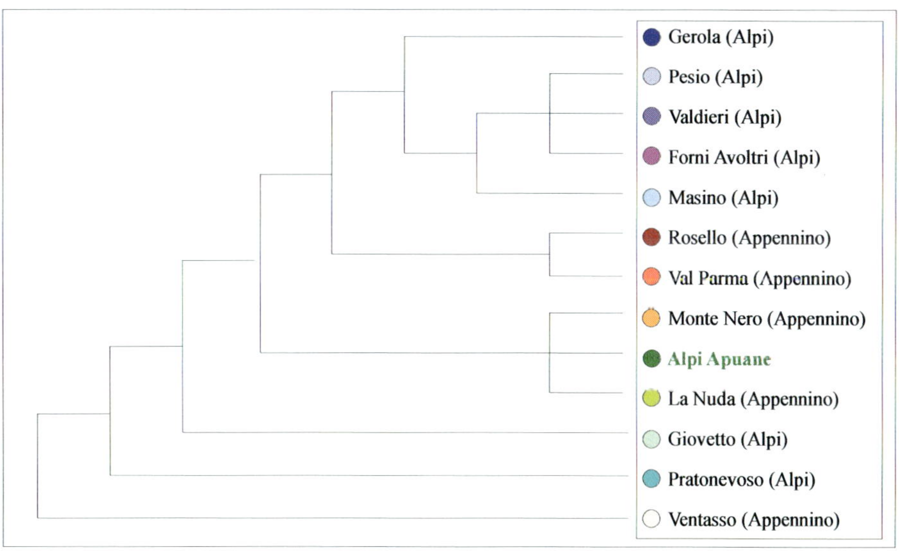

Fig. 5a: Phylogenetic tree of some Alpine and Apennine *Abies alba* populations, constructed using the neighbour-joining method, and based on a genetic distance matrix.

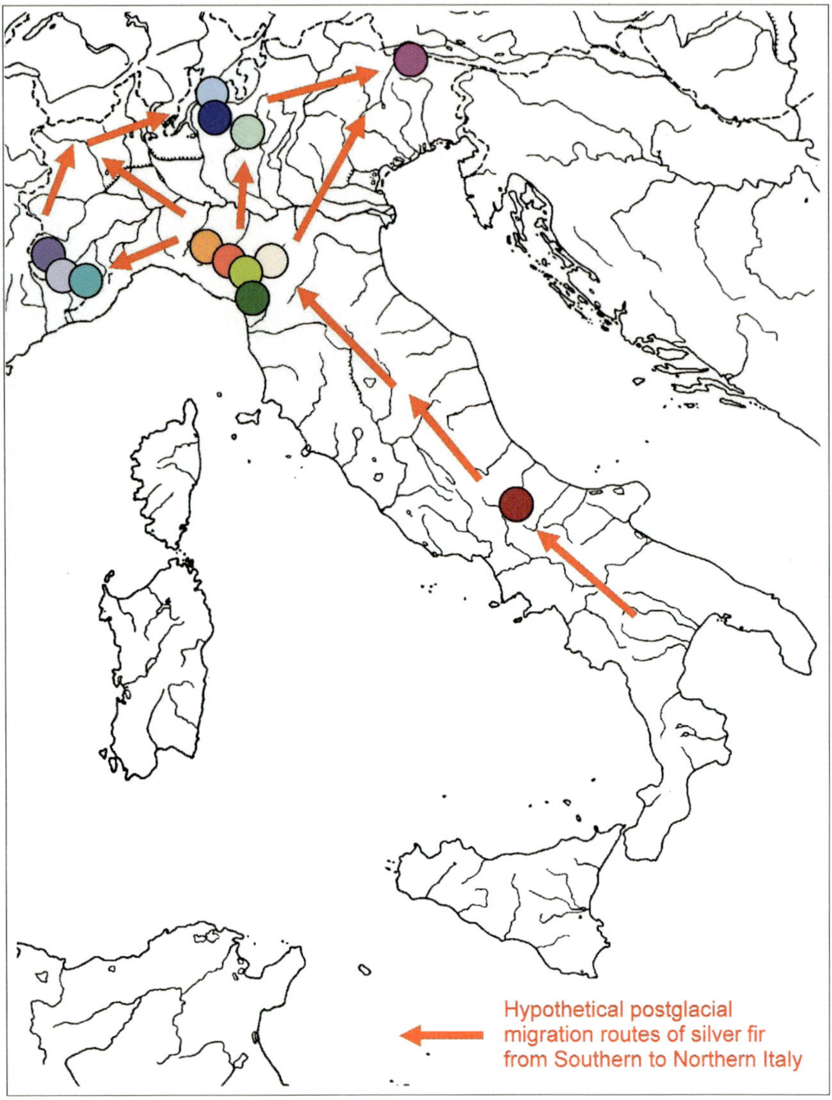

Fig. 5b: Similarity of geographically close populations along post-glacial migration routes from southern to northern Italy.

a) *conservation of vegetation habitats*: selective and timely thinning of surrounding woods has been performed to increase the natural conditions for the silver fir population (Fig. 6); thus, favourable ecological conditions have been created in order to foster natural regeneration of this autochthonous conifer as well as limit the expansion of beech (*Fagus sylvatica*);

b) *creation of a new silver fir nucleus*: seed collection and propagation of new plants (Fig. 7) in the Forest Garden of Camporgiano (Lucca) have enabled us to obtain some reliable material for an experimental reintroduction of *Abies alba* into a habitat located very close to the relict population. The place chosen – situated near a recently restructured alpine shelter – is an abandoned ancient marble quarry, already cleared and remodelled using environmental engineering techniques (Fig. 8).

Fig. 6: *In situ* beech thinning carried out to favour silver fir specimens.

Conclusions

The aim of experimental work on the Apuan Alps populations of *Abies alba* has been to support the safeguarding of a relict habitat, mentioned as a priority on the 92/43/CEE Directive, in which it is denominated as "Apennine beech forests with *Abies alba*, and beech forests with *Abies nebrodensis*". Recently [6], the same Apuan population, along with those of Verna (Arezzo), have been reported as the only two nuclei of *Abies alba* worthy of inclusion in the

Regional checklist for seed conservation of the specimens of forest interest in Tuscany.

Fig. 7: *Ex situ* plants produced after seed collection.

Fig. 8: Silver fir reintroduction on a suitable, easily accessible and information-provided area.

REFERENCES

1. Ambrosi A.C., 1957 - Osservazioni sulla voce dialettale [piel(l)a] (= A*bies alba* Mill.) in Lunigiana. Giornale Storico della Lunigiana, n.s., **8** (1-2): 44-46.
2. Ansaldi M., Bartelletti A., Tomei P.E., 1988 - L'abete bianco (*Abies alba* Miller) sulle Alpi Apuane. Atti Soc. Tosc. Sci. Nat., mem., ser. B, **95**: 41-49.
3. Bertoloni A., 1819 - Flora Alpium Apuanarum, Bologna.
4. De Stefani C., 1883 - Monografia agraria sul circondario di Castelnuovo di Garfagnana. In: Atti della Giunta per l'inchiesta agraria e sulle condizioni della classe agricola. Provincia di Massa Carrara, 10 (2), Roma.
5. Longo B., 1926 - L'*Abies alba* Miller nelle Alpi Apuane. Bull. Soc. Bot. Ital., **120** (7-9): 120.
6. Lumini M.B., 2004 - Il germoplasma delle specie forestali della Toscana. A.R.S.I.A.: 31-34. Regione Toscana, Firenze.
7. Vendramin G.G., Anzidei M., Madaghiele A., Bucci G., 1998 - Distribution of genetic diversity in *Pinus pinaster* Aiton as revealed by chloroplast microsatellites. Theoretical and Applied Genetics, **97**: 456-463.

DUNE SLACKS – AN ENDANGERED HABITAT IN THE CONFLICT BETWEEN NATURE CONSERVATION AND GROUNDWATER MANAGEMENT[1]

Richard POTT
Institut für Geobotanik, Universität Hannover, Nienburger Str. 17,
30167 Hannover (GERMANY); e-mail: pott@geobotanik.uni-hannover.de

Abstract: This paper presents preliminary results of an interdisciplinary research project entitled "Sustainable groundwater management in hydrogeological and ecological sensitive areas of the North Sea Coast"[2]. For years different and contrary concepts of the water supply and distribution authorities and nature conservation had led to conflict situations, which could result in the endangering especially of wet biotopes. These plant communities of wet dune slacks are strongly related to the water table, hence they can tolerate fluctuations only over a certain range without damage. Therefore, the main goal of the project is to develop a useful guide for sustainable groundwater management. Within the project, detailed investigations were carried out on the Frisian islands of Norderney and Langeoog as well as on the mainland in the Harlinger Land at the boundary between coastal marshland and the Pleistocene hinterland of N.W. Germany. Using a variety of scientific methods (phytosociology, groundwater-monitoring, GIS mapping, cellular automata model), it was possible to construct a comprehensive picture of threshold values or tolerance ranges of the water table for the characteristic vegetation units of the conservation areas and adjacent landscapes.

Introduction

In the North Sea Coast area, the Wadden Sea and its islands are a globally unique habitat and one of the last almost natural ecological systems of Central Europe. Due to their privileged wilderness position, the islands of the Wadden Sea have been of special interest for a long time. On the one hand this reflects tourism, because the short distance to a natural coastal landscape has been and still is the basis for tourism, which at the same time represents the most important source of financial income for the coastal population. The necessity for water management in conjunction with tourism to make drinking water available at seasonally different intensities, causes however a different demand on the resource groundwater, so that the excessive extraction of groundwater may endanger the dune slacks ecosystem and ultimately their groundwater-dependent vegetation.

On the other hand, the East Frisian Islands belong to the national park of "Niedersächsisches Wattenmeer" (Fig. 1). The mission of the national park is, among others, the preservation of a species-rich flora and fauna and the minimization of anthropogenic influences [2, 12, 20]. In order to be able to evaluate these influences, it is important to have a comprehensive knowledge of the occurrence and ecological condition of groundwater-dependent vegetation.

---

[1] Opening lecture given by Richard Pott after a publication from Petersen *et al.* [27].

[2] Sponsored mainly by Volkswagenstiftung Hannover.

Since the vegetation units of the wet dune slacks (= hygroseries), the most threatened plant communities of the coastal vegetation, are involved [3, 24, 29, 37,], this habitat "is shaped by threatened or strongly endangered" biotope types [4, 29, 30].

Fig. 1: Satellite photo of the Wadden-Sea National Park.

For these reasons a new interdisciplinary research project entitled "Sustainable groundwater management in hydrogeological and ecological sensitive areas of the North Sea Coast" started in 1999. Project partners are hydrogeologists (Prof. Wolff, TU Braunschweig), plant ecologists (Prof. Janiesch, Univ. Oldenburg), socio-economists (Prof. Magoulas, Univ. Hanover) and geobotanists (Prof. Pott, Univ. Hanover). The main goal is to supply the water management with some sort of "guide" for the sustainable management of groundwater.

Investigation area

The aggravated ecological problems of groundwater management in the area of conflict between nature protection and water supply and their effects can be examined in an almost exemplary way on the East Frisian Islands. With the urban-moulded Norderney and the much more rather village-structured Langeoog, two islands differing in their economic and population structure have been selected as the study area. Additionally, a protected area on the mainland in the Harlinger Land, at the boundary between coastal marshland and the Pleistocene hinterland, has been included as a study area for reasons of comparison (Fig. 1).

Work procedures and first results

Strategies for sustainable groundwater management within the coastal area have been compiled within the context of this interdisciplinary project. The work procedures and first results of the geobotany working group can be arranged into four sections:

The selection of "indicators" – vegetation units and installation of the observation surfaces in connection with groundwater measuring levels

For the selection of the vegetation units, among other things an estimate is necessary of the degree of dampness of the plant communities, e.g. the relative indicator values of Ellenberg *et al.* [6]. It concerns empirically raised numbers of high indicator value, which were verified in the course of the time by measurements. After Ellenberg *et al.* [6] the dampness number is the best secured pointer value, basing on numerous investigations and observations concerning the relationship between plants and groundwater levels, so that a relatively good estimate is possible of the requirements of the plant communities.

Such differences are clarified in a succession scheme of the dune slack vegetation of the Wadden Sea islands [26] (Fig. 2). Apart from the degree of dampness, soil acidity is a further crucial factor in the structure of this habitat [10, 16]. These phenomena are based on data from 719 soil samples from all 17 Wadden Sea islands, off the Netherlands as far up as Denmark, as well as the computation of average dampness numbers of a total of 2775 vegetation relevés. Thus in the ecological scheme, the natural vegetation development is emphasized within the hygroseries (Fig. 2). Because the majority of the groundwater catchment areas of the North Sea islands are covered with dune vegetation (xeroseries = plant communities of the dry dunes), special "indicator"- units of the dry dunes were selected besides the rare plant communities of the dune slacks [28, 30].

The vegetation units examined were selected according to the following criteria:

- they should be characteristic elements of the groundwater catchment areas;
- they should form the main part of the vegetation of these groundwater catchment areas;
- they should differ in their dependence on the groundwater;
- they should occur both in the groundwater catchment areas and in reference areas of the islands uninfluenced by groundwater usage.

On consideration of the criteria specified above, altogether five vegetation units are suitable, and therefore comprise the emphasis of our investigations:

1. *Caricetum trinervi-nigrae*;
2. *Empetro-Ericetum*;
3. *Betula pubescens* communities;
4. *Hieracio-Empetretum*;
5. *Corynephorion* communities (above all *Violo-Corynephoretum*).

In the habitat of these plant communities on the islands of Norderney and Langeoog, as well as on the mainland, a total of 47 long-term observation surfaces with groundwater measuring levels were installed.

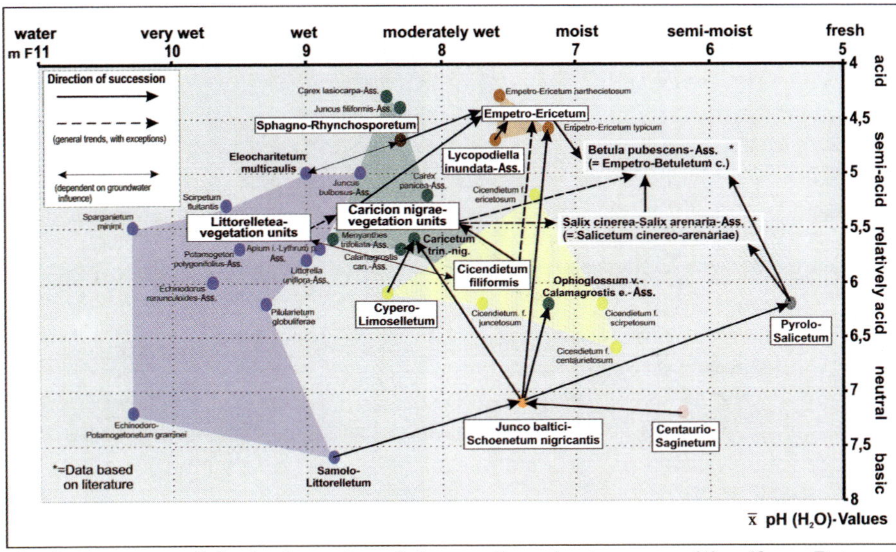

Fig. 2: Ecogram of the hydroseries of dune-valley plant communities (from Petersen [26]).

Ascertainment of the "*status quo*" of the vegetation of the groundwater catchment areas of the Norderney and Langeoog water companies, as well as a comparison with older and historical vegetation maps supported by GIS

The groundwater catchment areas of the Norderney and Langeoog waterworks were mapped phytosociologically in 2004 at the scale 1:3.000. For the area on the mainland, a vegetation map was revised by Hofmann [14], which was produced in the years 1996/1997. The basis of the mapping is always sampling by vegetation relevés of the plant communities present. In order to achieve a high resolution and accurate rendition of all areas, EDP-supported representation and evaluation techniques were linked with traditional mapping technology. By means of the employment of GIS (ArcView) it was possible to blend different data records with one another. These are based on the following parameters (Fig. 3):

- digital basic map (scale 1:3.000);
- digital aerial photographs: black-and-white and colour infrared digital elevator model (basis was a laser-scan flight, which made resolution possible up to 1 m);
- digitized maps of available vegetation maps (e.g. Fromke [8] for Langeoog).

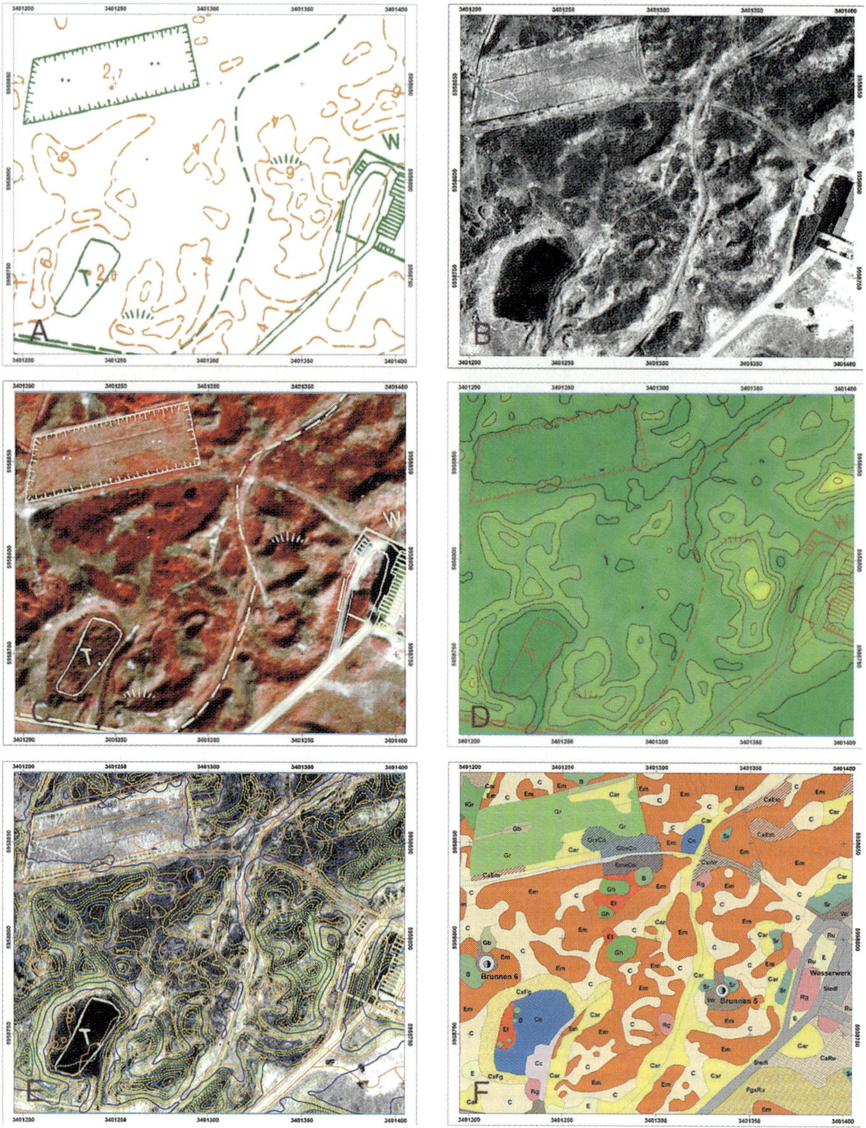

Fig. 3: a. German Topographic Base Map 1:5,000 (DKG 5); b. Black & white aerial photos (rectified and georeferenced); c. Infrared aerial photos (rectified and georeferenced); d. Digital model of the topographic relief (isochromes: 1 m – elevation contour lines: 5 m); e. Vegetation units delineation on the digital model of the topographic relief; f. Vegetation map 1: 5,000 (on the topographical basis DKG 5).

By correlation of these digital mappings with our area, high-resolution data and precise vegetation maps could be made (Fig. 3). This actual condition of the vegetation of a groundwater catchment area thus creates a current basis for the exact localization and differentiation of groundwater-dependent and -independent vegetation. Thereby the basis for a "guide system" of sustainable groundwater management is created, as well as a future-oriented monitoring system in connection with the long-term observation areas and groundwater measuring levels.

During the interpretation of vegetation changes, it is to be noted however that it usually concerns natural succession, which is induced by changes of local conditions such as acidification, demineralization and enrichment with organic matter. In addition, it has to be considered that anthropogenic influences, for example groundwater drawdown by intensive utilization, which changes natural succession processes, can therefore lead to the loss of plant communities of the wet dune slacks and to the occurrence of vegetation units of dry dunes or other vegetation areas (Fig. 4).

During the interpretation of the genesis of the actual condition, above all, older and historical vegetation maps are helpful. For Norderney, vegetation maps are available for the year 1949 by Tüxen & Neumann, as well as one for 1990 [13] and a current vegetation relevé [11]. These vegetation maps were digitized by means of GIS and, on the basis of these data, a calculation of areas has been carried out.

For the interpretation of current and historical vegetation data, figures of the total utilization by the water-company Norderney were available [18, 19]. Public utility Norderney GmbH conveys from two waterworks ("Ort" and "Weiße Düne" waterworks, or waterworks I and II) c. 1 million m^3/year of groundwater. The "Ort" waterworks was brought into operation at the end of the 19th century. The amount of water conveyed was 150,000–300,000 m^3/year at the beginning of the 20th century, increasing until 1958 to circa 500,000 m^3/year and reduced again to 150,000 m^3/year in 1997. The "Weiße Düne" waterworks was started up in 1959, and in 1960 reached a conveyed volume of scarcely 300,000 m^3/year. Problematic for the vegetation of the wet dune slacks surely was the clear increase to c. 800,000 m^3/year at the end of the 1970s.

Unfortunately no vegetation maps are available to document vegetation changes from these interesting periods in time. One must consider that during the summer months, during the optimal period for vegetation, up to two thirds of the total amount of conveyed groundwater is utilized [21]. Within the area of the "Weiße Düne" waterworks a groundwater drawdown of approx. 0.40–1.40 m occurred, according to Müller & Wolff [19], in the years 1988 and 1991. A drawdown of this order of magnitude has undoubtedly had a negative influence on groundwater-dependent plant communities of the wet dune slacks. The intensive conveyance of groundwater of the "Weiße Düne" waterworks is of concern above all to the nearby Bakenlegde dune slack. It is of great importance here that some dune slacks within the area of the "Weiße Düne" waterworks lie in protected zone I (=quiescent zone) of the national park Niedersächsisches Wattenmeer and the most sensitive parts of the landscape, where most of the national park's plant and animal species are located (Fig. 1) [12, 20, 25].

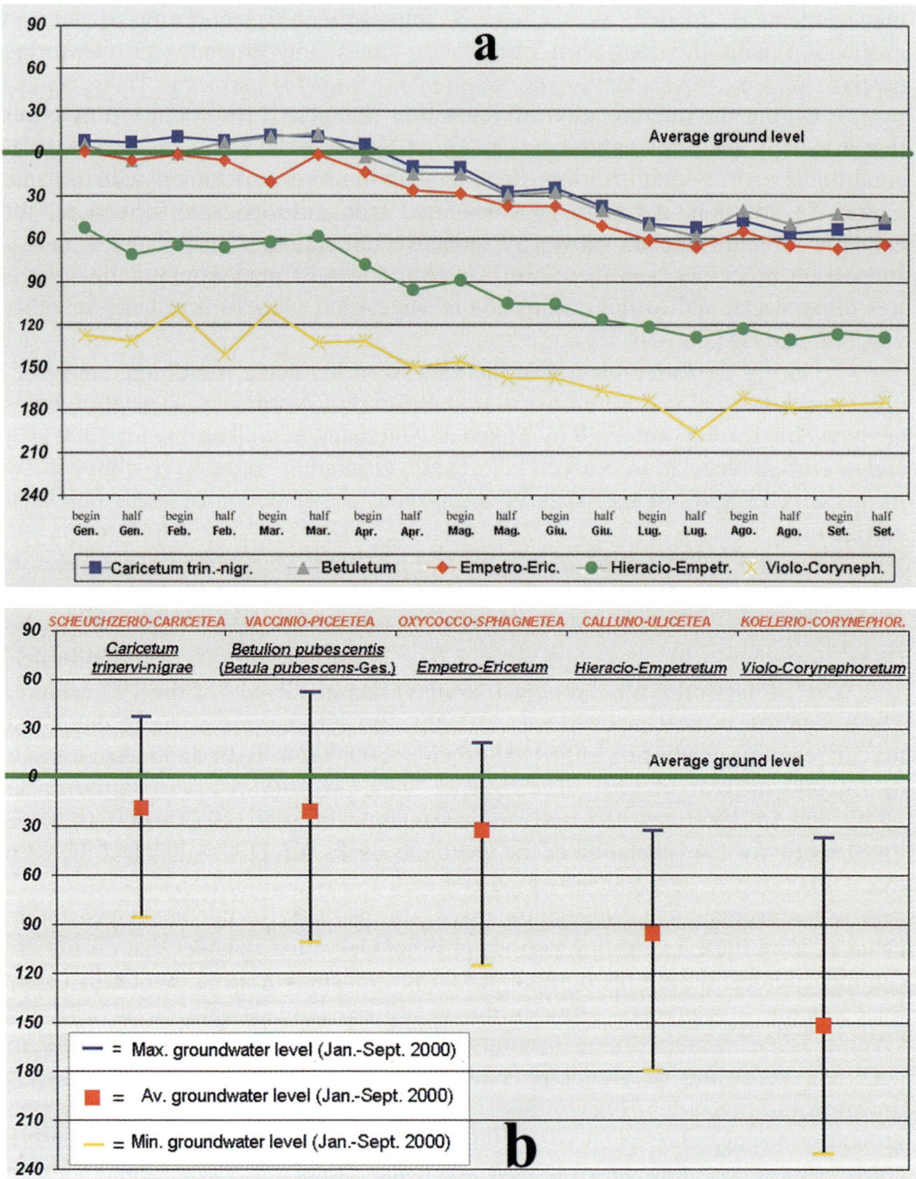

Fig. 4: a. Distribution of the average piezometric measurements performed from January to September 2000 in different plant community types. b. Variations in the average groundwater level by plant community type during the monitored time-span.

Fig. 4 clarifies further also that in this area obvious groundwater drawdowns are present. The comparison of the vegetation maps from 1949, 1990 and 2000 of the groundwater catchment area shows that overall the expansion of the vegetation units of the hygroseries has increased. However, the increase is limited, on the one hand, to the new increase area "Südstrandpolder" where salt-meadow stands have developed into wet biotopes; and on the other hand, mainly to species-poor vegetation types of general occurrence on the islands. A detailed analysis of all plant communities occurring in the hygroseries clarifies that the local increase on Norderney is for the most part due to the increase of *Scirpo-Phragmitetum, Betula pubescens* communities, *Caricetum trinervi-nigrae* and *Salix cinerea-Salix repens* communities (Fig. 5). Since *Scirpo-Phragmitetum* and *Salix cinerea-Salix repens* occur almost exclusively on the "Südstrandpolder", and this area lies almost outside the groundwater catchment area, they are relatively insignificant for this comparison.

Fig. 5: Dune valley on island with *Caricetum trinervi-nigrae* and *Salix cinerea-Salix repens* communities.

The increase of *Caricetum trinervi nigrae* and, above all, *Betula pubescens* communities can be explained by natural succession in wet dune slacks, whereby the *Betula pubescens* communities are to be evaluated within the succession series of the hygroseries clearly as a climax community (Fig. 3 and 5); [28, 26]. The example of the groundwater catchment area, and above all the Bakenlegde, suggests that the competitively strong, species-poor *Betula pubescens* communities may even have been promoted by groundwater drawdowns. It is of importance that *Betula* alters its habitat conditions to its favour where it has once established, because it extracts a relatively large amount of groundwater from the soil and thereby produces a competitive advantage over the wet- and moisture-preferring plant species and communities of the hygroseries. The propagation of the *Betula pubescens* communities has probably been driven partially by *Betula* growth [13, 25].

First groundwater threshold values of the selected vegetation units

The hydrological system of the freshwater lenses on the North Sea islands and the occurrence of plant communities in the wet dune slacks are directly dependent. The consideration of groundwater fluctuations is of great importance for the understanding of the frequently complicated succession conditions in dune slacks. Changes in groundwater level can occur also with the growth or retreat of the coastline at sea-lateral island sections. Due to the formation of new ranks of dunes, groundwater level can rise in the dune slacks and they thus become marshy. Short-term fluctuations, for example by excessive precipitation, have a smaller influence than seasonal fluctuations [7, 32, 33, 36]. Groundwater drawdown develops inevitably with the extraction of drinking water and must be held therefore within a controlled range of tolerance, especially as far as the wetlands of protected areas are concerned. For the regulation of these interferences in dunes and dune valley landscapes, vegetation-specific groundwater characteristic data are to be developed as some sort of early warning system.

Important basic data for the characterisation of the groundwater dependent on different plant-communities are:

- the absolute minimum value and/or the lowest point during/after a dry period;
- the absolute maximum value and/or the highest point during/after one wet period in combination with the duration of flooding (months);
- a "representative" middle groundwater level;
- a "representative" middle groundwater hydrograph curve.

Similar data can be found in Tüxen [34], Niemann [22, 23], Kloetzli [15], Grootjans *et al.* [9], DVWK [5] and Lammerts [16]. In Figure 7 the first results are visible, based on data collected from 38 groundwater measuring levels. The installation of the measuring levels was finished at the end of 1999, so that groundwater measuring data are available starting from January 2000. Data collection takes place at the beginning and in the middle of the month. Additionally the data measured for many years of the respective water companies are to be evaluated.

The minimum value of the groundwater level is of great importance for the characterization of a vegetation type, but is of no worthwhile value for sustainable groundwater management. For the maximum value the duration of flooding is important, less the absolute value. This is, however, a key factor of vegetation composition, and thus an essential condition for stature of the respective plant community. The middle groundwater level or range around this value is to be aimed at as the control value of the groundwater conveying, whereby the characteristic middle groundwater hydrograph curves are to be considered. Figure 4 shows the first results of the determination of the groundwater threshold values of the five selected vegetation units – the clear differentiation of the groundwater basic data between the plant communities of the dunes and dune slacks is remarkable. While the vegetation units of the hygroseries differ only slightly, a stronger gradient is present within the xeroseries, e.g. for *Hieracio-Empetretum* and *Violo-Corynephoretum*.

For *Caricetum trinervi-nigrae*, *Betula pubescens* communities and *Empetro-Ericetum*, a maximum flooding level within the range 20–50 cm is estimated. The crucial factor here is not so much the flood level, but rather the duration of continuous flooding. The minimum value for these vegetation units varies around a groundwater corridor distance of c. 90 cm (86–116 cm). For these plant communities can thus be specified a range of groundwater levels from 0 to 90 cm corridor distance. The middle groundwater corridor distance lies for all three plant communities within c. 30 cm. Estimation of the duration of flooding cannot yet be made due to the few available data (January–September 2000). In order to attain a finer differentiation of the plant communities of the wet dune slacks, a partitioning can take place into a wet and a dry variant of the respective vegetation units. According to DVWK [5], a classification of predominantly groundwater-affected vegetation types for the mainland range is indicated.

Here quite comparable values are present, i.e. for these three plant communities of the hygroseries a maximum and minimum value with groundwater distances are specified within the range 60–100 cm. However no middle groundwater level, but only a middle range is indicated. This phenomenon is justified with the very heterogeneous data available, whereby according to the DVWK [5] the middle groundwater level is to lie within this range. For an analysis of the characteristic groundwater data of plant communities, the seasonal groundwater condition process must be naturally determined, represented by so-called groundwater hydrograph curves (Fig. 4). Only the combination of these benchmark figures makes it possible to develop both nature protection and practice-relevant statements about an appropriate indicator system by the evaluation of series of measurements of at least several years. Therefore, the specified groundwater results, which are based only on one measuring period, must be evaluated initially as provisional. Yearly ranges of the water levels can vary strongly because of annually changing weather conditions and have therefore to be considered.

Moist years or moist summers show up as reduced groundwater fluctuation ranges and deep, long-persistent flooding. More crucially, however, are dry years or dry summers with hardly tolerable groundwater drawdown for the vegetation [5]. Therefore series of measurements should at least contain a wet, a dry and, if possible, a further year with average weather conditions [17]. The past weather data point to 2000 as an average year, so that the force of expression of the specified data will probably be seen to be confirmed in the long-term.

"Perspective" - modelling of the vegetation development in wet dune slacks by the example of the island of Norderney

For the Bakenlegde dune slack on Norderney, in the central part of the drinking water catchment area of the "Weiße Düne" waterworks, a spatial vegetation model has been developed to measure the effects of future groundwater management on vegetation dynamics. At the outset, it must be stated that the deficits of classical ecological models exist mostly in the non-explicit space reference. In order to avoid this problem, more and more raster-based approaches are used for the simulation of the spatial and temporal dynamics of ecosystems. Furthermore raster-based models make the illustration of natural processes by interactions possible for a finite number of raster cells. Cellular automata offer an approach for explicit spatial simulation. Baltzer *et al.* [1] give an overview of the use of cellular automata within the range ecology, biology and environment, which consist after Weimar [35] of the following four components:

- A lattice of many finite raster cells;
- A finite condition quantity for the characterisation of the cells;
- A neighbourhood of limited range;
- A monitoring system, which specifies the changes of the cell conditions.

For the simulation of vegetation development in wet dune slacks, a lattice from square raster cells is used. Each cell is assigned a state vector, which contains the vegetation unit and the five location factors: soil moisture, soil acidity, organic matter, salinity and anthropogenic-zoogenic impairment (Fig. 6). The spatial propagation and interaction of the vegetation units are coupled over a probability monitoring system to the files of the location factors. The information about the connections between the location factors and the vegetation units originates for the most part from Petersen [26]. Data for the sensitivity of the vegetation units in relation to changes of location factors derive from the same source. The role of these sensitivity data is to describe which environmental conditions the vegetation units are able to tolerate.

The computation of the establishment-probability of the locations by vegetation units considers location factors, the occurrence of the vegetation units in the immediate neighbourhood and propagation potential over larger distances. The possible stages of succession are taken from the succession-figure of Petersen [26] (Fig. 5). The model is able to simulate new establishment of a dune slack landscape.

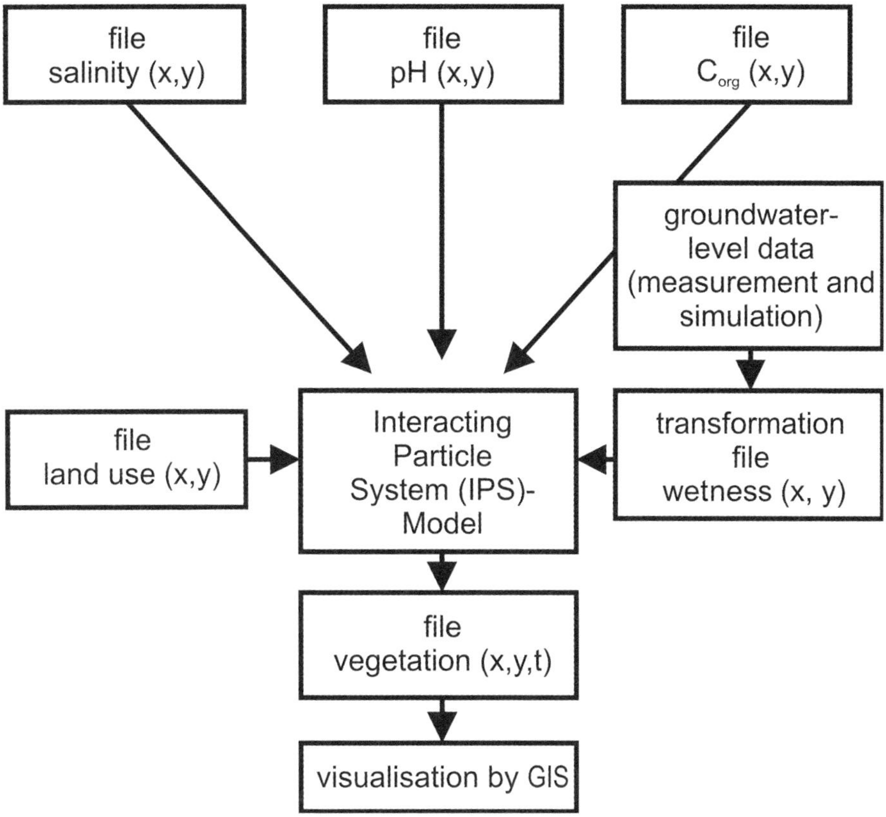

Fig. 6: IPS-model (Interacting Particle System) linked to a GIS (Geographic Information System) used to forecast possible scenarios corresponding to different types of groundwater management.

By still current analyses of sensitivity and validation, it attempts to adjust the system parameters optimally. The goal is to predict more exactly the effects of changes of the groundwater level on the vegetation development in wet dune slacks by the vegetation model. Because of the coupling of the vegetation model to geographical information systems, an effective visualization of the simulation results can be achieved (Fig. 7), which can show – for example – decision makers in water management and nature protection the effects of measures taken; and can contribute to better communication between these departments and, if necessary, the interested public.

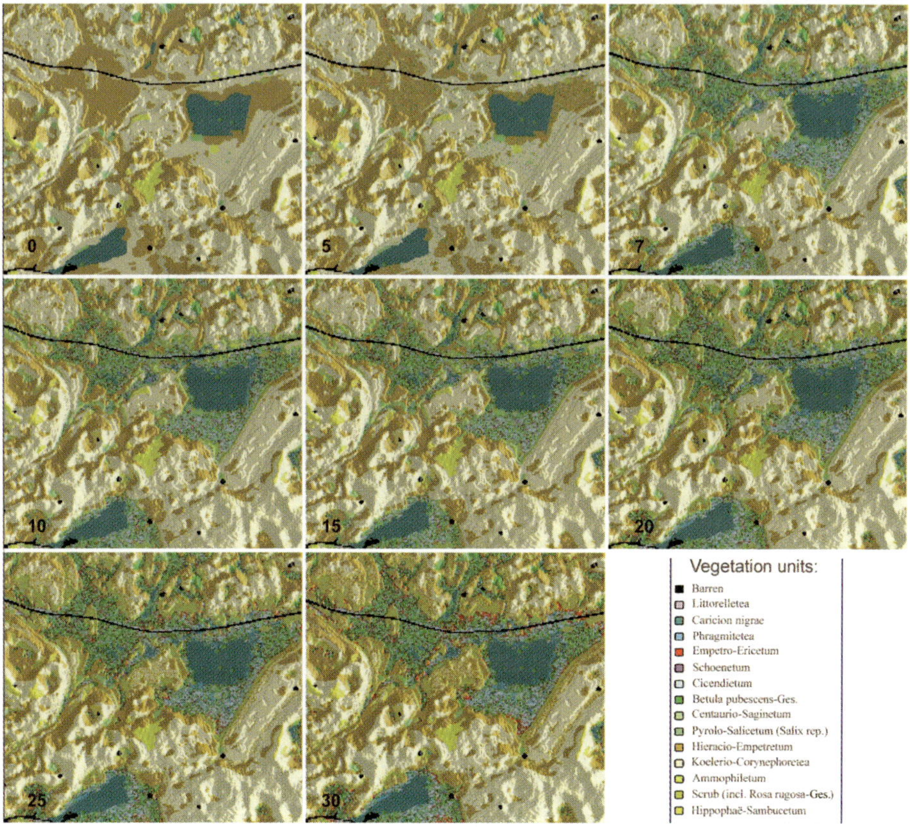

Fig. 7: Simulation of the effects of soil aridity increment in the next 30 years, obtained by the integration of the digital model of soil water regime with data on the ecological amplitude and the different plant communities (0, 5, 7, 10, 15, 20, 25, 30 years from the beginning of the simulation).

REFERENCES

1. Baltzer H., Braun P., Köhler W., 1998 - Cellular automata models for vegetation dynamics. Ecological Modelling, **107**: 113–125.
2. Buchwald K., 1990 - Nordsee – ein Lebensraum ohne Zukunft? 552 S., Göttingen.
3. Dierssen K., 1988 - Rote Liste der Pflanzengesellschaften Schleswig-Holsteins. Schriftenreihe Landesamt für Naturschutz und Landschaftspflege Schleswig-Holstein 6: 157 S., Kiel.
4. Drachenfels O. von, 1996 - Rote Liste der gefährdeten Biotoptypen in Niedersachsen. Naturschutz Landschaftspfl. Niedersachsen, 34: 148 S., Hannover.
5. DVWK, 1996 - Klassifikation überwiegend grundwasser-beeinflußter Vegetationstypen. Schriftenreihe des Deutschen Verbandes für Wasserwirtschaft und Kulturbau e.V. (DVWK) 112: 492 S., Bonn.

6. Ellenberg H., Weber H., Düll R., Wirth V., Werner W., Paulissen D., 1992 - Zeigerwerte von Pflanzen in Mitteleuropa. Scripta Geobot., Göttingen, **18**: 258 S.
7. Fahsold A., 1960 - Rhythmus des Grundwassers auf den Ostfriesischen Inseln. Gas- und Wasserfachmann, Bremen, **101** (2): 26-33.
8. Fromke A., 1996 - Vergleichende geobotanische Untersuchungen der Ostfriesischen Inseln Baltrum und Langeoog im Nationalpark „Niedersächsisches Wattenmeer" – 229 S., Diss. Univ. Hannover.
9. Grootjans A., Hartog P., Fresco L., Esselink H., 1991 - Succession and fluctuation in a wet dune slack in relation to hydrological changes. J. Veg. Sci., **2**: 545–554.
10. Grootjans A., Lammerts E., Beusekom C. van, 1995 - Kalkrijke duinvalleien op de Waddeneilanden. KNNV: 175 S., Utrecht.
11. Hahn D., 2005 - Neophyten auf den Ostfriesischen Inseln – Verbreitung, Ökologie und Vergesellschaftung. Diss. Univ. Hannover.
12. Helbing C.-D., 1991 - Zwischen Land und See. Nationalpark 2/91, 4 S., Wilhelmshaven.
13. Hobohm C., 1993 - Die Pflanzengesellschaften von Norderney. Arbeiten aus der Forschungsstelle Küste 12. Niedersächsisches Landesamt für Ökologie: 202 S.
14. Hofmann K., 2000 - Standortsökologie und Vergesellschaftung der *Utricularia*-Arten Nordwestdeutschland. – 130 S., Diss. Univ. Hannover.
15. Klötzli F., 1969 - Die Grundwasserbeziehungen der Streu- und Moorwiesen im nördlichen Schweizer Mittelland. Beitrag zur Geobotanischen Landesaufnahme der Schweiz: 52, 296 S., Bern.
16. Lammerts E., 1999 - Basiphilous pioneer vegetation in dune slacks on the Dutch Wadden Sea Islands. Proefschrift: 160 S., Groningen.
17. Laan D. van der, 1979 - Spatial and temporal variation in the vegetation of dune slacks in relation to the ground water regime. Vegetatio, **39** (l): 43–51.
18. Müller A., Müller H., Vries D. de, 1997 - Eine schonende Bewirtschaftung der Süßwasserlinsen auf den ostfriesischen Inseln am Beispiel der Insel Norderney. Zbl. Geol. Paläont. Teil I. (1/2): 111–124, Stuttgart.
19. Müller H., Wolff J., 1998 - Ermittlung einer schonenden Bewirtschaftung der Süßwasserlinsen auf den Ostfriesischen Inseln am Beispiel der Insel Norderney. 135 S., unveröffentlichter Endbericht, TU Braunschweig.
20. Nationalparkverwaltung Niedersächsisches Wattenmeer (ed.), 1992 - Nationalpark Niedersächsisches Wattenmeer. 52 S., Wilhelmshaven.
21. Neuhaus R., Petersen J., 1999 - Dunes. In: Jong F. de, Bakker J., Berkel C. van, Dankers N., Dahl K., Gätje C., Marencic H., Potel P. (eds.), Wadden Sea Quality Status Report. Wadden Sea Ecosystem No. 9. Common Wadden Sea Secretariat. Trilateral Monitoring and Assessment Group, Quality Status Report Group: 53–56, Wilhelmshaven.
22. Niemann E., 1963 - Beziehungen zwischen Vegetation und Grundwasser. Ein Beitrag zur Präzisierung des ökologischen Zeigerwertes von Pflanzen und Pflanzengesellschaften. Archiv f. Naturschutz u. Landschaftsforsch., B 3, H 1: 3-36, Berlin.
23. Niemann E., 1973 - Grundwasser und Vegetationsgefüge. Nova Acta Leopoldina, **6** (38): 172 S.
24. Nordheim H. von, Norden Andersen O., Thissen J., 1996 - Red Lists of Biotopes, Flora and Fauna of the Trilateral Wadden Sea Area. Helgoländer Meeresuntersuchungen 50 (Suppl.): 136 S., Hamburg.

25. Peters M., Pott R., 1999 - Natur & Tourismus auf Norderney. Abh. Westf. Museum f. Naturkde 61, 174 S., Münster.
26. Petersen J., 2000 - Die Dünentalvegetation der Wattenmeer-Inseln in der südlichen Nordsee. Eine pflanzensoziologische und ökologische Vergleichsuntersuchung unter Berücksichtigung von Nutzung und Naturschutz. 336 S., Husum.
27. Petersen J., Pott R., Richter O., 2001 - Dünentäler – Ein gefährdeter Lebensraum im Interessenkonflikt zwischen Naturschutz und Grundwasserbewirtschaftung. Zbl. Geol. Paläont. Teil I. (1/2): 93-110, Stuttgart.
28. Pott R., 1995 - Farbatlas Nordseeküste und Nordseeinseln. 288 S., Stuttgart.
29. Pott R., 1996 - Biotoptypen – Schützenswerte Lebensräume Deutschlands und angrenzender Regionen, Stuttgart, 448 S.
30. Pott R., 2003 - Die Nordsee. C.H. Beck, München.
31. Richter O., Söndgerath D., Belde M., Schwartz S., Schröder B., 1997 - Kopplung geographischer Informationssysteme (GIS) mit ökologischen Modellen für das Naturschutzmanagement. In: Kratz R., Suhling F. (eds.), GIS im Naturschutz - Forschung, Planung, Praxis, pp. 5–29. Verlag Westarp Wissenschaften, Magdeburg.
32. Streif H., 1987 - Geologie unseres Küstenraumes. Wilhelmshavener Tage, **1**: 5–12.
33. Streif H., 1990 - Das Ostfriesische Küstengebiet. Nordsee, Inseln, Watten und Marschen. Samml. Geol. Führer, Berlin, **57**: 376 S.
34. Tüxen R., 1954 - Pflanzengesellschaften und Grundwasser-Ganglinien. Angew. Pflanzensoz. (Stolzenau), **8:** 64–98.
35. Weimar J., 1998 - Simulation with cellular automata. Logos-Verlag, Berlin.
36. Westhoff V., 1991 - Die Küstenvegetation der Westfriesischen Inseln. Ber. d. Reinh.-Tüxen-Ges., Hannover, **3**: 269–290.
37. Westhoff V., Hobohm C., Schaminee J., 1993 - Rote Liste der Pflanzengesellschaften des Naturraumes Wattenmeer unter Berücksichtigung der ungefährdeten Vegetationseinheiten. Tuexenia, **13**: 109–140.

ON THE PROBLEMS OF NATURE CONSERVATION IN THE DESERT ZONE IN RUSSIA

Irina SAFRONOVA
Komarov Botanical Institute of the Russian Academy of Sciences, Prof. Popov Str. 2, **197376 St.-Petersburg (RUSSIA)**; e-mail: irinasaf@is1189.spb.edu

Abstract: At present, in the desert zone of Russia, nature monuments, national parks and reserves have been established by the efforts of scientists and public ecological organizations. But these protected areas are aimed at the conservation of azonal vegetation (meadows, forests and rare but not desert plants), not zonal deserts. Reserves are needed in order to conserve and restore deserts.

Introduction

The desert zone in Russia occupies a small territory bordering the north-western shore of the Caspian Sea. Its southern boundary runs from the Kuma river (45° N) along the eastern slope of the Ergeni Upland up to 47° N; from which turns to the north-east, crosses the River Volga near 48° N and then extends east just north of this parallel. The desert flora of the region is poor and includes about 200 species [1, 3]. Desert vegetation here has low diversity, reflecting the uniformity of the natural environment [6]. The Caspian desert region in Russia is traversed by the valley of the Volga with azonal vegetation such as meadows, scrub and forests, and a richer flora [2].

This paper addresses the problem of conservation of desert vegetation in Russia.

Vegetation outline

In the temperate deserts of Eurasia, the western part of which is the deserts of the Caspian Lowland, the main growth forms are dwarf semishrubs of *Chenopodiaceae*, *Asteraceae* and other families. In the Caspian region species of *Asteraceae* predominate, primarily dwarf to semishrubby wormwoods of the genus *Artemisia* – *Artemisia lerchiana*, *A. pauciflora* and *A. arenaria* [5].

The geographical range of *Artemisia lerchiana* consists of three isolated parts: the western in Romania (on the Dobrudzhi limestones), the central in the Crimea, while the largest, eastern part stretches from the River Don up to the Northern Cis-Aral area. Its northern limit reaches 51° N, the southern 43° N, the western 27° E, the eastern 66° E. Between the rivers Kuma and Ural, in the desert zone *A. lerchiana* is a eurytopic species.

Artemisia pauciflora extends from the River Don in the West (42° E) up to Lake Zaisan in the East (85° E). In the north this wormwood reaches 53° N, in the south 47° N. *A. pauciflora* is a halophyte restricted to saline soils and solonetz

The distribution of *Artemisia arenaria*, like that of *A. pauciflora*, is very extensive, from the River Volga in the west (45° N) up to Lake Zaisan in the east

(85° E). In the north *A. arenaria* reaches 50°30' N, in the south 45° N. *A. arenaria* is a typical psammophyte, growing on weakly stabilized sands.

The Caspian Lowland from the Kuma to the Ural river is the domain of *Artemisia lerchiana*. The desert communities of *A. lerchiana* occur here under diverse conditions. Especially widespread in the region are bluegrass wormwood (*Artemisia lerchiana*, *Poa bulbosa*) deserts, which comprise hemi-psammophytic and psammophytic ecotones of vegetation on sandy-loam and sandy soils. The *Artemisia lerchiana* communities with a grass component (*Stipa sareptana, S. lessingiana* or *Agropyron fragile*) are formed on hillocky fixed sands, as well as communities of *A. lerchiana* with *Ephedra distachya* or locally *Calligonum aphyllum*. On plains with loamy soils, *A. lerchiana* communities combine into a complex with the *A. pauciflora* communities confined to solonetz. The psammophyte deserts of *Artemisia arenaria*, usually with the shrubs *Calligonum aphyllum, Tamarix ramosissima* and *T. laxa*, are widespread on barkhan weakly stabilized sands.

Natural desert vegetation in Russia has been considerably altered. Dynamically unstable communities of annual weeds (*Alyssum turkestanicum, Anisantha tectorum, Ceratocarpus arenarius, Descurainia sophia, Eragrostis minor,* etc.) and perennial species (*Alhagi pseudalhagi, Anabasis aphylla, Artemisia scoparia*, *A. austriaca, Euphorbia seguierana, Xanthium spinosum,* etc.) dominate over large areas. Human impact plays a significant role in increasing the representation of grasses in the Caspian Deserts. With hay cutting, overgrazing and fires the structure of grass-wormwood deserts on sandy-loam and sandy soils changes radically: wormwoods disappear and grasses spread. As a result the illusion of occurrence of steppe communities arises, which does not correspond to the natural conditions of the region. Sands are being broken up fast by grazing and colonized by weed species. With overgrazing, vegetation is annihilated completely; only scattered plants occur – *Coryspermum arenarium*, *Leymus racemosus*, *Peganum harmala.*

On degraded sand massifs the restoration of natural communities is proceeding slowly; therefore they need phyto-ameliorative measures for their stabilization. For planting both native plants *Calligomun aphyllum*, *Krascheninnikovia ceratoides* are used, and plants newly introduced into cultivation. Some of them have acclimatized well and form communities previously absent in a particular region, for instance, communities of *Haloxylon aphyllum*.

Conclusion

Conservation of desert vegetation in Russia is very important. At present, nature monuments, national parks and reserves have been established by the efforts of scientists and public ecological organizations. But there are some problems. Basically they conserve azonal vegetation in the desert zone, not zonal deserts. There are many nature monuments for meadows, forests, and rare but not desert plants (*Nelumbo caspica*, *Trapa astrakhanica*, etc.) in the delta and the valley of the Volga [4]. In the individual biosphere reserves, for the most part the desert plant communities are not natural. So, reserves are necessary not only for the conservation but also the restoration of deserts.

REFERENCES

1. Baktasheva N.M., 2000 - Flora of Kalmykia and its analysis. Ed. Prof. Kamelin R.V., Elista.
2. Golub V.B., Laktionov A.P., Barmin A.N., Pilipenko V.N., 2002 - Vascular plants of the valley of Low Volga. Ed. Prof. Rozenberg G.S., Toljatti.
3. Neronov V.V., Ochirova N.N., 1998 - Vascular plants of the reserve "Black lands". Ed. Gubanov I.A., Moscow.
4. Pjatin V.A. (ed.), 1997 - Atlas of the Astrakhan region. Moscow.
5. Safronova I.N., 1996 - Species of *Artemisia* subgenus *Seriphidium* in the West Turan and their ecology. In: Hind D.J.N. (ed.), Proceeding of the International Compositae Conference. Royal Botanic Gardens, Kew, **2**: 105-110.
6. Safronova I.N., 2002 - On the Caspian subprovince of the Caharo-Gobian Desert Region. Botanical journal (St.-Petersburg), **87** (3): 57-62.

INDEX

A

B

C

M

N

O

P

Q

R

S

T

X

Y

Z

Printing: Krips bv, Meppel
Binding: Stürtz, Würzburg